geoENV VI – GEOSTATISTICS FOR ENVIRONMENTAL APPLICATIONS

Quantitative Geology and Geostatistics

VOLUME 15

The titles published in this series are listed at the end of this volume.

geoENV VI – GEOSTATISTICS FOR ENVIRONMENTAL APPLICATIONS

Proceedings of the Sixth European Conference on Geostatistics for Environmental Applications

Edited by

AMÍLCAR SOARES
Instituto Superior Técnico,
Lisbon, Portugal

MARIA JOÃO PEREIRA
Instituto Superior Técnico,
Lisbon, Portugal

and

ROUSSOS DIMITRAKOPOULOS
McGill University, Montreal, Canada

Springer

Editors

Amílcar Soares
Instituto Superior Técnico
Lisbon, Portugal

Maria João Pereira
Instituto Superior Técnico
Lisbon, Portugal

Roussos Dimitrakopoulos
McGill University
Montreal, Canada

ISBN: 978-90-481-7643-4 e-ISBN: 978-1-4020-6448-7

Softcover reprint of the hardcover 1st edition 2008

Cover illustrations by A. Soares

Printed on acid-free paper

springer.com

Contents

Part III Meteorology

Part IV Remote Sensing

Part V Soil

Foreword

geoENV: Ten Years Later

The environment has unquestionably become a key topic of focus and concern for today's society, encompassing themes that include sustainable development, climate change, reduction of biological diversity, and carbon emissions along with the need for new energy paradigms. These themes are no longer the exclusive domain of academic and scientific exploration. They are now high-priority issues for governments and environmental agencies of all industrialized countries because of the tremendous effects on the industrialized and, to an even grater extent, developing world. Quantifying and predicting global environmental impacts and risks encompasses political, social, economic as well as technical dimensions, and is now an integral part of strategic planning for both governments and international organizations.

Geostatistics has become an important set of technical tools for environmental problem-solving, in particular spatial and temporal assessment of uncertainty of physical/environmental phenomena and related natural resources. Geostatistics has been applied to a variety of fields from the characterization of desertification, degradation of soil, air and water quality, to the evaluation of health and pollutant space-time relationships in the field of environmental epidemiology, and the assessment of climatic and meteorology for predicting the dynamic of natural phenomena.

Geostatistics for Environmental Applications (geoENV), a series of bi-annual conferences, was started in 1996 with the ambitious goal of bringing together the disparate geostatisticical community to discuss ideas and methods regarding new and diverse applications in the environmental field. Thanks to everyone involved in the organization and scientific coordination of the conferences, first in Lisbon, then Valencia, Avignon, Barcelona, Neufchâtel and most recently in Rhodes, the geoENV international conferences and subsequent publication of selected papers have contributed to maintaining the high standards of scientific quality in approaching the diversity of new environmental modeling problems. Ten years after Lisbon, we are proud to see that geoENV has become a well-respected and well-supported scientific project.

This book marks the first decade of geoENV and reflects the status of the most up-to-date research in the field as presented in Rhodes. As in past years, scientists who approach environmental problems with different methodological perspectives than those encountered in our field, were invited to present their methodologies at geoENV conferences, with the goal of enriching our own field through cross-fertilization with related fields and their approaches, concepts, tools and developments. Ricardo Trigo, the keynote speaker in Rhodes, introduced us to new methods for modeling climate change and assessing corresponding impacts on natural resources and human health. The additional two keynote papers from Pierre Goovaerts and Philippe Renard presented the state of the art of geostatistical applications in analyzing public health data and stochastic hydrology, respectively. The 42 papers of this book were presented in oral sessions of Methods, Environment and Health, Soil, Hydrology, Remote Sensing and Meteorology.

We would like to thank to all the authors and reviewers for their outstanding efforts and technical contributions to the present volume.

Amílcar Soares
Maria João Pereira
Roussos Dimitrakopoulos

Organizing Committee geoENV2006

Amilcar Soares, IST, Lisbon, Portugal
Denis Allard, INRA, Avignon, France
Hélène Demougeot-Renard, FSS International, Neuchâtel, Switzerland
Jaime Gómez-Hernández, UPV, Valencia, Spain
Maria João Pereira, IST, Lisbon, Portugal
Pascal Monestiez, INRA, Avignon, France
Phaedon Kyriakidis, UC Santa Barbara, Santa Barbara, USA
Philippe Renard, University of Neuchâtel, Switzerland
Pierre Goovaerts, BioMedware, Inc., USA
Roland Froidevaux, FSS International, Geneva, Switzerland
Roussos Dimitrakopoulos, BRC, U. Queensland
Xavier Sanches-Vila, UPC, Barcelona, Spain

International Scientific Committee geoENV2006

Aldo Fiori
Alberto Guadagnini
Alexander Boucher
Alfred Stein
Amilcar Soares
Carla Nunes
Carol Gotway Crawford
Chantal de Fouquet
Chris Loyd
Denis Allard
Denis Marcotte
Dick Brus
Dimitri D'Or
Edzer Pebesma
Geoffrey M. Jacquez
Hans Wackernagel
Harrie-Jan Hendricks
Helene Demougeot
Henrique G. Pereira
Hirotaka Saito
Jacques Rivoirard
Jaime Gomez-Hernandez
Jean Paul Chilés
Jennifer Mckinley
Jorge Sousa
José Almeida
Júlia Carvalho
Laura Guadagnini
Marc Van Merveinne
Margaret Oliver
Maria João Pereira
Maria-Theresa Schafmeister
Mohan Srivastava
Monica Riva
Muray Lark
Oy Leuangthong
Pascal Monestiez
Patrick Bogaert
Peter Atkinson
Peter Kettlewell
Phaedon Kiriakidis
Philippe Renard
Pierre Goovaerts
Ricardo Garcia-Herrera
Richard Webster
Roland Froidevaux
Roussos Dimitrakopoulos
Souheil Ezzedine
Vicente Serrano
Xavier Emery
Xavier Sanchez-Vila

Contributors

P.M. Adler
UPMC – Sisyphe, 4 place Jussieu, 75252 Paris cedex 05, France
padler@ccr.jussieu.fr

P.M. Atkinson
School of Geography, University of Southampton, Highfield, Southampton, SO17 1BJ, UK
pma@soton.ac.uk

O. Baume
Department of Environmental Sciences, Wageningen University, The Netherlands
olivier.baume@wur.nl

M. Andrade-Bejarano
School of Biological Sciences, Statistics Section, Harry Pitt Building, The University of Reading, RG6 6FN, Reading, UK
snr02ma@reading.ac.uk

M. Bérengier
Section Acoustique Routière et Urbaine, Laboratoire Central des Ponts et Chaussées, 44341 Bouguenais, France
michel.berengier@lcpc.fr

L. Bloom
School of Engineering and Mathematics, Edith Cowan University, Perth, Western Australia
l.bloom@ecu.edu.au

G. Blöschl
Institute for Hydraulic and Water Resources Engineering, Vienna University of Technology, Karlsplatz 13/222, A-1040 Vienna, Austria
bloeschl@hydro.tuwien.ac.at

P. Bogaert
Department of Environmental Sciences and Land Use Planning, Université catholique de Louvain, Belgium
bogaert@enge.ucl.ac.be

A. Boucher
Dpt. of Geological and Environmental Sciences, Stanford University, CA, USA
aboucher@stanford.edu

B. Briand
Laboratory of Radioecological Studies for Marine and Terrestrial Ecosystems, Institute for Radioprotection and Nuclear Safety (IRSN), DEI/SESURE/LERCM, Cadarache, Bld 153, 13105 St-Paul-lez-Durance cedex, France

G. Buttafuoco
CNR – Institute for Agricultural and Forest Systems in the Mediterranean, Via Cavour, 87030 Rende (CS), Italy
buttafuo@ieif.cs.cnr.it

H. Caetano
CMRP – Centre for Modelling Petroleum Reservoirs, Mining and Georesources Department, Instituto Superior Técnico, Av. Rovisco Pais, 1049-001, Lisboa, Portugal
pcl42082@popsrv.ist.utl.pt

T.J. Calado
Centro Geofísico da Universidade de Lisboa, 1749-016 Lisboa, Portugal
mtcalado@fc.ul.pt

N. Caputi
Marine and Fisheries Research Laboratories, Department of Fisheries, North Beach, WA 6020, Australia
ncaputi@fish.wa.gov.au

G. Cardenas
Institut National de l'Environnement Industriel et des Risques, Parc Technologique ALATA, BP 2, 60550 Verneuil-en-Halatte, France
giovanni.cardenas@ineris.fr

F. Carrat
INSERM UMR-S 707, Paris, France
carrat@u707.jussieu.fr

J. Carvalho
Environmental Group of the Centre for Modelling Petroleum Reservoirs – CMRP, Instituto Superior Técnico, Lisboa, Portugal
jcarvalho@ist.utl.pt

A. Castrignanò
CRA – Agronomic Research Institute, Bari, Italy
annamaria.castrignano@entecra.it

J.-P. Chilès
Centre de Géosciences, École des Mines de Paris, 77305 Fontainebleau, France
jean-paul.chiles@ensmp.fr

M.F.E.S. Coelho
Instituto de Meteorologia, Lisboa, Portugal
Fatima.Coelho@meteo.pt

D. Cornford
Neural Computing Research Group, Aston University, Birmingham B4 7ET UK
d.cornford@aston.ac.uk

J.A. Corte-Real
Centro de Geofísica de Évora, Universidade de Évora, Portugal
jmcr@uevora.pt

A.C.M. Costa
ISEGI, Universidade Nova de Lisboa, Campus de Campolide, 1070-312 Lisbon, Portugal
ccosta@isegi.unl.pt

C.C. DaCamara
Centro Geofísico da Universidade de Lisboa, 1749-016 Lisboa, Portugal

A. Dassargues
Applied Geology and Mineralogy, Department of Geology-Geography, Katholieke Universiteit Leuven, Belgium; Hydrogeology and Environmental Geology, University of Liège, Belgium
alain.dassargues@ulg.ac.be

G.A. Degan
Dipartimento di Ingegneria Meccanica e Industriale; Università degli studi Roma Tre, Rome, Italy
g.alfarodegan@uniroma3.it

H. Demougeot-Renard
FSS International r&d 1956, Av. Roger Salengro, 92370 Chaville,
France
demougeot.renard@fssintl.com

M. Dentz
Department of Geotechnical Engineering and Geosciences, Technical University of Catalonia, Gran Capità S/N, 08034 Barcelona, Spain
marco.dentz@upc.es

R. Dimitrakopoulos
Department of Mining and Materials Engineering, McGill University, Montreal, Qc, Canada H3A 2A7
roussos.dimitrakopoulos@mcgill.ca

D. D'Or
FSS International r&d 1956, Av. Roger Salengro, 92370 Chaville, France
dimitri.dor@fssintl.com

G. Dubois
European Commission – DG Joint Research Centre, Institute for Environment and Sustainability, Ispra, Italy
gregoire.dubois@jrc.it

V. Durand
Laboratory of radioecological studies for marine and terrestrial ecosystems, Institute for Radioprotection and Nuclear Safety (IRSN), DEI/SESURE/LERCM, Cadarache, Bld 153, 13105 St-Paul-lez-Durance cedex, France

R.M. Durão
Centro de Geofísica de Évora, Universidade de Évora, Portugal
rddurao@fc.ul.pt

S.N. Elogne
Department of Mineral Resources Engineering, Technical University of Crete, Chania 73100, Greece
elogne@mred.tuc.gr

D. Fasbender
Department of Environmental Sciences and Land Use Planning,
Université catholique de Louvain, Belgium
fasbender@enge.ucl.ac.be

D. Fernàndez-Garcia
Polytechnic University of Valencia, Ingeniería Hidráulica y
Medio Ambiente, Camino de Vera s/n., 46022 Valencia, Spain
dafernan@dihma.upv.es

A. Furon
Dept. of Land Resource Science
University of Guelph, Guelph, Ontario, Canada

I. García
IFAPA, Centro "Las Torres-Tomejil", Junta de Andalucía. Ctra.
Sevilla-Cazalla km 12.2, 41200 Alcalá del Río (Seville), Spain
ivan.garcia.ext@juntadeandalucia.es

M. Garcia
FSS International r&d 1956, Av. Roger Salengro, 92370 Chaville,
France
michel.garcia@fssintl.com

B. Gauvreau
Section Acoustique Routière et Urbaine, Laboratoire Central
des Ponts et Chaussées,
44341 Bouguenais
France
benoit.gauvreau@lcpc.fr

C. Giasi
Department of Environmental and Civil Engineering, Polytechnic of Bari, Italy
c.giasi@poliba.it

J.J. Gómez Hernández
Polytechnic University of Valencia, Department of Hydraulic Engineering and Environment
Camino de Vera s/n., 46022 Valencia, Spain
jaime@dihma.upv.es

E. Gommeren
Openbare Afvalstoffenmaatschappij voor het Vlaamse Gewest (OVAM), Stationstraat 110, 2800 Mechelen, Belgium
egommere@ovam.be

P. Goovaerts
BioMedware, Inc. 516 North State Street, Ann Arbor, MI 48104 – 1236, USA
goovaerts@biomedware.com

C. Gouveia
CGUL, Departamento de Física, Faculdade de Ciências, Universidade de Lisboa, Portugal; Escola Superior de Tecnologia, Instituto Politécnico de Setúbal, Portugal
cgouveia@est.ips.pt

A. Guadagnini
Dipartimento di Ingegneria Idraulica, Ambientale, Infrastrutture Viarie, Rilevamento (DIIAR), Politecnico di Milano, Piazza L. Da Vinci 32, 20133 Milano, Italy
alberto.guadagnini@polimi.it

A. Horta
Environmental Group of the Centre for Modelling Petroleum Reservoirs – CMRP, Instituto Superior Técnico, Lisboa, Portugal
ahorta@ist.utl.pt

D.T. Hristopulos
Department of Mineral Resources Engineering, Technical University of Crete, Chania 73100, Greece
dionisi@mred.tuc.gr

M. Huysmans
Applied Geology and Mineralogy, Department of Geology-Geography, Katholieke Universiteit Leuven, Belgium
marijke.huysmans@geo.kuleuven.be

J.R. Ilarri
Universidad Politécnica de Valencia, Instituto de Ingeniería del Agua y Medio Ambiente, Camino de Vera s/n., 46022 Valencia, Spain
jrodrigo@upv.es

C. Innocenti
Central Institute for Marine Research, Via di Casalotti, 300, 00166 Roma, Italy.
c.innocenti@icram.org

Y.O. Isselmou
France Télécom R & D, RESA/FACE, Issy-les-Moulineaux, France; Ecole des Mines, CG – Geostatistics, Fontainebleau, France

C. Jégat
INSERM UMR-S 707, Paris, France; Ecole des Mines de Paris – Geostatistics group, Fontainebleau, France
cyrille.jegat@mines-paris.org

J.A. Jiménez
IFAPA, Centro "Las Torres-Tomejil", Junta de Andalucía. Ctra. Sevilla-Cazalla km 12.2, 41200 Alcalá del Río (Seville), Spain

F. Junker
Département Analyse Mécanique et Acoustique, Electricité de France, France
fabrice.junker@edf.fr

M. Kangas
Marine and Fisheries Research Laboratories, Department of Fisheries, Northbeach, Western Australia 6020
mkangas@fish.wa.gov.au

C. Kaiser
Institute of Geography, University of Lausanne, CH-1015 Switzerland

M. Kanevski
Institute of Geomatics and Analysis of Risk, University of Lausanne, CH-1015 Switzerland

C. Lajaunie
Ecole des Mines de Paris – Geostatistics group, Fontainebleau, France
christian.lajaunie@ensmp.fr

D. Lippiello
Dipartimento di Ingegneria Meccanica e Industriale; Università degli studi Roma Tre, Rome Italy
d.lippielo@uniroma3.it

S. Mackie
Xstrata, Brisbane, Qld, Australia

F. Marconi
Central Institute for Marine Research, via di Casalotti, 300 – 00166 Roma, Italy

D. Marcotte
École Polytechnique de Montréal, Département des génies civil, géologique et des mines, C.P. 6079, Succ. Centre-ville, Montréal, Qc, Canada, H3C 3A7
denis.marcotte@polymtl.ca

G. Martínez
IFAPA, Centro "Las Torres-Tomejil", Junta de Andalucía, Ctra. Sevilla-Cazalla km 12.2, 41200 Alcalá del Río (Seville), Spain

H. Mata-Lima
CMRP – Centre for Modelling Petroleum Reservoirs, Mining and Georesources Department, Instituto Superior Técnico, Av. Rovisco Pais, 1049-001, Lisboa, Portugal
Departamento de Matemática e Engenharias, Universidade da Madeira, 9000-390 Funchal, Portugal
hlima@uma.pt

M. Van Meirvenne
Dept. Soil Management and Soil Care, Ghent University, Coupure 653, 9000 Gent, Belgium
marc.vanmeirvenne@ugent.be

T. Meklit
Dept. Soil Management and Soil Care, Ghent University, Coupure 653, 9000 Gent, Belgium
meklit.tariku@ugent.be

C. Mercat-Rommens
Laboratory of radioecological studies for marine and terrestrial ecosystems, Institute for Radioprotection and Nuclear Safety (IRSN), DEI/SESURE/LERCM, Cadarache, Bld 153, 13105 St-Paul-lez-Durance cedex, France
catherine.mercat-rommens@irsn.fr

J.-M. Metivier
Laboratory of radioecological studies for marine and terrestrial ecosystems, Institute for Radioprotection and Nuclear Safety (IRSN), DEI/SESURE/LERCM, Cadarache, Bld 153, 13105 St-Paul-lez-Durance cedex, France
jean-michel.metivier@irsn.fr

J.M.M. Avalos
Faculty of Sciences, University of Coruña, A Zapateira 15071, A Coruña, Spain
jmirasa@udc.es

U. Mueller
School of Engineering and Mathematics, Edith Cowan University, Perth, Western Australia
u.mueller@ecu.edu.au

E. Mumelter
Central Institute for Marine Research, Via di Casalotti, 300, 00166 Roma, Italy
e.mumelter@icram.org

J.L. Muriel
IFAPA, Centro "Las Torres-Tomejil", Junta de Andalucía. Ctra. Sevilla-Cazalla km 12.2, 41200 Alcalá del Río (Seville)
Spain
josel.muriel@juntadeandalucia.es

J. Negreiros
Instituto Superior de Línguas e Administração, Estrada da Correia 53, 1500-210 Lisbon, Portugal
c8057@isegi.unl.pt

P. Pasquier
Golder Associates, 9200, L'Acadie blvd., Montréal, Québec, Canada, H4N 2T2

D. Patriarche
Gaz de France
361, Avenue du President Wilson, BP 33, 93211 Saint Denis La Plaine Cedex, France
delphine.patriarche@gazdefrance.com

A.P. González
Faculty of Sciences, University of Coruña, La Zapateira 15071, La Coruña, Spain

E.J. Pebesma
Geosciences Faculty, Utrecht University, The Netherlands
e.pebesma@geo.uu.nl

E. Perdrix
Ecole des Mines de Douai, France
perdrix@ensm-douai.fr

F. Perea
IFAPA, Centro "Las Torres-Tomejil", Junta de Andalucía. Ctra. Sevilla-Cazalla km 12.2, 41200 Alcalá del Río (Seville),
Spain

J.M.C. Pereira
Cartography Centre, Tropical Research Institute, Lisboa, Portugal; Department of Forestry, Instituto Superior de Agronomia, Lisboa, Portugal
jmocpereira@sapo.pt

E. Pili
Commissariat à l'Energie Atomique, Département Analyse Surveillance Environnement, BP 12, 91680 Bruyères-le-Châtel, France
eric.pili@cea.fr

M. Pinzari
Dipartimento di Ingegneria Meccanica e Industriale; Università degli studi Roma Tre, Rome Italia
pinzari@uniroma3.it

G. Raspa
Dipartimento di Ingegneria Chimica, dei Materiali delle Materie Prime e Metallurgia; La Sapienza Università di Roma, Dept. ICMMPM, via Eudossiana, 18 – 00184 Roma, Italy
giuseppe.raspa@uniroma1.it

M. Riva
Dipartimento Ingegneria Idraulica, Ambientale, Infrastrutture Viarie, Rilevamento (DIIAR), Politecnico di Milano, Piazza L. Da Vinci 32, 20133 Milano, Italy
monica.riva@polimi.it

M. Rivest
École Polytechnique de Montréal, Département des génies civil, géologique et des mines, C.P. 6079, Succ. Centre-ville, Montréal, Qc, Canada, H3C 3A7
martine.rivest@polymtl.ca

A. Russo
CMRP, Instituto Superior Técnico, Av. Rovisco Pais, 1049-001 Lisboa, Portugal
arusso@ist.utl.pt

A. Salmeri
Central Institute for Marine Research, Via di Casalotti, 300, 00166 Roma, Italy
a.salmeri@icram.org

X. Sanchez-Vila
Department of Geotechnical Engineering and Geosciences, Technical University of Catalonia, Gran Capità S/N, 08034 Barcelona, Spain
xavier.sanchez-vila@upc.edu

E. Sevens
Openbare Afvalstoffenmaatschappij voor het Vlaamse Gewest (OVAM), Stationstraat 110, 2800 Mechelen, Belgium
erwin.sevens@ovam.be

J.O. Skøien
Department of Physical Geography, Utrecht University, P.O. box 80115, 3508 TC Utrecht, The Netherlands
jskoien@geo.uu.nl

A. Soares
Environmental Group of the Centre for Modelling Petroleum Reservoirs – CMRP, Instituto Superior Técnico, Lisbon, Portugal
asoares@ist.utl.pt

W. Tabbara
Supélec, DRE-LSS, Gif-sur-Yvette, France
walid.tabbara@lss.supelec.fr

F. Tack
Dept. Soil Management and Soil Care, Ghent University, Coupure 653, 9000 Gent, Belgium
filip.tack@ugent.be

D.M. Tartakovsky
Department of Mechanical and Aerospace Engineering, University of California, San Diego, La Jolla, CA 92093, USA; Theoretical Division, Los Alamos National Laboratory, Los Alamos, NM 87545, USA
dmt@ucsd.edu

J.-F. Thovert
Laboratoire de Combustion et Détonique, SP2MI, BP 179, 86960 Futuroscope Cedex, France
thovert@lcd.ensma.fr

R.M. Trigo
CGUL, Departamento de Física, Faculdade de Ciências, Universidade de Lisboa, Portugal; Departamento de Engenharias, Universidade Lusófona, Lisboa, Portugal
rmtrigo@fc.ul.pt

D. Tuia
Institute of Geomatics and Analysis of Risk, University of Lausanne, CH-1015 Switzerland
devis.tuia@unil.ch

K. Vanderlinden
IFAPA, Centro "Las Torres-Tomejil", Junta de Andalucía. Ctra. Sevilla-Cazalla km 12.2, 41200 Alcalá del Río (Seville), Spain
karl.vanderlinden.ext@juntadeandalucia.es

H.S. Vargas
CMRP – Centre for Modelling Petroleum Reservoirs, Mining and Georesources Department, Instituto Superior Técnico, Av. Rovisco Pais, 1049-001, Lisboa, Portugal
hugo.vargas@total.com

S. Verstraete
Dept. Soil Management and Soil Care, Ghent University, Coupure 653, 9000 Gent, Belgium
samuel.verstraete@ugent.be

H. Wackernagel
Centre de Géosciences, École des Mines de Paris, 77305 Fontainebleau, France
Ecole des Mines de Paris – Geostatistics group, Fontainebleau, France
hans.wackernagel@ensmp.fr

C. Wagner-Riddle
Dept. of Land Resource Science, University of Guelph, Guelph, Ontario, Canada
cwagnerr@uoguelph.ca

J. Wiart
France Télécom R & D, RESA/FACE, Issy-les-Moulineaux, France
joe.wiart@francetelecom.fr

B.E. Wohlberg
Theoretical Division, Los Alamos National Laboratory, Los Alamos, NM 87545, USA
brendt@lanl.gov

Part I
Environment and Health

Geostatistical Analysis of Health Data: State-of-the-Art and Perspectives

P. Goovaerts

Abstract The analysis of health data and putative covariates, such as environmental, socio-economic, behavioral or demographic factors, is a promising application for geostatistics. It presents, however, several methodological challenges that arise from the fact that data are typically aggregated over irregular spatial supports and consist of a numerator and a denominator (i.e. population size). This paper presents an overview of recent developments in the field of health geostatistics, with an emphasis on three main steps in the analysis of aggregated health data: estimation of the underlying disease risk, detection of areas with significantly higher risk, and analysis of relationships with putative risk factors. The analysis is illustrated using age-adjusted cervix cancer mortality rates recorded over the 1970–1994 period for 118 counties of four states in the Western USA. Poisson kriging allows the filtering of noisy mortality rates computed from small population sizes, enhancing the correlation with two putative explanatory variables: percentage of habitants living below the federally defined poverty line, and percentage of Hispanic females. Area-to-point kriging formulation creates continuous maps of mortality risk, reducing the visual bias associated with the interpretation of choropleth maps. Stochastic simulation is used to generate realizations of cancer mortality maps, which allows one to quantify numerically how the uncertainty about the spatial distribution of health outcomes translates into uncertainty about the location of clusters of high values or the correlation with covariates. Last, geographically-weighted regression highlights the non-stationarity in the explanatory power of covariates: the higher mortality values along the coast are better explained by the two covariates than the lower risk recorded in Utah.

1 Introduction

Since its early development for the assessment of mineral deposits, geostatistics has been used in a growing number of disciplines dealing with the analysis of data

P. Goovaerts
BioMedware, Inc. 516 North State Street, Ann Arbor, MI 48104-1236, USA
e-mail: goovaerts@biomedware.com

A. Soares et al. (eds.), *geoENV VI – Geostatistics for Environmental Applications*, 3–22

distributed in space and/or time. One field that has received little attention in the geostatistical literature is medical geography or spatial epidemiology, which is concerned with the study of spatial patterns of disease incidence and mortality and the identification of potential "causes" of disease, such as environmental exposure or socio-demographic factors (Waller and Gotway 2004). This lack of attention contrasts with the increasing need for methods to analyze health data following the emergence of new infectious diseases (e.g. West Nile Virus, bird flu), the higher occurrence of cancer mortality associated with longer life expectancy, and the burden of a widely polluted environment on human health.

Individual humans represent the basic unit of spatial analysis in health research. However, because of the need to protect patient privacy publicly available data are often aggregated to a sufficient extent to prevent the disclosure or reconstruction of patient identity. The information available for human health studies thus takes the form of disease rates, e.g. number of deceased or infected patients per 100,000 habitants, aggregated within areas that can span a wide range of scales, such as census units, counties or states. Associations can then be investigated between these areal data and environmental, socio-economic, behavioral or demographic covariates. Figure 1 shows an example of datasets that could support a study of the impact of demographic and socio-economic factors on cervix cancer mortality. The top map shows the spatial distribution of age-adjusted mortality rates recorded over the 1970-1994 period for 118 counties of four states in the Western USA. The corresponding population at risk is displayed in the middle map, either aggregated within counties or assigned to 25 km^2 cells. The bottom maps show two putative explanatory variables: percentage of habitants living below the federally defined poverty line, and percentage of Hispanic females. Indeed, Hispanic women tend to have elevated risk of cervix cancer, while poverty reduces access to health care and to early detection through the Pap smear test in particular (Friedell et al. 1992). These socio-demographic data are available at the census block level and were assigned to the nodes of a 5 km spacing grid for the purpose of this study (same resolution as the population map).

A visual inspection of the cancer mortality map conveys the impression that rates are much higher in the centre of the study area (Nye and Lincoln Counties), as well as in one Northern California county. This result must however be interpreted with caution since the population is not uniformly distributed across the study area and rates computed from sparsely populated counties tend to be less reliable, an effect known as "small number problem" and illustrated by the top scattergram in Fig. 1. The use of administrative units to report the results (i.e. counties in this case) can also bias the interpretation: had the two counties with high rates been much smaller in size, these high values likely would have been perceived as less problematic. Last, the mismatch of spatial supports for cancer rates and explanatory variables prevents their direct use in the correlation analysis.

Unlike datasets typically analyzed by geostatisticians, the attributes of interest are here measured exhaustively. Ordinary kriging, the backbone of any geostatistical analysis, thus seems of little use. Yet, I see at least three main applications of geostatistics for the analysis of such aggregated data:

1. Filtering of the noise caused by the small number problem using a variant of kriging with non-systematic measurement errors.
2. Modeling of the uncertainty attached to the map of filtered rates using stochastic simulation, and propagation of this uncertainty through subsequent analysis, such as the detection of aggregate of counties (clusters) with significantly higher or lower rates than neighboring counties.
3. Disaggregation of county-level data to map cancer mortality at a resolution compatible with the measurement support of explanatory variables.

Goovaerts (2005a, 2006a,b) introduced a geostatistical approach to address all three issues and compared its performances to empirical and Bayesian methods which have been traditionally used in health science. The filtering method is based on Poisson kriging and semivariogram estimators developed by Monestiez et al. (2006) for mapping the relative abundance of species in the presence of spatially heterogeneous observation efforts and sparse animal sightings. Poisson kriging was combined with p-field simulation to generate multiple realizations of the spatial distribution of cancer mortality risk. A limitation of all these studies is the assumption that the size and shape of geographical units, as well as the distribution of the population within those units, are uniform, which is clearly inappropriate in the example of Fig. 1. The last issue of change of support was addressed recently in the geostatistical literature (Gotway and Young 2002, 2005; Kyriakidis 2004). In its general form kriging can accommodate different spatial supports for the data and the prediction, while ensuring the coherence of the predictions so that disaggregated estimates of count data are non-negative and their sum is equal to the original aggregated count. The coherence property needs however to be tailored to the current situation where aggregated rate data have various degree of reliability depending on the size of the population at risk (Goovaerts, 2006b).

This paper discusses how geostatistics can benefit three main steps of the analysis of aggregated health data: estimation of the underlying disease risk, detection of areas with significantly higher risk, and analysis of relationships with putative risk factors. An innovative procedure is proposed for the deconvolution of the semivariogram of aggregated rates and the disaggregation of these rates, accounting for heterogeneous population densities and the shape and size of administrative units. The different concepts are illustrated using the cervix cancer data of Fig. 1.

2 Estimating Mortality Risk from Observed Rates

For a given number N of entities v_α (e.g. counties), denote the observed mortality rates as $z(v_\alpha) = d(v_\alpha)/n(v_\alpha)$, where $d(v_\alpha)$ is the number of recorded mortality cases and $n(v_\alpha)$ is the size of the population at risk. Let us assume for now that all entities v_α have similar shapes and sizes, with a uniform population density. These entities can thus be referenced geographically by their centroids with the vector of spatial

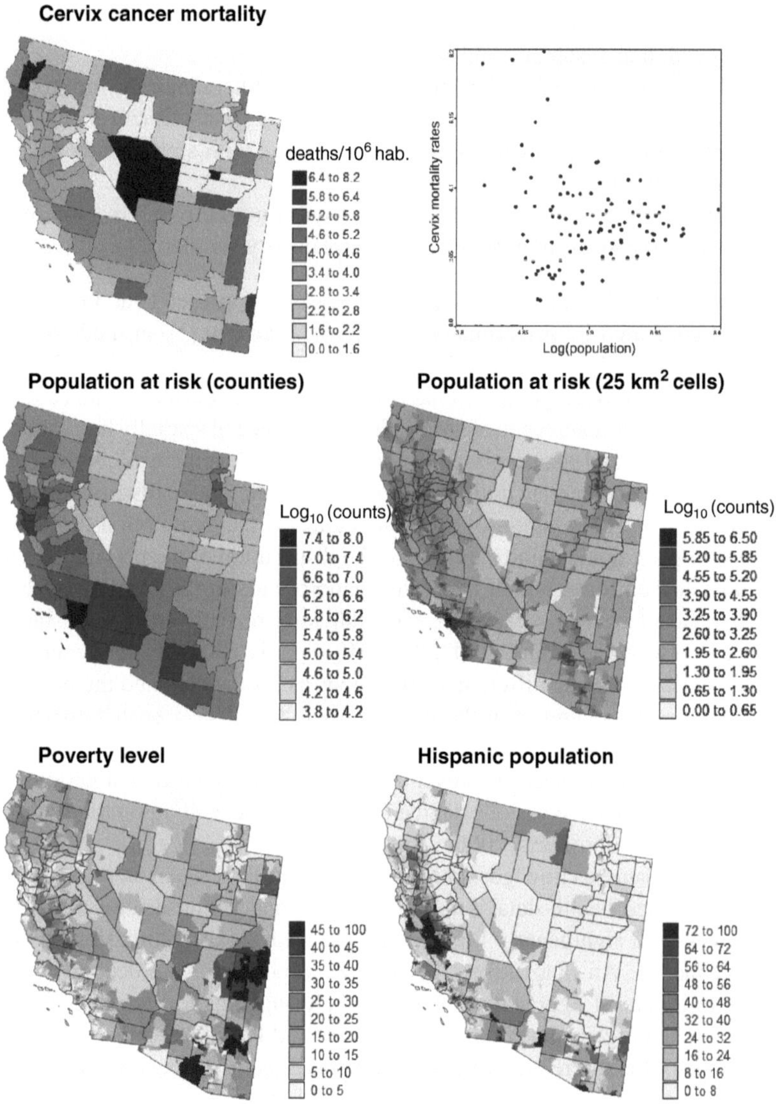

Fig. 1 Geographical distribution of cervix cancer mortality rates recorded for white females over the period 1970–1994, and the corresponding population at risk (aggregated within counties or assigned to 25 km^2 cells). Scatterplot illustrates the larger variance of rates computed from sparsely populated counties. Bottom maps show two putative risk factors: percentage of habitants living below the federally defined poverty line, and percentage of Hispanic females

coordinate's $\mathbf{u}_\alpha = (\mathrm{x}_\alpha, \mathrm{y}_\alpha)$. The disease count $d(\mathbf{u}_\alpha)$ is interpreted as a realization of a random variable $D(\mathbf{u}_\alpha)$ that follows a Poisson distribution with one parameter (expected number of counts) that is the product of the population size $n(\mathbf{u}_\alpha)$ by the local risk $R(\mathbf{u}_\alpha)$, see Goovaerts (2005a) for more details.

In Poisson kriging (PK), the risk over a given entity v_α is estimated as a linear combination of the kernel rate $z(\mathbf{u}_\alpha)$ and the rates observed in (K-1) neighboring entities:

$$\hat{r}_{PK}(\mathbf{u}_\alpha) = \sum_{i=1}^{K} \lambda_i(\mathbf{u}_\alpha) z(\mathbf{u}_i) \tag{1}$$

where $\lambda_i(\mathbf{u}_\alpha)$ is the weight assigned to the rate $z(\mathbf{u}_i)$ when estimating the risk at $\mathbf{u}_\alpha$. The K weights are the solution of the following system of linear equations:

$$\begin{aligned} &\sum_{j=1}^{K} \lambda_j(\mathbf{u}_\alpha)\left[C_R(\mathbf{u}_i - \mathbf{u}_j) + \delta_{ij}\frac{m*}{n(\mathbf{u}_i)}\right] + \mu(\mathbf{u}_\alpha) = C_R(\mathbf{u}_i - \mathbf{u}_\alpha) \quad \mathrm{i} = 1,\ldots,K \\ &\sum_{j=1}^{K} \lambda_j(\mathbf{u}_\alpha) = 1 \end{aligned} \tag{2}$$

where δ_{ij}=1 if $\mathbf{u}_i$=$\mathbf{u}_j$ and 0 otherwise, and $m*$ is the population-weighted mean of the N rates. The addition of an "error variance" term, $m*/n(\mathbf{u}_i)$, for a zero distance accounts for variability arising from population size, leading to smaller weights for less reliable data (i.e. measured over smaller populations). The prediction variance associated with the estimate (1) is computed using the traditional formula for the ordinary kriging variance:

$$\sigma^2_{PK}(\mathbf{u}_\alpha) = C_R(0) - \sum_{i=1}^{K} \lambda_i(\mathbf{u}_\alpha) C_R(\mathbf{u}_i - \mathbf{u}_\alpha) - \mu(\mathbf{u}_\alpha) \tag{3}$$

The computation of kriging weights and kriging variance (Equations (2) and (3)) requires knowledge of the covariance of the unknown risk, $C_R(\mathbf{h})$, or equivalently its semivariogram $\gamma_R(\mathbf{h})$=$C_R(0)$- $C_R(\mathbf{h})$. Following Monestiez et al. (2006) the semivariogram of the risk is estimated as:

$$\hat{\gamma}_R(\mathbf{h}) = \frac{1}{2\sum_{\alpha=1}^{N(\mathbf{h})} \frac{n(\mathbf{u}_\alpha)n(\mathbf{u}_\alpha+\mathbf{h})}{n(\mathbf{u}_\alpha)+n(\mathbf{u}_\alpha+\mathbf{h})}} \sum_{\alpha=1}^{N(\mathbf{h})} \left\{ \frac{n(\mathbf{u}_\alpha)n(\mathbf{u}_\alpha+\mathbf{h})}{n(\mathbf{u}_\alpha)+n(\mathbf{u}_\alpha+\mathbf{h})} \left[z(\mathbf{u}_\alpha) - z(\mathbf{u}_\alpha+\mathbf{h})\right]^2 - m* \right\} \tag{4}$$

where the different pairs $[z(\mathbf{u}_\alpha) - z(\mathbf{u}_\alpha+\mathbf{h})]$ are weighted by the corresponding population sizes to homogenize their variance.

2.1 Area-to-Area (ATA) Poisson Kriging

In the situation where the geographical entities have very different shapes and sizes, areal data can not be simply collapsed into their respective polygon centroids. Following the terminology in Kyriakidis (2004), ATA kriging refers to the case where both the prediction and measurement supports are blocks (or areas) instead of points. The PK estimate (1) for the areal risk value $r(v_\alpha)$ thus becomes:

$$\hat{r}_{PK}(v_\alpha) = \sum_{i=1}^{K} \lambda_i(v_\alpha) z(v_i) \tag{5}$$

The Poisson kriging system (2) is now written as:

$$\begin{aligned} &\sum_{j=1}^{K} \lambda_j(v_\alpha) \left[\bar{C}_R(v_i, v_j) + \delta_{ij} \frac{m*}{n(v_i)} \right] + \mu(v_\alpha) = \bar{C}_R(v_i, v_\alpha) \quad \mathrm{i} = 1,\ldots,K \\ &\sum_{j=1}^{K} \lambda_j(v_\alpha) = 1. \end{aligned} \tag{6}$$

The main change is that point-to-point covariance terms $C_R(\mathbf{u}_i - \mathbf{u}_j)$ are replaced by area-to-area covariances $\bar{C}_R(v_i, v_j) = \mathrm{Cov}\{Z(v_\mathrm{i}), Z(v_\mathrm{j})\}$. Like in the traditional block kriging, those covariances are approximated by the average of the point support covariance $C(\mathbf{h})$ computed between any two locations discretizing the areas v_i and v_j :

$$\bar{C}_R(v_i, v_j) = \frac{1}{\sum_{s=1}^{P_i} \sum_{s'=1}^{P_j} w_{ss'}} \sum_{s=1}^{P_i} \sum_{s'=1}^{P_j} w_{ss'} C(\mathbf{u}_s, \mathbf{u}_{s'}) \tag{7}$$

where P_i and P_j are the number of points used to discretize the two areas v_i and v_j, respectively. For the example of Fig. 1 a grid with a spacing of 5 km was overlaid over the study area, yielding a total of 11 to 2,082 discretizing points per county depending on its area. The high-resolution population map in Fig. 1 clearly shows the heterogeneous distribution of population within counties. To account for spatially varying population density in the computation of the area-to-area covariance, the weights $w_{ss'}$ were identified to the product of population sizes within the 25 km^2 cells centred on the discretizing point $\mathbf{u}_s$ and $\mathbf{u}'_s$:

$$w_{ss'} = n(\mathbf{u}_s) \times n(\mathbf{u}_{s'}) \text{ with } \sum_{s=1}^{P_i} n(\mathbf{u}_s) = n(v_i) \text{ and } \sum_{s'=1}^{P_j} n(\mathbf{u}_{s'}) = n(v_j) \tag{8}$$

The kriging variance for the areal estimator is computed as:

$$\sigma_{PK}^2(v_\alpha) = \bar{C}_R(v_\alpha, v_\alpha) - \sum_{i=1}^{K} \lambda_i(v_\alpha) \bar{C}_R(v_i, v_\alpha) - \mu(v_\alpha) \tag{9}$$

where $\bar{C}_R(v_\alpha, v_\alpha)$ is the within-area covariance that depends on the form of the geographical entity v_α and decreases as its area increases. Thus, ignoring the size of the prediction support in the computation of the kriging variance (3) can lead to a systematic overestimation of the prediction variance of large blocks.

2.2 Area-to-Point (ATP) Poisson Kriging

A major limitation of choropleth maps is the common biased visual perception that larger rural and sparsely populated areas are of greater importance. A solution is to create continuous maps of mortality risk, which amounts to perform a disaggregation or area-to-point interpolation. At each discretizing point $\mathbf{u}_s$ within an entity v_α, the risk $r(\mathbf{u}_s)$ can be estimated as the following linear combination of areal data:

$$\hat{r}_{PK}(\mathbf{u}_s) = \sum_{i=1}^{K} \lambda_i(\mathbf{u}_s) z(v_i) \tag{10}$$

The Poisson kriging system is similar to system (6), except for the right-hand-side term where the area-to-area covariances $\bar{C}_R(v_i, v_\alpha)$ is replaced by the area-to-point covariance $\bar{C}_R(v_i, \mathbf{u}_s)$. The latter is approximated by a procedure similar to the one described in equation (7). A critical property of the ATP kriging estimator is its coherence, that is the aggregation of the P_α point risk estimates within any given entity v_α yields the areal risk estimate $\hat{r}_{PK}(v_\alpha)$:

$$\hat{r}_{PK}(v_\alpha) = \frac{1}{n(v_\alpha)} \sum_{s=1}^{P_\alpha} n(\mathbf{u}_s) \hat{r}_{PK}(\mathbf{u}_s) \tag{11}$$

Condition (11) differs from the constraint commonly found in the geostatistical literature (Kyriakidis, 2004) in that: 1) the observation $z(v_\alpha)$ is uncertain, hence it is the reproduction of the PK risk estimate $\hat{r}_{PK}(v_\alpha)$ that is imposed, and 2) the incorporation of the population density in the computation of the areal covariance implies that it is the population-weighted average of the point risk estimates, not their arithmetical average, that satisfies the coherence condition. The constraint (11) is satisfied if the same K areal data are used for the estimation of the P_α point risk estimates. Indeed, in this case the population-weighted average of the right-hand-side covariance terms of the K ATP kriging systems is equal to the right-hand-side covariance of the single ATA kriging system:

$$\begin{aligned} \frac{1}{n(v_\alpha)} \sum_{s=1}^{P_\alpha} n(\mathbf{u}_s) \bar{C}_R(v_i, \mathbf{u}_s) &= \frac{1}{n(v_\alpha)} \sum_{s=1}^{P_\alpha} n(\mathbf{u}_s) \left[\frac{1}{n(v_i)} \sum_{s'=1}^{P_i} n(\mathbf{u}_{s'}) C(\mathbf{u}_{s'}, \mathbf{u}_s) \right] \\ &= \bar{C}_R(v_i, v_\alpha) \, , \end{aligned} \tag{12}$$

per relations (7) and (8). Therefore, the following relationship exists between the two sets of ATA and ATP kriging weights:

$$\lambda_i(v_\alpha) = \frac{1}{n(v_\alpha)} \sum_{s=1}^{P_\alpha} n(\mathbf{u}_s)\lambda_i(\mathbf{u}_s) \quad i = 1, \ldots, K \tag{13}$$

which ensures the coherence of the estimation.

2.3 Deconvolution of the Semivariogram of the Risk

Both ATA and ATP kriging require knowledge of the point support covariance of the risk $C(\mathbf{h})$, or equivalently the semivariogram $\gamma(\mathbf{h})$. This function cannot be estimated directly from the observed rates, since only aggregated data are available. Derivation of a point support semivariogram from the experimental semivariogram of areal data is called "deconvolution", an operation that is frequent in mining and has been the topic of much research (Journel and Huijbregts, 1978). However, in typical mining applications all blocks (areas) have the same size and shape, which makes the deconvolution reasonably straightforward. Goovaerts (2008) proposed an iterative approach to conduct the deconvolution in presence of irregular geographical units. This innovative algorithm starts with the derivation of an initial deconvoluted model $\gamma^{(0)}(\mathbf{h})$; for example the model $\gamma_{\mathrm{R}}(\mathbf{h})$ fitted to the areal data. This initial model is then regularized using the following expression:

$$\gamma_{regul}(\mathbf{h}) = \bar{\gamma}^{(0)}(v, v_h) - \bar{\gamma}_h^{(0)}(v, v) \tag{14}$$

where $\bar{\gamma}^{(0)}(v, v_h)$ is the area-to-area semivariogram value for any two counties separated by a distance h. It is approximated by the population-weighted average (7), using $\gamma^{(0)}(\mathbf{h})$ instead of $C(\mathbf{h})$. The second term, $\bar{\gamma}_h^{(0)}(v, v)$, is the within-area semivariogram value. Unlike the expression commonly found in the literature, this term varies as a function of the separation distance since smaller areas tend to be paired at shorter distances. To account for heterogeneous population density, the distance between any two counties is estimated as a population-weighted average of distances between locations discretizing the pair of counties:

$$Dist(v_i, v_j) = \frac{1}{\sum_{s=1}^{P_i} \sum_{s'=1}^{P_j} n(\mathbf{u}_s)n(\mathbf{u}_{s'})} \sum_{s=1}^{P_i} \sum_{s'=1}^{P_j} n(\mathbf{u}_s)n(\mathbf{u}_{s'}) \left\|\mathbf{u}_s - \mathbf{u}_{s'}\right\| \tag{15}$$

Note that the block-to-block distances (15) are numerically very close to the Euclidian distances computed between population-weighted centroids (Goovaerts, 2006b). The theoretically regularized model,$\gamma_{regul}(\mathbf{h})$, is compared to the model fitted to experimental values, $\gamma_{\mathrm{R}}(\mathbf{h})$, and the relative difference between the two curves, denoted D, is used as optimization criterion. A new candidate point-support semivariogram $\gamma^{(1)}(\mathbf{h})$ is derived by rescaling of the initial point-support model $\gamma^{(0)}(\mathbf{h})$, and then regularized according to expression (14). Model $\gamma^{(1)}(\mathbf{h})$ becomes the new optimum if the theoretically regularized semivariogram model $\gamma^{(1)}_{regul}(h)$ gets closer to the model fitted to areal data, that is if $D^{(1)} < D^{(0)}$. Rescaling coefficients are then updated to account for the difference between $\gamma^{(1)}_{regul}(h)$ and $\gamma_{\mathrm{R}}(\mathbf{h})$, leading to a new candidate model $\gamma^{(2)}(\mathbf{h})$ for the next iteration. The procedure stops when the maximum number of allowed iterations has been tried (e.g. 35 in this paper) or the decrease in the D statistic becomes negligible from one iteration to the next. The use of lag-specific rescaling coefficients provides enough flexibility to modify the initial shape of the point-support semivariogram and makes the deconvolution insensitive to the initial solution adopted. More details and simulation studies are available in Goovaerts (2006b, 2008).

2.4 Application to the Cervix Cancer Mortality Data

Figure 2 (top graph, dark gray curve) shows the experimental and model semivariograms of cervix cancer mortality risk computed from aggregated data using estimator (4) and the distance measure (15). This model is then deconvoluted and, as expected, the resulting model (light gray curve) has a higher sill since the punctual process has a larger variance than its aggregated form. Its regularization using expression (14) yields a semivariogram model that is close to the one fitted to experimental values, which validates the consistency of the deconvolution.

The deconvoluted model was used to estimate aggregated risk values at the county level (ATA kriging) and to map the spatial distribution of risk values within counties (ATP kriging). Both maps are much smoother than the map of raw rates since the noise due to small population sizes is filtered. In particular, the high risk area formed by two central counties in Fig. 1 disappeared, which illustrates how hazardous the interpretation of the map of observed rates can be. The highest risk (4.081 deaths/100,000 habitants) is predicted for Kern County, just west of Santa Barbara County. ATP kriging map shows that the high risk is not confined to this sole county but spreads over four counties, which is important information for designing prevention strategies. By construction, aggregating the ATP kriging estimates within each county using the population density map of Fig. 1 (right medium graph) yields the ATA kriging map.

The map of ATA kriging variance essentially reflects the higher confidence in the mortality risk estimated for counties with large populations. The distribution of population can however be highly heterogeneous in large counties with contrasted urban and rural areas. This information is incorporated in the ATP kriging variance map that shows clearly the location of urban centers, such as Los Angeles, San Francisco, Salt Lake City, Las Vegas or Tucson. The variance of point risk estimates is much larger than the county-level estimates, as expected.

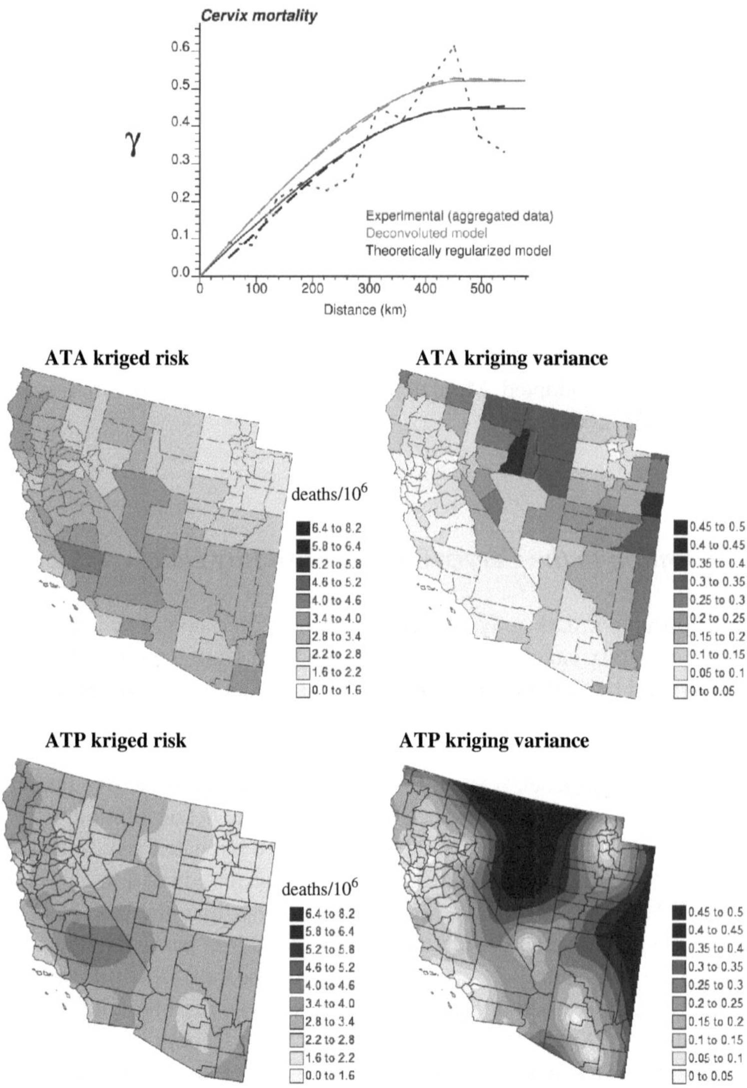

Fig. 2 Experimental semivariogram of the risk estimated from county-level rate data, and the results of its deconvolution (top curve). The regularization of the point support model yields a curve (black dashed line) that is very close to the experimental one. The model is then used to estimate the cervix cancer mortality risk (deaths/100,000 habitants) and associated prediction variance at the county level (ATA kriging) or at the nodes of a 5 km spacing grid (ATP kriging)

3 Detection of Spatial Clusters and Outliers

Mapping cancer risk is a preliminary step towards further analysis that might highlight areas where causative exposures change through geographic space, the presence of local populations with distinct cancer incidences, or the impact of different cancer control methods.

3.1 Local Cluster Analysis (LCA)

The local Moran test aims to detect the existence of local clusters or outliers of high or low cancer risk values (Goovaerts, 2005b). For each county, the so-called LISA (Local Indicator of Spatial Autocorrelation) statistic is computed as:

$$LISA(v_\alpha) = \left[\frac{z(v_\alpha) - m}{s}\right] \times \left(\sum_{j=1}^{J(v_\alpha)} \frac{1}{J(v_\alpha)} \times \left[\frac{z(v_j) - m}{s}\right]\right) \quad (16)$$

where $z(v_\alpha)$ is the mortality rate for the county being tested, which is referred to as the "kernel" hereafter; $z(v_j)$ are the rates for the $J(v_\alpha)$ neighboring counties that are here defined as units sharing a common border or vertex with the kernel v_α (1-st order queen adjacencies). All values are standardized using the mean m and standard deviation s of the set of risk estimates. Since the standardized values have zero mean, a negative value for the LISA statistic indicates a negative local auto-correlation and the presence of spatial outlier where the kernel value is much lower (higher) than the surrounding values. Cluster of low (high) values will lead to positive values of the LISA statistic.

In addition to the sign of the LISA statistic, its magnitude informs on the extent to which kernel and neighborhood values differ. To test whether this difference is significant or not, a Monte Carlo simulation is conducted, which traditionally consists of sampling randomly and without replacement the global distribution of rates (i.e. sample histogram) and computing the corresponding simulated neighborhood averages. This operation is repeated many times (e.g. $M = 999$ draws) and these simulated values are multiplied by the kernel value to produce a set of M simulated values of the LISA statistic for the entity v_α. This set represents a numerical approximation of the probability distribution of the LISA statistic at v_α, under the assumption of spatial independence. The observed statistic (Equation 16) is compared to the probability distribution, enabling the computation of the probability of not rejecting the null hypothesis of spatial independence. The so-called p-value is compared to the significance level chosen by the user and representing the probability of rejecting the null hypothesis when it is true (Type I error). Every county where the p-value is lower than the significance level is classified as a significant spatial outlier (HL: high value surrounded by low values, and LH: low value surrounded by high values) or cluster (HH: high value surrounded by high values, and LL: low value surrounded by low values). If the p-value exceeds the significance level, the county is declared non-significant (NS).

Figure 3 (left top map) shows the results of the LCA of the observed cervix cancer mortality rates. Only two counties are declared significant HL outliers, a result that must be interpreted with caution given their small population sizes. Indeed, these two counties become non-significant when the analysis is conducted on the map of kriged risks, see Fig. 3 (right top map). Accounting for population size in the analysis reveals a cluster of low risk values in Utah, which likely reflects cultural or religious influence on sexual practices resulting in reduced transmission of human papillomavirus. Yet, the smoothing effect of kriging tends to enhances spatial autocorrelation in the risk map, with the risk of inflating artificially cluster sizes. For example, the one-county HH cluster detected in the middle of the mortality map grows to become an aggregate of seven counties on the map of kriged risks. Another weakness is that the uncertainty attached to the risk estimates (i.e. kriging variance) is ignored in the analysis.

3.2 Stochastic Simulation of Cancer Mortality Risk

Static maps of risk estimates and the associated prediction variance fail to depict the uncertainty attached to the spatial distribution of risk values and do not allow its propagation through local cluster analysis. Instead of a unique set of smooth risk estimates $\{\hat{r}_{PK}(v_\alpha), \alpha = 1, \ldots, N\}$, stochastic simulation aims to generate a set of L equally-probable realizations of the spatial distribution of risk values, $\{r^{(1)}(v_\alpha), \alpha = 1, \ldots, N; l = 1, \ldots, L\}$, each consistent with the spatial pattern of the risk as modeled using the function $\gamma_R(\mathbf{h})$. Goovaerts (2006a) proposed the use of p-field simulation to circumvent the problem that no risk data (i.e. only risk estimates), hence no reference histogram, is available to condition the simulation. The basic idea is to generate a realization $\{r^{(1)}(v_\alpha), \alpha = 1, \ldots, N\}$ through the sampling of the set of local probability distributions (ccdf) by a set of spatially correlated probability values $\{p^{(1)}(v_\alpha), \alpha = 1, \ldots, N\}$, known as a probability field or p-field. Assuming that the ccdf of the risk variable is Gaussian, each risk value can be simulated as:

$$r^{(l)}(v_\alpha) = \hat{r}_{PK}(v_\alpha) + \sigma_{PK}(v_\alpha) y^{(l)}(v_\alpha) \tag{17}$$

where $y^{(l)}(v_\alpha)$ is the quantile of the standard normal distribution corresponding to the cumulative probability $p^{(l)}(v_\alpha)$. $\hat{r}_{PK}(v_\alpha)$ and $\sigma_{PK}(v_\alpha)$ are the ATA kriging estimate and standard deviation, respectively. The L sets of random deviates or normal scores, $\{y^{(l)}(v_\alpha), \alpha = 1, \ldots N\}$, are generated using non-conditional sequential Gaussian simulation with the distance metric (15) and the semivariogram of the risk, $\gamma_R(\mathbf{h})$, rescaled to a unit sill; see Goovaerts (2006a) for a detailed description of the algorithm.

Figure 3 (medium row) shows two realizations of the spatial distribution of cervix cancer mortality risk values generated using p-field simulation. The simulated maps are more variable than the kriged risk map of Fig. 2, yet they are smoother than the map of potentially unreliable rates of Fig. 1. Differences among realizations depict the uncertainty attached to the risk map. For example, Nye County in the center of the map, which has a very high mortality rate (recall Fig. 1) but low

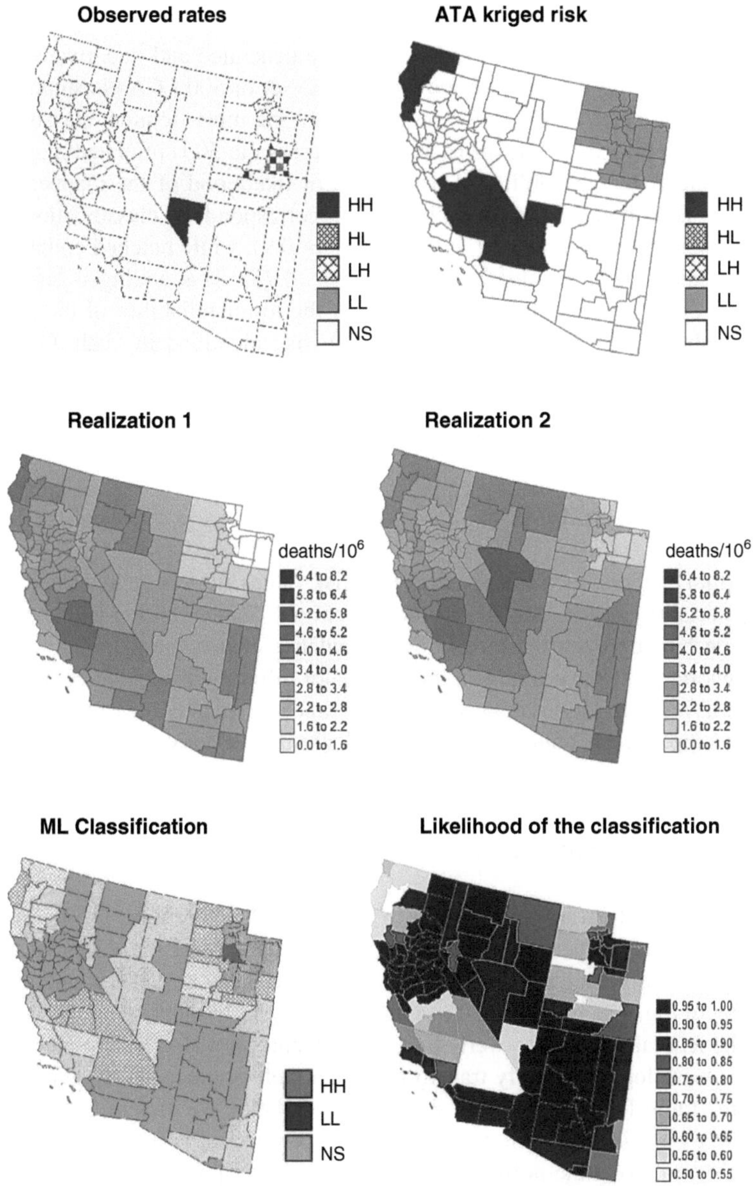

Fig. 3 Results of the local cluster analysis conducted on cervix cancer mortality rates and estimated risks (top maps); see legend description in text. Middle maps show two realizations of the spatial distribution of cervix cancer risk, while the bottom map shows the most likely (ML) classification inferred from 500 realizations. The intensity of the shading increases as the classification becomes more certain (i.e. the likelihood increases)

population, has a simulated risk that is small for realization 1 but large in the next realization. Five hundreds realizations were generated and underwent a local cluster analysis. The information provided by the set of 500 LCAs is summarized at the bottom of Fig. 3. The color code indicates the most frequent classification (maximum likelihood = ML) of each county across the 500 simulated maps. The shading reflects the probability of occurrence or likelihood of the mapped class, see Fig. 3 (right bottom graph). Solid shading corresponds to classifications with high frequencies of occurrence (i.e. likelihood > 0.9), while hatched counties denote the least reliable results (i.e. likelihood < 0.75). This coding is somewhat subjective but leads to a clear visualization of the lower reliability of the clusters of high values relatively to the cluster of low risk identified in Utah. Only one county south of Salt Lake City is declared a significant low-risk cluster with a high likelihood (0.906).

4 Correlation Analysis

Once spatial patterns, such as clusters of high risk values, have been identified on the cancer mortality map, a critical step for cancer control intervention is the analysis of relationships between these features and putative environmental, demographic, socioeconomic and behavioral factors. The major difficulty is the choice of a scale for quantifying correlations between variables that are typically measured over very different supports, e.g. counties and census blocks in this study.

4.1 Ecological Analysis

The most straightforward approach is to aggregate the finer data to the level of coarser resolution data, resulting in a common geographical scale for the correlation analysis. For example, Fig. 4 shows the county-level kriged risk and the two covariates of Fig. 1 aggregated to the same geography: percentage of habitants living below the federally defined poverty line, and percentage of Hispanic females. Both variables were logarithmically transformed, and their product defines the interaction term. Table 1 (first two rows) shows the correlation coefficient between each of the three covariates and the mortality rates before and after application of Poisson kriging. Filtering the noise due to the small number problem clearly enhances the explanatory power of the covariates: the proportion of variance explained (R^2) increases by almost one order of magnitude (6.2% to 48.8%) and all correlation coefficients become highly significant. The uncertainty attached to the risk estimates can be accounted for by weighting each estimate according to the inverse of its kriging variance, leading to slightly larger correlation coefficients and R^2 (Table 1, 3rd row).

So far the significance of the correlation coefficient is tested using the common assumption of independence of observations, which is clearly inappropriate for most

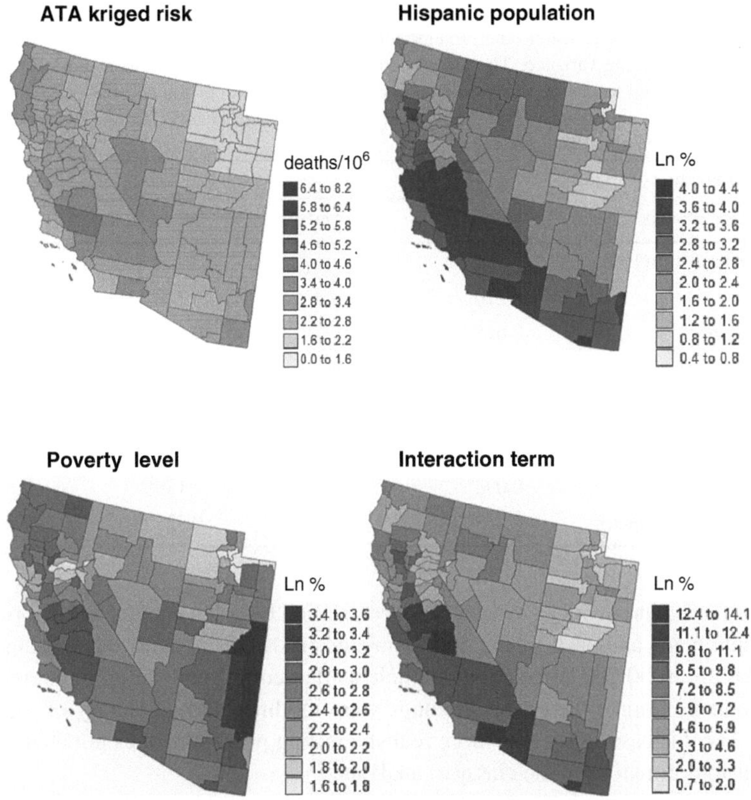

Fig. 4 Maps of cancer mortality risk estimated by Poisson kriging and the logtransformed values of three putative covariates aggregated to the county-level for conducting the ecological analysis

spatial datasets. Instead of computing the correlation between each covariate and the smoothed risk map, the correlation was quantified for each of the 500 risk maps generated by p-field simulation in Section 3.2. This propagation of uncertainty leads to a range of correlation coefficients and R^2 that can be fairly wide, see Table 1 (4th row). Next, this distribution must be compared to the one expected under the assumption of no correlation between mortality risk and each covariate. This reference distribution was obtained empirically in 2 steps. First, the maps of covariates were modified using the spatially ordered shuffling procedure proposed by Goovaerts and Jacquez (2004). The idea is to generate a standard normal random field with a given spatial covariance, e.g. the covariance of the demographic variable in this paper, using non-conditional sequential Gaussian simulation. Each simulated normal score is then substituted by the value of same rank in the distribution of proportion of Hispanic females. To maintain the correlation among covariates, all three covariate maps were modified simultaneously. The operation was repeated 100 times, yielding

Table 1 Results of the correlation analysis of cervix cancer mortality rates and kriged risks with two putative covariates, as well as their interaction. Kriging estimates are weighted according to the inverse of their kriging variance. The use of neutral models allows one to incorporate the spatial uncertainty attached to cancer risk estimates into the computation of the correlation coefficients and testing of their significance (* = significant, ** = highly significant). The last two rows show the results obtained after disaggregation

Correlation with covariates				
Regression models	Hispanic	Poverty	Interaction	R^2(%)
County-level correlation				
Rates	0.210*	0.144	0.240**	6.2
ATA kriging	0.625**	0.473**	0.690**	48.8
ATA kriging (weighted)	0.641**	0.613**	0.729**	54.1
ATA kriging (neutral model)	0.247–0.703**	0.173–0.590**	0.347–0.716**	14.4–52.0
Point-level (25 km^2 cells) correlation				
ATP kriging	0.096**	−0.036**	0.188**	9.8
ATP kriging (weighted)	0.239**	0.090**	0.321**	14.0

100 sets of covariate maps. Second, the correlation between each of the re-ordered covariate maps and each of the 500 simulated risk maps is assessed, leading to a distribution of 50,000 correlation coefficients that corresponds to an hypothesis of independence, since the covariate maps were modified independently of the risk maps. For this case study, this more realistic testing procedure does not change the conclusions drawn from the classical analysis.

Correlations computed between health outcomes and risk factors averaged over geographical entities, such as counties, are referred to as 'ecological correlations'. The unit of analysis is a group of people, as opposed to individual-based studies that relies on data collected for each cancer case. A limitation of ecological analyses is the resolution available which might be too coarse to obtain a detailed view of geographical patterns in disease mortality or incidence. The aggregation may also distort or mask the true exposure/response relationship for individuals, a phenomenon called the *ecological fallacy*. The disaggregation performed by ATP Poisson kriging eliminates the need for using averaged values, and the correlation coefficients between both risk and covariates estimated at the nodes of the 5-km spacing grid are listed in Table 1 (last rows). The correlation is much weaker than for county-level data, which might be due to the noise in the map of socio-demographic variables and/or reflects the scale-dependence of the relationship.

4.2 Geographically-Weighted Regression

The analysis in Table 1 is aspatial and makes the implicit assumption that the impact of covariates is constant across the study area. This assumption is likely unrealistic for large areas which can display substantial geographic variation in demographic, social, economic, and environmental conditions. Several local regression techniques have been developed to account for the non-stationarity of relationships in space (Fotheringham et al., 2002). In geographically-weighted regression (GWR) the regression is performed within local windows centred on each observation or the nodes of a regular grid, and each observation is weighted according to its proximity to the centre of the window. This weighting avoids abrupt changes in the local statistics computed in adjacent windows. Local regression coefficients and associated statistics (i.e. proportion of variance explained, correlation coefficients) can then be mapped to visualize how the explanatory power of covariates changes spatially (Goovaerts, 2005c).

GWR regression was conducted using as dependent variable the mortality risk estimated by ATA and ATP kriging (20 km spacing grid). The centers of the local windows were identified to either the county population-weighted centroids or the nodes of the 5 km spacing grid. The window size was defined as the set of 50 closest observations for both county-level and point-level data. The weight assigned to each observation $\mathbf{u}_\alpha$ was computed as $C_{sph}(h_{0\alpha})/\sigma^2_{PK}(\mathbf{u}_a)$, where $C_{sph}(h_{0\alpha})$ is the value of the spherical covariance at a distance $h_{0\alpha}$ to the center $\mathbf{u}_0$ of the window, and $\sigma^2_{PK}(\mathbf{u}_a)$ is the kriging variance of the ATA or ATP kriged estimate. The range of $C_{sph}(h)$ was set to the distance between the center of the window and the most distant observation. Two statistics are displayed in Fig. 5: the proportion of variance explained within each window (left column) and the covariate with the highest significant correlation coefficient (right column).

The analysis of county-level data (Fig. 5, top maps) shows a clear SW-NE trend in the explanatory power of the local regression models: the higher mortality values along the coast are better explained by the two covariates than the lower risk recorded in Utah. In this state, none of the covariates displays significant correlation with cancer mortality. Poverty level is the best correlated covariate in Northern California while the interaction between economic and demographic variables is the most significant factor in Central California and in the South of the study area. The proportion of Hispanic females is the most significant covariate in a very small transition area between the coast where higher mortality rates and proportion of Hispanic females are observed and Utah where the same two variables have lower values. The computation of the GWR statistics over a regular grid allows one to visualize the within-county variability (Fig. 5, middle maps), yet the analysis is still based on county-level aggregates of socio-demographic variables which can be overly simplistic for some counties, recall Figure 1 (bottom maps). For example, the largest R^2 observed in the Northeast corner of the study area (Fig. 5, left bottom map) corresponds to the Eastern border of a county that display great variation for both proportion of Hispanic females and habitants below the poverty level. Differences between the GWR of county-level and point-support data are even more striking for the map of significantly correlated covariates. The pattern becomes much more complex and correlations are locally negative, see hatched areas in Figure 5 (right bottom map). These maps are mainly used for descriptive purpose and should guide further studies to interpret these local relationships.

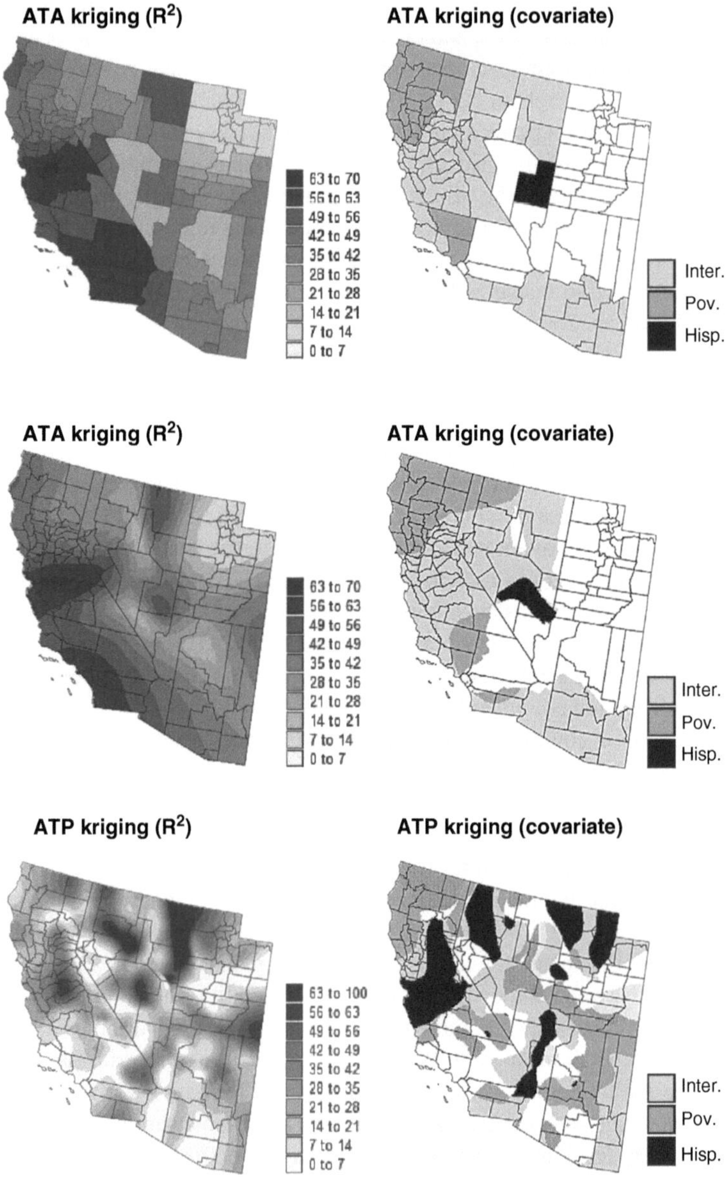

Fig. 5 Results of the geographically-weighted regression applied to the ATA and ATP kriged risk values. Left column displays the maps of the local proportion of variance explained, while the right maps show, for each county or node of the 5 km spacing grid, the covariate (Hispanic population, poverty level, and interaction) that has the highest significant correlation (hatched areas = negative correlation) with cancer mortality risk. The analysis of county-level data conducted at each node of the 5 km spacing grid is shown in the middle maps

Conclusions

The analysis of health data and putative covariates, such as environmental, socio-economic, behavioral or demographic factors, is a promising application for geostatistics. It presents, however, several methodological challenges that arise from the fact that data are typically aggregated over irregular spatial supports and consist of a numerator and a denominator (i.e. population size). Common geostatistical tools, such as semivariograms or kriging, thus cannot be blindly implemented in environmental epidemiology. This paper demonstrated how recent developments in other disciplines, such as ecology for Poisson kriging or remote sensing for area-to-point kriging, can foster the advancement of health geostatistics. Capitalizing on these results and an innovative approach for semivariogram deconvolution, this paper presented the first study where the size and shape of administrative units, as well as the population density, is incorporated into the filtering of noisy mortality rates and the mapping of the corresponding risk at a fine scale (i.e. disaggregation).

Like in other disciplines, spatial interpolation is rarely a goal per se; rather it is a step along the decision-making process. In epidemiology one main concern is to establish the rationale for targeted cancer control interventions, including consideration of health services needs, and resource allocation for screening and diagnostic testing. It is thus important to delineate areas with significantly higher mortality or incidence rates, as well as to analyze relationships between health outcomes and putative risk factors. The uncertainty attached to cancer maps needs however to be propagated through this analysis, a task that geostatisticians have been tackling for several decades using stochastic simulation. Once again the implementation of this approach in epidemiology faces specific challenge, such as the absence of measurements of the target attribute. This paper introduced the application of p-field simulation to generate realizations of cancer mortality maps, which allows one to quantify numerically how the uncertainty about the spatial distribution of health outcomes translates into uncertainty about the location of clusters of high values or the correlation with covariates. Last, this study demonstrated the limitation of a traditional aspatial regression analysis, which ignores the geographic variations in the impact of covariates.

The field of health geostatistics is still in its infancy. Its growth cannot be sustained, or at least is meaningless, if it does not involve the end-users who are the epidemiologists and GIS specialists working in health departments and cancer registries. Critical components to its success include the publication of applied studies illustrating the merits of geostatistics over current methods, training through short courses and updating of existing curriculum, as well as the development of user-friendly software. The success of mining and environmental geostatistics, as we experience it today, can be traced back to its development outside the realm of spatial statistics, through the close collaboration of mathematically minded individuals and practitioners. Health geostatistics will prove to be no different.

Acknowledgments This research was funded by grant R44-CA105819-02 from the National Cancer Institute. The views stated in this publication are those of the author and do not necessarily represent the official views of the NCI.

References

Fotheringham AS, Brunsdon C, Charlton M (2002) Geographically weighted regression: the analysis of spatially varying relationships. Wiley, Chichester

Friedell GH, Tucker TC, McManmon E, Moser M, Hernandez C, Nadel M (1992) Incidence of dysplasia and carcinoma of the uterine cervix in an Appalachian population. J Nat Cancer Inst 84:1030–1032

Goovaerts P (2005a) Geostatistical analysis of disease data: estimation of cancer mortality risk from empirical frequencies using Poisson kriging. Int J Health Geogr 4:31

Goovaerts P (2005b) Detection of spatial clusters and outliers in cancer rates using geostatistical filters and spatial neutral models. In: Renard Ph, Demougeot-Renard H, and Froidevaux R (eds) geoENV V-Geostatistics for Environmental Applications. Springer-Verlag, Berlin, Germany, pp 149–160

Goovaerts P (2005c) Analysis and detection of health disparities using Geostatistics and a space-time information system. The case of prostate cancer mortality in the United States, 1970–1994. Proceedings of GIS Planet 2005, Estoril, May 30-June 2

Goovaerts P (2006a) Geostatistical analysis of disease data: visualization and propagation of spatial uncertainty in cancer mortality risk using Poisson kriging and p-field simulation. Int J Health Geogr 5:7

Goovaerts P (2006b) Geostatistical analysis of disease data: accounting for spatial support and population density in the isopleth mapping of cancer mortality risk using area-to-point Poisson kriging. Int J Health Geogr 5:52

Goovaerts (2008) Kriging and semivariogram deconvolution in presence of irregular geographical units. Math Geology 40, in press

Goovaerts P, Jacquez GM (2004) Accounting for regional background and population size in the detection of spatial clusters and outliers using geostatistical filtering and spatial neutral models: the case of lung cancer in Long Island New York. Int J Health Geogr 3:14

Gotway CA Young LJ (2002) Combining incompatible spatial data. J Am Stat Assoc 97:632–648

Gotway CA, Young LJ (2005) Change of support: an inter-disciplinary challenge. In: Renard Ph, Demougeot-Renard H, and Froidevaux R (eds) geoENV V - Geostatistics for environmental applications. Springer-Verlag, Berlin, Germany, pp 1–13

Journel AG, Huijbregts CG (1978) Mining Geostatistics. Academic Press, London

Kyriakidis P (2004) A geostatistical framework for area-to-point spatial interpolation. Geogr Anal 36:259–289

Monestiez P, Dubroca L, Bonnin E, Durbec JP Guinet C (2006) Geostatistical modelling of spatial distribution of *Balenoptera physalus* in the Northwestern Mediterranean Sea from sparse count data and heterogeneous observation efforts. Ecol Modell 193:615–628

Waller LA, Gotway CA (2004) Applied Spatial Statistics for public health data. John Wiley and Sons, New Jersey

Early Detection and Assessment of Epidemics by Particle Filtering

C. Jégat, F. Carrat, C. Lajaunie and H. Wackernagel

Abstract Influenza infects from 5% to 20% of the population during an epidemic episode, which typically lasts a few weeks, and, as the conditions leading to an outbreak are not well understood, the moment of its start is difficult to foresee. The early detection of an epidemic would however make it possible to limit its impact by adopting appropriate actions—this is particularly desirable in account of the threat of a major pandemic. As a side-product of the forecasting system an estimation of the total number of infected people in each region at every time step is provided.

We first present the classical epidemiological model for contagious diseases and explain the way regionalization has been incorporated into it. Then we describe the assimilation technique that is used to process the data, which are daily reported cases of influenza-like illness. Tests on simulated data are presented to illustrate the efficiency of the process and finally real observational data from the French Sentinelles network are processed.

Introduction: the *Sentinelles* Network Data

The UMR-S 707 of the INSERM and the Geostatistics Group of the Centre de Geosciences have been collaborating since several years on a prototype system for assimilating data from a sentinel network of voluntary general practitioners into a stochastic epidemiological model (Biboud, 2002; Bui, 2001) The model includes temporal evolution of the epidemic state and a basic regionalization. The improvements made with respect to those studies were mainly concerned regionalization aspects.

The observational data are provided by the French *Sentinelles* network. This network gathers 1200 physicians (among the 60000 general practitioners of metropolitan France) who connect themselves about once every week to the internet to report the number of cases of influenza-like-diseases (as well as 8 other diseases) they have observed among their patients.

H. Wackernagel
INSERM UMR-S 707, Paris, France; Ecole des Mines de Paris - Geostatistics group, Fontainebleau, France
e-mail: hans.wackernagel@ensmp.fr

A. Soares et al. (eds.), *geoENV VI – Geostatistics for Environmental Applications*, 23–35

The observed cases may give an idea of the evolution of the epidemic, but they are subject to many uncertainties. First, the diagnosis leading a physician to report a case is only based on the symptoms of the disease and usually no virologic analysis is carried out. The physician mainly looks for the characteristic symptoms of influenza: sudden fever above 39°C, breathing troubles and muscle soreness. Now, these symptoms could also the consequence of other diseases and it is also possible that some patients could have different symptoms, or no rcaction at all, and still bc contagious.

Furthermore, an unknown proportion of the population does not consult a doctor when they have got influenza. In this study, given the lack of precise data and according to estimations made at the INSERM, this proportion is assumed to be 50%.

Finally, doctors do not regularly connect themselves to the network, and sometimes forget to report an absence of cases, and yet this would also be an important information. This irregularity also may lead to an accumulation of cases on the day of each connection, whereas the patients that were reported may have consulted already a few days earlier.

The cases have therefore been redistributed over the period prior to correction to take this irregularity into account. A *Sentinelles'* doctor is considered as active if the gap between two consecutive connections is inferior or equal to 12 days. For example, if a doctor connects on a Friday, and his last connection was on the Monday of the same week, he will be considered active on every day from Tuesday to Friday, and if he reports 6 cases, they will be spread out uniformly over these 4 days, that is 1.5 cases per day. This uniform distribution may appear unrealistic since it is more probable that a case reported one day has consulted the same day or the day before, rather than several days earlier. However, this reallocation, even though imperfect, seems acceptable and has been adopted for this study.

In this study, the time scale will be the day, in order to enable an early detection of the epidemic, and the geographic scale will be the French administrative regions (all 21 of the metropolitan, continental regions). Figure 1 shows the number of cases reported in four different regions of France for the last 5 years, as well as the number of active doctors.

1 The Epidemic Model

The SIR model, widely used to model epidemic diseases (Daley and Gani, 1999; Andersson and Britton, 2000) partitions the populations into three categories:

- *Susceptible*: the people free from the disease and without specific immunity,
- *Infected*: the people infected by the virus and who are still infectious,
- *Removed*: the people who have recovered from their illness, are no longer infectious and are immunized.

We will consider a discrete time-space structured model. The population in each region x at time t is partitioned into by $S_t(x)$, $I_t(x)$ and $R_t(x)$. The total population

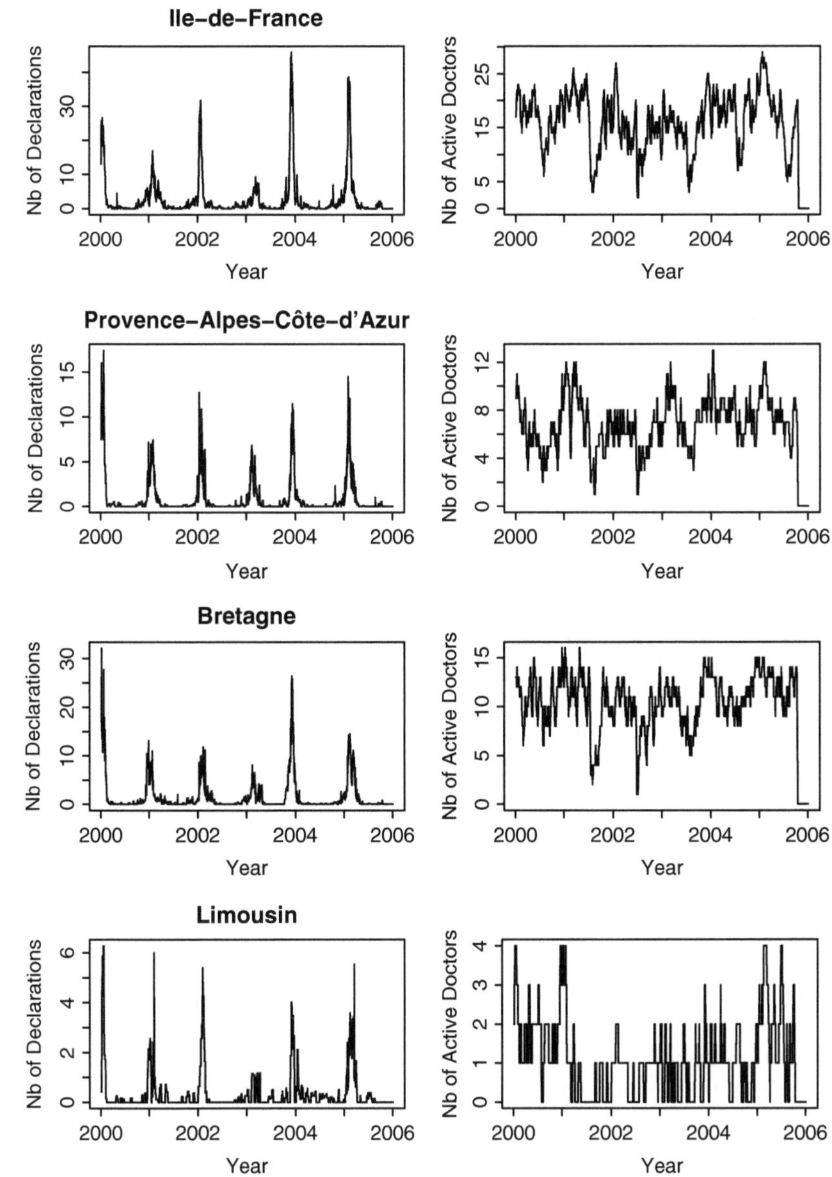

Fig. 1 Daily declarations of the *Sentinelles* network and number of active doctors over the last 5 years, in four French regions

in each region $N(x)$ is assumed constant during the course of an epidemic, so we have $\forall t$: $N(x) = S_t(x) + I_t(x) + R_t(x)$.

The new cases in a region x result from contacts between susceptibles in given region x and infectives in the other regions y. If the contact rate per time unit between x and y is denoted by $a(x,y)$, and if each $S-I$ contact produces a contamination with probability τ, each susceptible avoids contamination during a time interval with probability:

$$q(x,t) = \prod_{y} \left(1 - \tau \frac{I(y)}{N(y)}\right)^{a(x,y)}$$

Therefore, under this simple model which assumes independence of contamination events, the number of new infectives at x is a binomial variable with parameters $S(x)$ and $1-q(x,t)$.

The observed behavior of real epidemics, in which bursts follow smooth periods, suggests a sharp increase of the probability of contamination τ during epidemic episodes. This increase could be due to the income of a new virus strain. Thus we modeled τ with independent two-state Markov chains for the different regions and write accordingly $\tau_t(x)$.

The next step in the modeling is about the recovery from disease. A Markovian structure, when only one category of infective is considered, would imply an exponential distribution of individual infectious periods. To obtain more realistic distributions, without sacrificing the Markovian property of the model, we further divided $I_t(x)$ into $I_t^0(x)$, ..., $I_t^6(x)$ according to time since infection. Recovery probability for each category in one time step is $p_{rec}^k; k = 1, \cdots, 7$, with $p_{rec}^7 = 1$, since it is assumed that infection cannot last more than 7 days.

Next we need to model how the declarations inform us on the epidemic state. We assume that each new infected subject is registered by *Sentinelles* with a known probability, which depends on the coverage of the region. Thus the number of declarations in x is conditionally binomial, with parameters $S_{t-1}(x) - S_t(x)$, $p_{dec}(x)$.

2 Regionalization of the Model

This section describes the calculation of the contact rate $a(x, y)$, which is a main parameter of the regionalized model.

2.1 The Origin-destination Matrix

A common tool in population flux modeling is the *origin-destination matrix*. This matrix is defined by: $M = (M_{xy})$ with $M_{x,y}$ being the number of people moving from x to y per time unit.

There are several ways to calculate this matrix. To describe precisely the population fluxes between regions, the ideal way would be to have access to the road and rail traffic data. Unfortunately, these data are not easy to find and we will finally use a mathematical model of the fluxes instead.

2.2 The Gravitational Model

The *gravitational model*, inspired by Newton's laws, assumes that the flux from x to y is proportional to the populations in x and y and also inversely proportional to a power of the distance between the regions. We thus write

$$T_{xy} = k \frac{P_x^\lambda P_y^\alpha}{d_{xy}^\beta}$$

with:

T_{xy} the number of people moving from x to y,
P_x the population at x,
d_{xy} the distance between x and y (based on their main cities),
k a proportionality constant depending on the time step,
λ the force of emission of the source,
α the force of attraction of the destination, mainly on the basis of its economic activities,
β the transport friction, depending on the efficiency of the transport system.

The last three parameters should be adjusted with observed data, which however lacked in this study, and so λ and α were set to 1, whereas β was set to 2.

2.3 Contact Probabilities

The *origin-destination matrix* can be set up in different ways. One of them assumes that the people effectively move, leading to a revision of the total population within each region at each step. This approach may be acceptable in the case of important population fluxes, for example on holidays, but it is far too rough to describe daily contacts between regions, which are mainly caused by commuting.

We choose a different approach and define the *contact matrix*. We assume that people from two regions are in contact and therefore can infect each other, but at the end of each time step, everyone is back home. The population within each region therefore remains constant. We further assume that the disease does not affect the population movements.

The contact matrix is generated according to the following choices:

- The total number of contacts for a single person (we set it to 100 in this study).
- We then determine the number of contacts each person has with people in the same region. In the present setting, the model assumes that the more a region is populated, the more it is attractive, and the lower is the within-region contact rate. This led to the formula: $N_{xx} = K/Pop(x)$ with $K = 0.000003$.
- The contacts with people in other regions are split on the basis of the origin-destination matrix obtained from the gravity model. Whatever the constant k,

this does not affect the final result since everything is driven by the choice of the number of total contacts and the rate of internal contacts.

- Lastly, the matrix has to be made symmetric. The numbers of total contacts are allowed to vary from a region to another, but they remain close.

3 Data Assimilation with Interacting Particle Simulation

Data assimilation is a term used by meteorologists and oceanographers to refer to the integration of real time observational data into simulation models. Two ingredients are required for this, one being a mechanistic model of the time evolution of the process. This model, which is not necessarily deterministic, is usually based on the physics of the process. The other ingredient is an observational equation, which should realistically model what the observed data tells us about the state of the process.

Sequential data assimilation is often based on some form of Kalman filtering (Bertino et al., 2003) In this paper, we present the application of an alternative method, known as *particle filtering* (Doucet et al., 2001; Oudjane, 2000; Cauchemez, 2005) which can handle highly non-linear dynamics at the expense of performing a large number of simulations in parallel. Approximations of the state conditional distribution are obtained sequentially and are based on populations of simulations (i.e. the particles).

3.1 The Principles of Particle Filtering

Let N_d be the number of regions and N_s the number of simulations. To notation, we denote the state vector at time t by:

$$X_t = \{\tau_t(x)\,,\, I_t(x)\,; x = 1 \ldots N_d\}$$

where I is itself structured according to the elapsed time since contagion. The simulated particles at time t are indexed by an exponent and so X_t^i refers to the ith simulation. Writing $D_{0:t} = D_0, \cdots, D_t$ the observational data up to time t, we look for an approximation of the conditional distribution

$$f(X_t\,|D_{0:t})$$

of the current state given the data. Note that this is a less demanding objective than that of approximating the whole trajectory. Bayes formula, which tells how the exact conditional distribution should be updated, gives a clue on how this should be done. We have:

$$\begin{aligned} f(X_t|D_{0:t}) &= f(X_t|D_t, D_{0:t-1}) \\ &= \frac{f(X_t, D_t|D_{0:t-1})}{\int f(x_t, D_t|D_{0:t-1})\, dx_t} \end{aligned}$$

Now, using two conditional independence properties of the model:

$$\begin{aligned} f(X_t, D_t|D_{0:t-1}) &= \int f(X_t, x_{t-1}, D_t|D_{0:t-1})\, dx_{t-1} \\ &= \int f(D_t|X_t, x_{t-1}, D_{0:t-1})\, f(X_t\,|x_{t-1}, D_{0:t-1}) \\ &\qquad f(x_{t-1}|D_{0:t-1})\, dx_{t-1} \\ &= \int f(D_t|X_t)\, f(X_t\,|x_{t-1})\, f(x_{t-1}|D_{0:t-1})\, dx_{t-1} \end{aligned}$$

we get the update equation:

$$f(X_t|D_{0:t}) = \frac{\int f(D_t|X_t)\, f(X_t\,|x_{t-1})\, f(x_{t-1}|D_{0:t-1})\, dx_{t-1}}{\int f(D_t|x_t)\, f(x_t\,|x_{t-1})\, f(x_{t-1}|D_{0:t-1})\, dx_t, dx_{t-1}}$$

In these equations:

- $f(X_t|x_{t-1})$ represents the system's *evolution* distribution. The term:

$$f(X_t\,|x_{t-1})\, f(x_{t-1}|D_{0:t-1})$$

  can be interpreted as a one step free evolution of the system, starting from a random initial state, generated according to the conditional distribution at $t-1$.
- $f(D_t|X_t)$ represents the *observational process*. It can be interpreted as the likelihood of X_t relatively to data D_t.

The updated distribution can thus be viewed as a reweighted version of the predictive one. The assimilation method follows this algorithm.

To describe the sequence, we assume that at $t-1$ an approximation of $f(X_{t-1}|D_{0:t-1})$ is available in the form of an empirical distribution of N_s particles:

$$\widehat{f_{t-1}} = \frac{1}{N_s}\sum_{i=1}^{N_s} \delta_{X^i_{t-1}}$$

The steps of the algorithm are then as follows:

Prediction step. Free evolution of each particle, according to the model:

$$X^i_{t-1} \rightarrow \widetilde{X^i_t}$$

When, as it is the case for the present SIR model, this evolution is stochastic, independent paths are generated.

Conditioning on D_t. This involves the likelihood calculation, $f(D_t|\widetilde{X^i_t})$ and its normalization, which yields the following probability masses:

$$\omega_i = \frac{f(D_t|\widetilde{X_t^i})}{\sum_j f(D_t|\widetilde{X}_t^j)}$$

At this point, we have the following approximation of the conditional distribution:

$$\widetilde{\mathrm{f}}(X_t|D_{0:t}) = \sum_i \omega_i \, \delta_{\widetilde{X}_t^i}$$

Resampling. The iteration is not yet complete, for we do not have an approximation having the form of an empirical distribution. The idea is to draw a new set of N_s particles independently from $\widetilde{\mathrm{f}}(x_t|D_{0:t})$. This is just a weighted resampling of the set $\widetilde{X}_t^1, \cdots, \widetilde{X}_t^{N_s}$ favoring the particles having the highest likelihood.

There are numerous possible variations of this algorithm. For example the independent resampling, which is not very efficient, can be replaced by better schemes regarding the representation of $\widetilde{\mathrm{f}}$ by empirical distributions.

4 Results

In order to test the efficiency of the model, we have to test the algorithm on simulated data first. The simulation algorithm randomly generates a time series of Markov chain states based on the transition matrix. Then, at each time step, in each region, given this state, the algorithm generates simulated new cases and recovered, computes the susceptible, infected and removed and finally generates the reported cases given the new cases. We then only keep the reported cases and assimilate them. If the algorithm is efficient, we should recover a Markov-chain state time-series similar to the one that generated the cases. We can also compare the estimation of the number of infected to the simulation.

4.1 Results on Simulated Data

In order to test the efficiency of the algorithm, we assimilate simulated observational data (of which we know all the parameters, especially the Markov chain states for all t), hoping to find back the same Markov chain states.

Figure 2 shows the original Markov chain, the simulated cases, the estimated Markov chain and the ratio of the simulated number of infected against the estimated number of infected for two regions.

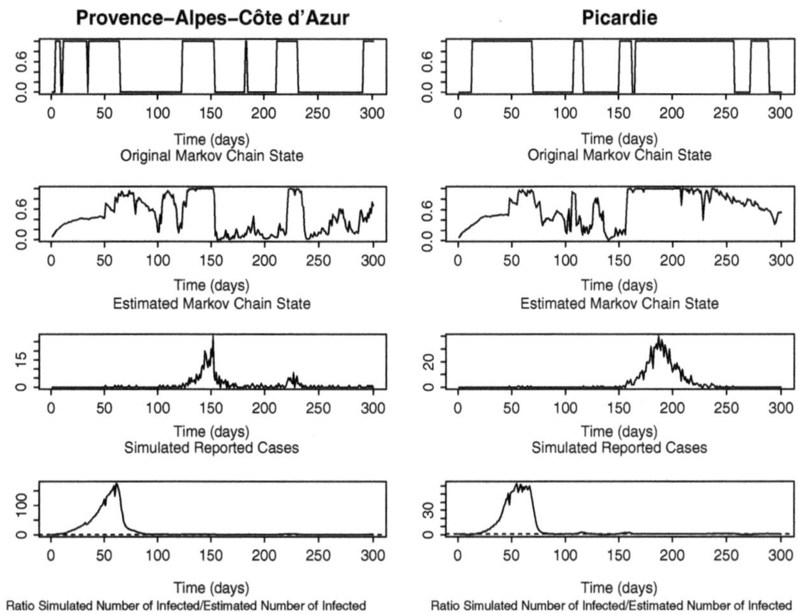

Fig. 2 Assimilation of simulated data (from top to bottom): original Markov chain, estimated Markov chain, simulated declarations, ratio of simulated/estimated number of infected

4.1.1 First Facts

Problems at the Beginning of the Time Period

There are two estimation problems at the beginning. During the time when there are no reported cases, or only few (less than two per day per region), the algorithm is in trouble for finding in which state the Markov chain is, and it underestimates the number of infected.

The underestimation is due to the fact that, in spite of the increasing number of infected in the simulation, the declaration probability is so low that there are no declarations for a long time. Therefore, the algorithm finds out only quite late (at the time of the first reported case) that there are more infected than estimated. Then it catches up on the number of infected by only keeping the particles in an epidemic dynamic and the number of infected is correctly estimated.

In particular, the likelihoods of the particles do not enable to decide between the two types of particles: because the number of infected is low, and because the declaration probability is low, even particles in an epidemic dynamic do not generate enough infected individuals to have a reported case. In that situation both types of particles are just as likely, and the algorithm cannot decide between them.

As said previously the algorithm has to catch up on the number of infected in the early stage of an epidemic. Therefore, only particles in an epidemic dynamic are kept, so that the number of infected progressively increases. Sometimes this is right, at other times the number has increased only because of a contagion within other regions.

4.1.2 Sensitivity Due to Parameter Variations

In this section we discuss the influence of the parameters which are most difficult to estimate.

Transition Probability

In the case of a simple two-state Markov chain model, where we chose to have the same transition probability from 0 to 1 than from 1 to 0, there is only one parameter. It is not actually a parameter which has to be known precisely. If we assume that this probability is totally unknown in reality, then in the assimilation algorithm that probability is only the proportion of particles generated with the same state as the previous particles. Therefore, even if only a few particles are in the right state, these are the ones which are kept in the resampling step. We only have to choose a probability high enough to generate a small number of particle at each step. On the other hand, a too high probability would generate too many epidemic particles at the beginning and lead to an overestimation of the number of infected individuals—even more than it already does. The good news is, however, that in the assimilation of simulations this probability actually has shown to have very little influence on the results.

Contamination Probabilities τ_i

Every contamination probability has to be calculated with reference to a single one. Indeed, in the case of a simple model, an epidemic can start if each infected can infect more than one susceptible. In this model, the number of susceptibles contaminated by an infected is defined as $R = \tau \cdot C \cdot D$, τ being the contamination probability of an S-I contact, C the number of contacts per person, and D the average length of the infection. The probability τ_l above which the epidemic can start is therefore $\tau_l = \frac{1}{C \cdot D} = 0.0018$ in our case, where $C = 100$ and $D = 5.5$. Tests show that even though our model is more complex, this probability does correspond to a case when the epidemic neither starts nor is completely extinct: the number of infected stays constant.

Next, the high and low values of τ must be chosen on each side of τ_l.

The low value does not have a great influence on the results. It only affects how fast the infection stops when dynamics are non-epidemic. However, to avoid total extinction, randomly chosen infected are introduced.

The high value is very hard to estimate from the data. If the state of the Markov chain were known at each time step, the probability could be estimated thanks to the total number of infected and the length of the epidemic. But given that we know neither the state, nor the probability corresponding to the state, this estimation is much harder.

A first rough estimation can be done, prior to the assimilation, by simulating a time series with constant epidemic dynamics. In that case we can avoid to choose a τ for which the epidemic is too strong or too weak.

Another approximate way to adjust this value is simply to look at the estimated Markov chain state. If the value stays high all the time, it is likely that the probability was chosen too low. If the value is oscillating throughout the epidemic period, it is is likely that the probability was set too high.

The Number of Initial Cases

The number of cases in the non-epidemic period is very hard to estimate. Actually, we do not know if the start of an epidemic is caused by a small number of infected remaining from a previous epidemic, or whether it is due to infection coming from abroad. Since we start the assimilation at a time when we know, almost for sure, that the epidemic has not yet started anywhere, this number of infected represents a very small proportion of the population. Moreover, given that we inject randomly a few infected to prevent the extinction of the infection, even if that number is very small, there will be a compensation from this injection. And if it is too high (which can only mean that it generates at least one declaration whereas it should not), it will tend to rapidly decrease.

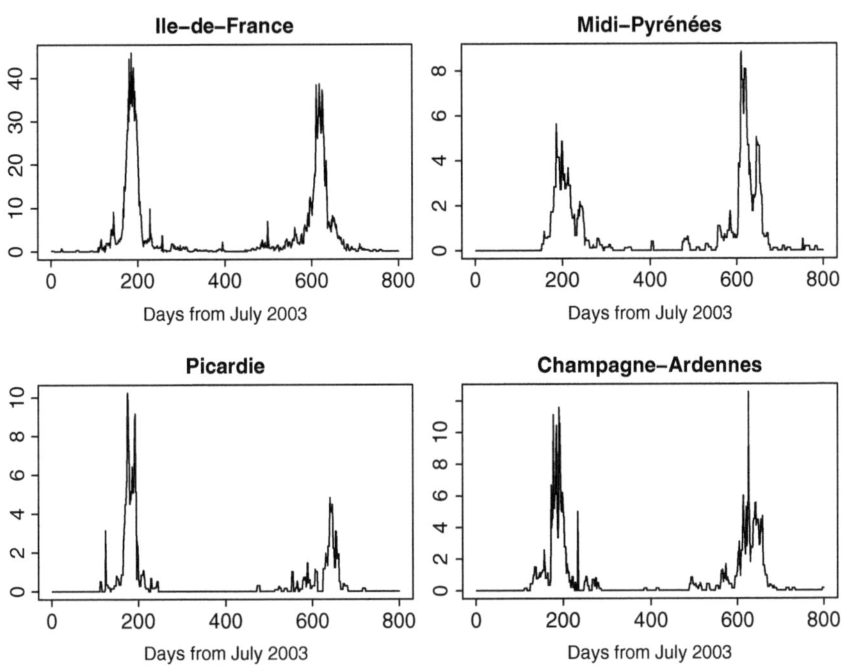

Fig. 3 Reported cases from July 2003 to September 2005

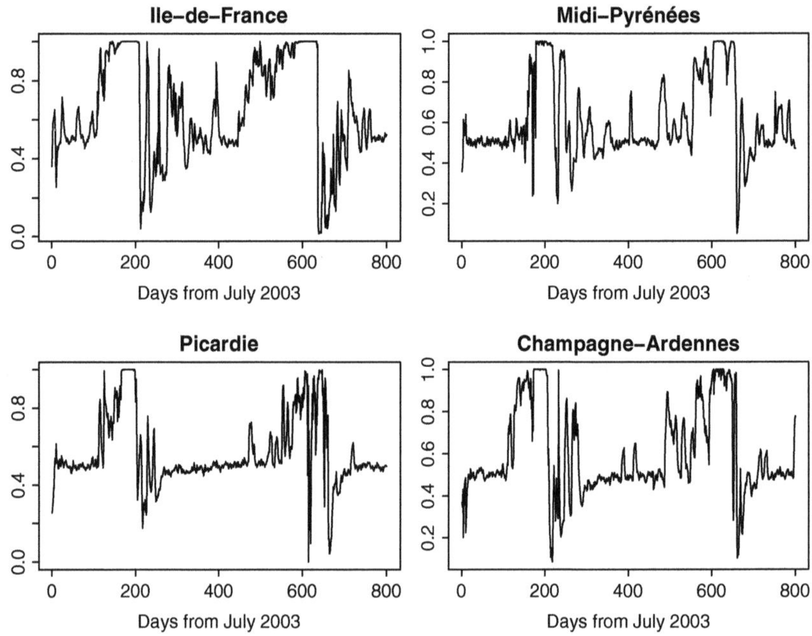

Fig. 4 Assimilation of 2003–2005 data: estimation of the two-state Markov chain

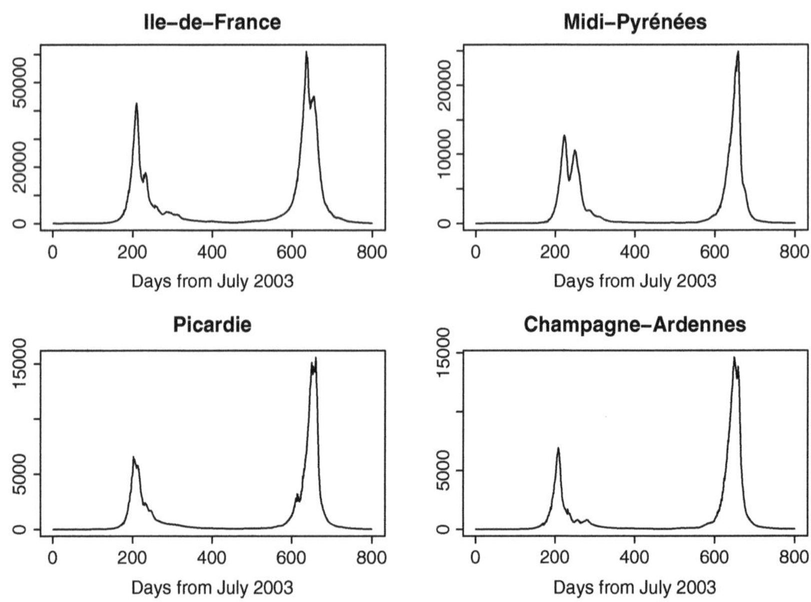

Fig. 5 Assimilation of 2003–2005 data: estimation of number of infected individuals

Other Parameters

The declaration probability (mainly based on the proportion of active doctors) and the recovery probability are relatively well known. This is to say that, given the confidence with which we know them, variations inside that confidence interval did not have a great influence on results.

4.2 Results on Real Data

Finally, we confronted the algorithm to real data taken between July 2003 and July 2005 as shown on Fig. 3. The data were assimilated into a two-state Markov chain model with $\tau_1 = 0.0006$ and $\tau_2 = 0.0027$. Results on Fig. 4 show the seemingly hesitating oscillation of the Markov chain between its two states in non-epidemic periods, but overall the algorithm appears to react fairly quickly to an increase in the number of declarations.

An important side-product of the present forecasting procedure is that it provides an estimation of the total number of infected people in each region at every time step. Fig. 5 displays the time evolution of the estimated number of infected individuals during the two epidemics.

References

Andersson H, Britton T (2000) Stochastic epidemic models and their statistical analysis. Lecture notes in statistics, vol 151. Springer-Verlag, New York, p 137

Bertino L, Evensen G, Wackernagel H (2003) Sequential data assimilation techniques in oceanography. *Int Stat Rev* 71:223–241

Biboud E (2002) Modélisation des épidémies de grippe. Technical Report S-436, Centre de Géostatistique, Ecole des Mines de Paris, Fontainebleau

Bui VAP (2001) Modélisation par chaînes de Markov cachées d'une épidémie de grippe. Technical Report S-420, Centre de Géostatistique, Ecole des Mines de Paris, Fontainebleau

Cauchemez S (2005) Estimation des Paramètres de Transmission dans les Modèles Epidémiques par Échantillonnage de Monte Carlo par Chaîne de Markov. PhD thesis, Université Pierre et Marie Curie, Paris.

Daley DJ, Gani J (1999) Epidemic modelling: an introduction. Cambridge series in mathematical biology. University Press, Cambridge, p 213

Doucet A, de Freitas N, Gordon N (eds) (2001) Sequential Monte Carlo Methods in Practice. Springer-Verlag, New York

Oudjane N (2000) Stabilité et Approximations Particulaires en Filtrage Non-linéaire: Application au Pistage. PhD thesis, Université de Rennes I, Rennes

Improvement of Forecast Noise Levels in Confined Spaces by Means of Geostatistical Methods

G. A. Degan, D. Lippiello, M. Pinzari and G. Raspa

Abstract The aim of this paper is to improve, through the application of geostatistical methods, the results provided by forecasting models for determining the noise level in confined spaces. After a brief glance at the legislative aspect of assessing noise risk in working environments, the importance is stressed of the space-time discretization of the work cycle for the purpose of the aforesaid assessment under the most general conditions, and hence of the need to have the sound pressure level map available for every time element. Attention was then shifted to the spatial reconstruction of the sound pressure levels by means of forecasting models. In particular, in a case prepared ad hoc, the discrepancy was evidenced, by means of a set of regular grid measurements, between forecasting and reality, characterized by a strong bias varying in size from one point to another. The external drift method, applied using a few measurements of the primary variable and the output of the model as the external drift, proved to be very effective in removing the bias.

Introduction

Noise is one of the most common physical agents having an impact on persons under working conditions. From the legislative standpoint, Directive 2003/10/CE adopted in Italy by Decree Law 195/2006 makes it compulsory for the employer, as the one responsible for the productive conditions, to carry out an assessment of the noise exposure risk every four years and to renew this or to integrate it in concomitance with any variation in the company's production cycle. This assessment regards all those involved in production organization and is an operation of basic importance for determining prevention and protection measures, in order to guarantee the safety and health of the workers. In particular the regulations establish the exposure and action limits, i.e. the reference values in terms of the exposure level referred to the eight working hours and peak pressure, corresponding to which the actions and measures for which the employer is responsible are laid down.

G. A. Degan
Dipartimento di Ingegneria Meccanica e Industriale; Università degli studi Roma Tre, Rome, Italy
e-mail: g.alfarodegan@uniroma3.it

A. Soares et al. (eds.), *geoENV VI – Geostatistics for Environmental Applications*, 37–44

Should conditions so require, an appropriate campaign of measurements is foreseen, aimed at calculating the values concerned, as required by the standard ISO 1999:90.

Regarding the choice of methodologies for the assessment concerned, the legislation refers to the specific choices of those responsible. Precisely on this aspect the international community resolved to carry out an in-depth, systematic inquiry for the purpose of identifying the salient steps of the methodological approach (Behar and Plener, 1984; Malchaire and Piette, 1997; Alfaro Degan and Pinzari, 2002). On the basis of the investigations it can be stated that the assessment of the risk should be centred on determining the sound pressure level, the duration of exposure and the space-time characteristics thereof. Precisely the space-time variability of the physical agent in working activity is, in this specific case, one of the most critical factors, requiring a considerable amount of field surveys for carrying out the assessment.

The present study thus falls into this context, with the intention of providing a methodological contribution towards evaluating the exposure to noise of workers in confined surroundings under conditions of space-time variability of sound pressure.

Assessment of the Noise Exposure Level of Workers

The noise exposure level is defined in standard ISO 1999:90, on the basis of which the daily personal exposure, to be compared with the legally allowed limits as per the preceding point, is given by:

$$L_{ep,d} = L_{A,eq,T_E} + 10 \log \frac{T_E}{T_0} \tag{1}$$

where

$$L_{A,eq,Te} = 10 \cdot Log_{10} \left[\frac{1}{T_e} \int_0^{T_e} \left(\frac{p_A(t)}{p_0} \right)^2 dt \right] \tag{2}$$

which represents the equivalent level of the A-weighted sound pressure against the time *Te*, referred to more simply below as the *equivalent level.* In the formula $p_A(t)$ is the sound pressure reaching the worker's ear at time t, p_0 is a reference value equal to 20 μPa and T_e and T_0 are, respectively, the worker's exposure time to noise and the duration of the standard working day ($T_0 = 8$ hours).

Hence the assessment of the exposure implies knowing the sound pressure value at the worker's ear. In cases of a large number of workers and of their mobility within the place of work or of an outstanding space-time variability in the sound field, due for example to the presence of a number of independent sound sources, risk assessment becomes difficult. These considerations indicate the advisability of a discrete approach to the problem based on the subdivision of the working environment into spatial elements S_i according to their functionality with respect

Table 1 Matrix of the *equivalent level*

	T_1	T_j	T_n
S_1			
S_i		$L_{a,eq}(i,j)$	
S_m			

to carrying out the activities, and to the discretization of working hours into time intervals T_j of constant duration T, duly defined (Table 1).

The above-mentioned spatial elements constitute the functional spaces (Alfaro Degan and Pinzari, 2001) and represent the minimum space necessary for the performance of the single activity. The formalization of this approach is defined by means of a matrix, which may be stated to represent the space-time discretization of the sound pressure level.

The element of the matrix represents the *equivalent level* of the functional space in the interval $T_j = t_{j+1} - t_j$. To determine this it is necessary to know the sound pressure level in the time interval Tj at the nodes of a grid.

The worker's exposure to noise is instead controlled by the indicator matrix $\{I_k(i,j)\}$, whose element assumes the value of one if the k-th operator occupies the i-th position in the time interval $t_{j+1} - t_j$, otherwise it assumes a zero value. The exposure time associated with the k-th operator is therefore:

$$T_E(k) = T \sum_i \sum_j I_k(i, j) \tag{3}$$

The *equivalent level* of exposure of operator k, mobile within the productive unit, is thus calculated on the basis of the relation:

$$L_{A,eq,T_E(k)} = 10Log \frac{1}{\sum_i \sum_j I_k(i, j)} \sum_i \sum_j I_k(i, j) \cdot 10^{0.1 L_{A,eq,T}(i,j)} \tag{4}$$

The evaluation procedure is then completed by applying the relation (1) which provides the daily exposure level of operator k to be compared with the regulation limits.

Determination of the Sound Pressure Level in Confined Spaces

To define the sound pressure level in space, there are two possible approaches: the first one foresees the use of forecasting models for confined spaces, while the second one is based on measuring the sound pressure level by means of a targeted campaign of measurements.

In the case of a fixed work station it is convenient to assess noise exposure by means of a certain number of measurements with an appropriate distribution of exposure time. In this regard Lajaunie et al. (1999) developed a method to determine the optimal sampling.

When instead during the carrying out of their activity workers occupy various stations and the noise level varies in both space and time, having recourse to forecasting models is decisive. Under such conditions in fact a campaign of measurements would require the simultaneous monitoring of various points of the space and hence different measuring instruments.

The use of models becomes indispensable when a new working situation has to be designed or in the case of variations or restructuring of the production cycle of an existing situation.

There are three categories in the current panorama of models for forecasting the noise level in confined spaces of large size: the semi-reverberating model (Beranek, 1971), the image sources model (Hodgson, Warnock, 1992) and the ray tracing model (Krokstadt et al. 1968; Farina, 1995).

The semi-reverberating model is based on the reverberating noise level theory which uses the following well known formula for calculating the sound pressure L_p:

$$L_p = L_W + 10Log\left[\frac{Q_\vartheta}{4\pi r^2} + \frac{4}{R}\right] \tag{5}$$

where L_w and Q_θ are respectively the power level and the directivity of the source, r is the source-receiver distance and

$$R = \frac{(1-\bar{\alpha})}{\bar{\alpha}}S$$

is the value of the energy absorbed by all the reflecting surfaces having an overall extent of S and an average sound-absorbing value equal to $\bar{\alpha}$.

The image sources model considers the contribution of multiple reflections as generated by the same number of virtual sources, called *image sources*. Having fixed the number of reflections, the algorithm identifies the position of the image sources and calculates the contribution to the sound field at the receiving point regarding them as free field sources.

The ray-tracing model is based on the assumption of launching from the source a very large number of rays with a certain initial energy depending on the directivity of the source. The rays are then followed in their rebounds on the surrounding surfaces with specular and diffuse laws.

The semi-reverberating model is the one of most immediate application in the case of rooms of simple geometry and for that reason it has been chosen for the example to be shown in the next point. The input of this model consists of the sound power level of the source and its directivity, the sound-absorbing characteristics of the constituent materials (floor, walls, ceiling, etc.) and the room geometry. The output consists of the sound pressure level at the nodes of a fixed-side grid.

As known, the output of models generally diverges from reality and often in a biased way, which is due both to the approximation of the algorithm on which the model is based, and to the inexact correspondence of the input parameters to real ones. The divergences from reality constitute the limit of models, the use of which remains, from what has been stated above, in any case essential. An example is set out below showing how it is possible to correct the bias by means of the geostatistical external drift method (Chilès and Delfiner, 1999) on the basis of a limited number of measurements.

Example of Forecasting the Sound Pressure Level and Correction of the Bias

The following example refers to forecasting the sound pressure level by using a semi-reverberating field model in a large room (a lecture hall of the Faculty of Engineering in the University of Rome Three, measuring 16.5 × 7 × 4.2 metres). The noise level concerned was obtained using an artificial isotropic sound source, whose characterization, in terms of the level of sound power produced, was effected according to the standard UNI-EN 3746 and supplied the value of $L_W = 88.25$ dB. Considering the presence of walls of porous calcareous material and irregular pattern, a ceiling consisting of soundproofing panels and a tiled floor, the better choice of the average sound absorption coefficient relating to the structural components was 0.25. Forecasting of sound pressure levels has been carried out at the nods of the grid shown in Fig. 1a, from which the map shown in Fig. 1b has been derived.

To be able to assess the correspondence between forecast and actual fact, phonometric measurements were performed at said 112 points. The result of the comparison is represented by the scatterplot in Fig. 2.

As may be observed, the scatterplot evidences an underestimation of the sound pressure level which is more accentuated in the higher values. The divergence between measurements and forecasts can be explained by the fact that, beyond a certain distance from the sound source, the reverberating field prevails over the direct field and from this point onwards the forecast becomes reliable only if the room is perfectly similar to a reverberating chamber.

To correct the bias it was decided to use the external drift method to be applied using only a few measurements of the sound pressure as primary variable and the model output, in terms of sound pressure, as external drift. After a number of attempts it was realized that, for the same number of points, the location of the most effective measurements for the purpose of correcting the bias was along the directions of the pressure gradients

Taking this into account, the two configurations shown in Fig. 3 were taken into consideration, one (configuration a) with 6 measurements located in the direction of the main gradient coinciding with that of the longest side of the hall, and the other one (configuration b) obtained from the preceding one with the addition of two further measurements in the direction of the second gradient.

The results obtained, compared with the measurements, are shown in the scatterplots in Fig. 4.

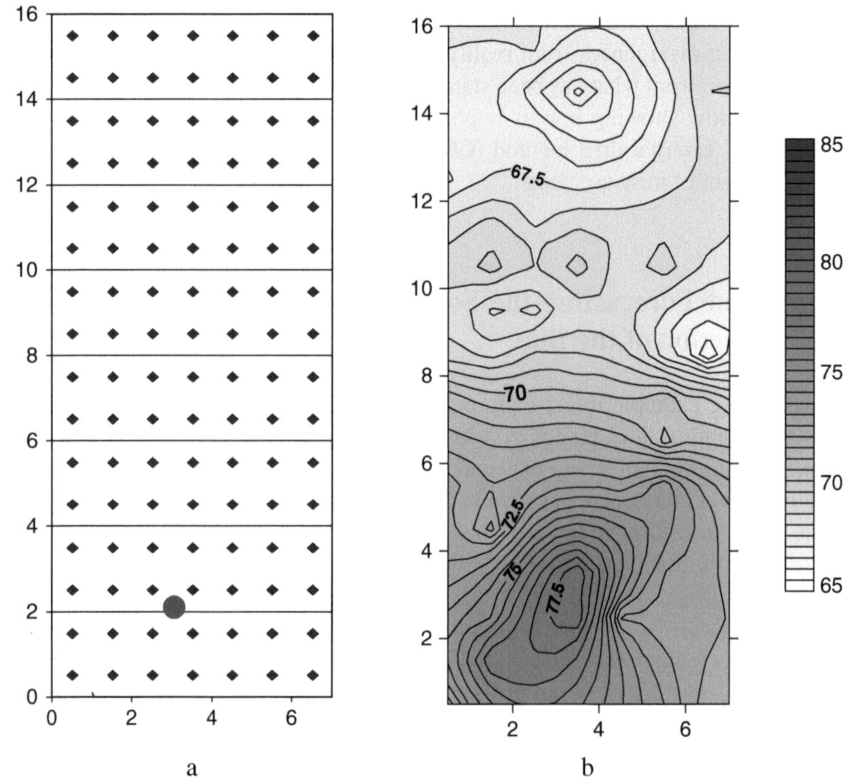

Fig. 1 Position of the source and forecasting point grid **(a)**; map of the sound pressure level reconstructed on the basis of the model response at the forecasting points **(b)**

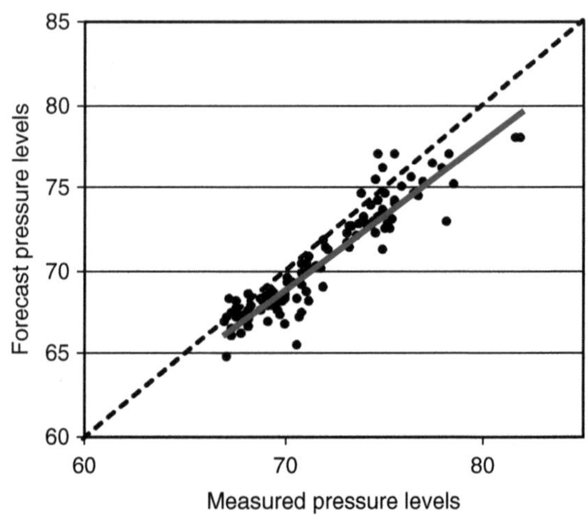

Fig. 2 Scatterplot and linear regression between measured and forecast sound pressure levels

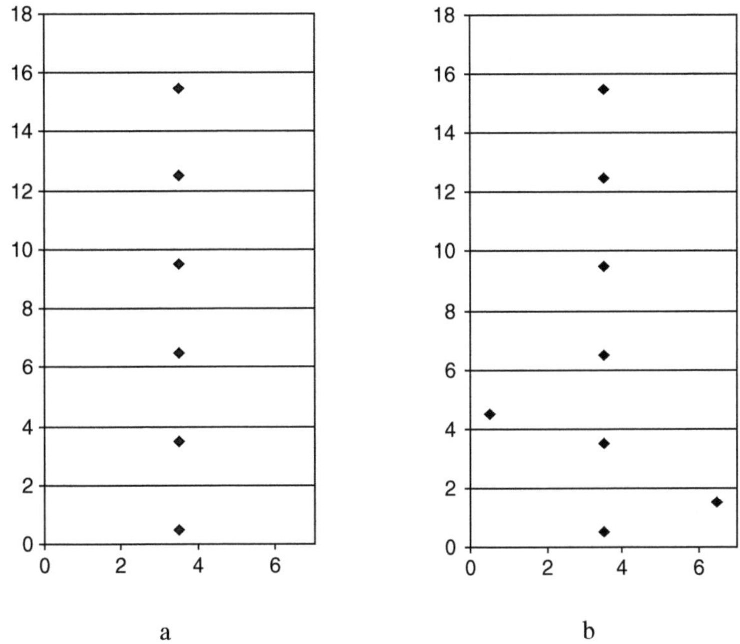

Fig. 3 Position of sound pressure measurements chosen for kriging with external drift: 6 measurements (**a**); 8 measurements (**b**)

It can be seen that with the 8 measurements located as in Fig. 3b the kriging with external drift has in fact eliminated the bias, but has not reduced the random error. The variance of the kriging with external drift does not take into account the random error contained in the drift because the forecasting models do not provide

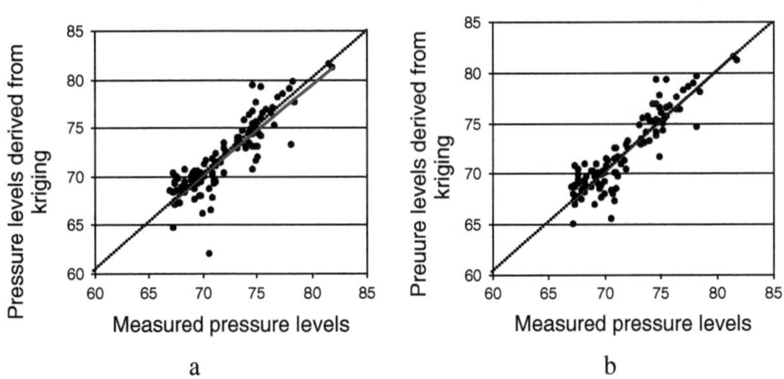

Fig. 4 Scatterplots and linerar regressions between measured pressure levels and the output of the kriging with external drift: 6 measurements of the primary variable (**a**); 8 measurements of the primary variable (**b**)

the accuracy of the output. When calculating the risk, therefore, the random error can only be obtained from analogous preceding practical applications.

Conclusions

Use of a model is in many cases essential for the space-time characterization of the sound pressure level. Models, even with appropriate refinements and duly calibrated, can be subject to bias, at times a considerable one and varying in size from point to point. In order to eliminate this bias, the external drift method has obtained good results. The critical aspect is represented by the selection of the points at which to conduct the measurements, which in this example has been made allowing for the gradient of the sound pressure level. Lastly, it should be observed that, as the use of this method implies simultaneous measurements and therefore the same quantity of instruments, the number of measurements has to be restricted, which can constitute a limit in the complex applications for which the problem remains open.

References

Alfaro Degan G, Pinzari M (2001) "The Functional analysis space Tecnique (FAST) in risk analysis". Safety and reliability ESREL 2001, vol 2. pp 1139–1146

Alfaro Degan G, Pinzari M (2002) "Risk evaluation: noise exposure". SRA-E 2002, Berlin, Germany

Behar A, Plener R (1984) "Noise exposure-sampling strategy and risk assessment". American Ind Hyg Assoc J 45:105–109

Beranek LL (1971)"Sound in Small Spaces". In: Beranek LL (ed) Noise and Vibration Control, McGraw-Hill, New York

Chilès JP, Delfiner P (1999) Geostatistics: modeling spatial uncentainty. Wiley Series in Probability and Statistics

Farina A (1995) "Pyramid Tracing vs. Ray Tracing for the simulation of sound propagation in large rooms". Computational Mechanics Publications, Brebbia CA, Southampton (GB), pp 109–116

Hodgson M, Warnock A (1992) "Noise in rooms". In: Beranek LL, Vèr IL (eds) Noise and Vibration Control Engineering. John Wiley & Sons, New York

Krokstadt A, Strom S, Sorsdal S (1968) "Calculating the acoustical room response by the use of a ray tracing technique". J Sound Vib 8:118

Lajaunie C, Wackernagel H, Thiéry L, Grzebyk M (1999) "Sampling multiphase noise exposure time series". In: Gomez-Hernandez J, Soares A, Froidevaux R (eds) geoENV II: Geostatistics for Environmental Applications, Kluwer, Amsterdam, pp 101–112

Malchaire J, Piette A (1997) "A comprehensive strategy for the assessment of noise exposure and risk of hearing impairment". British Occupational Hygiene society vol 41. no 4, pp 467–484

Geostatistical Modeling of Environmental Sound Propagation

O. Baume, H. Wackernagel, B. Gauvreau, F. Junker, M. Bérengier and J.-P. Chilès

Abstract The atmospheric influence on sound propagation along the ground surface is a critical issue for estimating the noise impact of industrial plants or road networks. Indeed, sound refraction in the surface layer has a dramatic impact on the geographical acoustic exposure. Many analytical and numerical models and studies based on the laws of physics are available in scientific papers whereas very few works in statistical analysis have been attempted. However several important practical issues need to be considered. Among these, time and space representativity of "in situ" measurements, sampling design, influence of meteorological and ground parameters on acoustic exposure show to be a few challenges. They need to be investigated with statistical tools taking into account space and time autocorrelation.

A new protocol which includes ground impedance monitoring, spatial micrometeorological and acoustical characterization has been applied to an experimental campaign from June to August 2005 in a case of sound propagation from a point sound source on a grassy flat ground. The first geostatistical study on such a multivariable experimental database is presented. It adresses both the issue of modelling space varying impedance properties of an homogeneous meadow and the issue of modelling the acoustic field itself. This latter includes an analysis of the spatial variogram of the acoustic field residual calculated from a basic physical model as an external drift.

Introduction

Long Range Sound Propagation in the surface layer of the atmosphere is mainly influenced by the ground impedance, the ground topography and the micrometeorological conditions (see for instance Wiener and Keast (1959), Attenborough et al. (1980) or Rasmussen (1982)). In this field of research realistic numerical models derive from the linearized Euler's equations or the Helmholtz Equation

O. Baume
Department of Environmental Sciences, Wageningen University, The Netherlands
e-mail: olivier.baume@wur.nl

A. Soares et al. (eds.), *geoENV VI – Geostatistics for Environmental Applications*, 45–57

(Ostashev, 1999). Particularly relevant results include the refraction of sound propagation in the form of vertical sound speed profiles and turbulence energy spectrum (Daigle et al., 1983; Wilson and Thomson, 1994; Dallois et al., 2001). But these advanced models have a heavy cost in CPU-time, especially when attempting to express the results of these models in more practical terms (1/3 octave bands e.g. or even more global noise level). Then the calculation time does not allow a time analysis of the influence of micro-meteorological conditions and ground characteristics on the environmental impact of a sound source.

On the other hand, the measurement of the noise impact of a road or an industrial plant is basically held with a poor space sampling design and a short time duration without in-situ representativity estimation. An estimate of the sound level is thus calculated with very empirical methods which do not take into account the total range of micro-meteorological conditions. New statistical work is needed to complete our knowledge for predicting in time and space the noise level emitted from a specific source. The geostatistical tools seem promisingly rich enough to support the needs of statistical modelling.

Laboratoire Central des Ponts et Chaussées, Électricité de France and the Geostatistics Group of the Centre de Géosciences (École des Mines de Paris) collaborate to develop new statistical methods for time and space estimation of the impact of a noise source. A new campaign with a space distribution of the sensors and a specific time and frequency sampling design which was set up by acousticians was held in 2005 from June to August. The database is proposed for a first geostatistical study in Long Range Sound Propagation.

First, we describe the protocol and focus on the data design. Second, the ground properties are studied at a day-time scale and a geostatistical model is proposed. The same variogram model is also used in conditional simulations. Third, we apply a universal kriging procedure to acoustic measurements. It includes the output of a simple physical model which gives a first order approximation of propagation of sound around a point source in a homogeneous field. Finally a linear drift component is included and it appears that it reflects the influence of the micro-meteorological conditions on the acoustic field.

The Measurement Campaign

The Experimental Protocol

Micro-meteorological towers have been used to determine the wind speed and temperature vertical gradients of the experimental site. These measurements will not be included into the geostatistical model but will be used separately for interpretation of the results.

Acoustical Configuration

In acoustics nothing is more sensitive to various micro-meteorological events on a flat ground than an omni-directional source. Its main advantage is to observe the time variability of acoustical time series which become very characteristic of weather conditions (Baume et al., 2005). Besides, its drawback is that its geometrical spreading for a constant signal imposes a fast decay in space and extraneous noise becomes tricky quite close to the source. The Brüel & Kjaer 4296 omnidirectional source emitting a pink noise level worth 115 dB is located 2 meters above the ground. As shown on the left picture of Fig. 1, microphones were settled at 2 and 4 meter high along 3 main axes corresponding to sound propagation planes. The distance between 2 acoustical masts is 25 m.

The acoustical sensors allow to record equivalent integrated sound levels (denoted L_{eq}) for periods greater than 1 second. A 10 to 15 minutes period is generally chosen because the average of corresponding micro-meteorological conditions are relevant to characterize mean refraction of the surface layer of the atmosphere. This was demonstrated in a former study using correspondence analysis (Séchet and Zouboff, 1995).

Ground Effect

Very special care was taken to characterize the ground properties during this campaign. Its influence on sound levels increases when propagation conditions become favorable between the source and the receivers. Indeed as sound energy is focused along the ground (downward curvature of the sound rays) more energy can be absorbed by a porous surface.

One monitoring station was designed to allow almost continuous data acquisition (every 4 hours) of the ground impedance at one specific point in space (emphasized on the right map of Fig. 1 by a black triangle). Furthermore a parallel campaign was held to explore the spatial variation of the ground impedance by performing measurements at 13 locations of the experimental area, which were repeated 13 times (13 days between early June and late August 2005). The global impedance protocol is drawn on Fig. 1 (right).

Validation of the Database Used for the Spatial Analysis

We want to stress the care that was taken in validating the database. For example the rainy and strong wind periods needed to be suppressed because of the microphones'sensitivity. Moreover an average acoustical spectrum of the source at several distances was determined in order to get rid of extraneous noise such as a few sound events that can appear anywhere (trains passing by in the valley, sparse road trafic, crickets). For acoustical data, the validation is applied to 1/3 octave bands ranging from 50 Hz to 8 kHz of 1s integrated L_{eq}.

For the ground impedance database each measured value of the spatial campaign is an average of 10 successive measurements which avoids the largest errors that are to be made with this protocol. After the campaign, the impedance value is converted into a phenomenological parameter called the *specific air flow resistivity* by fitting a ground model called the Delany and Bazley model (Delany and Bazley, 1970).

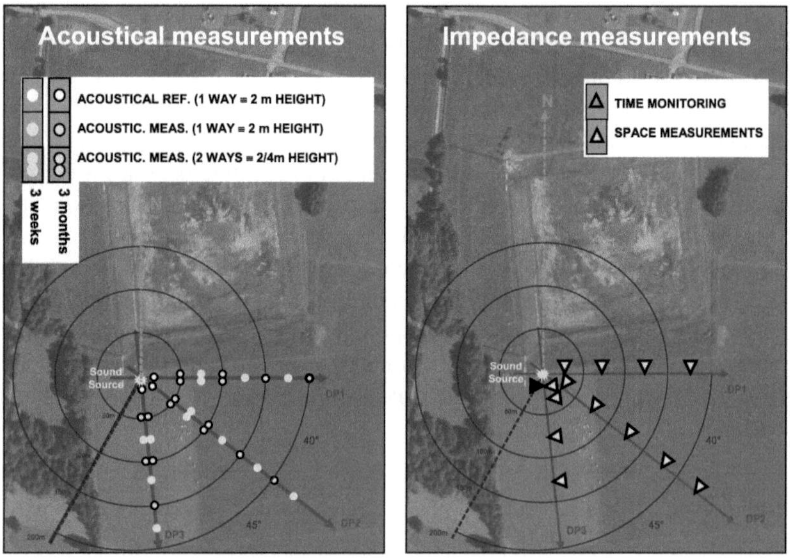

Fig. 1 Spatial distribution of acoustical (left) and ground impedance measurements (right) during the Lannemezan campaign in 2005

The specific air flow resistivity result is then obtained by averaging the Delany and Bazley model fit that was found by 3 different people. For this type of grassy ground, the best model to use is a 2 parameter model which includes the specific air flow resistivity (denoted σ_{ground}) and the thickness of the porous layer of the ground (denoted e) (Rasmussen, 1982).

More details on the validation process of the data during this campaign can be found in Junker et al. (2006).

Data Analysis

Ground Properties Analysis

The *Time Monitoring* measurement helped to provide evidence that the short period fluctuations are much smaller than the large fluctuations. The first is due to the day and night thermal gap and to the effect of the dew on the porous layer of the ground while the second is related to the general weather conditions. Thus we focused our analysis on the large time scaled evolution of the spatial measurements of σ_{ground} along the 3 propagation axes DP1, DP2 and DP3. We first studied the autocorrelation of the data and then applied ordinary kriging and conditional simulations with the variogram models assuming second order stationarity of the random function.

The interest of kriging is to visualize the zones of the field where the ground is more reflective to sound energy (i.e. σ_{ground} is higher) and those where the ground is more absorbing. The conditional simulations are useful in providing impedance fields as input data of numerical models of sound propagation.

Modelling the Autocorrelation of the Specific Air Flow Resistivity Data

As we had a fairly low number of samples there is quite some uncertainty about the true variogram of impedance. The experimental variogram of Fig. 2 illustrates how low the number of pairs is. Our fit of the variogram was guided by our knowledge of the experimental uncertainties.

From our experience on impedance measurements and the Delany and Bazley model we know that experimental uncertainties are about 20 $kNsm^{-4}$. This uncertainty shows as a measurement error in the ordinary kriging model which imposes a nugget effect worth 200 (semi-variance). See for instance the model we chose for June 20th at 14h00 (Fig. 2(a)). The map of ordinary kriging performed with a unique neighborhood is shown on Fig. 2(b) and provides a smooth picture of the actual ground impedance. It represents a spatial variability that is greater than the uncertainties in the measurements. While the weather conditions evolved in such a way that the meadow dried from late June to late August, we found different variogram models – spherical, stable (power $3/2$) and cubic – with a range varying from 60 m to 180 m. This precluded the use of a mean model over time.

Use of Conditional Simulations

Contrarily to kriging, conditional simulations do not provide a smooth picture of the spatial variability. Especially when a nugget effect is imposed in the variogram model, the simulated surfaces of the specific air flow resistivity is very erratic. As an example we show on Fig. 2(c) a corresponding conditional simulation of the model of June 20th. The gaussian assumption was made to simulate the target random function.

As the conditional simulations bring a realistic behavior of the parameter, they have an important application potential as input data for numerical models of Long Range Sound Propagation: these up to now did not have access to the spatial evolution of the ground properties. By repeated runs of the LRSP model with different members of a geostatistical simulation ensemble the sensitivity of sound to ground impedance can be assessed.

Results and Discussion

Air flow resistivity is the main parameter for characterizing the acoustical properties of a porous ground like a crop field, a meadow or even a field covered with snow. A complete measurement campaign is very demanding with respect to both human and material resources. The Lannemezan 2005 campaign is a step forward in the spatial characterization of environmental sound propagation. The experimental variograms resulted from a low number of measurements in space and were quite shaky. It was then possible to fit a variogram model which included a nugget effect to take into account measurement uncertainties. The resulting maps from ordinary kriging

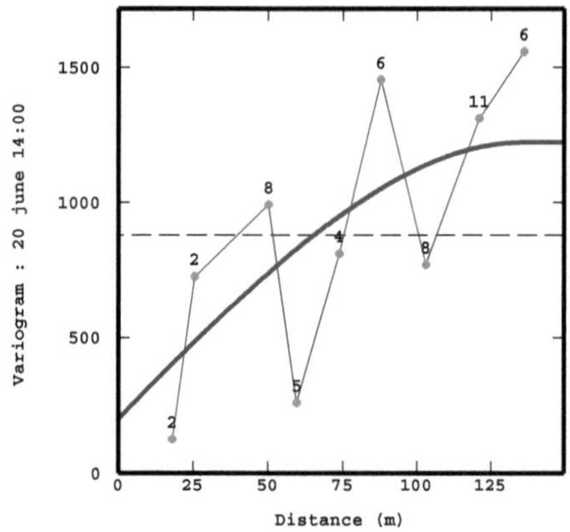

(a) Experimental variogram and variogram model

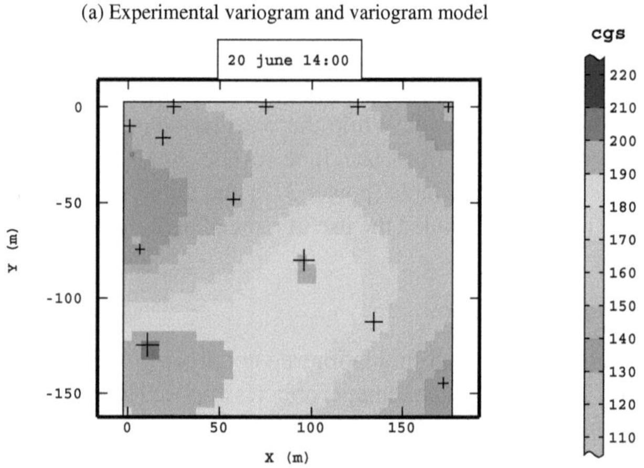

(b) Ordinary kriging estimation

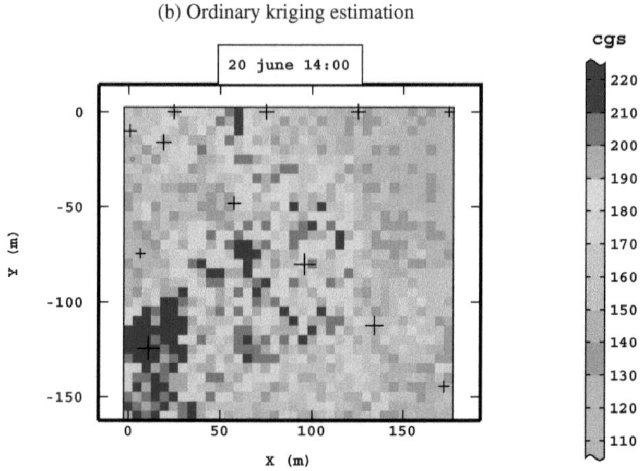

(c) Example of a conditional simulation

Fig. 2 Example: June 20th at 14:00 – nugget (sill 200) and spherical model (range 136 m and sill 1022); the number of pairs is written for each class of distance estimation give a new information about slow space variability of the specific air flow resistivity. The weather conditions have an impact on the zones which are more reflecting at the beginning of the campaign and then more absorbing as they get dryer and more porous.

The conditional simulations are ideal as a basis for sensitivity analyses of the influence of σ_{ground} on acoustical results. Such a study should be an outcome of our work. *In the next section a kriged mean value of the acoustic properties will be used (see details below).*

Sound Propagation Analysis

This is to our knowledge the first study in geostatistical modelling of outdoor sound propagation. The purpose is to better understand the impact of different models on the interpolation for different 1/3 octave bands and to give a first interpretation in terms of the ground and micro-meteorological influence on the acoustic propagation.

Case Study

To avoid the effects linked to the directivity of the source, the measured L_{eq} (in 1/3 octave bands) are transformed into attenuation levels between the reference microphones located 10 meters away from the source and more remote microphones (2 meter height) for each propagation direction (see Fig. 1). A particular day, June 22nd, is selected because many samples have complete spatial information. One 1/3 octave band is examined: 1 kHz which is the central frequency in the log scale of the audible sounds.

Universal Kriging

Sound propagation is a non-stationary phenomenon which includes a geometrical spreading and atmospheric absorption. A simple first-order approximation explains the main component of the non-stationarity. So we orient the work to the development of a universal kriging procedure that includes a physical model in the external drift. Such an approach has already been used in air pollution modelling and is known in meteorology as data assimilation (see e.g. Wackernagel et al. (2004)). Besides, the external drift method is presented in Chilès and Delfiner (1999) and applied in other multivariate contexts in Wackernagel (2003).

The method can be split into 4 steps. The first step is the calculation of a first order physical model. The second step is a least squares fit of the model to the experimental data. We take 2 different options: one is a least squares fit without linear drift while the other includes a linear drift. The third step consists in modelling the residuals and the fourth includes the resulting variogram model to a kriging computation known as kriging with external drift.

Step 1: calculation of the first order physical model

We propose to use the Embleton physical model (Embleton, 1983). It takes into account a mean spatial value of the specific air flow resistivity rather than a simulated field as it is used to calculate a first order approximation of the acoustic attenuation. The general form of the acoustic pressure above a porous half-space at location **x** is given by

$$p(\mathbf{x}) = p_d(\mathbf{x}) + Q.p_r(\mathbf{x}), \tag{1}$$

where p_d is the contribution of the real source, p_r the contribution of the image source and Q the complex form of the reflection coefficient of the porous ground. Q explicitly depend on the impedance Z_{ground}. In the Embleton model, the Delany and Bazley approach is taken to express the impedance. The following is used

$$Z_{ground} = Z_0\left(1 + 9.08\left(\frac{f}{\sigma_{ground}}\right)^{-0.754} + j11.9\left(\frac{f}{\sigma_{ground}}\right)^{-0.732}\right)$$

where Z_0 is the air impedance.

The relation 1 is frequency dependent and the computation is made for a minimum of 7 frequencies for a 1/3 octave band. The Fig. 3(a) displays a map of the output from this model which will subsequently be used as a drift component for the assimilation of the entire day of data. It shows an isotropic propagation around the omni-directional source located in the upper left corner.

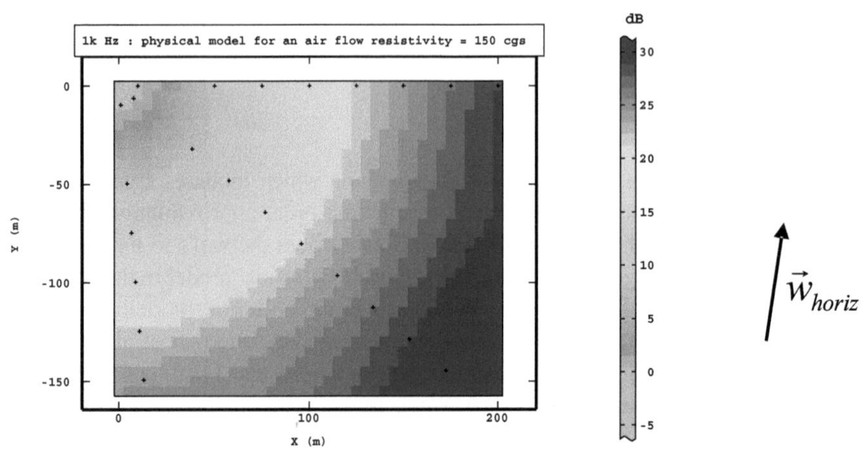

(a) (b)

Fig. 3 (a) Acoustic attenuation calculated with the Embleton model for a mean $\sigma_{\text{ground}} = 150$ cgs (b) Indicative horizontal wind direction for the sample June 22nd at 23:45 – wind speed at 10 meter high was 3.6 m.s^{-1}

Step 2: least squares fit to the data

The physical model output is fitted by least squares to the data at 23h45 and the result is shown on Fig. 4(a). An option is to include also a linear drift component and the corresponding map is displayed on Fig. 4(b). The linear drift is an assumption made from the acousticians experience that the acoustic field is highly anisotropic following the wind direction (Wiener and Keast, 1959). We use this assumption to refine the method and then, the linear drift is calculated from acoustic measurements independently from wind data. In the least squares fit process example, the off-set component just lowers the attenuation whereas the linear drift gives an image of anisotropy due to the mean horizontal wind speed which was observed in the mean time (Fig. 3(b)).

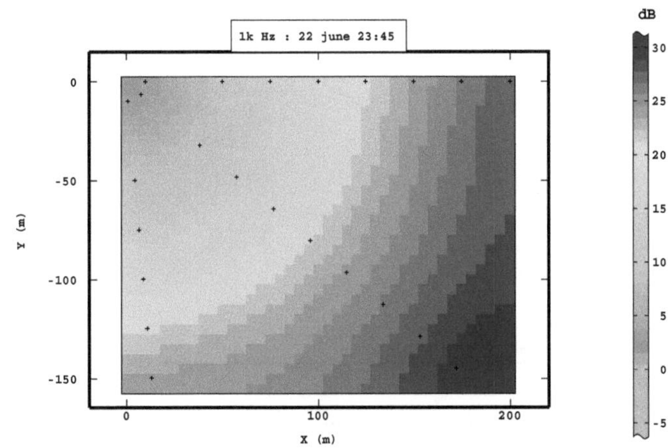

(a) *Without linear drift*

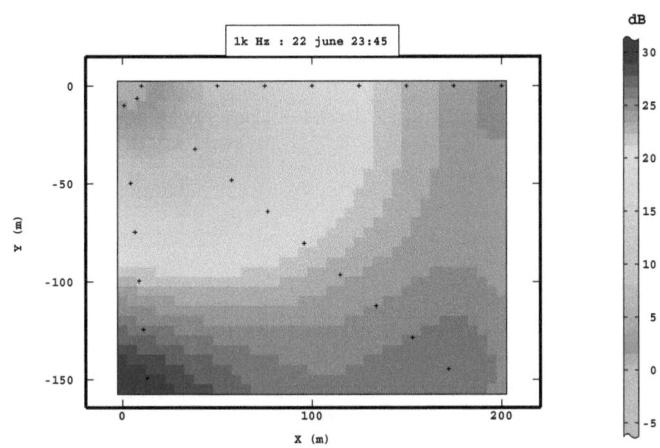

(b) *Including a linear drift*

Fig. 4 *Least squares fit of the Embleton model to the measurements at 1 kHz, June 22nd at 23h45*

Step 3: variograms of residuals

The residuals between the measurements and the fitted drift (optionnally including a linear drift component) are computed and corresponding variograms are displayed on Fig. 5. The variogram of the residuals of the plain least-squares fit (Fig. 5(a)) shows a parabolic shape at the origin that cannot be reproduced well using a single cubic model. At large distances it does not stabilize at a sill, but continues to increase. The option of including a linear drift component to catch non-stationarity that is not explained by the Embleton model has the effect that the variance of the

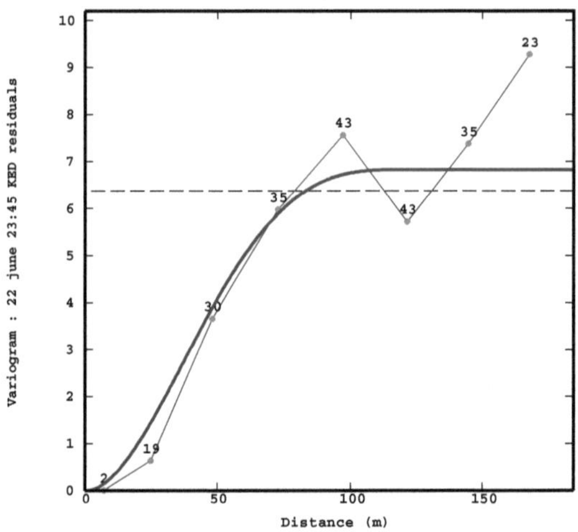

(a) *Cubic model (range 125 m and sill 6.8).*

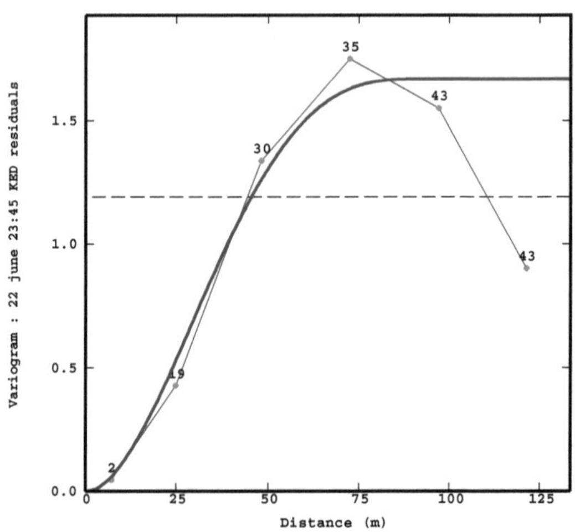

(b) *Cubic model (range 96 m and sill 1.66).*

Fig. 5 Variogram of residuals for least squares fit of external drift (a) without linear drift and (b) including a linear drift residuals is divided by a factor of 6 (Fig. 5(b)). Furthermore, the behavior of the variogram at the origin is fairly well compatible with a cubic model and the decrease beyond a distance of 75 is typical for the experimental variogram of residuals.

Step 4: external drift kriging

The external drift kriging was performed using optionally a linear drift component. The resulting maps are displayed on Fig. 6. The incorporation of a linear drift component strongly increases the anisotropy of the estimated attenuation with respect to the source in the top-left corner.

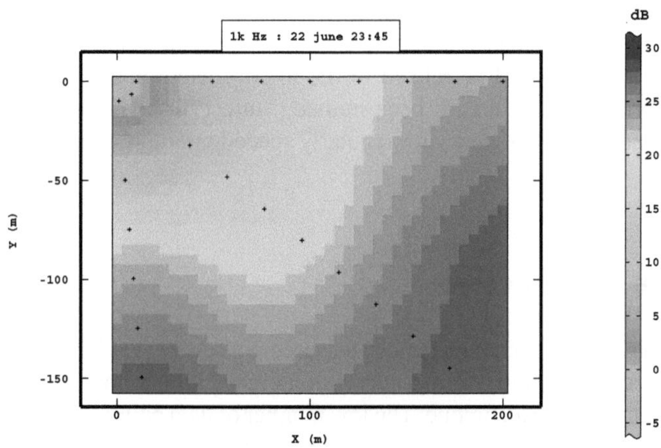

(a) *Without linear drift.*

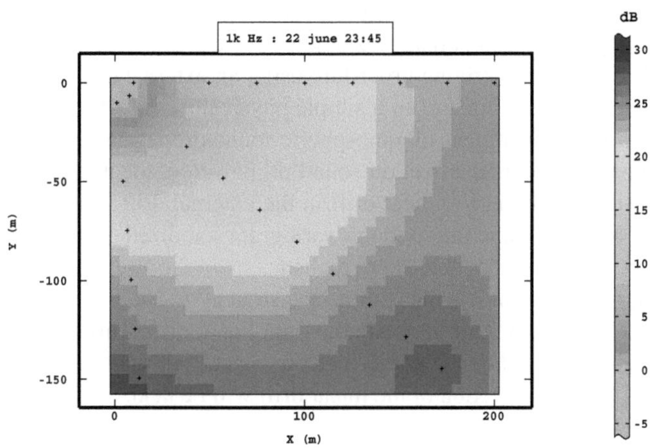

(b) *Including a linear drift.*

Fig. 6 Kriging with external drift

Results and Discussion

The computations shown here for the measurements taken at 23h45 were repeated for the whole day of June 22nd at 1 kHz. It turns out that the linear drift component obviously reflects the effect of the micro-meteorology on the sound propagation. In particular the wind direction and intensity seem to be related to the shape of the map of acoustic attenuation. In our example, the wind is strong enough to bring anisotropy in mean wind vector direction (South $\Rightarrow$ North direction parallel to the Y axis).

For low frequencies (< 500 Hz), acoustical propagation is generally less influenced by micro-meteorology due to ground effects. So the inclusion of a linear drift to fit the Embleton model to the measurements has a slighter effect. On the contrary, at high frequencies (e.g. 1 kHz) a linear drift is needed to obtain a good anisotropy on the kriging map when the wind blows.

The residuals variogram is hardly ever interpretable in terms of physical phenomenon. Its shape should be related to the micro-meteorological turbulence for which autocorrelation has already been studied (Stull, 1988). The study of this relation requires another set of data with regularly spaced microphones.

Conclusion and Outcomes

The experimental data from impedance measurements during the Lannemezan 2005 campaign were analysed to obtain a succession of specific air flow resistivity maps with an ordinary kriging technique. The low number of data points in space let us use a nugget effect to model the experimental uncertainties. The resulting conditional simulations will give a first opportunity to use realistic field of σ_{ground} as an input data of numerical models in Long Range Sound Propagation.

An entire day of acoustical data was studied for the first time with geostatistical tools. We developed a kriging procedure using an external drift. This drift was chosen to be a fit by least squares of a simple physical model which takes into account the unvarying phenomena of atmospheric sound propagation. When the wind conditions have a high order effect on sound propagation, the anisotropy must be modelled by the inclusion of a linear drift in the external drift. Then, the physical model is well reoriented and the residuals variogram stabilized to a sill. Finally, the output of the kriging calculation is more realistic.

We still have to determine which is the best strategy to use when the number of space measurements is too low to obtain the right linear drift. This lack of information must be compensated by the micro-meteorological data and further analysis on its correlation with the direction of the linear drift will be held.

Acknowledgments The authors of the present paper would like to acknowledge the French Ministry of Environment and Sustainable Development for its financial support to the Lannemezan 2005 experimental campaign.

References

Attenborough K, Hayek SI, Lawther JM (1980) Propagation of sound over a porous half space. J Acous Soc Am 68(5):1493–1501

Baume O, Gauvreau B, Junker F, Wackernagel H, Bérengier M, Chilès J-P (2005) Statistical exploration of small scale variation in acoustic time series taking into account micro-meteorological conditions. Budapest, Hungary. Forum Acusticum

Chilès J-P, Delfiner P (1999) Geostatistics, modelling spatial uncertainty. Wiley series on probability and statistics. Wiley, New York

Daigle GA, Piercy JE, Embleton TFW (1983) Line-of-sight propagation through atmospheric turbulence near the ground. J Acoust Soc Am 74(5):1505–1513

Dallois L, Blanc-Benon P, Juvé D (2001) Long range sound propagation in a turbulent atmosphere within parabolic equation. *Acta Acustica* 87(6):659–669

Delany ME, Bazley EN (1970) Acoustical properties of fibrous absorbant materials. *Applied Acoustics*, 3

Embleton TFW (1983) Effective flow resistivity of ground surfaces determined by acoustical measurements. *J Acoust Soc Am* 74(4):1239–1244

Junker F, Gauvreau B, Cremezi-Charlet C, Gérault C, Ecotière D, Blanc-Benon P, Cotté B (2006) Classification de l'inuence relative des paramètres physiques affectant les conditions de propagation à grande distance : campagne expérimentale de lannemezan 2005. Tours, France. Congrès Francais d'Acoustique

Ostashev V (1999) Acoustics in Moving Inhomogeneous Media. E & FN Spon

Rasmussen KB, (1982) A note on the calculation of sound propagation over impedance jumps and screens. Journal of Sound and Vibration 84(4):598–602

Séchet E, Zouboff V (1995) Application des méthodes factorielles à la caractérisation des effets météorologiques sur la propagation du bruit à grande distance. Bulletin de liaison Laboratoire des Ponts et Chaussées, 198(référence 3908)

Stull RB (1988) An introduction to boundary layer meteorology. Kluwer Academic, Dordrecht, The Netherlands

Wackernagel H (2003) Multivariate geostatistics, 3rd edn. Springer-Verlag, Berlin

Wackernagel H, Lajaunie C, Blond N, Roth C, Vautard R (2004) Geostatistical risk mapping with chemical transport model output and ozone station data. *Ecological Model* 179:177–185

Wiener FM, Keast DN (1959) Experimental study of sound over ground. J Acoust Soc Am, 31(6):724–733

Wilson DK, Thomson DW (1994) Acoustic propagation through anisotropic, surface layer turbulence. J Acoust Soc Am 96(2):1080–1095

Geostatistical Estimation of Electromagnetic Exposure

Y. O. Isselmou, H. Wackernagel, W. Tabbara and J. Wiart

Abstract The electromagnetic environment in urban areas is growing increasingly complex. Sources of electromagnetic exposure like TV, FM, GSM, Wifi and others are spreading continuously and in the case of Wifi their geographical locations cannot be cataloged exhaustively anymore. Furthermore, the complexity of any highly urbanized environment and the lack of information about the dielectric properties of buildings lead to complex configuration so that a precise deterministic modeling of the electromagnetic exposure at any a given location of interest is probably out-of-reach.

On the other hand there is a growing demand to assess the human exposure induced by these wireless communications. In a project between France Télécom R & D, Ecole des Mines and Supélec the application of geostatistical methods in this context is being explored.

Geostatistics provides the right framework for setting up such exposure maps and its spatial statistical model yields an estimate of exposure as well as an associated error (De Doncker et al., 2006).

The project consists of three phases: geostatistical evaluation of data generated by the numerical model EMF Visual (both in free space and with the addition of obstacles), statistical analysis of measurements performed in the area of the Quartier Latin in Paris and, finally, joint evaluation of an urban area both by statistical and deterministic numerical modeling.

The paper reports about the third phase of this ongoing project, in which the spatial variation is modeled using the *variogram*, followed by a spatial regression known as *kriging*. The paper presents results about using a kriging algorithm that integrates numerical model output as an *external drift*.

Y. O. Isselmou
France Télécom R & D, RESA/FACE, Issy-les-Moulineaux, France

A. Soares et al. (eds.), *geoENV VI – Geostatistics for Environmental Applications*, 59–70

1 Introduction

The wireless communication systems and cellular handset are nowadays intensively used worldwide. To support these equipments thousand of base stations or access points operating at different frequencies and using different communication protocol have been implemented. These antennas have induced a public concern about exposure to electromagnetic fields. To check the compliance with the relevant limits such as those of ICNIRP (1998) international bodies such as CENELEC, IEC, CEPT or IEEE have developed accurate methods to assess exposure. The exposure to the electromagnetic field induced by TV, GSM, FM... is weak as shown in ANFR (2004), yet the public is looking for the exposure assessment and not only for the compliance. Numerical tools such as EMF Visual have been developed, but even if the locations of the antennas are public (e.g. www.cartoradio.fr or www.sitefinder.radio.gov.uk) the characteristics of emission (type of antenna, tilt, azimuth, power emitted) are considered as quite sensitive data. Because of that, simulations are not always possible, and an exposure assessment based on field measurements is needed. A frequency selective "in situ" exposure assessment can be carried out using either a personal dosimeter or a monitoring station (e.g. EME SPY and INSITE Station by Antennessa), but tools have to be developed to spatially interpolate the data.

In this paper we propose a new approach to mapping electromagnetic exposure which resembles methods that are common practice in numerical weather forecasting (Daley, 1991). The approach is subdivided into two main steps.

First, a guess of the electromagnetic field is constructed by a numerical model, using the known antenna positions as well as likely parameters for their emission characteristics. This provides a possible, yet inaccurate first picture of the propagation of the electromagnetic signal in the field.

Second, the surface generated with the numerical model, the *guess field*, is combined with the measurements of radio-electric exposure to base station emissions in order to provide a corrected picture of the electromagnetic field. The combination of the data with the numerical model output is performed using the geostatistical method of *kriging*, which is preceded by a statistical analysis and modeling of the residual between the guess field and the observations.

2 Electromagnetic Field Strength Assessment

The wireless communications services operate at different frequencies, for instance the frequency band used by FM, TV3, GSM 900 downlink, GSM 1800 downlink, UMTS downlink and WIFI are respectively in the MHZ [88 108], [174 223], [925 960], [1805 1880], [2110 2170], [2400 2500] bands. The wireless systems use also different communication protocols and techniques (Wiart et al., 2000) such as Time Division Multiplex Access (TDMA), Discontinuous Transmission (DTX), Code Division Multiplex Access (CDMA) and Adaptive Power Control (APC), therefore the power emitted by these systems are variable. Moreover, the signal emitted by

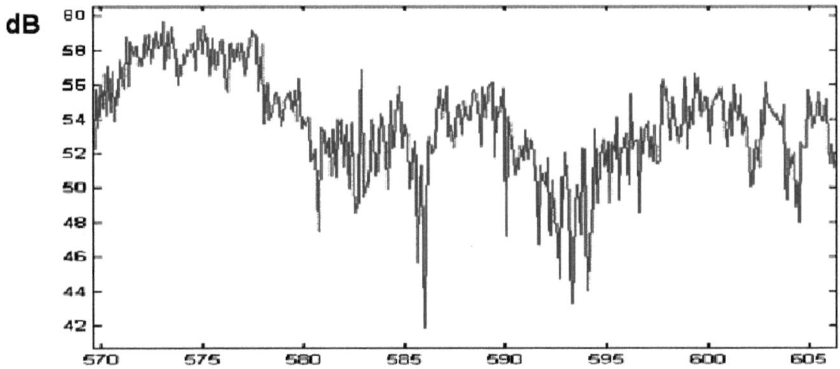

Fig. 1 Electric field strength (dBμv/m) along a line

an antenna operating at a given frequency is reflected and diffracted by walls and edges of buildings. Therefore the field received is a combination in phase and in counter-phase of these contributions, which create fading (see Fig. 1)

The exposure assessment has to take into account such variations (Larchevêque et al., 2005) and measurements performed at different locations cannot be compared point to point.

3 Measurement System

Different frequency-selective measurement systems exist: on the one hand, integrated equipments based on hard filters such as the *personal dosimeter*, on the other hand, the classical *spectrum analyzer*. These equipments are in either case connected to an antenna.

In the case of a personal dosimeter the frequency bands under test are predetermined and consist of the main services occupying the spectrum, such as TV, FM, GSM, UMTS, WIFI. The dosimeter made available for this study (see Fig. 2) uses hardware filters and was able to measure 9 bands (a newer version will be able to analyze 12 bands). This included GSM 900, GSM 1800 and UMTS downlink (i.e emitted by base stations). The axial isotropy of the device is greater than 1 dB at 66% and greater than 2 dB at 95%; the minimum and maximum detection is respectively 0.05 V/m and 5 V/m; the resolution is 0.01 V/m.

The memory of the dosimeter is able to record up to 7000 measurements with a minimum sampling duration of 3 seconds. Figure 3 shows a set of 220 measurements performed along streets in Paris with a sampling duration of 10 seconds. Since the sensitivity of the equipment is 0.05 V/m, the field strength below or equal to this limit is recorded at 0.05 and is not significant.

The spectrum analyzer is a versatile equipment: its output consists in a trace that represents the spectrum sampled at N points in a frequency interval (*fstart*,

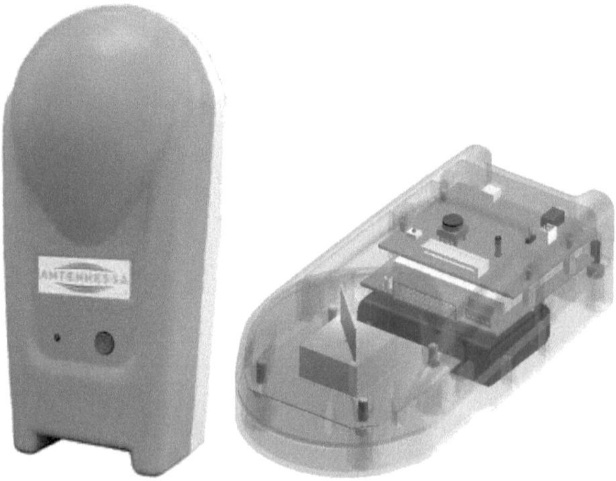

Fig. 2 View of the dosimeter used in this study and its internal structure

fstop) that can be defined as needed. The spectrum analyzer is more sensitive and its resolution is smaller, but the equipment is less portable.

Since measurements can be performed with both methods we have carried out a comparison between spectrum analyzer and dosimeter measurements. This study

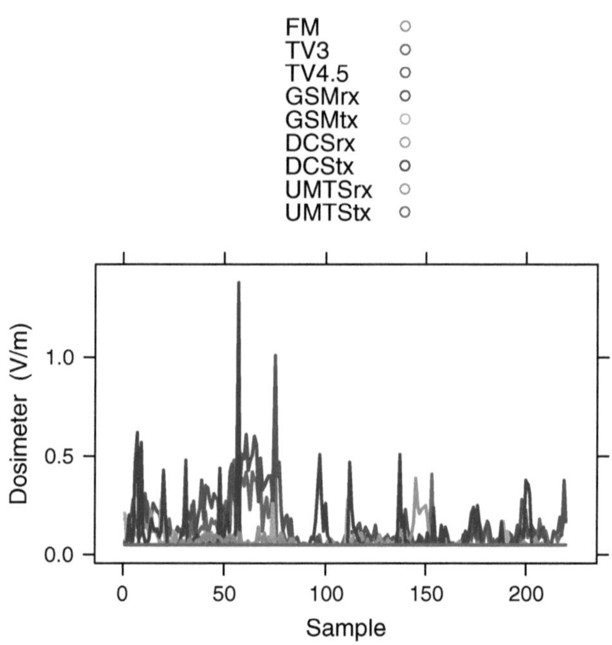

Fig. 3 Example of 220 measurements (sampling duration 10s) performed with the dosimeter along streets in the Quartier Latin in Paris

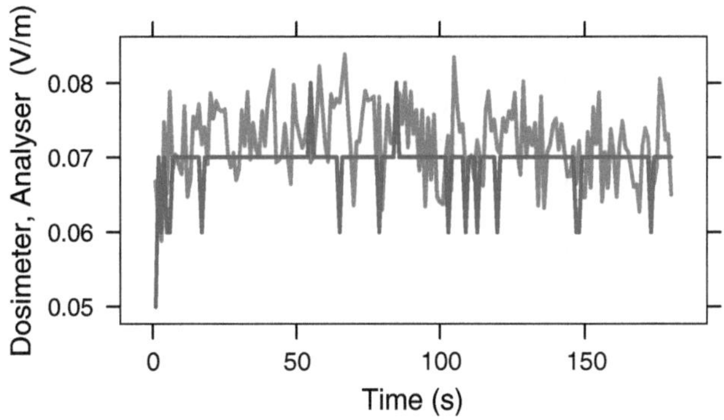

Fig. 4 Comparisons between measurement performed in laboratory, at 900 MHz, with a dosimeter and a spectrum analyzer

has been performed in laboratory and has been focused on GSM 900 downlink, known as *GSM900rx*, since this will be the frequency of interest in this study. The Fig. 4 reveals that there is a good agreement between the measurements from the two devices. As explained previously, the measurements performed using a spectrum analyzer are more precise, so the measurements exhibit only a small fading. The measurements from the dosimeter have a limited resolution as compared to the spectrum analyzer. The comparison further shows that measurements performed with a dosimeter match well those stemming from the spectrum analyzer.

4 Data from the Quartier Latin and EMF Visual Simulation

The first author took sample dosimeter measurements in the Quartier Latin area of Paris. The sequence of the measurements for the different frequency bands was displayed on Fig. 3. We will concentrate on the GSMrx band, which corresponds to the radio-electric exposure related to emissions from GSM base stations.

The histogram of the 220 GSMrx values is displayed on Fig. 5 and it has a typical right-skew shape. The tail has been colored in blue and it can be noted that three values single out on the right, which we may qualify as *outliers*.

The geographical locations of the 220 samples can be seen on Fig. 6. The symbols are proportional in size to the values of GSMrx exposure measured at each location. We note that two of the three outliers are located next to very low values so that we may expect strong small-scale spatial variation.

Taking account of the base station locations and of the reflectors (buildings) in that section of Paris, as displayed on Fig. 7, a simulation of electromagnetic propagation was performed with the software EMF Visual. The geographical map of the output is shown on Fig. 8 and the corresponding histogram on Fig. 9. As the characteristics of the antennas were not disclosed to us, the parameters were set to likely values and we

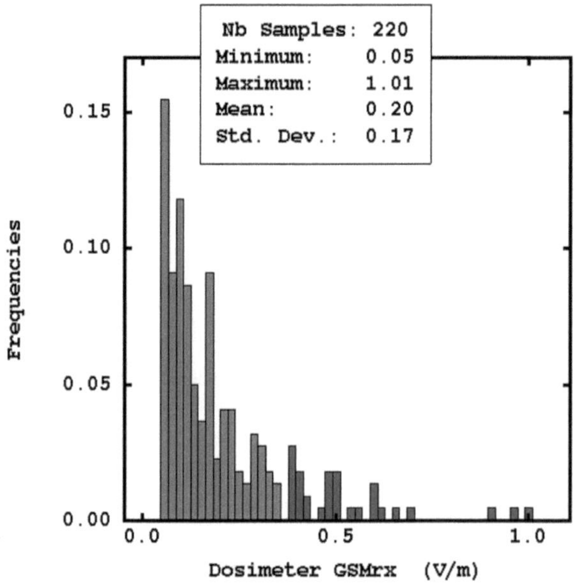

Fig. 5 Histogram of dosimeter GSMrx data; There are 3 outliers on the right

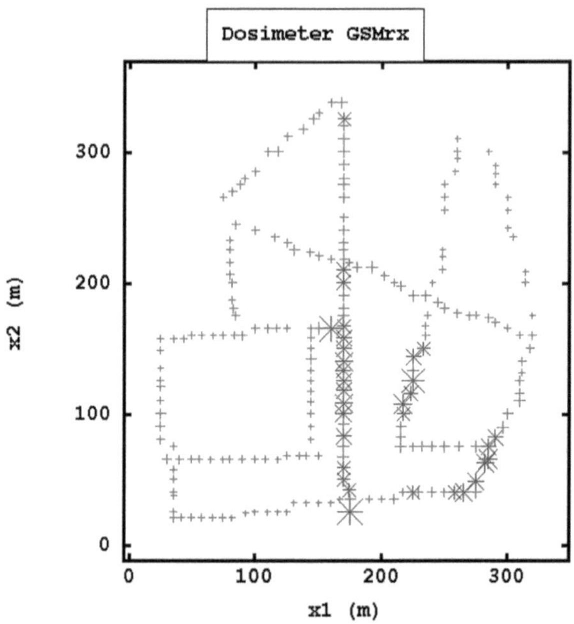

Fig. 6 Map of sampling locations of dosimeter GSMrx data in the Quartier Latin; the diagonal line of samples was taken along boulevard Saint Germain. The stars correspond to values in the tail of the histogram; the symbol sizes are proportional to the values. Two of the 3 outliers are next to low values

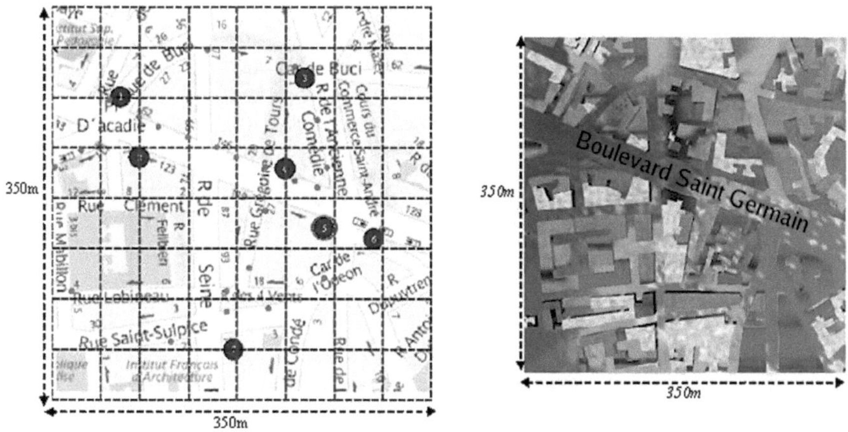

Fig. 7 The EMF Visual simulation is taking account of the position of the base stations (dots on the left) and of the reflectors (buildings)

therefore qualify the result as a *guess field* (like in meteorological applications (Chilès and Delfiner, 1999)).

The histograms of the data and the model output are both right-skew as they reflect/reproduce the same type of physical phenomenon. However, one cannot expect them to be perfectly compatible due to the qualitatively different sources of error,

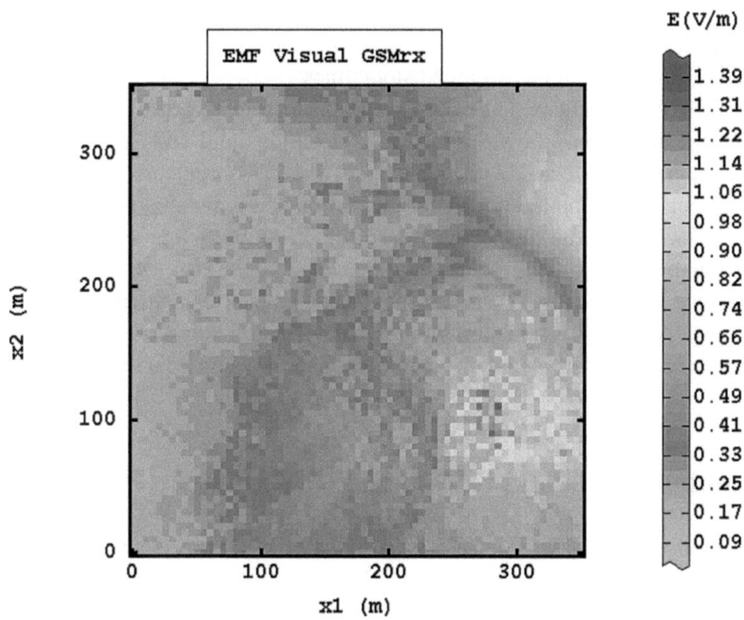

Fig. 8 Guess field generated with EMF Visual using known antenna positions and reflectors, setting likely values for their characteristics

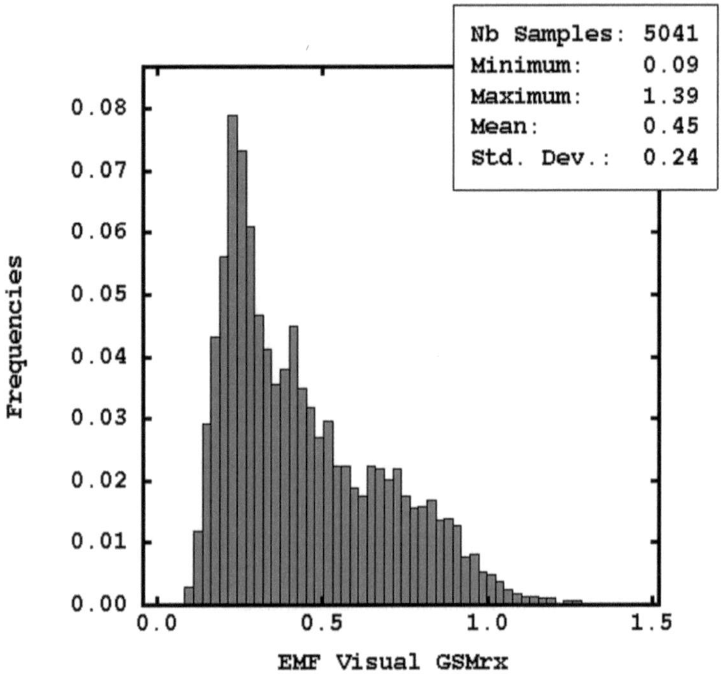

Fig. 9 Histogram of EMF Visual GSMrx values

to the non-stationarity of the field and, in particular, to the sampling design of the dosimeter data with samples arranged along lines.

The EMF Visual guess field provides an inaccurate sketch of the electromagnetic propagation from known point source locations and with known obstacles. This needs to be corrected with the help of the measurements. Conversely, the measurements do not include explicit information about the location of the sources and the reflectors, so the inclusion of the EMF Visual output is a way of bringing that information into the interpolation problem.

A balance has to be found between the data and the model output: this will result from the statistical analysis and modeling of the spatial variation of the *residuals*, i.e. the differences between the measurements and the numerical model output.

5 Combining Data and Guess Field by Kriging

The method we apply is known in geostatistics as *kriging with external drift* (KED) (Wackernagel, 2003; Chilès and Delfiner, 1999) and has been applied to many problems. Kriging is a spatial regression based on a statistical model of the spatial variation. The use of numerical model output as external drift is fairly recent in geostatistics (Wackernagel et al., 2004) but is common practice in meteorology and oceanography. KED is implemented in the following steps:

1. fit the numerical model output to data by least-squares (LS-FIT),
2. compute the differences (residuals) between data values and corresponding LS-FIT values,
3. compute and model the auto-correlation of the residuals using the *variogram*,
4. final KED estimate based on the residuals' variogram model and the LS-FIT as external drift.

After performing the first two steps, which are elementary, we wish to characterize the spatial correlation of the residuals. This is performed in geostatistics with the variogram $\gamma(\mathbf{h})$, ie the expectation of the squared differences of pairs of values Y that are $\mathbf{h}$ apart,

$$\gamma(\mathbf{h}) = \frac{1}{2} E\left[(Y(\mathbf{x}+\mathbf{h}) - Y(\mathbf{x}))^2\right] \tag{1}$$

where $\mathbf{x}$ and $\mathbf{x}+\mathbf{h}$ are two locations in geographical space. In practice, the variogram can be computed for different directions by calculating mean-squared differences of data values for pairs of points belonging to given distance and angle classes.

The variogram of GSMrx dosimeter measurements was computed into perpendicular directions of space and is displayed on Fig. 10 for the NS and EW directions.

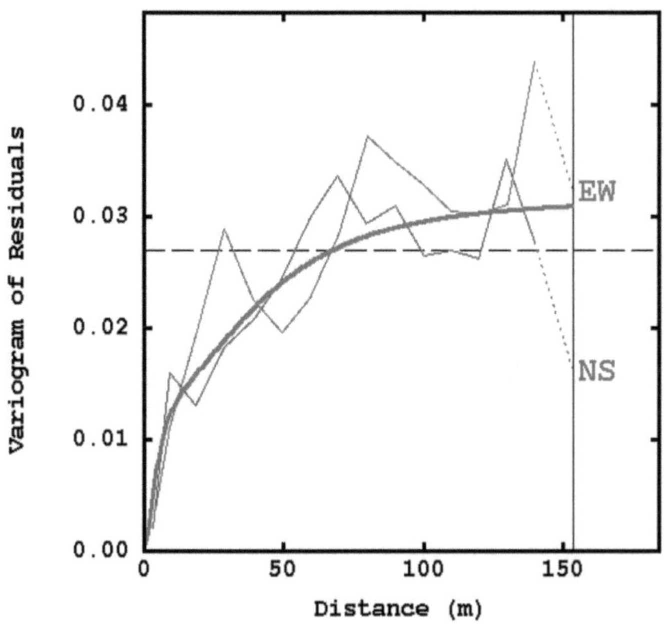

Fig. 10 Variogram of residuals between samples and EMF Visual model output. The dashed horizontal line represents the variance. The thin lines are respectively the experimental variograms computed in the North-South (NS) and East-West (EW) directions. The thick line represents the isotropic variogram model consisting of: a small measurement-variance term, a short-range (16m) and a long-range (140m) Cauchy-type variogram structure

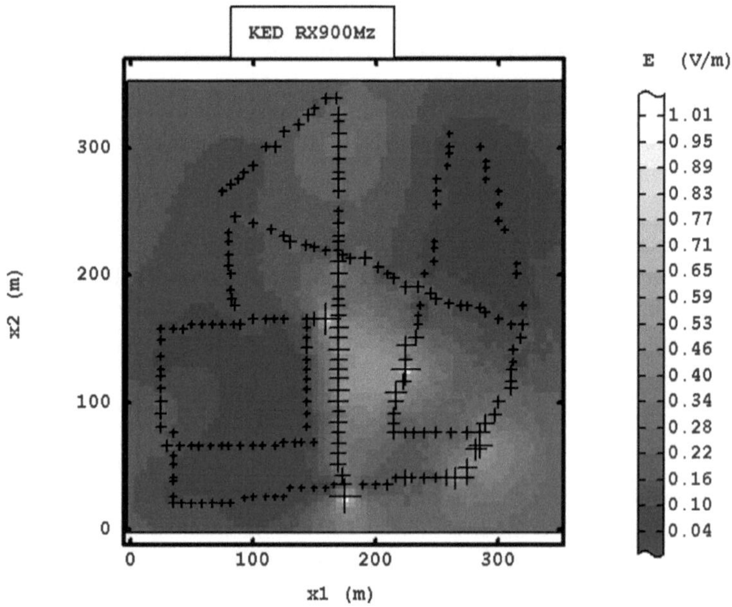

Fig. 11 Kriged map of GSMrx exposure: the "hot spots" are due to the 3 outliers

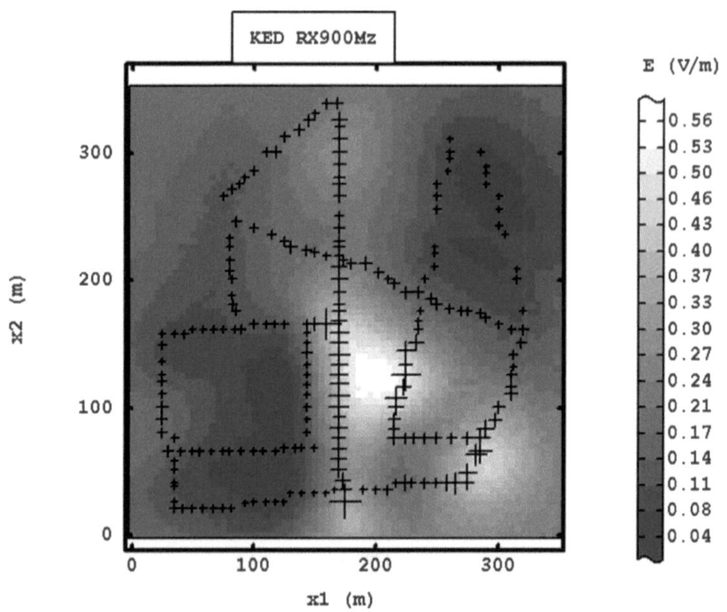

Fig. 12 Kriged map of GSMrx exposure, filtering small-scale variation: large-scale phenomena are emphasized

The directional variograms obviously overlay, so that we could fit an *isotropic* variogram model. We use a nested variogram model consisting of the sum of a term corresponding to the variance of measurement-error and of two Cauchy-type variogram terms, which are commonly used in exploration geophysics to model the spatial variation of magnetic and gravimetric data (Spector and Grant, 1970; Chilès and Delfiner, 1999).

The map of the GSMrx values estimated by KED is shown on Fig. 11. The influence of the three dosimeter outliers is quite important and leads to three small-scale "hot spots" on the map, which are eventually due to fading.

As an alternative representation we propose to filter small-scale variation (see Wackernagel (2003) for details on methodology) by removing the variation associated both with the measurement variance and the short-range Cauchy-type structure. The resulting map is displayed on Fig. 12: it emphasizes large-scale phenomena and in particular the areas of higher exposure.

6 Conclusion

The purpose of this paper is to provide a first example of application of geostatistical methods to radio-electric exposure mapping. The methodology proposed so far is simple as it belongs to the realm of linear geostatistics.

Important aspects that have been played down in this first application are the pronounced right-skew shape of the distribution of GSMrx dosimeter values (Fig. 5), the censoring by the dosimeter of values below 0.05 V/m and the status of outliers. These questions will need a careful treatment and will lead to the application of more sophisticated methodology.

Acknowledgments Geostatistical calculations were performed with the software package Isatis (www.geovariances.fr).

References

ANFR (2004) Panorama du rayonnement électromagnétique en France. Technical report, Agence Nationale des Fréquences, Maisons-Alfort. Available on: www.anfr.fr

Daley R (1991) Atmospheric Data Analysis. Cambridge University Press, Cambridge

De Doncker P, Dricot JM, Meys R, Hélier M, Tabbara W (2006) Electromagnetic fields estimation using spatial statistics. Electromagnetics 26:111–122

Chilès JP, Delfiner P (1999) Geostatistics: Modeling Spatial Uncertainty. Wiley, New York

ICNIRP (1998) Guidelines for limiting exposure to time-varying electric, magnetic, and electromagnetic fields (up to 300 GHz). Health Phys, 74(4):494–522

Larchevêque E., Dale C., Wong M-F, Wiart J. (2005) Analysis of electric field averaging for in situ radiofrequency exposure assessment. *IEEE Trans on Vehicular Tech* 54(4): 1245–1250

Spector A, Grant FS (1970) Statistical models for interpreting aeromagnetic data. *Geophysics* 35:293–302

Wackernagel H (2003) Multivariate Geostatistics: an Introduction with Applications, 3rd edn Springer-Verlag, Berlin

Wackernagel H, Lajaunie C, Blond N, Roth C, Vautard R (2004) Geostatistical risk mapping with chemical transport model output and ozone station data. Ecological Model 179:177–185

Wiart J., Dale C., Bosisio A. V., Le Cornec A (2000) Analysis of the influence of the power control and discontinuous transmission on RF exposure with GSM mobile phones. IEEE Trans on EMC 42(4):376–385

How Spatial Analysis Can Help in Predicting the Level of Radioactive Contamination of Cereals

C. Mercat-Rommens, J.-M. Metivier, B. Briand and V. Durand

Abstract The study was devoted to the identification of the spatial parameters that contribute mainly to the assessment of the vulnerability of cereals in the context of accidental discharges of radioactivity into the environment. Linking an agronomical model and a radioecological model highlighted first that the flowering date was the main parameter, since it determines the beginning of an exponential transfer of contaminants from the leaves of cereal plants to the edible part, the grain. Secondly, yield also appeared to be an important parameter as it allows the quantification of the number of contaminated products. The spatial statistical analysis performed on the yield data allowed the creation of vulnerability maps with clear spatial trends, which can facilitate the management of risks associated with radioactive contamination of cereals during the post-accidental phase.

1 Introduction and context

The radiological consequences of radioactive releases, as shown, in particular, by feedback relating to the Chernobyl accident, highlight the fact that the consequences of industrial pollution on man and the environment depend not only on the extent and nature of this pollution but also on the territory that is polluted. Whether expressed in economic, toxic or health risk terms, these consequences can be more or less detrimental depending on the features of the environment affected (environmental parameters) and the usage of this environment by man (human parameters). Urban, farming, forest, river, lake, sea or mountain environments show different pollution sensitivity levels and, within these major environmental categories, the response or reaction to a pollution event is determined by different natural or human factors, specific to the ecosystem in question. For example, in a farming region,

C. Mercat-Rommens
Laboratory of radioecological studies for marine and terrestrial ecosystems, Institute for Radioprotection and Nuclear Safety (IRSN), DEI/SESURE/LERCM, Cadarache, Bld 153, 13105 St-Paul-lez-Durance cedex, France
e-mail: catherine.mercat-rommens@irsn.fr

A. Soares et al. (eds.), *geoENV VI – Geostatistics for Environmental Applications*, 71–82

the type of production is a significant sensitivity factor. For the same surface area affected by a given pollution event, wheat and milk products will show very different respective contamination levels and responses over time. The persistence of this contamination in successive crops will also strongly depend on the soil characteristics. Generally, a territory's specific sensitivity to the pollution will be determined by the features inherent in its ecosystem, which have an influence on pollutant transfer. The same effect is observed with human factors such as farming methods (use of fertilizers, irrigation, sowing periods) or zootechnical practices (animal feeding regimes, animals housed indoors or kept outdoors). A territory's radioecological sensitivity is therefore dictated by two components: environment and human factors.

Although we can establish that a territory may be susceptible to the pollution it receives, it remains difficult to compare overall sensitivity between different territories. Radioecological sensitivity is a concept for the evaluation of the intensiveness of a territory's response to a pollution event. Since 2003, the French Institute for Radioprotection and Nuclear Safety (IRSN) has been running an inter-organizational project entitled SENSIB (an acronym referring to radioecological sensitivity) (Mercat-Rommens and Renaud 2005). Its aim is to develop a standardized tool with a single scale of values to describe and compare the sensitivity of various environments to radioactive pollution, thereby providing a classification of the territory based on its intrinsic features (annual rainfall, soil type, agricultural practices, dietary habits etc.).

2 Objectives and Methods

This paper focuses solely on the agricultural aspects of the SENSIB project. The object is to develop a method for classifying agricultural areas by their sensitivity to atmospheric radioactive pollution. The factors likely to increase or reduce the consequences of the pollution event need to be identified, characterized and ranked, in order to develop a system of indices to be used for operational classification. The present study focused first on the sensitivity of winter wheat and then generalized the main results to other French cereals. The main objective was to establish whether a uniform, localized deposit would entail an identical contamination of cereals on a national scale. If not, we want to know the quantity of cereals that will be contaminated and where in France, according to the date of the accidental release. Three sources of spatial variability are taken into account (weather conditions, soil type and agricultural practices (fertilizer, irrigation and sowing dates according to variety)).

The study used the ASTRAL model ("Assistance Technique en Radioprotection post-Accidentelle"), a computing code developed by the IRSN, which enables the assessment of radionuclide transfers to the terrestrial food chain following an accidental atmospheric discharge (Mourlon and Calmon 2002). The aim is to determine the effect of a regionalization of the ASTRAL model parameters on a specific activity in farm produce. In this study, the agricultural produce analyzed was the winter wheat plants exposed to an accidental deposit of atmospheric cesium 137 (^{137}Cs), and the ^{137}Cs concentration in the grain at the time of harvest. The STICS software ("Simulateur mulTIdisciplinaire pour des Cultures Standard") from INRA (Brisson et al. 1998, Brisson and Mary 2002) was used in order to identify regional (environmental and agronomical) factors likely significantly to influence contaminant transfers in agricultural products. This model provides a day-by-day estimate of the leaf area index, a variable that could be correlated with radioecological parameters of the ASTRAL model. Coupling the radionuclide transfers in the food chain model (ASTRAL) and the STICS model will enable the spatial variability of the radioecological sensibility of agricultural products to be quantified.

3 Models Used for the Study

3.1 The ASTRAL Radioecological Model

ASTRAL comprises a calculation module involving geographical and radioecological databases (radionuclide transfer parameters in the environment), enabling the assessment of the impact of radioactive deposits on agricultural produce (specific activities), agronomic resources (areas and quantities affected) and populations (doses received) in areas affected by a possible nuclear accident.

In the case of cereals, only leaf transfer was studied, as this transfer pathway is predominant over root transfer for the first year following accidental release. Activity at time t after deposit, C, is thus determined by the initial contamination, the retention capacity of plants taking into account the weather conditions on the day of deposit, plant growth (activity dilution), radioactive decay during the period between deposit and harvest:

$$C = \frac{D \times TLF}{Yld} \times \left[Kr \times RC_{dry} + (1 - Kr) \times RC_{wet}\right] \times e^{-\lambda r \times t} \qquad (1)$$

D: Total deposit on plant ($Bq.m^{-2}$)
TLF: translocation factor, which defines the proportion of radionuclide migrating from the leaves to the grain (-)
Kr: Dry deposit as a proportion of total deposit
Yld: Crop yield (fresh $kg.m^{-2}$) at harvest
RC_{dry}, RC_{wet}: Retention ratio in dry and rainy weather, respectively
λr: Radioactive decay constants (d^{-1})
t: Time elapsed since deposit (d)

The parameter Kr indicates the proportion of dry deposit within a given total deposit. It is likely to be a regionalized value as it varies according to rainfall intensity at the time of the deposit. However, it is impossible to predict weather conditions in accidental situations. In fact, this is a risk factor exclusively linked to the event. For this study, a value of Kr equal to 1 (entirely dry deposit) or 0 (entirely wet deposit) was therefore used in the ASTRAL simulations.

The retention ratio (also referred to as "interception ratio") is the dimensionless ratio of activity taken up by vegetation and total activity deposited on 1 m^2 (Chamberlain, 1970). In the ASTRAL model, the retention value depends on the surface area occupied by plant cover at the time of the deposit. It corresponds to the total deposited activity fraction intercepted by the above-ground part of the plants. The retention ratio, RC_{dry}, can be estimated on the basis of the leaf area index, LAI, defined as the surface area likely to collect aerosols, i.e. the leaf area per soil surface unit. The German model, ECOSYS-87 (Müller and Pröhl 1993), proposes a calculation method for retention in dry weather conditions based on the LAI. It makes the assumption that a radioactive deposit, on any surface area, is calculated as the product of a deposition velocity by the radionuclide concentration value in air, and that deposition velocity on the plant depends on represented foliar development. The variable chosen in order to characterize the plant development stage is leaf area index, and the following equation defines the retention ratio in dry weather:

$$RC_{dry} = \frac{\frac{LAI}{LAI_{max}}}{\frac{LAI}{LAI_{max}} + \frac{Vg_s}{Vgi_{max}}} \tag{2}$$

LAI: Leaf area index
LAI_{max}: Maximum leaf area index
Vg_{max}: Maximum deposit rate ($m.s^{-1}$)
Vg_s: Deposit rate on soil ($m.s^{-1}$), a constant for all plant types

In the case of deposit in rainy weather (cumulated rainfall over the deposit period in excess of 1mm, according to the ASTRAL model assumptions), the studies by Angeletti and Levi (1977), followed by those by Hoffman (1989) demonstrated that interception (represented by retention ratio in damp conditions, RC_{wet}), essentially correlates to the biomass, represented by the LAI, and to the contaminated rainfall (P):

$$RC_{wet} = \left(LAI \times \frac{S_2}{P}\right) \times \left(1 - 2^{-\frac{P}{3 \times S_2}}\right) \tag{3}$$

S_2: Saturation coefficient (mm), dependent on the radionuclide and the plant
P: Rainfall (mm)

3.2 *The STICS Agronomical Model*

STICS is a daily crop growth model developed by INRA (Brisson and Mary 2002). Its input variables relate to climate, soil and technical management. Its output variables relate to production (quantity and quality), environment and soil property changes under the influence of cultivation. STICS was designed as an operational simulation tool for agricultural conditions. Its main objective is to simulate the consequences of variations in environment and farming methods on production from an agricultural plot, over the course of the year. Crop growth is generally appraised through its above-ground biomass and nitrogen content, its leaf area index, as well as the number and biomass (and nitrogen content) of harvested organs.

The soil is considered as a series of horizontal layers, each characterized by its water, mineral nitrogen and organic nitrogen reserves. Interactions between soil and crops occur via the roots, which are defined by a distribution of root density in the soil profile. The model simulates the system's carbon, water and nitrogen balances and allows the calculation both of agricultural variables (yield, fertilizer consumption) and environmental variables (water and nitrate loss) in various agricultural situations.

3.3 Coupling Both Models

This study assumes that plant cover is not at the same stage at any given time t, across the entire French territory. The ASTRAL model's radioecological parameters of retention (RC_{dry} and RC_{wet}) change in accordance with the STICS model's agronomic parameter LAI. This enables the ASTRAL model's outputs (radioactive contamination of the crops), to take regional diversity in cultivation conditions into account.

4 The Winter Wheat Study

In order to study the influence of regional variability, we tried to define the simulation scenarios able to illustrate the full variability of agricultural conditions in France. 12 kinds of weather conditions (Fig. 1) and two varieties of winter wheat (one early variety, Talent and one late variety, Allure) were chosen. The values for the interception ratios during dry and wet weather were then calculated by STICS for each of the 24 scenarios.

The results showed that the process of interception during dry weather conditions can be correctly represented by a single curve because the effect of regional climatic variability is low (Fig. 2). However, if regional variability leads to a low effect on the magnitude of the interception, it can influence the contamination of the crop because of a shift during the year. With respect to wet interception, regional variability influences the magnitude of transfer and also causes a shift between the curves according to region.

The flowering date is another parameter which varies in space and which highly influences the contamination process of grain by the translocation process (Fig. 3).

When we combined the various regional sources of variability in a simulation of an accidental fallout of radionuclides on various French regions, it appeared

Fig. 1 Localization of the 12 stations of the study

that the date fallout occurs is the main point of contamination of the crop. If the fallout occurs before the flowering date, grain contamination is negligible because translocation is very low, even if the interception process is very high. If the fallout occurs after the flowering date, the grain is highly contaminated. That is why during

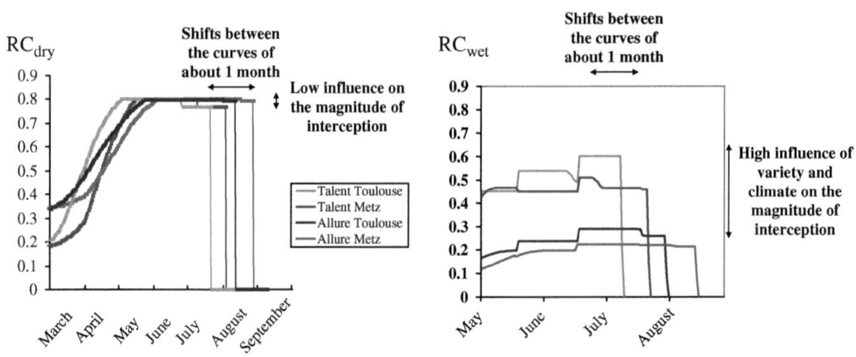

Fig. 2 Examples of evolution of the interception ratio during dry and rainy weather for both varieties and for two very different climatic conditions (stations of Metz and Toulouse)

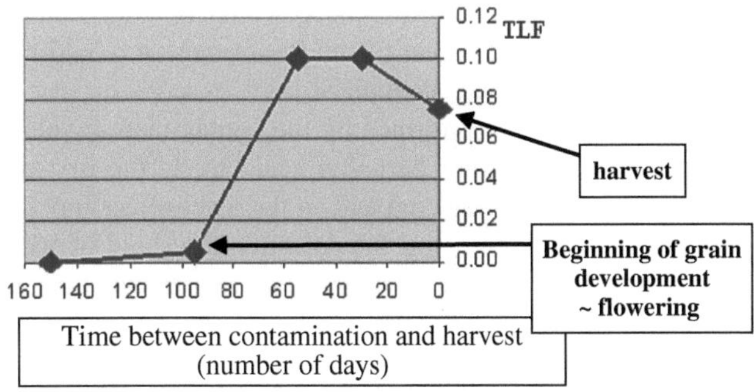

Fig. 3 Evolution of the translocation process in ECOSYS (Müller and Pröhl 1993)

spring and early summer, some French regions could totally avoid the contamination because flowering has not happened whereas other regions would be seriously affected. It depends mainly on the day of flowering, which determinates the exponential increase of transfer of radionuclides from leaves to crop (Fig. 3). As an example, the Fig. 4 shows the results for both varieties, Talent and Allure, and for two stations (Perpignan and Rouen which are located respectively in Southern and Northern France – cf. Fig. 1).

From the Fig. 4, we can see that an accidental fallout before May would not contaminate the grain, either in the South of France or in the North. Any fallout in May and at the beginning of June would cause contamination of the winter wheat produced in the Perpignan region, whereas that produced in the Rouen region would avoid the contamination because the flowering stage would not have started. From late June, and until the harvest in Perpignan (beginning of July) both produce types

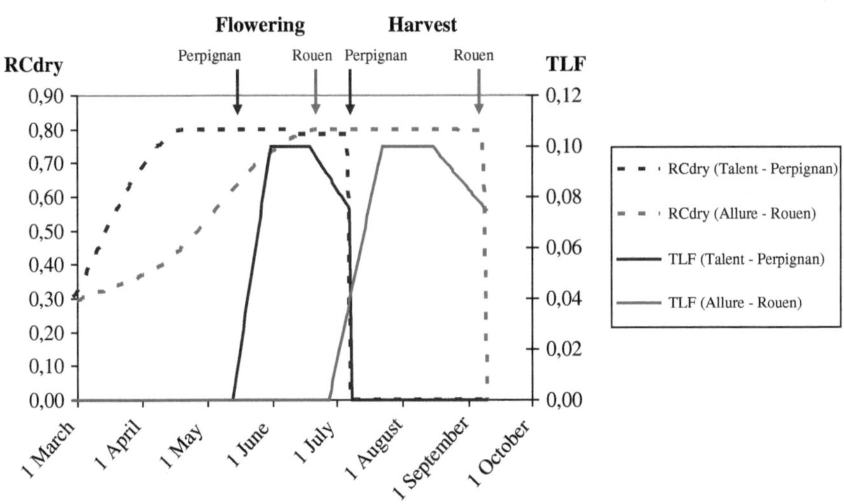

Fig. 4 Results for interception and translocation for 2 stations

and consequently the entire territory of France would be contaminated. Then in August and September, only the wheat of the Rouen region would be vulnerable because the harvest would not have taken place.

These modeling results were confirmed by the contamination values measured in wheat crops in France after the Chernobyl accident. In the first days of May, very few varieties of wheat had flowered and so the level of cesium 137 measured remained low in grain, very often below the detection threshold of 50 Bq per fresh kg, in most French regions. The ten samples of wheat, in which cesium 137 was significantly detected after the Chernobyl accident, came from regions located in the south of France (French administrative regions labeled "Drôme", "Ardèche", "Bouches-du-Rhône" and "Vaucluse").

5 Generalization to Other French Cereals

The previous study of winter wheat shows that the radioecological sensitivity of grains relies primarily on the flowering date. This flowering date indicator may also account for the level of contamination of other French cereals because the translocation process acts in the same way for all cereals. That is why we proposed - as a tool for managing potential contaminated cereal fields - an early calculation of the flowering date of all cereals, combined with an estimation of regional cereal production. The first indicator, "the flowering date", will answer the question "where in France would cereals be contaminated?", while the second indicator, "the yield", will answer the question "what quantity of cereals would be contaminated?".

The determination of the main factor for the contamination can be made from spatial parameters such as rainfall and temperature, which mainly influence the date flowering takes place. This work is being done by the French Institute "Arvalis-Institut du vegetal" which is conducting a reconstruction of the spatial variability of the flowering dates of the main French cereals (winter wheat, barley, corn and triticale) from meteorological data available from 309 stations in France. This reconstruction will be made for each location in the country and respectively, for early and late varieties and will also consider the variability in time of flowering dates by taking annual changes of climatic parameters into account.

In parallels, we studied the effect of the spatial distribution of the 309 stations cross-referenced with the variability of potential cereal production, represented by yield. The objective was to determine the area where cereal production is more (or less) intensive and subsequently the degree of vulnerability of cereal production in the event of an accidental fallout occurring after the flowering date. The 309 meteorological stations are distributed fairly homogeneously throughout French territory (except in Corsica). We calculated the Voronoï polygons corresponding to the stations with GeoStatisticalAnalyst of ArcGis9.1 (Fig. 5).

The Voronoï polygons represent the area of influence of each meteorological station. The Voronoï polygons were then cross-referenced with the agricultural statistical data available for the total cereal producing area (Agreste 2000) and for

Fig. 5 Voronoï polygons

the total yield of cereals (Agreste 1988), in order to reconstruct the occupation of agricultural lands by cereals and the yield expected in each polygon (Fig. 6).

Then we performed a local cluster analysis on yield to detect any potential geographical clusters (if Voronoï polygons are significantly correlated with surrounding polygons) in order to propose a judicious classification of the territory by aggregating the polygons. We performed a spatial statistical analysis with the LISA statistics of the software TerraSeer-STIS (Jacquez *et al.* 2005):

$$LISA(u) = \underbrace{\left[\frac{z(u) - m}{s}\right]}_{term\,1} \times \underbrace{\left(\sum_{j=1}^{J(u)} \frac{1}{J(u)} \times \left[\frac{z(uj) - m}{s}\right]\right)}_{term\,2} \tag{4}$$

u: one of the 309 Voronoï polygons constructed from the 309 meteorological stations,
z(u): yield reconstituted in the u polygon,
m: mean of the 309 yield values,
s: standard deviation for the 309 yield values,
J(u): number of neighbors.

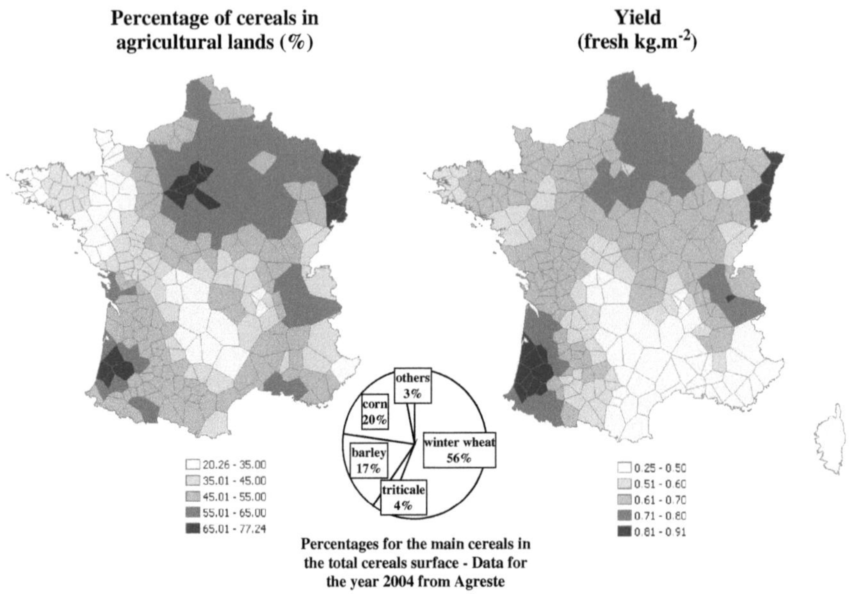

Fig. 6 Spatial patterns for French production of cereals as a percentage of the cereal share of arable land (%) and yield (kg.m^{-2})

This statistic is based on the local Moran test (Moran 1950) (Anselin 1995) for the detection of local spatial autocorrelation.

LISA $<$ 0: when term 1 is negative and term 2 is positive (or the opposite). It happens when the value z(u) is higher than the mean, whereas the values for the neighboring polygons are globally lower than the mean. In this case, we have an outlier named High-Low, which identifies an anomaly. The opposite (when a value is lower than the local mean whereas neighboring values are globally higher than the local mean) is named a Low-High outlier and is also considered as an anomaly.

LISA $>$ 0: when the value z(u) and most of the surrounding values go in the same direction (higher or lower) compared to the local mean. In these cases, we have clusters respectively High-High or Low-Low. The spatial correlation of a polygon with its neighbors can then justify propositions of spatial aggregation, a first step towards a classification of the territory.

In addition to the sign of the LISA statistic, its magnitude informs on the extent to which the value of a polygon differs from or corresponds to its neighbors' values.

The LISA statistic was applied for the 1st order queen neighbors, which are defined as polygons sharing a common border or vertex with the u polygon. We tested the LISA significance by applying a Monte Carlo randomization (10 000 calculation) and a significance level of 5% (Moran test). The results are presented in Fig. 7.

No outlier was detected. Good agreement was observed between the results of LISA statistics and the French agricultural patterns. As shown by Fig. 6, the main

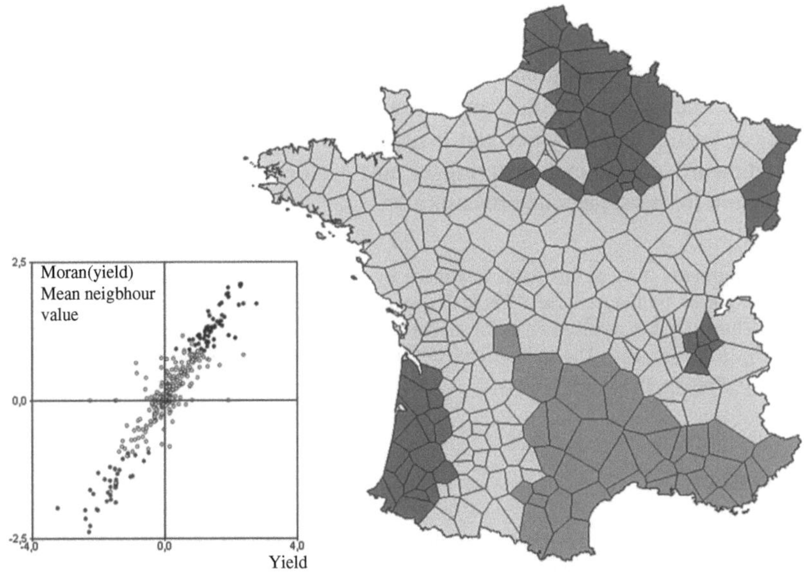

Fig. 7 Results of local cluster analyses for global yield of cereals. High-High clusters are shown in red and Low-Low clusters are shown in blue

regions for cereal production in terms of surface devoted to cereal growing are the Center and the North of France (wheat and barley), the Southwest (corn) and the East (corn). We observed in Fig. 7 that the yield is high (High-High cluster) for almost all of these regions, except the Center, whereas the Southeast region is less productive (Low-Low cluster). A small area with a High-High cluster also appeared in the North of the Alps, which corresponds to a more secondary region for cereal production.

If we cross-reference knowledge about the yield and flowering dates for winter wheat reconstructed by the STICS model, we observe that the yield is higher for the Northern part of France while flowering dates are later. This result can provide information for improving the dimensioning of countermeasures from a spatial point of view (where are the vulnerable areas?), but also from a temporal point of view (at which period of the year?) and a magnitude point of view (what quantities of contaminated cereals are expected?).

6 Conclusion

The study was devoted to the identification of the spatial parameters that contribute mainly to the radioecological sensibility of cereals. The flowering date appeared to be the main parameter because it determines the beginning of an exponential transfer of contaminants from the leaves of the cereal plants to the edible part, the grain. This parameter, the flowering, is especially useful in decision-making for at least two reasons: 1. It is a stage in the development of cereals, easily visualized on fields; 2. An early calculation of flowering dates based on the spatial variability induced by weather conditions, soil types and farming practices is achievable for all cereals. The spatial statistical analysis allows the creation of vulnerability maps with clear spatial trends, which can facilitate the management of risks associated with radioactive contamination during the post-accident phase for cereals, and guide the decision for possible countermeasures.

Acknowledgments Part of this work was made by Aude Delboe in the framework of the agricultural engineer training course ("Ecole Nationale Supérieure d'Agronomie de Montpellier", France). This study has received financial support from ADEME (French Agency for Environment and Energy Management). The authors wish to thank Pierre Goovaerts for his prompt and accurate answers to several e-mails of questions.

References

Agreste (1988) Recensement général de l'agriculture. Agreste(ed), France

Agreste (2000) CD ROM Recensement agricole 2000 - La fiche comparative., Eds Agreste France

Angeletti L, Levi E (1997) Etudes comparatives des facteurs de transfert de l'eau, de l'iode et du strontium sur le ray-grass et le trèfle. Rapport CEA-R-4960, Commissariat à l'Energie Atomique, Saclay, France

Anselin L (1995) Local Indicators of Spatial Association — LISA, Geographical Analysis Vol 27. pp 93–115

Brisson N, Mary B, Ripoche D, Jeuffroy MH, Ruget F, Nicoullaud B, Gate P, Devienne F, Antonioletti R, Dürr C, Richard G, Beaudoin N, Recous S, Tayot X, Plénet D, Cellier P, Machet JM, Meynard JM, Delécolle R (1998) STICS: a generic model for the simulation of crops and their water and nitrogen balances. I. Theory and parameterization applied to wheat and corn. Agronomie, 18: 311–346

Brisson N, Mary B (2002) STICS : Notice concepts et formalismes. Version 5.0. (eds) INRA, France

Chamberlain AC (1970) Interception and retention of radioactive aerosols by vegetation. Atmos Environ 4: 57–78

Hoffman FO (1989) Pasture grass interception and retention of 131I, 7Be and insoluble microspheres deposited in rain. OAK-RIDGE Laboratory, Environmental Sciences Division.

Jacquez GM, Goovaerts P, Rogerson P (2005) Space-Time intelligence Systems: Technology, applications and methods. J Geo Syst 7: 1–5

Mercat-Rommens C, Renaud P (2005) From radioecological sensitivity to risk management: the SENSIB project, second international conference radioactivity in the environment, nice.

Moran PAP (1950) Notes on continuous stochastic phenomena, Biometrika 37: 17–23

Mourlon C, Calmon P (2002) ASTRAL: a code for assessing situations after a nuclear accident. 12th annual meeting of SETAC Europe, Vienna 37: pp 12–16

Müller H, Pröhl G (1993) ECOSYS-87: a dynamic Model for assessing radiological consequences of nuclear accidents. Health Phy 64(3): 232–252

Stochastic Modelling Applied to Air Quality Space-Time Characterization

A. Russo, R. M. Trigo and A. Soares

Abstract Atmospheric pollution directly affects the respiratory system, aggravating several chronicle illnesses (e.g. bronchitis, pulmonary infections, cardiac illnesses and cancer). This pertinent issue concerns mainly highly populated urban areas, in particular when meteorological conditions (e.g. high temperature in summer) emphasise its effects on human health.

The proposed methodology aims to forecast critical ozone concentration episodes by means of a hybrid approach, based on a deterministic dispersion model and stochastic simulations. First, a certain pollutant's spatial dispersion is determined at a coarse spatial scale by a deterministic model, resulting in an hourly local trend. Afterwards, spatial downscaling of the trend will be performed, using data recorded by the air quality (AQ) monitoring stations and an optimization algorithm based on stochastic simulations (Direct sequential simulation and co-simulation). The proposed methodology will be applied to ozone measurements registered in Lisbon. The hybrid model shows to be a very promising alternative for urban air quality characterization. These results will allow further developments in order to produce an integrated air quality and health surveillance/monitoring system in the area of Lisbon.

1 Introduction

Research activities focusing on possible associations between climate variability/climate change, atmospheric pollution and health went through an exceptional boost on the past few years. The respiratory system is directly affected by atmospheric pollution (e.g. bronchitis, pulmonary infections, cardiac illnesses and cancer). This pertinent issue concerns mainly highly populated urban areas, in particular when meteorological conditions (e.g. high temperature in summer) emphasise its effects on human health. The ominous consequences resulting from population's

A. Russo
CMRP, Instituto Superior Técnico, Av. Rovisco Pais, 1049-001 Lisboa, Portugal
e-mail: arusso@ist.utl.pt

A. Soares et al. (eds.), *geoENV VI – Geostatistics for Environmental Applications*, 83–93

exposure to air pollution (Cobourn et al., 2000; De Nevers, 2000; Kolehmainen et al. 2000; Seinfeld, 1986) as well as the association of the occurrence of meteorological extreme events to poor AQ episodes, have been greatly discussed and explored during the past few years (Díaz et al., 2005; Díaz et al., 2004; García-Herrera et al., 2005; Monteiro et al., 2005a, 2005b; WHO, 2004).

Recently, some studies relating health, meteorological and environmental parameters have been formulated through neural network modelling. This methodology has revealed itself as a sufficiently innovative and efficient technique (Hitlermann et al., 1998; Lippman, 1989). A variety of other methodologies can be applied as an alternative to neural network modelling (Sokhi et al., 2006). Usually, urban atmospheric models are based on simple approaches such as box or Gaussian models (Kitwiroon et al., 2002; Middleton, 1997). These simple modelling approaches can provide swift solutions. Nevertheless, they rely on significant simplifying assumptions and do not fully describe the processes and interactions that control the transport and behaviour of pollutants in the atmosphere (Sokhi et al., 2006). Eulerian models, such as UAM, ROM, HASTE, CMAQ, enable a more realistic and detailed description of the atmosphere (San José et al., 2002; Smyth et al., 2006; Sokhi et al., 2006; Wang et al., 2002). This new generation of atmospheric modelling tools now applies an integrated approach to the model domain by solving the Navier–Stokes partial differential equation system by using state-of-the-art numerical methods whilst benefiting from advances in recent computer platforms (Sokhi et al., 2006). The use of such Eulerian grid models has progressively increased over the past few years for a range of air quality research applications, including development of strategies to control photochemical smog (Sokhi et al., 2006).

Although the referred models take into account meteorological conditions and observed data for a certain time period, they have some limitations as it regards the purpose of this study. These models are able to provide mean values of a certain pollutant's concentration at coarse scale grids. However, health problems (*e.g.* asthma crisis) are greatly associated with extreme episodes of pollution that are not revealed by the mean trends of those models. On the other hand, given the high spatial variability of such phenomena, the coarse scale is, most of the times, insufficient to characterize locally pollutants' extreme values.

Spatial-temporal geostatistical models (Kyriakidis and Journel, 1999; Nunes and Soares, 2005) can constitute a complementary tool for the AQ characterization at a regional level.

2 Objectives

The main objective of this work is to produce an integrated air quality and health surveillance/monitoring system, which will include all relevant information, in order to analyze and follow-up any possible associations between environmental factors and health issues in the area of Lisbon. The proposed methodology aims to forecast critical ozone concentration episodes by means of a hybrid approach, based on a deterministic dispersion model and stochastic simulations.

Spatial-temporal AQ models will be developed using meteorological, environmental and health information at different time and spatial scales as input. A hybrid approach, based on a deterministic dispersion model and stochastic simulations, will be applied in order to characterize a certain pollutant spatial dispersion at a fine scale. The proposed methodology can be divided in two basic steps:

i. In a first step, a deterministic simulation model – Community Multiscale Air Quality (CMAQ), model used and developed by the U.S. Environmental Protection Agency (EPA) for forecasting urban AQ – is used to calculate a trend of a certain pollutant's concentration at a coarse grid of 4×4 km. This dynamic model takes into account the meteorological conditions as well as the pollutant's concentrations measured at the existing monitoring stations.
ii. In a second step, the trend previously determined by CMAQ is downscaled through an iterative optimization procedure. This optimization procedure is based on direct sequential simulations and co-simulations.

3 Methodology

3.1 Coarse Spatial Scale

CMAQ system is a powerful third generation air quality modelling and assessment tool, designed to support air quality modelling for various applications (Sokhi et al., 2006). CMAQ is a three-dimensional grid-based AQ model that can be applied to simulate concentrations of tropospheric ozone, acid deposition, visibility, fine particulate and other air pollutants, involving complex atmospheric pollutant interactions on regional and urban scales. The target grid resolutions and domain sizes for CMAQ range spatially and temporally over several orders of magnitude. CMAQ temporal flexibility allows the evaluation of longer-term (annual to multi-year) pollutants' behaviour as well as short-term transport from localized sources. CMAQ was used in order to calculate a trend of a certain pollutant's concentration at a coarse grid of 4×4 km.

3.2 "Downscaling" of the CMAQ Local Trend

The intention underneath the downscaling of CMAQ's local trend is to characterize the spatial dispersion of a pollutant concentration at a fine scale (500×500 m), in order to reproduce for a certain time period CMAQ's local trend and the monitoring stations' measurements, and also spatial patterns revealed by correlograms determined by monitoring stations observations (Fig. 2).

Let us consider the variable Z(x) at a punctual scale (fine grid), characterized by the variogram and cumulative distribution function: γ_1(h) and F_Z(z). Z_v(x) is the up-scaled value of CMAQ model that one wishes to be reproduced at the final coarse grid scale.

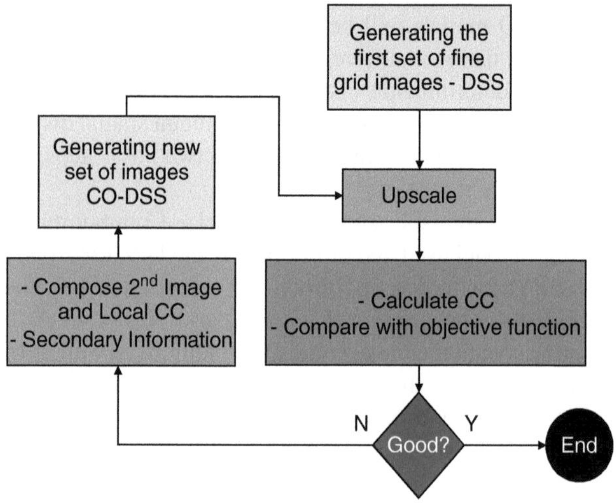

Fig. 1 Methodology

For the selected time period (one day), CMAQ is run in order to produce a trend with 24 layers (1 per hour). Each layer is a trend map of the study area for a particular hour.

The proposed iterative optimization methodology can be described as follows:

i. Application of Direct Sequential Simulation (DSS) algorithm with local means (CMAQ trend) to generate a multiple set of fine scale images.
ii. The set of generated images will be used, after being up-scaled spatially to a 4×4 km grid, to calculate a match with the objective function (CMAQ trend):

$$Z_v^s(x) = \sum_{i=1}^{16} Z^s(x_i) \text{ and } \mathrm{Z_v(x)}$$

This match can be calculated, for example, through the use of a correlation coefficient (CC) between the CMAQ trend and the simulated images. The "best" parts of each image, revealed by the correlation coefficient, are selected and composed in a unique fine grid image for the next iterative step.

For each spatial cell of each simulation, the match with the trend at the coarse grid scale is computed. The correlation coefficient between the 24 simulated values and the trend value is used. For each of the 20 realizations, a 2D map of correlation coefficients at the coarse grid scale is thus obtained. One composite 3D image at the fine scale is build by selecting for each 2D coarse grid cell the realization among the 20 where the match is the best. The 24 series of fine cells included in the coarse cell are retrieved and used in the composite image.
iii. A co-simulation of Z(x) (Soares, 2000) is performed using the composed best image as secondary information and the equivalent correlation coefficients to derive

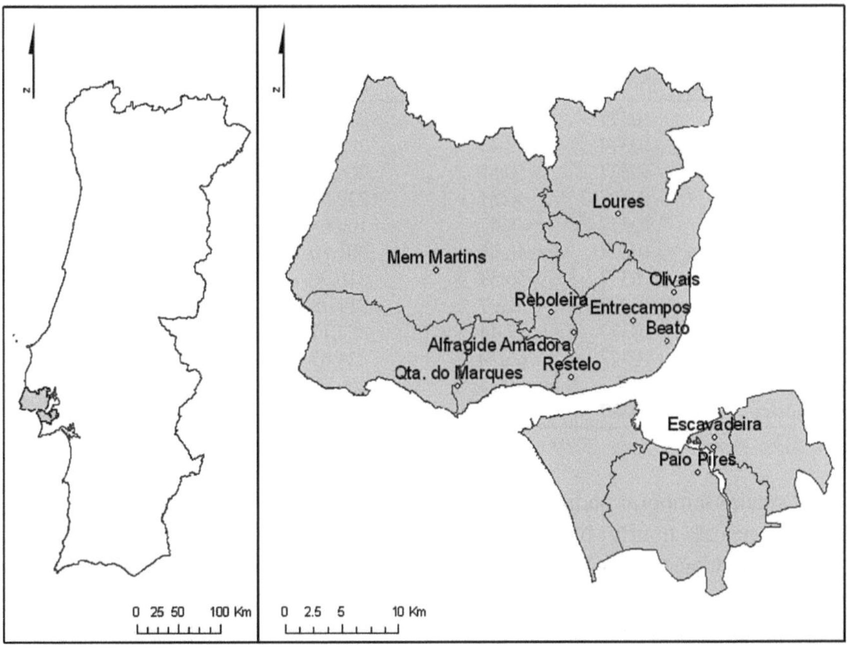

Fig. 2 Case study area

the co-regionalization model of collocated co-kriging system. The equivalent correlation coefficients used correspond to the coefficients associated to the best 3D coarse grid cells among the 20 realizations where the match between the trend and the simulations are the best. After generating a new set of co-simulated images the process return to step ii) until the match with CMAQ trend is satisfactory. The final output of the methodology is a unique composite 3D image.

Note that at each step the composed "best" image used as secondary information does not have the spatial structure of Z(x). The variogram model is imposed to the next generation of co-simulated images. It is important to stress out that the space of uncertainty, that we achieve with the proposed algorithm, is defined by the set of equi-probable images (co-simulated), where each one of them reproduce the coarse grid trend given by the deterministic model.

4 Case Study

The proposed methodology intends to characterize critical ozone (O_3) concentration episodes. In order to test the proposed methodology a case study is presented: the city of Lisbon and its surroundings (Fig. 2). The selection of Lisbon as a case study relates with its geographical location and its social relevance. "Great" Lisbon occupies about 2.750 km^2 and is inhabited by 2.1 million people.

Table 1 O3 statistics. O_3 critical values for population's information (PI) (hourly average exceeds 180 mg/m^3) and for alert (A) (hourly average exceeds 240 mg/m^3)

Monitoring Stations	N Samples	Mean (mg/m^3)	Max (mg/m^3)	PI < O_3 < A	O_3 > A
Hospital Velho	10377	45.32	232.23	11	0
Paio Pires	10377	52.43	288.19	17	1
Beato	10377	50.85	208.55	5	0
Olivais	10377	46.81	222.94	2	0
Entrecampos	10377	38.47	166.64	0	0
Alfragide-Amadora	10377	46.96	201.80	3	0
Reboleira	10377	56.59	216.90	7	0
Loures	10377	52.97	211.78	10	0
Restelo	10377	52.44	257.17	1	2
Mem-Martins	10377	66.70	224.63	4	0
Quinta do Marques	10377	61.75	269.99	1	2
Escavadeira	10377	48.47	186.00	1	0

O_3 spatial-temporal samples were colleted at twelve AQ monitoring stations (Fig. 2) on an hourly basis for a period of 24 months (from 1/1/2003 to 31/12/2004). Considering O_3 critical values for population's information (PI) (hourly average exceeds 180 mg/m^3) and for alert (A) (hourly average exceeds 240 mg/m^3) and the statistical results for O_3, we can conclude that exceeding values are a very small portion of the complete sample (Table 1). A random day (24 hours) was selected for DSS simulation in order to determine a local mean spatial-temporal trend.

5 Results and Discussion

5.1 Exploratory O_3 Data Analysis

Spatial and temporal O_3 variograms (Fig. 3, Fig. 4) were determined for the complete set of monitoring stations and for a period of 10377 hours. The histogram (Fig. 5) was determined for the complete set of monitoring stations and for a random day.

As the 12 monitoring stations are the only available spatial data, it is assumed that the variogram calculated with this information reflects the spatial pattern of the average behaviour for the 24 hours period. $C(\mathrm{h}, t) = C(|\mathrm{h}|)$ was adopted as space-time model, where the generalized distance $|\mathrm{h}| = \sqrt{x^2 + y^2 + t^2}$ is based on a simple metric of two spatial dimensions plus the time component (Dimitrakopoulos and Luo, 1994; Kyriakidis and Journel, 1999; Soares, 2000). The space-time variogram model is an anisotropic spherical model with a spatial range of 5000 m and time range of 7.5 hours.

Just for an illustrative purpose a synthetic CMAQ grid of 4 km $\times$ 4 km $\times$ 24 hours was built covering the entire area. A coarse scale grid of 8 km $\times$ 8 km spatial blocks $\times$ 24 periods of time is considered as our reference trend of O_3 for that period (Fig. 6).

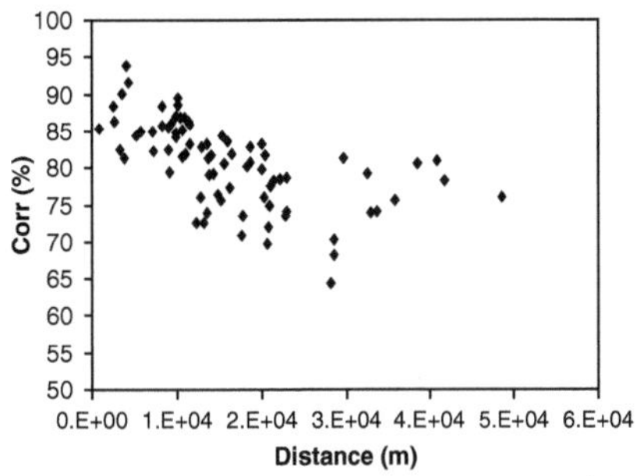

(a)

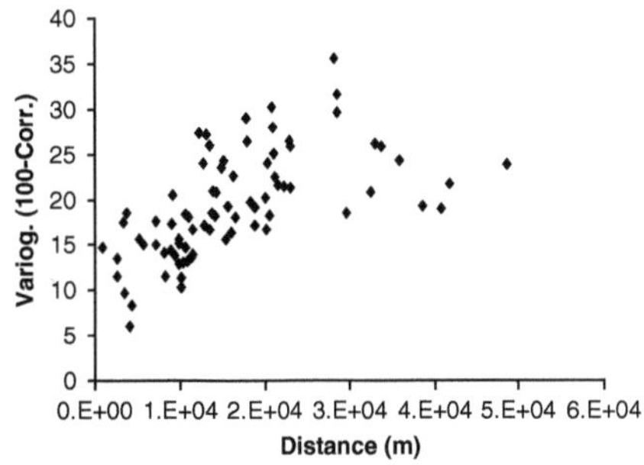

(b)

Fig. 3 (a) Correlation between a 24 hours sample of O_3 on each pair of monitoring stations and the distance between the two; (b) Variogram between a 24 hours sample of O_3 on each pair of monitoring stations and the distance between the two

5.2 "Downscaling" of the Local Trend

The iterative methodology described previously (*c.f.* Section 3.2) was applied to perform the downscaling of the coarse grid trend in order to reproduce the space-time variogram model, the monitoring stations' measurements for the reference period and also to reproduce the coarse grid trend behaviour.

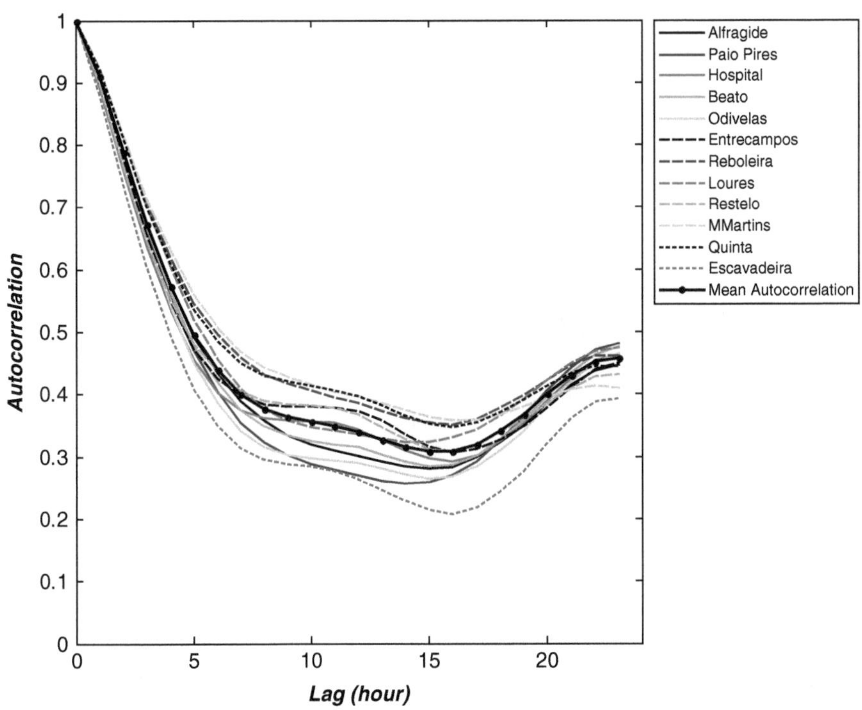

Fig. 4 Autocorrelation for each monitoring station and mean autocorrelation

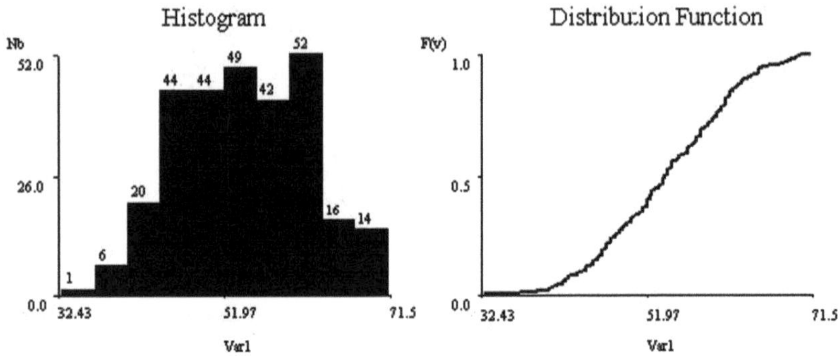

Fig. 5 O_3 Histogram

The first set of 20 fine scale grid images was generated by DSS (Fig. 7 (a)). Each of the 20 simulated images was filtered (up-scaled) to a $4\,\text{km} \times 4\,\text{km} \times 24$ hours grid (Fig. 7 (b)) in order to be compared with the O_3 trend represented in Fig. 6 (objective).

Afterwards, the mean correlation between each 24 hours block of each simulated image and the reference image (trend) is calculated. The best blocks (higher correlation with reference image) compose the best secondary image, for the next iterative step (Fig. 8).

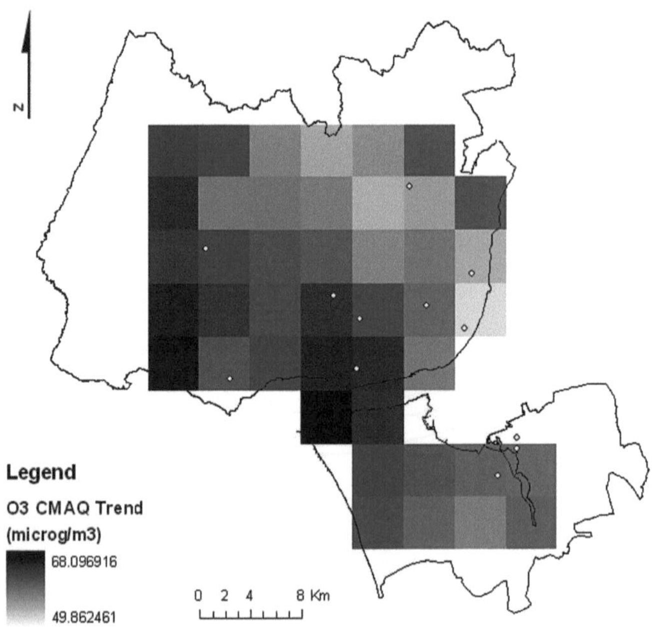

Fig. 6 Reference image (Example for hour 1)

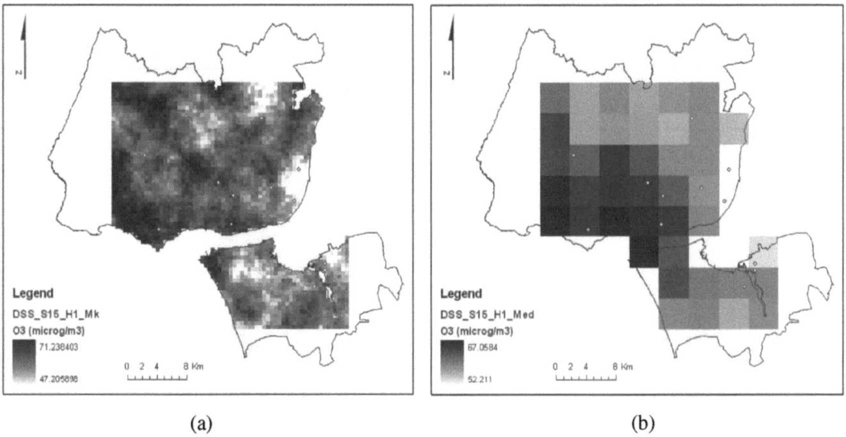

Fig. 7 (a) Example of hour 1 of one of the 20 simulated images; (b) Example of hour 1 of one of the filtered 20 simulated images (respectively)

The best image was used as secondary information for the co-DSS for the next generation of images. The correlation coefficients determined previously were used to build local co-regionalization models.

This procedure was repeated until, after the third co-simulation, the correlation between average 24 blocks of the simulated images and the respective 24 hours

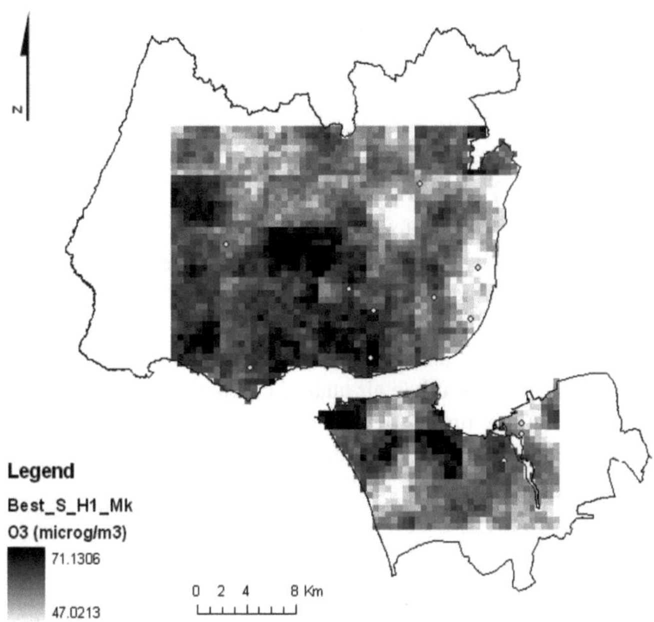

Fig. 8 Example of best secondary image

Table 2 Simulation (S) and co-simulation (Co-S) correlation results

Correlation (%)	S	1st Co-S	2nd Co-S	3rd Co-S	4th Co-S
Maximum	98.7	99.1	99.3	99.4	99.5
Minimum	34.6	82.4	85.3	88.5	90.0

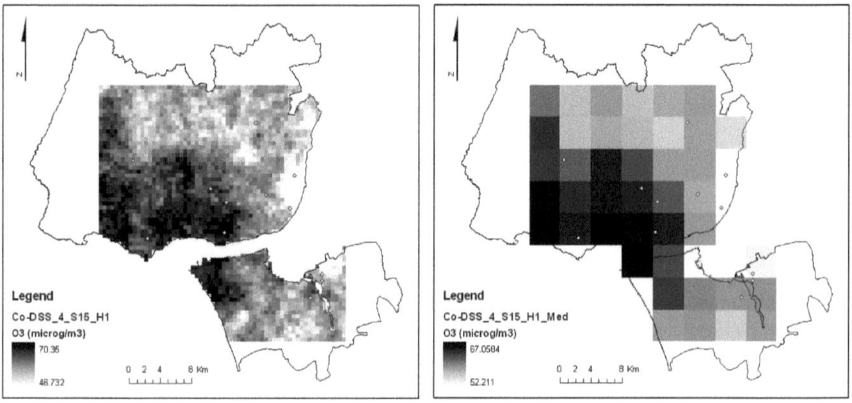

Fig. 9 (a) Example of hour 1 of the best co-simulated images; (b) Example of hour 1 of the up-scaled filtered best co-simulated images

block of the reference image reached a satisfactory match for this illustrative case study. (Table 2, Fig. 9 (a)).

The best co-simulation (co-simulation 15) is chosen because it's the one that has the higher correlation with the reference image. The trend downscaling using data recorded by the AQ monitoring stations was reached only with 4 co-simulations, converging rapidly (Table 2).

6 Conclusions

The proposed methodology transforms a coarse grid air quality trend into maps with local extreme values, which are a valuable tool for the evaluation of space-time relationships with health effects.

The hybrid model, coupling a deterministic air quality trend model, conditioned by meteorological conditions and space-time stochastic simulations shows to be a very promising alternative for urban air quality characterization.

These results will allow further developments in order to produce an integrated air quality and health surveillance/monitoring system in the area of Lisbon.

References

Cobourn W.G., Dolcine L., French M. and Hubbard M.C. (2000), A Comparison of Nonlinear Regression and Neural Network Models for Ground-Level Ozone Forecasting, Journal of the Air & Waste Management Association, Volume 50, 1999–2009

De Nevers N. (2000) Air Pollution Control Engineering, 2nd edn. McGraw-Hill

Díaz J., García-Herrera R., Trigo R.M., Linares C., Valente M.A. and Hernadéz E. (2005) "The impact of summer 2003 heat wave in Iberia: how should we measure it?". International Journal of Biometeorology, DOI: 10.1007/s00484-005-0005-8

Díaz J., Linares C., López C., García-Herrera R. and Trigo R.M. (2004) "Relationship between Environmental Factors and Infant Mortality in Madrid, 1986-1997". Journal of Occupational and Environmental Medicine 6 (8), 768–774

Dimitrakopoulos R. and Luo X. (1994). "Spatiotemporal Modelling: Covariances and Ordinary Kringing Systems". Geostatistics for the Next Century. Dimitrakopoulos Ed.. Kluwer Academic Pub., 88–93

García-Herrera R., Díaz J., Trigo R.M. and Hernandez E. (2005) "Extreme summer temperatures in Iberia: health impacts and associated synoptic conditions". Annales Geophysicae , 23, 239–251

Hitlermann T., Stolk J. and van der Zee S. (1998), Asthma severity and susceptibility to air pollution, European Respiratory Journal, 11, 686–693

Kitwiroon N., Sokhi R.S., Luhana L.. and Teeuw R.M. (2002). Improvements in air quality modelling by using surface boundary layer parameters derived from satellite land cover data. Water, Air and Soil Pollution: Focus 2 (5–6), 29–41

Kolehmainen M, Martikainen H and Ruuskanen J (2000) Neural networks and periodic components used in air quality forecasting. Atmospheric Environment, 35, 815–825

Kyriakidis P and Journel A. (1999), Geostatistical space-time models: a review. Mathematical Geology, 31(6), 651–685

Lippman M. (1989), Health effects of ozone: A critical review. J. Air Waste Manage. Assoc., 39, 672–695

Middleton D.R. (1997). A new model to forecast urban air quality: BOXURB. Environmental Monitoring and Assessment 52, 315–335

Monteiro A., Lopes M., Miranda A., Borrego C. and Vautard R. (2005a), Air Pollution Forecast In Portugal: A Demand From The New Air Quality Framework Directive. International Journal of Environment and Pollution, 25, (1–4): 4–15 2005

Monteiro A., Vautard R., Borrego C. and Miranda A. (2005b), Long-Term Simulations Of Photo Oxidant Pollution Over Portugal Using The CHIMERE Model, Atmospheric Environment 39, 17, 3089–3101

Nunes C. and Soares A. (2005), "Geostatistical Space-Time Simulation Model", Environmetrics, 16: 393–404

San José R., Pérez J.L., Blanco J.F., Barquín R. and González R.M. (2002), An operational version of MM5–CMAQ modelling system over Madrid City. Forth Symposium on the Urban Environment, 20–24 May, American Meteorology Society

Seinfeld J.H. (1986) Atmospheric Chemistry and Physics of Air Pollution. John Wiley & Sons, New York, 738

Smyth S.C., Jiang W., Yin D., Roth H. and Giroux E. (2006), Evaluation of CMAQ O_3 and PM2.5 performance using Pacific 2001 measurement data. Atmospheric Environment 40, 2735–2749

Soares A. (2000) *Geoestatística Aplicada às Ciências da Terra e do Ambiente*, IST PRESS, 206p

Soares A., (2002) "Stochastic Modelling of Spatio-Temporal Phenomena in Earth Sciences". *Geoinformatics*. Chapter of Encyclopedia of Life Support Systems (EOLSS). Ed. Atkinsson P. Developed under the Auspices of the UNESCO, Eolss Publishers, Oxford, UK

Sokhi R.S., San José R., Kitwiroon N., Fragkoua E., Pérez J.L. and Middleton D.R. (2006), Prediction of ozone levels in London using the MM5–CMAQ modelling system, Environmental Modelling & Software 21, 566–576

Wang X., Hu Y., Russell A., Mauzerall D. and Zhang Y. (2002), Using Models3/CMAQ to simulate regional air quality in China. 2002 Models-3 User's Workshop, 21–23 October, EPA, Research Triangle Park

WHO (2004), Heath and Global Environmental Change Series, Heat waves: risks and responses, No. 2 e Systematic review of health aspects of air quality in Europe, http://www.euro.who.int

Automatic Mapping Algorithm of Nitrogen Dioxide Levels from Monitoring Air Pollution Data Using Classical Geostatistical Approach: Application to the French Lille City

G. Cardenas and E. Perdrix

Abstract This work aims to test a new method of automatic mapping of gaseous nitrogen dioxide in an urban area. These maps have to be realised based on the data provided by automatic on-site monitoring stations, taking into account their scarcity that could hamper a correct spatial interpolation of the pollution levels. In the first part of this study, we propose a new methodology to generate additional data, based on several previous field campaigns performed by passive sampling. Among these passive sampling sites some of them (henceforth called "virtual stations") are time-correlated to a given fixed station (called "reference station"). In the second part of this study, we have tested the suitability of our method for the automated generation of variograms on a case study as well as the quality of the estimations calculated based on these data. For mapping, geostatistical methods were applied, particularly the cokriging one. This multivariable method exploits the additional information given by auxiliary variables; in the case of nitrogen dioxide, variables depicting the area, such as the population density or the emissions inventory, may therefore be used. In order to take into account the uncertainty of the data generated in the virtual stations, we included in the variance-covariance kriging matrix an additional component called the variance of measurement error (VME); a methodology to calculate this component is described. Finally, the resulting maps are well detailed and do show the main features of the pollution due to nitrogen dioxide on the considered domain.

1 Introduction

In France, the air quality monitoring networks have to regularly publish the levels of pollutants in their respective zone. Several methods of cartography have been developed, since maps have proved to be an efficient way to present information about air quality to the public.

G. Cardenas
Institut National de l'Environnement Industriel et des Risques, Parc Technologique ALATA, BP2, 60550 Verneuil-en-Halatte, France
e-mail: giovanni.cardenas@ineris.fri

A. Soares et al. (eds.), *geoENV VI – Geostatistics for Environmental Applications*, 95–101

In order to produce them, geostatistical methods of interpolation (Kriging) are often used. Their main advantage is to integrate the spatial correlation of the pollutant through the calculation of the spatial covariance function based on the available data.

Usually, maps are realised with measurements provided by passive samplers. Because of their low cost, these data are abundant and allow a large spatial coverage. However, they provide a time average pollutant's concentration over quite a long period (7 or 14 days).

On another hand, the data given by fixed automatic monitoring stations consisting of every 15-minutes measurements are used to monitor the alert thresholds of given pollutants such as ozone or nitrogen dioxide. However the number of fixed automatic stations is generally insufficient to allow a good spatial interpolation of the pollution levels.

In order to produce a daily pollution maps, it is compulsory, firstly, to add additional data to the measurements given by the fixed monitoring stations and, secondly, to conceive an automatic mapping algorithm able to integrate both type of data.

2 Methodology

The aim is to set up a method in order to i) generate additional data and ii) build up an algorithm for automatic mapping, based on the automatic modelling of the spatial covariance of NO2.

2.1 Temporal Generation of Additional Data

Our method to produce additional data is based on the existence of several previous field campaigns done by passive sampling on an urban area.

These field campaigns are used in the way that sites of passive sampling (called "virtual stations") are time-correlated to a given fixed station (called "reference station") and put together.

In these determined "virtual stations" it is possible, by applying this method, to estimate the concentration of nitrogen dioxide, at any given time, from the value measured at the correlated reference station.

This method is an alternative to generate additional data. However for the urban areas lacking of sufficient campaigns done by passive sampling, other methods must be applied. Another possibility may be, for instance, the generation of additional data from mobile campaigns and forecast models (models such as ANACOVA), etc.

2.2 Automatic Mapping

The previously explained method enables to estimate the concentration of nitrogen dioxide at the virtual stations; this dataset is then used to run a geostatistical interpolation of the concentrations, in order to get a map of the pollution levels.

As a reminder, among the mostly used geostatistical methods in the field of air quality, one may cite: the monovariable methods, made up of ordinary kriging and simple kriging or with known mean, and the multivariable methods which aim to make the most of the additional information brought by auxiliary variables.

The multivariable methods are an adaptation of kriging to different assumptions in order to take into account the auxiliary variable; more precisely, for the estimation of the nitrogen dioxide concentrations, the consideration of auxiliary variables depicting the area such as the relief, the population density or the emission inventory allow to produce more detailed maps of the spatial distribution of nitrogen dioxide.

In the field of air quality, the most often used multivariable methods are cokriging, kriging with external drift and a variation of this latter which is kriging of the residues (Cardenas and Malherbe, 2002; Chilès and Delfiner, 1999). In order to select the better approach, the classical approach consists in comparing the results from the different methods. Table 1 shows the steps of a geostatistical study aiming to map nitrogen dioxide concentrations.

The main steps of the procedure are related to the search of the spatially-correlated auxiliary function with pollution and (subsequently) to the calculation of variograms and fitting of the respective models. With robustness in mind, pre-existing configurations have been tested to build the mapping algorithm, therefore taking advantage of the available knowledge of the phenomenon.

1. Search of the auxiliary function: the goal of this step is to find a known function in the whole domain correlated with the nitrogen dioxide concentration. This auxiliary function may be constituted of variables describing the area such as the population density, the relief and the emission inventory. We propose to primarily study all the possible configurations of these auxiliary variables and to only retain the best-correlated ones to the concentrations. Then this selected function will be used daily in the mapping algorithm.

Table 1 Steps of a geostatistical study

Steps of a geostatistical study	
Action	Results
1. Study of the available auxiliary data: relief, emissions, population density, meteorological data, etc.	Choice of the best correlated auxiliary function to the concentrations.
2. Calculation of the anisotropic and isotropic variograms of the concentration and the auxiliary variable.	Search of possible anisotropies, analysis of the auxiliary variable quality.
3. Fitting of the variogram's model(s)	Choice of the basic structures and the range of the models.
4. Test of the models by cross-validation.	Validation of the models and choice of the estimating method: ordinary kriging, cokriging, kriging with external drift or kriging of the residues (Gallois et al., 2005).

Table 2 Automation of the procedure

Automation of the procedure	
Action	**Result**
1. Creation of the dataset based on the data from the "Temporal generation of additional data method" (reference stations and the correlation parameters: slope and Y-axis intercept).	Dataset of the daily concentrations of nitrogen dioxide.
2. Creation of the algorithm enabling the automatic execution of daily maps of the estimation of nitrogen dioxide.	The algorithm must fulfill the following tasks: 1. Calculation of the correlation cloud with the pre-determined auxiliary variable. 2. Calculation of the experimental variograms. 3. Based on these experimental variograms: fitting of the sills of the pre-selected basic structures. 4. Cross-validation. 5. Using the pre-selected kriging method: interpolation of the concentrations at a grid cell. 6. Realisation of estimated maps and of the map of estimation variance using a pre-determined color scale. 7. Export of the results: • Export of the resulting maps in an image format bmp or jpeg. • Export of the results files in ascii format or grid format for a sub-sequent use in a GIS.
3. Results analysis	Here are some useful information to determine the quality of the obtained estimations. 1. Correlation cloud between the concentration and the auxiliary variable (correlation coefficient). 2. Experimental variograms and fitted models. 3. Cross-validation statistics: error and relative error variance. 4. Correlation cloud: concentrations at the virtual stations *versus* estimated concentrations (correlation coefficient). 5. Estimation and estimation variance statistics.

However the function may differ with seasons (one for summer and another for winter). Indeed several studies have shown that the correlation between nitrogen dioxide and the other variables describing the area are not as good in summer as for other seasons. On another hand, the existence of a decreasing relationship between the temperature and the concentration of NO2 is well-known, which is why the cartography of NO2 is more interesting in winter than in summer where concentrations are lower, from the point of view of air quality monitoring.

2. To fit the variograms, the basic structures (nugget effect, choice of the model: spherical, exponential or gaussian, etc.) have been pre-selected from the analysis of the available information such as the existing passive sampling campaigns .
3. The best method of estimation, among the four available ones (ordinary kriging, cokriging, kriging with external drift and kriging of the residues (Gallois et al., 2005) has been draw out from the previous estimations of pre-existing campaigns by passive sampling.
   Table 2 shows the built-up procedure to develop the mapping algorithm.

3 Case study

These methods have been tested on data from the city of Lille and its suburbs. The dataset consists of 16 fixed NO2 monitoring stations distributed on the estimation area and 15 measurement campaigns by passive sampling, lasting 14 days each, done during the years 1998/1999 and 2003/2004. Moreover, the following auxiliary variables are available: NOx emission inventory and population density.

3.1 Selection of the Virtual Stations

The virtual stations have been selected applying the "Temporal generation of additional data method". They represent 65 passive sampling sites which are very satisfactorily time-correlated to the measurements done at two fixed reference stations: the suburban station "Halluin caillou" and the urban station "Roubaix Château".

3.2 Calculation of the VME

The data extracted from the virtual stations are, by construction, marred with errors. Geostatistics are able to take into account this uncertainty through the variance of measurement error (VME), term that can be added to the variance-covariance matrix during kriging. MEV is calculated from the difference or residue between the estimated values and the ones measured during the 15 campaigns.

$$R(x) = NO_2\text{Estimated} - NO_2\text{Measured}$$

Finally for each virtual station, variance of residues is calculated, showing the "dispersion" of the residues around their mean value. Figure 1 shows the main statistics of this variable. The detailed analysis of this figure leads to the existence of a group of 10 passive sampling sites (virtual stations) related to a clear "urban environment" where it is more difficult to get accurate estimations; these are the stations 62, 132, 137, 54, 14, 356, 94, 144, 161 and 50 where the mean value of the 15 campaigns is superior to 30 $\mu g/m^3$.. These stations may be representative of a more local environment, influenced, for example, by traffic emissions.

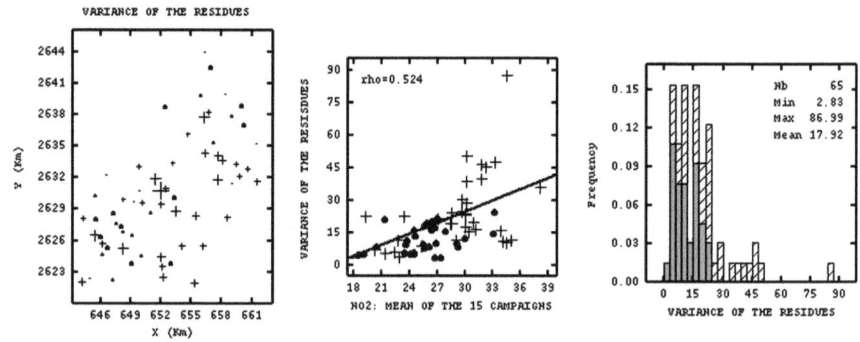

Fig. 1 Statistics of the Variance of the Residues or MEV
Note: Figure on the left: Geographical localisation of the virtual stations (the symbols size are proportional to the variance of the residues).
Figure on the center: Correlation cloud between the variance of the residues and the mean of the 15 nitrogen dioxide campaigns (the straight line represent the linear regression).
Dots represent the virtual stations linked to the periurban fixed station Halluin Caillou.
Cross in green represent the virtual stations linked to the urban fixed station Roubaix Château.
Figure on the right: histogram of the variance from the residues at the virtual stations.
Grey bars represent the virtual stations linked to the periurban fixed station Halluin Caillou.
Hatched bars represent the virtual stations linked to the urban fixed station Roubaix Château.

3.3 Choice of the Estimation Method

After a detailed study of 15 campaigns by passive sampling, we decided to perform the daily estimation by means of the collocated cokriging method with VME of the virtual stations included, and with a combination of emission inventory and population density as auxiliary variable.

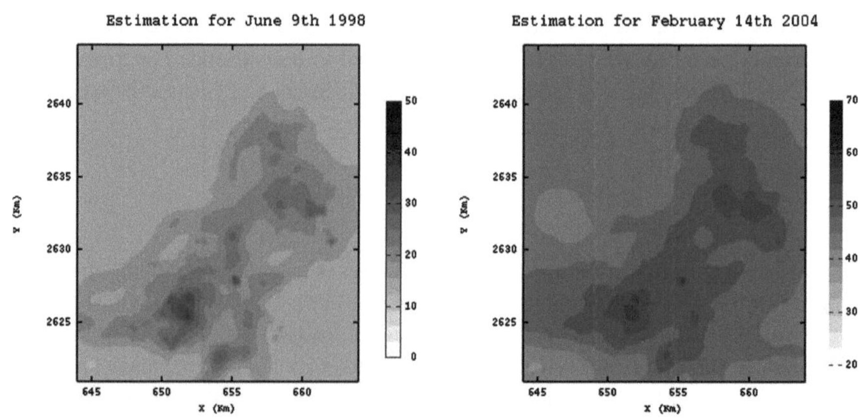

Fig. 2 Daily estimation maps of NO_2 for June 9th 1998 and February 14th 2004
Note: The scale is μgm^{-3}. The maximum values are located in the town center of Lille and Roubaix cities

Moreover, a different variogram model is fitted for each season: for summer, an isotropic model is constituted by a nugget effect and a gaussian structure with a 5 km range and, for winter a nugget effect and a gaussian structure with a 8 km range. The sill of the structures are fitted automatically.

3.4 Results

As an example, two daily estimations are shown: one in June 9th 1998 (summer period) and the other one in February 14th 2004 (winter period).

In order to evaluate the quality of our estimations, the daily means estimations relative to the winter period from February 5th to 18th 2004 and to the summer period from May 27th to June 9th 1998 were calculated and therefore, compared to the estimations got by kriging from the passive samplers measurements.

In spite of the use of auxiliary variables one may notice, on the example, a smoothing of the estimated results: high values have been under-estimated by 27% in winter and 37% in summer; in the same way, low values have been over-estimated by 35% in winter and 47% in summer. (The quality criteria is fixed to 50% for this kind of method).

4 Conclusions

A way to improve our method seems to be the reduction of the smoothing effect, maybe by working on "Temporal generation of additional data method" (as they do a first smoothing of the high values) and by creating more virtual stations.

Finally from a methodological point of view, we conclude that in order to carry out of daily maps of nitrogen dioxide, one has to perform as a starting point, a manual analysis of the pre-existing campaigns in order to pre-define some variables and parameters (best correlated auxiliary variable, structure and range of the variogram).

References

Cardenas G. Malherbe L. (2002). Application des méthodes géostatistiques à la cartographie de la qualité de l'air, STIC et Environnement, Rouen, 19–20 juin 2003

Chilès JP, Delfiner P (1999) Geostatistical: modelling spatial uncertainty, Wiley Series in Probability and Mathematical Statistics, p. 695

Gallois D, de Fouquet C, Le Loc'h G, Malherbe L, Cardenas G (2005) Mapping annual nitrogene dioxide concentrations in urban areas. Geostatistics Banff 2004. Springer, 2005, pp. 1087–1096

King Prawn Catch by Grade Category from an Economic and a Stock Management Perspective

U. Mueller, L. Bloom, M. Kangas and N. Caputi

Abstract We give a geostatistical analysis of western king prawn logbook data collected from the Shark Bay prawn fishing fleet in Western Australia for the 2000 and the 2004 fishing seasons, aggregated into total catch, together with three weight sub-classes and grouped into lunar months. For each of the two years we discuss both the spatial correlation between the weight classes and the spatial correlation for corresponding months in the two years under consideration. Finally, we use a cost function that takes account of the different weight classes to compare the financial return by location between 2000 and 2004.

1 Introduction

In this paper we consider king prawn catch data from the Shark Bay Prawn Managed Fishery in Western Australia and carry out a comparative statistical analysis by weight classes *Small*, *Medium* and *Large* for the 2000 and 2004 fishing seasons. Firstly we analyse the numerical distribution of catch by weight class and investigate any change from 2000 to 2004 and link this to differences in fishery management practice between these years. Secondly we analyse the spatial distribution of catch by weight class and investigate differences between 2000 and 2004 in the areas where the more commercially valuable large prawns are caught. As well as spatial maps we calculate Spearman's rank correlation coefficient, Tjøstheim's index of spatial association, and cross-semivariograms. We consider both total prawn catch accumulated by location over the entire fishing season, as well as results for selected lunar months.

The Shark Bay Prawn Managed Fishery is Western Australia's most significant prawn fishery. Trawling takes place between March and October each year and fishing occurs nightly, except for a closure period around the full moon of each month (Kangas and Sporer, 2000). In addition the fishing ground is subject to a number of area closures throughout the fishing season to protect smaller prawns

U. Mueller
School of Engineering and Mathematics, Edith Cowan University, Perth, Western Australia
e-mail: u.mueller@ecu.edu.au

A. Soares et al. (eds.), *geoENV VI – Geostatistics for Environmental Applications*, 103–113

and the spawning stock. Changes in management practice between 2000 and 2004 include targeting the larger prawns and avoidance of smaller less valuable prawns. The general migration pattern of smaller prawns is in a northerly direction from the southern breeding grounds and annual research surveys of size distributions were used to implement area closures to optimise size. Our study provides an assessment of the success or otherwise of this targeting strategy as evidenced between the 2000 and 2004 fishing seasons.

2 Data Description

The data considered here are prawn catch logbook data from 2000 and 2004. The logbooks are completed on a voluntary basis by the prawn fishers. The information collected comprises the start latitude and longitude for each trawl, the effort (minutes trawled) and the catch in kilogram by grade and species. There are six grade categories (number of prawns per pound: U10, U15, U20, U25, U20-30 and U30+). In our analysis we use the weight classes *Large,* consisting of king prawns in categories U10 and U15, *Medium,* U20 king prawns, and *Small,* comprising those graded as U25, U20-30 and U30+. High catches and workload during the fishing can lead to incomplete records. Typically there are no coordinates for about 10% of the records, and for approximately 65% there is no information by trawl shot but rather on a coarser basis, such as the start latitude and longitude for the entire night together with the amount of prawns by grade. For our analysis the catch locations were converted to nautical miles and a local coordinate system with origin at 24° south latitude and 113° east longitude. Records without coordinates were eliminated from the data sets and the remaining records were aggregated to catch per square nautical mile. Two time spans were chosen, the entire fishing season and a fishing month. In order to compare the spatial patterns, only locations common to both years were considered.

The within-season time series for the total catch per month and weight class in kg and the overall effort expended in hours for 2000 and 2004 are shown in Fig. 1. The month with the maximum catch in both years was May, and in both cases the weight class whose contribution is greatest is that of medium sized prawns. This peak in May is associated with the opening of an extended nursery area in the southeast of Shark Bay. The overall catch for 2000 was higher than that for 2004. There has been little change in the shape of the time series for medium and small prawns but there is a marked change for large prawns with a clear maximum in May 2004, but a much flatter distribution in 2000.

The catch composition has changed between the two years. In 2000 medium sized prawns comprised at least 50% of the catch in the first four months of the fishing season. In 2004 the split between the three weight classes was more even in the first four months. The most marked shift is towards prawns in the larger grade categories in 2004, with a minimum contribution of 35% and a maximum of 62%, compared to a minimum of 23% and a maximum of 55% in 2000.

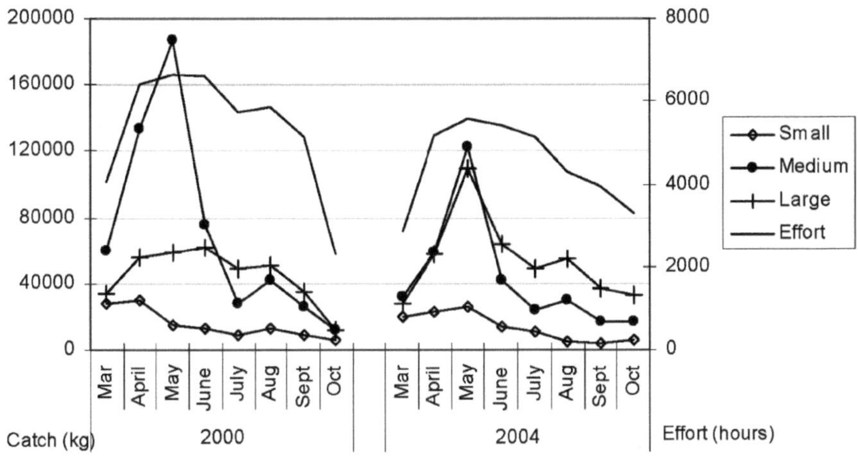

Fig. 1 Total catch per month by weight class and effort by month

The spatial distribution for the total catch over the entire year for 2000 and 2004 is shown in Fig. 2. There are two distinct fishing regions.

The region south of latitude −90 Nm is referred to as Denham Sound, the region north of latitude −90 Nm and south of −70 Nm as Peron and that north of −70 Nm as Red Cliff. For both years there is a region of high catch in the Peron fishing ground east of 26 Nm longitude. The western part of Peron appears to show greater variability, as does Red Cliff. For Denham Sound, high catches are concentrated in

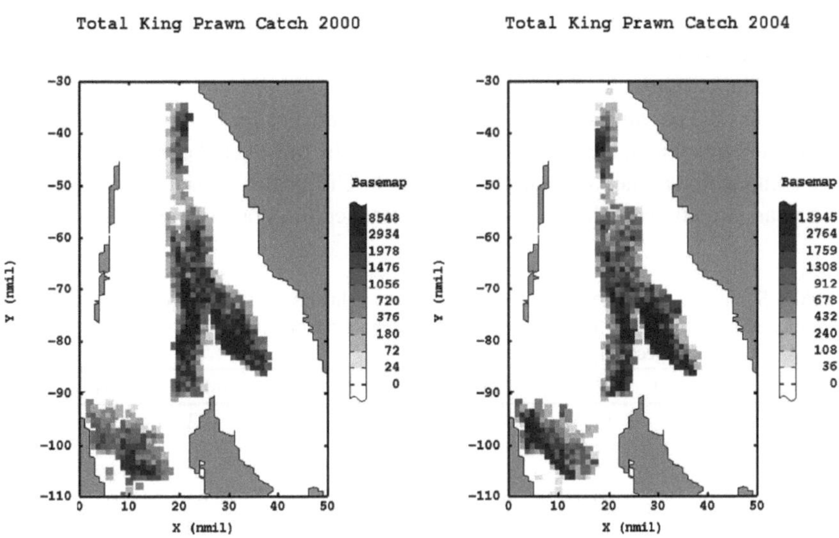

Fig. 2 Location maps (levels in deciles in kg) of the total king prawn catch in 2000 and 2004

the south-west in both years, and there is a region of high catch in the west in 2004. The spatial distribution of the effort for both years exhibits similars features.

Since the time spent fishing differs markedly between grid locations the effort needed to be accounted for in the analysis. The total amount caught per Nm^2 was converted to a standardised catch by dividing the total catch by the total effort expended in hours within the 1 Nm by 1 Nm window. The standardised catch data for the years 2000 and 2004 are moderately to strongly positively skewed (see Table 1). The relative ordering for the skewness coefficients is the same in both years with standardised small prawns being the most skewed followed by medium sized prawns and then large prawns.

Except for the large prawns the standardised catches in 2000 have greater variances. The overall means are comparable but the individual means again reflect changes in the fishing practice, whereby in 2004 the contribution of large prawns to the overall catch is almost 50% compared to 35% in 2000. The proportions of small prawns are comparable. Spearman's rank correlation coefficient for the entire region between the two years indicates a (statistically) significant positive correlation between the small and medium catch ($r = 0.2$) and medium and large catch ($r = 0.43$) in 2000. In 2004 there is significant positive correlation between any pair of variables with $r = 0.63$, $r = 0.12$ and $r = 0.53$ for small and medium, small and large and medium and large respectively. Between the years all three weight classes have significant rank correlation coefficients ($r = 0.09$, $r = 0.52$ and 0.21 for small, medium and large respectively).

The spatial distributions for the various weight classes are shown in Fig. 3 and Fig. 4. For all three weight classes, the distributions in Denham Sound have changed considerably. For small prawns low catch values are more widely dispersed in 2000 than in 2004 when the higher catches are concentrated in the east of the northern fishing region. There is a region of low catch in the central part of the Red Cliff fishing ground in 2000 which is not present in 2004. The pattern for medium and large prawns is similar between −90 Nm latitude and −50 Nm latitude, but there is little similarity north of −50 Nm.

May has the overall highest catch. Summary statistics for the standardised catch in May are given in Table 2. As already noted in the raw data, there are shifts in the size composition of the total standardised catch. The means for the total catch for the two years are comparable while the mean for medium prawns was approximately

Table 1 Summary statistics of standardised catch by weight class in 1 Nm^2 blocks, 2000 and 2004. StSm00, StSm04, StM00, StM04, StL00 and StL04 denote standardised small, medium, large catch in 2000 and 2004 respectively

	StSm00	StM00	StL00	StT00	StSm04	StM04	StL04	StT04
Mean	2.78	13.70	8.86	25.34	3.06	9.63	12.54	25.23
Variance	7.22	92.98	14.59	149.10	5.97	37.78	24.98	125.56
Minimum	0.00	0.00	0.00	0.00	0.00	0.00	0.00	3.47
Median	2.18	10.93	8.46	22.62	2.56	8.02	12.02	22.83
Maximum	24.00	54.44	25.34	75.67	17.21	35.10	33.40	78.81
Skewness	2.60	1.49	0.62	1.04	1.40	1.38	0.82	1.19

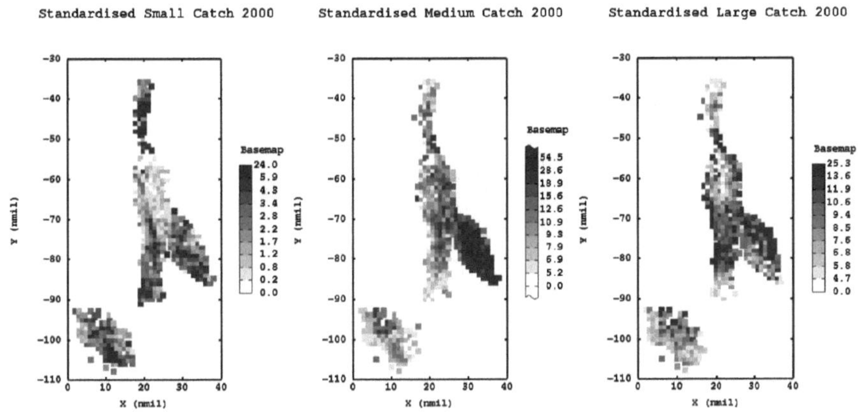

Fig. 3 Location maps (levels in deciles in kg/h) for standardised catch by weight class, 2000

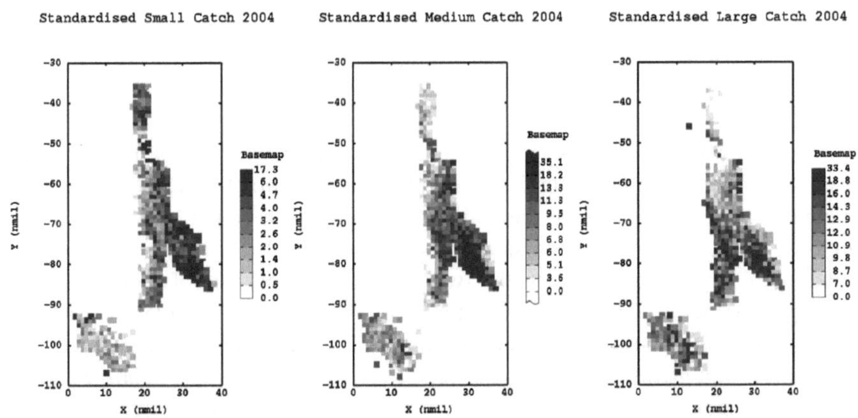

Fig. 4 Location maps (levels in deciles in kg/h) for standardised catch by weight class, 2004

30% lower in 2004. In contrast the mean for large prawns in 2000 was only approximately 60% of the 2004 mean. Apart from small prawns the data are moderately positively skewed, with the total catch in both years having the least skew.

Spearman rank correlation coefficients show a significant positive correlation between small prawn catch and the other two weight classes within the same year.

Table 2 Statistics of standardised catch by weight class, May 2000 and May 2004

	StSm00	StM00	StL00	StT00	StSm04	StM04	StL04	StT04
Mean	2.73	30.41	11.39	44.53	4.65	21.50	18.87	45.02
Variance	8.10	173.18	66.52	285.40	10.46	142.30	92.05	442.33
Minimum	0.00	2.30	0.00	4.61	0.00	0.00	1.02	1.02
Median	2.11	29.83	9.67	45.02	4.22	21.82	18.23	45.03
Maximum	19.10	66.78	35.20	93.05	15.43	59.83	48.38	115.01
Skewness	2.15	0.12	0.71	0.05	1.00	0.38	0.58	0.20

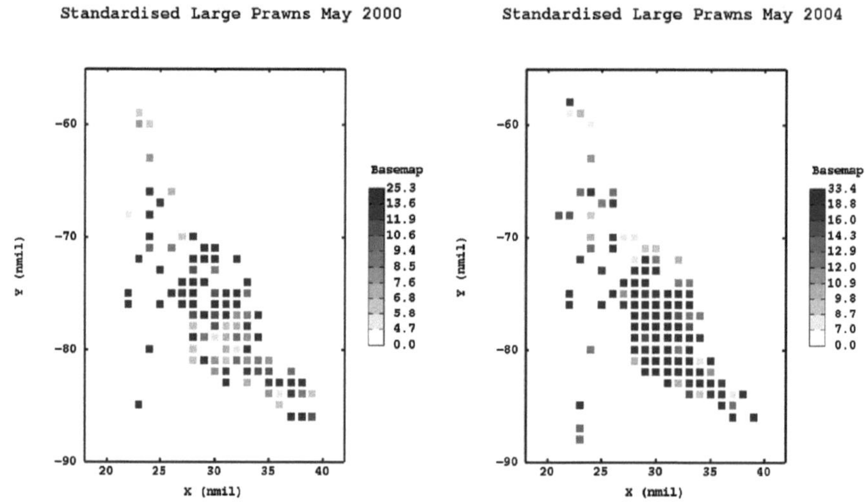

Fig. 5 Location maps (levels in deciles in kg/h) of standardised large prawn catch in May

In 2004 this was also the case for medium and large prawn catches. Between the years, Spearman's coefficient indicates significant correlation for the medium catch. In contrast to the situation for the entire year, there are no other significant associations. Location maps for the standardised prawn catch for May show that the locations of high catch between the two years differ considerably for large and small prawns, while the distributions for medium prawns are similar (see Fig. 5 for large prawns). These changes may indicate differences in the timing of recruitment and/or the growth of prawns between the years. In addition, subtle changes in real-time management between the two years may have resulted in finer scale area openings that vary between the years and may have influenced the fishing pattern.

3 Assessment of Spatial Correlation

Consideration of the spatial maps suggests that locations of high catch do not change significantly across the years. When considering monthly intervals, the spatial maps indicate that whether or not there is a change may depend on the weight class. This is also supported by changes in the statistical significance already observed when going from the annual time scale to the monthly time scale. To further test these observations several approaches were used. Experimental semivariograms and cross-semivariograms were computed to assess the similarities in the spatial structure for the weight classes between years. In particular, we check whether or not a linear model of coregionalisation can be fitted. To obtain a numerical measure Tjøstheim's index of spatial association (Tjøstheim, 1978; Hubert and Golledge 1982) was calculated. This index was used to check if it is possible to predict the position of the location ranked i for one variable from knowledge of the location ranked i for the other variable. Tjøstheim's index A is defined as

$$A = \sum_{i=1}^{n} [x_F(i)x_G(i) + y_F(i)y_G(i)]$$

where the two variables F and G are observed over the same n locations and ranked from 1 to n, and $(x_F(i), y_F(i))$ and $(x_G(i), y_G(i))$ denote the location of rank i on F and G respectively, and where the coordinates of the locations are standardised in such a way that

$$\sum_{i=1}^{n} x_F(i) = \sum_{i=1}^{n} x_G(i) = \sum_{i=1}^{n} y_F(i) = \sum_{i=1}^{n} y_G(i) = 0$$

and

$$\frac{1}{n}\sum_{i=1}^{n} x_F^2(i) = \frac{1}{n}\sum_{i=1}^{n} x_G^2(i) = \frac{1}{n}\sum_{i=1}^{n} y_F^2(i) = \frac{1}{n}\sum_{i=1}^{n} y_G^2(i) = 1.$$

Ties in the ranking were broken by ordering locations of equal rank first in ascending order of x then in ascending order of y. Under randomization of ranks, the index has a normal distribution with $\mathrm{E}(A) = 0$ and $\mathrm{var}(A) = (1 + r_{xy}^2)/(2(n-1))$. The quotient $A/\sqrt{\mathrm{var}(A)}$ is a test statistic to assess whether or not there is any spatial relation between the two variables.

3.1 Variography

Consideration of the semivariogram surfaces of the standardised catch for the entire fishing season indicates anisotropy for medium and total catch with major direction N-S, while large prawn catch and small prawn catch are isotropic. The anisotropy may in part be attributed to the shape of the northern fishing region. The data for Denham Sound exhibit isotropy when considered separately while for medium prawns and total catch the northern region exhibits the same spatial pattern as the overall region. However, the small prawn catch is anisotropic with major direction N-S in 2000 but is isotropic in 2004 and large prawn catch shows anisotropy in the direction of azimuth $-60°$ in 2004 and is isotropic in 2000. For large prawns the shapes of the semivariograms are spherical, but this is not the case for the cross semi-variogram whose shape appears gaussian. In contrast, for the standardised medium and small prawn catch it is possible to fit linear models of coregionalisation, whether the entire, the northern region or Denham Sound is considered. For example the model for standardised small king prawn catch in the entire region is given by

$$\gamma(\mathbf{h}) = \begin{bmatrix} 3 & 0 \\ 0 & 2.3 \end{bmatrix} nug(\mathbf{h}) + \begin{bmatrix} 1.8 & 0.2 \\ 0.2 & 0.8 \end{bmatrix} spher_{3.3}(\mathbf{h}) + \begin{bmatrix} 2.3 & 0.7 \\ 0.7 & 3.5 \end{bmatrix} spher_{17}(\mathbf{h})$$

where *nug* and $spher_a$ denote the nugget effect and spherical model of range a respectively. The model fit is shown in Fig. 6. In the light of results from survey data for assessing Atlantic cod stock (Warren, 1997), which showed that an increase in relative nugget preceded the total exhaustion of the species, we note here that the relative nugget for the entire fishing season was essentially unchanged from 2000 to 2004 for both small and large prawns.

Apart from October, for which there were few samples in 2000, the monthly experimental semivariograms for corresponding weight classes in the two years have similar features, but with varying relative nugget. For both small and large prawns the relative nugget increased in May, June and August, was unchanged in April and July and decreased in March and September. For May the experimental semivariograms can be fitted by a nugget plus one spherical structure. (see Table 3 for the chosen parameters).

The quality of the fit is generally good as is illustrated by the model fit for the standardised total catch shown in Fig. 7.

The situation is more complex for the associated cross semivariograms. For March, April, September and October they show little structure. In May the cross semivariograms do not show a spherical shape so that a model of coregionalisation cannot be fitted for any of the pairs of variables between years. The cross-semivariogram of the model that produces an appropriate fit for the semivariograms of the standardised large catch is shown in Fig. 8. The behaviour exhibited here is similar for the entire year. The cross-semivariogram shows small values until a lag spacing of 10 Nm is reached, at which time a steep rise is visible.

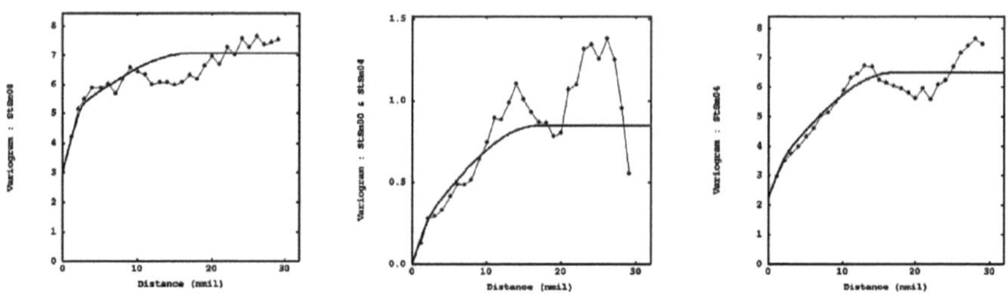

Fig. 6 Model fit for the linear model of coregionalisation for standardised small prawns

Table 3 Semivariogram model parameters, May 2000 and 2004

	2000			2004		
	Nugget	Range	Sill	Nugget	Range	Sill
StSmall	5	9.9	3	7.4	7.4	2.3
StMedium	50	5.8	102	72	8.1	67
StLarge	24	11.2	43	52	14.2	39
StTotal	47	6.6	208	240	12.6	230

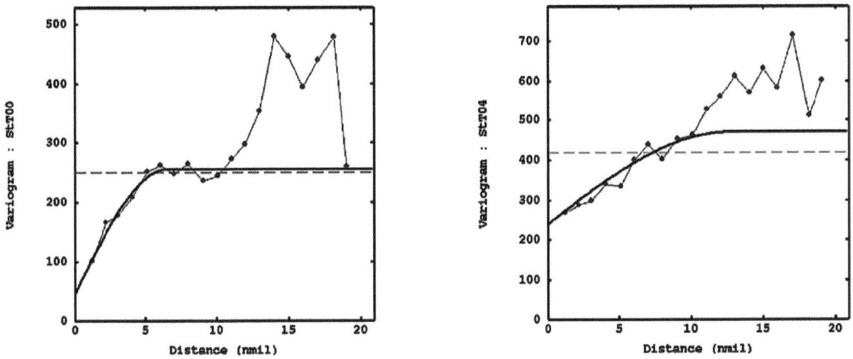

Fig. 7 Model fit for standardised total catch in May 2000 (**left**) and 2004 (**right**)

There are instances in individual months where a linear model of coregionalisation provides a good fit. Medium prawns in July (see Fig. 9) or large prawns in June provide an example where such a model does apply.

3.2 Spatial Rank Correlation

Tjøstheim's index of spatial association (see Table 4) shows that, taken across the entire year, locations of high rank for medium sized prawns in 2004 tend to be close to those of similar rank in 2000. There is no such relationship for standardised small or large prawn catch between the two years. In 2000 there was no association between the locations of high rank for small prawns with those of large or medium sized prawns, while in 2004 locations of high small prawn and medium prawn catch were associated. On the other hand, in both years there was a weak association between medium and large prawns. For May the only Tjøstheim index that is statistically significant is the one measuring the association between medium and small prawns in 2004 (A = 0.23).

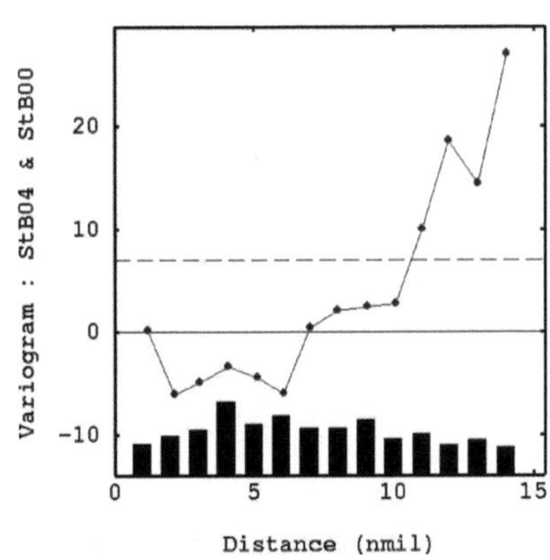

Fig. 8 Experimental cross semivariogram for standardised large prawn catch

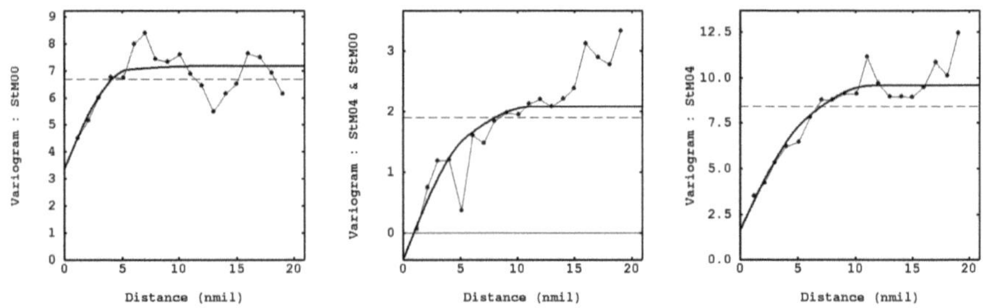

Fig. 9 Linear model of coregionalisation for standardised medium prawn catch, July 2000/2004

The commercial value of the catch varies with the weight class. An analysis of the prices for prawns indicates that if the value for medium sized prawns in set to 1 monetary unit per kg, then the appropriate values for small and large prawns are 0.75 and 1.25 respectively. The location maps for May are shown in Fig. 10.

Table 4 Tjøstheim's indices for standardised small, medium and large catch in 2000 and 2004, * indicates the value is significant at the 0.05 level

Tjøstheim	StSm00	StM00	StL00	StSm04	StM04	StL04
StSm00	–	0.00	0.02	−0.01	–	–
StM00		–	0.07*	–	0.11*	–
StL00			–	–	–	0.02
StSm04				–	0.17*	−0.04
StM04					–	0.10*
StL04						–

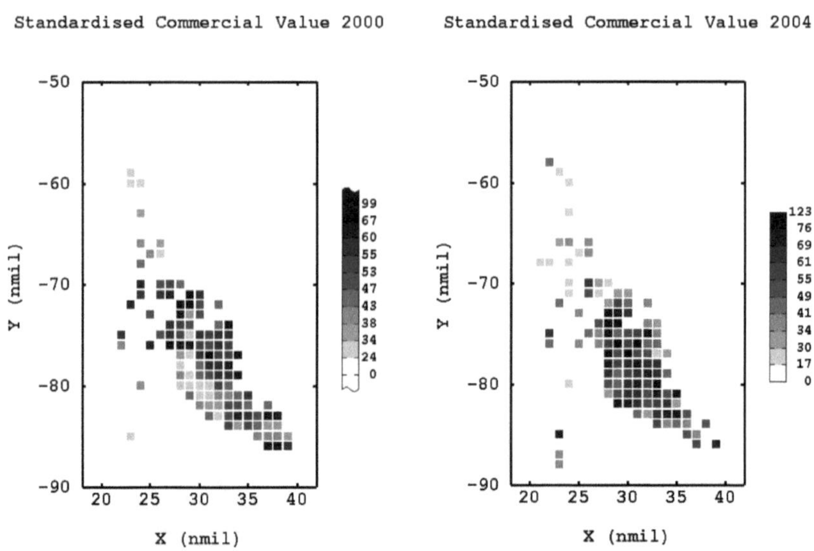

Fig. 10 Standardised commercial value of king prawn catch, May 2000 and 2004

The spatial distribution of the economic value of the catch for the entire year in Peron and Red Cliff has changed little between 2000 and 2004, but there were considerable changes in Denham Sound. This is also confirmed by the value of Tjøstheim's index for Peron and Red Cliff whose value of 0.13 is significant at the 0.01 level of significance, while that for Denham Sound (0.01) is not significant. For May Tjøstheim's index for the standardised commercial values is −0.12 indicating a lack of coincidence of locations with equally ranked financial returns for the two years, this may be visually confirmed in Fig. 10.

4 Conclusions

In summary, while annual variation in the timing and strength of recruitment of king prawns may result in shifts in overall size distributions from year to year, the spatial structure of the catch distribution is not affected. Also, there are no major changes in the spatial continuity for the individual weight classes from year to year. There are slight changes in the range and variations in the relative nugget but the shapes of the semivariograms do not change. Similarly, apart from the start and the end of the fishing season, the semivariogram structures in corresponding months of the two years are comparable. However, there is no evidence that equally ranked values for different weight classes occur at the same locations within one year and high values of large catch and high values of small catch are not co-located due to the clear migration pattern of smaller prawns in Shark Bay moving from south to north. The higher proportion of larger individuals retained in 2004 indicates that targeting larger prawns as a strategy is working, with the management arrangements protecting small prawns as intended. This in turn will result in economic benefits to the industry due to the higher value of larger prawns.

References

Hubert LJ and Golledge RG (1982) Comparing rectangular data matrices, *Environ Plan* 14: 1087–1095

Kangas M and Sporer E (2000). Shark Bay prawn managed fishery status report, in Penn J (ed) State of the fisheries report 2000/2001, //http:www.fish.wa.gov.au, pp 40–43

Tjøstheim D (1978). A measure of association for spatial variables. *Biometrika* 65: 109–114

Warren W (1997). Changes in the within-survey spatio-temporal structure of the northern cod (Gadus morhua) population, 1985–1992, *Can Fish Aquat Sci* 54 (Suppl. 1), 139–148

Part II
Hydrology

Machine Learning Methods for Inverse Modeling

D. M. Tartakovsky, A. Guadagnini and B. E. Wohlberg

Abstract Geostatistics has become a preferred tool for the identification of lithofacies from sparse data, such as measurements of hydraulic conductivity and porosity. Recently we demonstrated that the support vector machine (SVM), a tool from machine learning, can be readily adapted for this task, and offers significant advantages. On the conceptual side, the SVM avoids the use of untestable assumptions, such as ergodicity, while on the practical side, the SVM out performs geostatistics at low sampling densities. In this study, we use the SVM within an inverse modeling framework to incorporate hydraulic head measurements into lithofacies delineation, and identify the directions of feuture research.

1 Introduction

Heterogeneous aquifers typically consist of multiple lithofacies, the spatial arrangement of which significantly affects flow and transport in the subsurface. The identification of these lithofacies is complicated by the sparsity of data and by the lack of a clear correlation between identifiable geologic indicators and attributes (e.g., hydraulic conductivity and porosity). This so-called zonation problem has been studied by Sun and Yeh (1985), Carrera and Neuman (1986), Eppstein and Dougherty (1996), Tsai and Yeh (2004), among others.

Data which are used in geomaterials classification procedures are typically obtained from core samples that often disturb soils and are by necessity sparse, thus contributing to predictive uncertainty associated with the location of different geomaterials. Within a stochastic framework, this uncertainty is quantified by treating a formation's properties as random fields that are characterized by multivariate probability density functions or, equivalently, by their joint ensemble moments. Geostatistics has become an invaluable tool for estimating facies distributions at points

D. M. Tartakovsky
Department of Mechanical and Aerospace Engineering, University of California, San Diego, La Jolla, CA 92093, USA; Theoretical Division, Los Alamos National Laboratory, Los Alamos, NM 87545, USA
e-mail: dmt@ucsd.edu

A. Soares et al. (eds.), *geoENV VI – Geostatistics for Environmental Applications*, 117–125

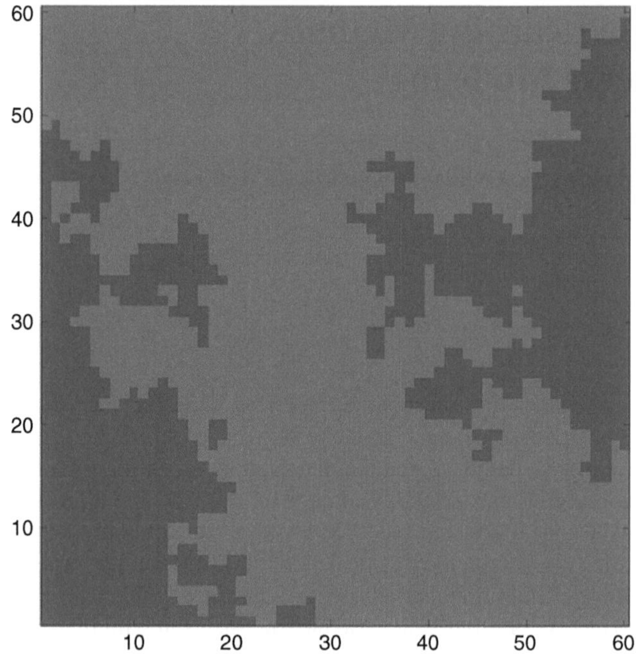

Fig. 1 Flow domain consisting of two contrasting geologic facies. A highly conducting material is shown in red and a low conducting material in blue

in a computational domain where data are not available, as well as for quantifying the corresponding uncertainty (Guadagnini et al., 2004).

Recently we (Tartakovsky and Wohlberg, 2004; Wohlberg et al., 2006; Guadagnini et al., 2006) demonstrated that Support Vector Machine (SVM) techniques, a subset of machine learning algorithms, provide a viable alternative to geostatistical frameworks by allowing one to delineate lithofacies in the absence of sufficient data parameterization, without treating geologic parameters as random and, hence, without the need for the ergodicity assumption. This has been done by using both well and poorly differentiated parameter data. For additional information on the use of the SVM and other machine learning techniques in environmental applications, we refer the interested reader to Kanevski and Maignan (2004).

In this study, we use machine learning within an inverse modeling framework to incorporate hydraulic head measurements into lithofacies identification. We apply the approach to a synthetic case of steady-state flow through a domain consisting of two materials separated by highly irregular boundaries (see Fig. 1). For simplicity, the hydraulic conductivity of each material is assumed to be constant.

2 A Problem of Facies Delineation

Consider the problem of reconstructing a boundary between two homogeneous geologic facies from either parameter data $K_i = K(\mathbf{x}_i)$ or system state data $h_i = h(\mathbf{x}_i, t)$ or both. Without loss of generality, we assume that both data sets are collected at the same N locations $\mathbf{x}_i = (x_i, y_i)^T$, where $i \in \{1, \ldots, N\}$. Such problems are ubiquitous in subsurface hydrology since the geologic structure of the subsurface plays a crucial role in fluid flow and contaminant transport. A typical example is the problem of locating permeable zones in the aquiclude that separates two aquifers, the upper aquifer contaminated with industrial pollutants, and the lower aquifer used for municipal water supplies (Guadagnini et al., 2004).

Parameter data can include measurements of hydraulic conductivity, electric resistivity, cumulative thickness of relevant geologic facies, and grain sizes. We will call the problem of estimating the internal boundaries between geologic lithofacies from such data *a forward facies delineation problem*. System state data consist of measurements of hydraulic head, flow rate, concentration, etc. We will call the problem of estimating the internal boundaries between geologic lithofacies from such data *an inverse facies delineation problem*. Often both data types are available at the same locations, e.g., when observation and/or pumping wells are outfitted with flow-meters.

2.1 Forward Facies Delineation Problem

Since parameter data characterize geologic materials, they allow one to label the points where they are taken by mean of the indicator function

$$I_i \equiv I(\mathrm{x}_i) = \begin{cases} +1 & \mathrm{x}_i \in M_1 \\ -1 & \mathrm{x}_i \in M_2, \end{cases} \tag{1}$$

where M_1 and M_2 are the two facies. This step typically involves an analysis of a data histogram, which is often nontrivial, since a typical geologic facies is heterogeneous. Here we assume that the available parameter data $\{K(\mathrm{x}_i)\}_{i=1}^N$ are well differentiated, so that the process of assigning the values of the indicator function to points $\{\mathrm{x}_i\}_{i=1}^N$ does not introduce interpretive errors. This assumption can be relaxed to account for poor differentiation of data (Guadagnini et al., 2004).

While it is customary to employ geostatistics for facies delineation, we (Tartakovsky and Wohlberg, 2004; Wohlberg, 2006) showed that the SVM, a tool from the statistical learning theory, can be readily adapted for this task and offers significant advantages. On the conceptual side, the SVM avoids the use of untestable assumptions, such as ergodicity. On the practical side, the SVM outperforms geostatistics at low sampling densities.

The SVM also has an advantage over neural networks, another tool from the machine learning theory. This is because the SVM solves a convex optimization by minimizing the quadratic functional

$$\max_{\gamma} \left\{ \sum_{i=1}^{N} \gamma_i - \frac{1}{2} \sum_{i=1}^{N} \sum_{j=1}^{N} \gamma_i \gamma_j I_i I_j \mathcal{K}(\mathrm{x}_i, \mathrm{x}_j) \right\}, \tag{2}$$

where $K(\mathrm{x}_i, \mathrm{x}_j)$ is a given Mercer kernel, subject to the constraints

$$0 \le \gamma_i \le C \quad \text{and} \quad \sum_{i=1}^{N} \gamma_i I_i = 0. \tag{3}$$

This optimization problem has a well defined global minimum that is influenced by the choice of the fitting parameter C. If $\{\gamma_i^{\star}\}_{i=1}^{N}$ denote a solution of the optimization problem (2), then the indicator function $I(\mathrm{x})$ at any point x, and hence the boundary separating the two materials, is given by Schölkopf and Smola (2002), p.203

$$I(\mathrm{x}) = \operatorname{sign}\left(\sum_{i=1}^{N} \gamma_i^{\star} I_i \mathcal{K}(\mathrm{x}, \mathrm{x}_i) + b^{\star} \right), \tag{4}$$

where

$$b^{\star} = I_j - \sum_{i=1}^{N} \gamma_i^{\star} I_i \mathcal{K}(\mathrm{x}_j, \mathrm{x}_i) \tag{5}$$

for some j such that $\gamma_j > 0$.

2.2 Inverse Facies Delineation Problem

The incorporation of system state data into the SVM framework is challenging, since (i) one cannot assign the indicator function to such data and (ii) the relationship between the two data types is nonlinear. To be concrete, we consider steady-state saturated flow, so that the parameter K stands for hydraulic conductivity and the system state h is hydraulic head. We propose the following SVM-based algorithm to delineate geologic facies from parameter and system state data.

1. Use an SVM to reconstruct facies from parameter data $\{K(\mathrm{x}_i)\}_{i=1}^{N}$.
2. Use the resulting parameter field as an initial guess for the optimization problem

$$\min_{\gamma} \left\{ -\sum_{i=1}^{N} \gamma_i + \frac{1}{2} \sum_{i=1}^{N} \sum_{j=1}^{N} \gamma_i \gamma_j I_i I_j \mathcal{K}(\mathrm{x}_i, \mathrm{x}_j) + \lambda \sqrt{\frac{1}{N} \sum_{i=1}^{N} \left[h_i - h_s(\mathrm{x}_i) \right]} \right\}, \tag{6}$$

subject to the constraints (3) and fixed $\lambda > 0$. Here $h_s(\mathrm{x})$ is a computed system state, e.g., a numerical solution of the steady-state flow equation $\nabla \cdot (K \nabla h) = 0$ subject to appropriate boundary conditions. The hydraulic conductivity $K(\mathrm{x})$ is determined by the current state of $\{\gamma_i\}_{i=1}^{N}$.

The proposed approach aims to retain the maximization of the SVM margin based on conductivity data (Step 1), while minimizing the difference between the measured and computed heads. This balance is controlled by the choice of the parameter λ in (6). The higher its value, the more weight one assigns to the head measurements relative to the conductivity measurements, and vice versa.

In the proposed approach, hydraulic head data affect only the radial functions weights, which, in principle, might provide too few degrees of freedom. Indeed, one can expect the estimated facies boundary to be overly smooth, when a conductivity data set is small. However, if such a data set is complemented by a few hydraulic head measurements, fewer degrees of freedom are necessary to obtain a good correspondence.

Also, it is important to note that the proposed approach replaces the quadratic optimization in the SVM (2) with the nonlinear optimization (6). This nonlinearity arises from the nonlocal relationship between hydraulic conductivity K and hydraulic head h. This raises important questions of whether the SVM parameterization of the boundaries is adequate and the use of SVMs for facies delineation is appropriate. We begin to address these questions by analyzing the computational example provided below.

3 Computational Example

We employ the proposed SVM-based algorithm to reconstruct the boundaries between two geologic facies shown in Fig. 1 from N randomly selected data points $\{\mathrm{x}_i\}_{i=1}^N$. At these data points, both hydraulic conductivity K and hydraulic head h are sampled. The values of hydraulic head are obtained by solving the steady-state flow equation $\nabla \cdot (K \nabla h) = 0$ with hydraulic conductivity $K(\mathrm{x})$ distribution shown in Fig. 1. Flow is driven by the hydraulic heads $h = H_1$ and $h = H_2$ prescribed along the left and right vertical boundaries, respectively. The lower and upper horizontal boundaries are impermeable. This results in the reference hydraulic head distribution shown in Fig. 2.

The first step in the proposed algorithm consists of the use of an SVM to reconstruct the boundaries from $N = 100$ conductivity measurements. The location of these measurements, the reconstructed boundaries, and the corresponding hydraulic head distribution are shown in Fig. 3. We used the Gaussian kernel $\mathcal{K}(\mathbf{x}, \mathbf{x}') = \exp\left[-||\mathbf{x} - \mathbf{x}'||^2/(2\sigma^2)\right]$ with parameter $\sigma = 5$, and set SVM parameter $C = 1000$. (These fixed values were chosen for good facies delineation performance based on our previous experience (Wohlberg et al., 2006), but in a more realistic setting these would be chosen via a cross-validation method.)

Using this hydraulic conductivity field as an initial guess in the second step, and choosing $\lambda = 1000$ to provide appropriate weighting to each term, we obtain the reconstructed boundaries and the corresponding hydraulic head distribution in Fig. 4. Table 1 provides a quantitative comparison of these reconstructions. While the incorporation of hydraulic head measurements leads to only a marginal reduction in the boundary reconstruction errors, it significantly reduces the error in predictions of hydraulic head distribution. Moreover, the L^2 norm used in this comparison does not tell the whole story. A visual comparison of the reconstructed conductivity fields

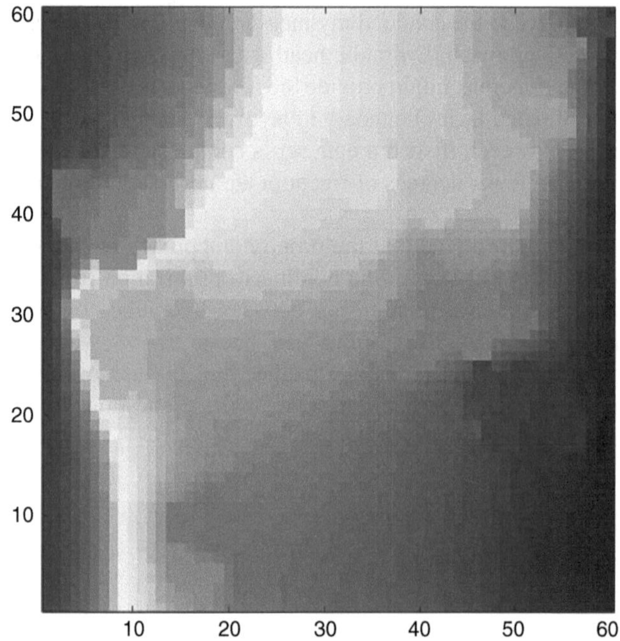

Fig. 2 Hydraulic head distribution in the flow domain shown in Fig. 1

(Figs. 3a and 4a) with the reference conductivity field (Fig. 1) reveals that the joint use of hydraulic conductivity and head data also noticeably improves the boundary reconstruction.

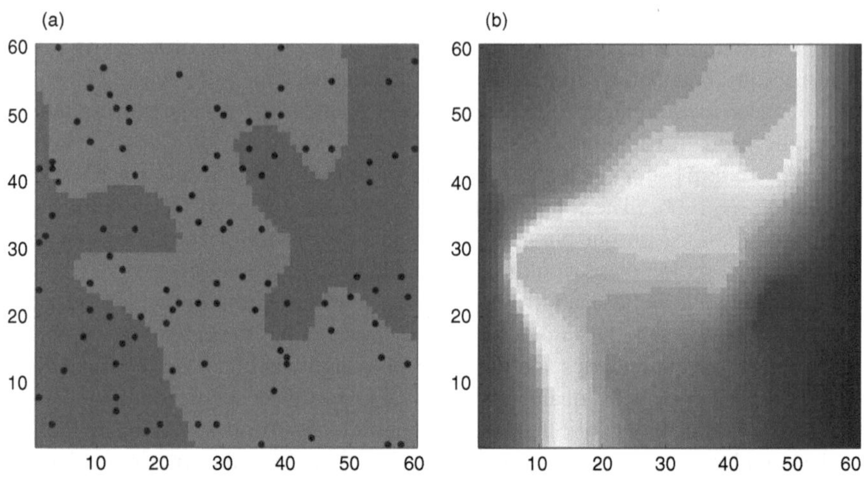

Fig. 3 The conductivity field reconstructed from $N = 100$ conductivity measurements, whose locations are shown by the black dots (a) and the corresponding hydraulic head distribution (b)

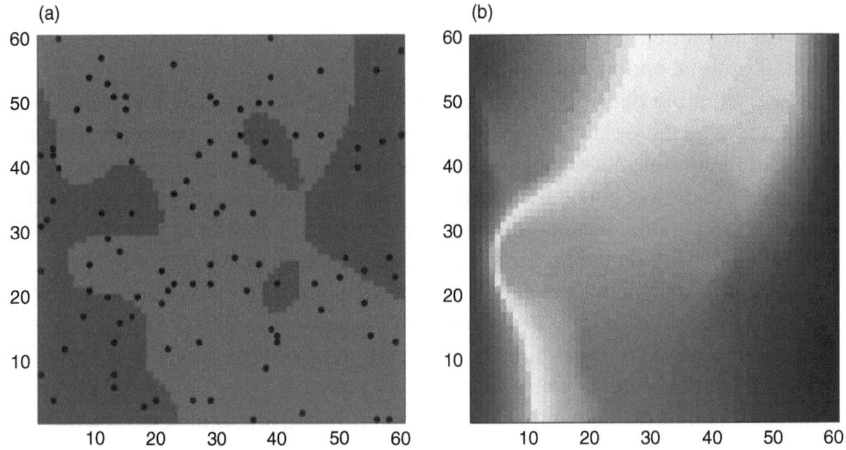

Fig. 4 The conductivity field reconstructed from $N = 100$ measurements of hydraulic conductivity and head, whose locations are shown by the black dots (a) and the corresponding hydraulic head distribution (b)

Table 1 Reconstruction errors (L^2 norm with respect to the reference field) resulting from the reliance on K data only and on combined K and h data

	Conductivity K	Head h
K data	111.83	52.04
K and h data	106.86	25.67

4 Summary and Discussion

Support Vector Machines (SVM), a tool from machine learning, provides a number of advantages over geostatistics. Previous applications of SVM focused on the delineation of lithofacies from measurements of properties of geologic materials (parameter data). Such data allow one to determine the membership of spatial locations where the measurements are made in a relatively straightforward fashion.

The task of identifying geologic units from system state data is significantly more challenging, since the membership of such data in a given unit is not identifiable from data analysis alone. We proposed an SVM-based approach that allows one to combine both types of data with the aim of improving the accuracy and robustness of the facies delineation. The preliminary results reported here demonstrate the potential of the proposed approach.

A number of issues remain open and remain the focus of our ongoing research. These include

- Since the proposed approach relies on a nonlinear optimization with many degrees of freedom, its utility and reliability depends critically on the selection of the optimization strategy. The nonlinear constrained optimization algorithm (function *fmincon* in the MATLAB Optimization Toolbox) used in the present analysis is known to converge to local, rather than global, minima and requires careful adjustment of optimization parameters. This poses the question of selection of optimal optimization strategies.

- As the sampling density (the number of elements in the numerical grid where data are available relative to the total number of elements) decreases, the last term in the optimization functional might become flat. Consequently, the SVM parameterization of boundaries might not guarantee an optimal performance with respect to hydraulic heads. This calls for a detailed analysis of the influence of sampling density on the performance of the proposed approach.
- Locations of data points are expected to play a significant role in the accuracy of facies delineation. To provide an unbiased assessment of the performance of the proposed approach, it is desirable to average the reconstruction errors resulting from several alternative placements of data points for the same sampling data. This task is impossible without a robust optimization algorithm (see above).

These issues and limitations might explain a relatively modest reduction (50%) of the reconstruction errors obtained with the SVM inversion, while geostatistical inverse methods, e.g., the inversion of stochastic moment equations based on the pilot point method (Hernandez et al., 2003, 2006), often yield an order of magnitude error reductions. However, it is important to emphasize that the success of this and other geostatistical inversion techniques depends heavily on the number and location of pilot points, the quality and quantity of data, the presence/absence of priors, and the way used to regularize the objective function (Alcolea et al., 2006). The results presented here reduce the bias by averaging over twenty possible locations of data points.

Finally, the proposed SVM inversion procedure might suffer from an inadequate number of degrees of freedom. To alleviate this problem, we are working on its extension that incorporates the ideas behind pilot point methods (but not their geostatistical implementation) into the SVM framework.

References

Alcolea A, Carrera J, Medina A (2006) Pilot points method incorporating prior information for solving the groundwater flow inverse problem. Adv Water Resour 29: 1678–1689

Carrera J, Neuman SP (1986) Estimation of aquifer parameters under transient and steady sate conditions. 3. Application to synthetic and field data. Water Resour Res 22: 228–242

Eppstein MJ, Dougherty DE (1996) Simultaneous estimation of transmissivity values and zonation. Water Resour Res 32: 3321–3336

Guadagnini L, Guadagnini A, Tartakovsky DM (2004) Probabilistic reconstruction of geologic facies. J Hydrol 294: 57–67

Guadagnini A, Wohlberg BE, Tartakovsky DM, De Simoni M (2006) Support vector machines for delineation of geologic facies from poorly differentiated data. In: Proceedings of the CMWR XVI conference, Copenhagen, June 2006

Hernandez AF, Neuman SP, Guadagnini A, Carrera J (2003) Conditioning mean steady state flow on hydraulic head and conductivity through geostatistical inversion. Stoch Environ Res Risk Assess 17: 329–338

Hernandez AF, Neuman SP, Guadagnini A, Carrera J (2006) Inverse stochastic moment analysis of steady state flow in randomly heterogeneous media. Water Resour Res 42: W05425, doi:10.1029/2005WR004449

Kanevski M, Maignan M (2004) Analysis and modelling of spatial environment data. EPFL Press, Marcel Dekker, Inc., Lausanne, Switzerland

Schölkopf B, Smola AJ (2002) Learning with Kernels. The MIT Press, Cambridge, MA, USA

Sun NZ, Yeh WWG (1985) Identification of parameter structure in groundwater inverse problem. Water Resour Res 21: 869–883

Tartakovsky DM, Wohlberg BE (2004) Delineation of geologic facies with statistical learning theory. Geophys Res Lett 31:L18502 doi:10.1029/2004GL020864

Tsai FTC, Yeh WWG (2004) Characterization and identification of aquifer heterogeneity with generalized parameterization and Bayesian estimation. Water Resour Res 40: W10102 doi:10.1029/2003WR002893

Wohlberg BE, Tartakovsky DM, Guadagnini A (2006) Subsurface characterization with support vector machines. IEEE Trans Geosci Remote Sens 44:47–57 doi:10.1109/TGRS.2005. 859953

Statistical Moments of Reaction Rates in Subsurface Reactive Solute Transport

X. Sanchez-Vila, A. Guadagnini, M. Dentz and D. Fernàndez-Garcia

Abstract Transport of reactive species in the subsurface is driven by mixing processes. Quantification of the mixing rate is, therefore, the basis for a proper characterization of the fate of pollutants in geochemically active environments. We consider the case of an anisotropic correlated random field, with perfect correlation in the horizontal plane, while the vertical integral scale is finite. Flow is uniform and takes place in the x-direction. Longitudinal constant dispersion is considered. Based on the analytical results of De Simoni et al. (2005) for the evaluation of reaction rates at the local scale, reaction is driven by local dispersion at any given point in space and time. Still, due to uncertainty in the advective velocity, reaction rates become a Spatial and Temporal Random Function. The aim of the work is to find the statistical moments of reaction rates, which in this particular configuration can be obtained exactly.

1 Introduction

Mixing has been recognized as the controlling process in several problems dealing with transport of reactive species in the subsurface. Mixing of two waters under perfect geochemical equilibrium with the natural porous medium would produce a local disequilibrium. A reaction will then take place to equilibrate the system. The types of reactions include precipitation/dissolution, adsorption/desorption, redox, and acid/base, amongst others. Appropriate quantification of the mixing rate is key for a proper characterization of solute spreading in geochemically active heterogeneous environments, with important implications in a number of environmental applications, including aquifer remediation schemes.

X. Sanchez-Vila
Department of Geotechnical Engineering and Geosciences, Technical University of Catalonia, Campus Nord D2-006, 08034 Barcelona, Spain
e-mail: xavier.sanchez-vila@upc.edu

A. Soares et al. (eds.), *geoENV VI – Geostatistics for Environmental Applications*, 127–139

Recently a methodology was presented to evaluate solute concentrations and reaction rates when homogeneous reactions (between aqueous species) or heterogeneous reactions (involving both aqueous species and the solid phase) take place under chemical equilibrium conditions (De Simoni et al., 2005). This methodology allows for the derivation of exact analytical expressions, applicable at the local scale, where mixing is mainly driven by diffusion or local dispersion. The salient question then becomes how to further elaborate on these results in order to obtain predictions of reaction rates in randomly heterogeneous media, together with a quantification of the associated uncertainty.

Our approach consists of the following steps. First, we define the geochemical problem. In this case, we consider a reaction of pure precipitation/dissolution, involving two aqueous species at equilibrium with an immobile solid mineral. Second, the flow and transport problems are formulated. These take place within a three-dimensional randomly heterogeneous, statistically uniform, hydraulic conductivity field, K, of infinite lateral extent. Hydraulic conductivity is isotropic at the local scale, but highly anisotropic in terms of correlation distances. The range of the conductivity variogram is very large (theoretically infinite) in the two horizontal directions, while it is finite along the vertical. This model is often employed to provide a depiction of a statistically stratified medium, where layering is not described in terms of a vertical sequence of disjoint materials, but rather as a continuous transition between different values of K, that are variably correlated along the vertical. A uniform head gradient is imposed parallel to the direction of layering. The (random) reaction rates are then computed exactly at the local scale as a function of the random K, following the methodology of De Simoni et al. (2005). It is then possible to compute the main statistics of the reaction rate, in terms of a simple quadrature in the probabilistic space from which hydraulic conductivity values are sampled. We concentrate on the first two statistical moments, and provide some ideas on the conditions where the ensemble values are actually a good representation of the real physical values.

2 Problem Statement

2.1 The Geochemical Problem

We consider the geochemical problem of the so-called bi-species system (Gramling et al., 2002; De Simoni et al., 2005). This involves the presence of two aqueous solutes, B_1 and B_2, which are in chemical equilibrium with a solid mineral, M_3. Without loss of generality we consider the case in which both stoichiometric coefficients are equal to one and the immobile solid mineral dissolves reversibly to yield species B_1 and B_2:

$$B_1 + B_2 \Leftrightarrow M_3. \tag{1}$$

In this particular case, the mass action law implies that the two aqueous species must satisfy at all points and all times the following condition

$$c_1 c_2 = K_{eq}, \tag{2}$$

where K_{eq} is the equilibrium constant, which is a function of temperature, pressure and chemical composition of the solution. In (2) we assume implicitly that activity coefficients are equal to 1. If at any given moment in space or time two waters satisfying (2) are put in contact, it is easily proven that the mixed water will not satisfy the equilibrium condition anymore (i.e., immediately it occurs that $c_{1,m}c_{2,m} > K_{eq}$, $c_{1,m}$ and $c_{2,m}$ respectively being the concentration of B_1 and B_2 in the mixed water). Under these circumstances precipitation takes place instantaneously and concentrations c_i $(i = 1, 2)$ are reduced in equal proportions, until (2) is again satisfied.

2.2 Geostatistical Model for Hydraulic Conductivity and Flow Set-up

We model hydraulic conductivity, K, as a three-dimensional stationary random space function, with mean $\langle K \rangle$ and variance σ_K^2. The two-point covariance of K is of axisymmetric anisotropy and is modeled with an exponential variogram $\gamma_K(\mathbf{h}) \equiv \gamma_K(h_1, h_2, h_3) = \sigma_K^2 \exp\left(-\left(\frac{h_1{}^2}{\lambda_1{}^2} + \frac{h_2{}^2}{\lambda_2{}^2} + \frac{h_3{}^2}{\lambda_3{}^2}\right)^{1/2}\right)$. Here, h_1, h_2, h_3 are separation distances along directions x,y, and z, respectively; λ_1, λ_2 $(= \lambda_1)$, λ_3 are measures of the corresponding correlation scales. The adopted assumption of stationarity implies that the probability density function (pdf) of K, p_K, be invariant under translation within the system.

We then consider one of the simplest models of heterogeneity, which is that of stratified formations. According to this model, K varies only in the vertical direction, z. Interest in this model has been motivated by its simplicity and by the recognition of the importance of layering upon solute transport in sedimentary formations (e.g. Matheron and de Marsily, 1980). Its simplicity allows grasping the key features of transport processes that can be recognized in more complex systems. From a practical viewpoint, we note that, although perfect layering is rarely found over large horizontal distances, the model can be applied to depict transport of contaminants over relatively short travel times.

In this work, layering is modeled in a geostatistical sense, i.e., we consider that λ_1, $\lambda_2 \to \infty$, while λ_3 is finite. We consider a saturated groundwater flow that is parallel to the direction of bedding. In other words, flow is driven by a constant horizontal gradient, $\boldsymbol{J}$, that is aligned along the x-direction ($\boldsymbol{J} = (J,0,0)$). We further assume the flow domain to be of large lateral extent, so that boundary effects can be disregarded. As a consequence, the steady-state Darcy's velocities at any point in the domain, $q(z)$, are given by

$$q(z) = -K(z)J. \tag{3}$$

Here, we note that J is negative, so that flow takes place in the direction of increasing x.

2.3 The Transport Equations

The mass balance equations for the two aqueous species are

$$\phi \frac{\partial c_i}{\partial t} = JK \frac{\partial c_i}{\partial x} + \phi D_L \frac{\partial^2 c_i}{\partial x^2} + w_e c_{e,i} - r \quad i = 1, 2, \tag{4}$$

where ϕ [–] is porosity, D_L [L^2T^{-1}] is the local scale longitudinal diffusion-dispersion value, w_e [T^{-1}] is an external recharge contribution, and $c_{e,i}$ [ML^{-3}] are the concentrations of species B_i in the recharging water. In this paper we concentrate on a diffusion coefficient which is constant and independent on local velocity. The problem could be made more general by adding an additional term accounting for a dispersion coefficient, but the main conclusions of the paper would not change.

Defining c_3 as the concentration of the mineral, the reaction rate, r [$ML^{-3}T^{-1}$] is incorporated as a sink/source term in the right hand side of (4), given as

$$\frac{\partial c_3}{\partial t} = r. \tag{5}$$

In (4) we have discarded the impact of the transverse dispersion, D_T, that causes mixing along the y-direction and between (statistically defined) layers. The reason is that we are interested in exploring the early-time behavior of the system, i.e., processes occurring within the regime for which $(tD_T)/\lambda_3^2 << 1$. In these scenarios, the effect of tranverse dispersivity has not yet developed in the system (Dentz and Carrera, 2003) and can be neglected in the governing equation.

Following the methodology of De Simoni et al. (2005), the system can be fully defined in terms of a conservative components, **u**, defined as stoichiometric combinations of the aqueous concentrations. In the bi-species system presented here, where the precipitation/dissolution reaction involves equal stoichiometric coefficients, a single component is needed, defined as

$$u = c_1 - c_2. \tag{6}$$

This component is advected and dispersed according to a transport equation that can be derived by subtracting the two transport equations in (4), leading to

$$\phi \frac{\partial u}{\partial t} = JK \frac{\partial u}{\partial x} + \phi D_L \frac{\partial^2 u}{\partial x^2} + w_e u_e. \tag{7}$$

The solution for u depends on the boundary and initial conditions of the problem to be solved. Combining (6) and the mass action law (2), it is possible to write the concentrations c_i explicitly, in terms of u (speciation process)

$$c_1 = \frac{u + \sqrt{u^2 + 4K_{eq}}}{2}; \quad c_2 = \frac{-u + \sqrt{u^2 + 4K_{eq}}}{2}. \tag{8}$$

Even though K_{eq} displays variations with temperature or salinity, in many groundwater-related geochemical processes we can assume that $K_{eq} \cong$ constant. In such a case, the aqueous concentrations are only functions of $u(i.e., c_i = f(u))$. Based on the method of De Simoni et al. (2005), it is possible to derive a closed-form expression for the reaction rate. The method consists of expanding (4) for one of the species, and then developing the spatial and temporal derivatives of c_i using the chain rule. We finally simplify the resulting expression by means of (7) and derive the following expression for the reaction rate,

$$r = \phi D_L \frac{\partial^2 c_2}{\partial u^2} \left(\frac{\partial u}{\partial x} \right)^2 + w_e \left(c_{e,2} - u_e \frac{\partial c_2}{\partial u} \right), \tag{9}$$

where the derivatives of c_2 with respect to u have the following expressions

$$\begin{aligned} \frac{\partial c_2}{\partial u} &= \frac{1}{2} \left(-1 + \frac{u}{\sqrt{u^2 + 4\,K_{eq}}} \right), \\ \frac{\partial^2 c_2}{\partial u^2} &= \frac{2\,K_{eq}}{(u^2 + 4\,K_{eq})^{3/2}}. \end{aligned} \tag{10}$$

2.4 Statistics of Reaction Rates

It is clear from (9) and (10) that the reaction rate is only a function of u and its spatial derivative, $u' = \partial u / \partial x$, while u itself is only a function of K. Thus, one can see that $r = r(u(K), u'(K)) = r(K)$. This implies that, if one can derive an analytical solution of (9) for the random rate, r, the moments of r can be obtained by integration in the probability space over which K is sampled.

Alternatively, it would be possible to obtain approximate expressions for the (statistical) moments of r, starting by approximating r in terms of a Taylor's expansion around the (constant) arithmetic mean of K, $K_A (\equiv \langle K \rangle)$

$$r(K) = r(K_A) + \sum_{i=1} \frac{1}{i!} \frac{d^i r}{dK^i}\bigg|_{K=K_A} (K - K_A)^i. \tag{11}$$

The leading terms for the first two moments of r can then be obtained after truncating (11) at second order. Thus, the (second-order) mean rate, $\langle r \rangle$, and the (second-order) variance of reaction rate, σ_r^2, are given respectively by

$$\langle r \rangle \approx r(K_A) + \frac{1}{2}\frac{d^2 r}{dK^2}\bigg|_{K=K_A} \sigma_K^2, \tag{12}$$

$$\sigma_r^2 = \left(\frac{dr}{dK}\bigg|_{K=K_A}\right)^2 \sigma_K^2. \tag{13}$$

An alternative way to find the derivatives involved in (12) and (13) is graphically. For a given set-up it would be possible to find the curve r vs. K. Then the derivatives can be derived graphically.

3 Application Example: 1-D Fixed-step Function

3.1 Explicit Random Solution for the Reaction Rates

The proposed methodology is here applied to the previously described setup problem. The initial and boundary conditions associated with the transport problem are described mathematically as follows:

$$\begin{aligned} &u\,(x, y, z, t = 0) = u^0 \quad x \geq 0 \\ &u\,(x = 0, y, z \in [-a, a], t) = u^0 + \Delta u^0 \quad t \geq 0 \\ &u\,(x = \infty, y, z, t) = u^0 \quad t \geq 0 \end{aligned} \tag{14}$$

The solution of the problem, in the absence of recharge $w_e = 0$, and along one-dimensional lines was provided by Ogata and Banks (1961). The normalized concentration of the component, $u_D = u/\Delta u^0$, can be written in terms of a dimensionless time, $t_D = -\frac{JKt}{\phi x}$ and the *Peclet* number, $\left(Pe = -\frac{JKx}{\phi D_L}\right)$:

$$u_D = \frac{u^0}{\Delta u^0} + \frac{1}{2}\left\{\exp\,(Pe)\,\mathrm{erfc}\left[\left(\frac{Pe}{4t_D}\right)^{1/2}(1+t_D)\right] + \mathrm{erfc}\left[\left(\frac{Pe}{4t_D}\right)^{1/2}(1-t_D)\right]\right\} \tag{15}$$

It is noted that *Pe* practically ranges between 1 and 100 for flow in aquifers. From (9) and (15), we can obtain an exact, random solution for the dimensionless rate, $r_D(x, y, z, t) = (rx^2)\,/\,(\phi\, D_L \Delta u^0) = f(K_{eqD}, u_D(K), Pe(K), t_D(K))$, as

$$\begin{aligned} r_D \equiv f\,(K) = \frac{1}{2}\frac{K_{eqD}}{\left(u_D^2 + 4\,K_{eqD}\right)^{3/2}} &\left\{Pe\, e^{Pe} \mathrm{erfc}\left[\left(\frac{Pe}{4t_D}\right)^{1/2}(1+t_D)\right]\right. \\ &\left. -\frac{2}{\sqrt{\pi}}\left(\frac{Pe}{t_D}\right)^{1/2}\exp\left(-\left(\frac{Pe}{4t_D}\right)(1+t_D)^2\right)\right\}^2 \end{aligned} \tag{16}$$

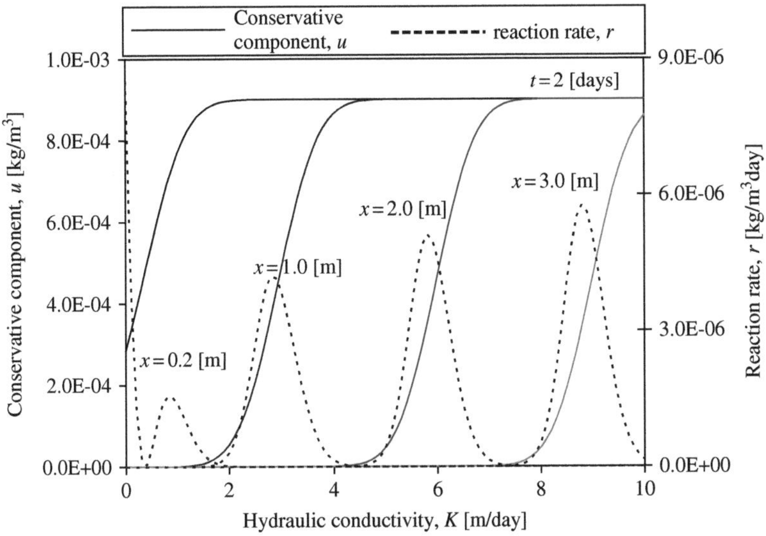

Fig. 1 Dependence of the (random) conservative component, u, and of the rate, r, on K at various distances from the injection line ($x = 0.2, 1.0, 2.0, 3.0\,\mathrm{m}$) and after a time $t = 2$ days has elapsed since injection. Constant parameters used in the evaluation of (15) and (16) are: $J / \phi = -0.167$, $D_L = 10^{-2}\mathrm{m}^2/\mathrm{day}$, $u^0 = 0.0\,\mathrm{kg/m}^3$, $\Delta u^0 = 9 \times 10^{-4}\,\mathrm{kg/m}^3$, $K_{eq} = 10^{-7}(\mathrm{kg/m}^3)^2$

Here, $K_{eqD} = K_{eq} / (\Delta u^0)^2$. Notice that r_D depends on K through u_D, Pe, and t_D, leading to a highly non-linear behavior. To illustrate this non-linearity, Fig. 1 depicts the dependence of the (random, dimensional) conservative component, u, and of the rate, r, on K at various distances from the injection line ($x = 0.2, 1.0, 2.0, 3.0\,\mathrm{m}$) and after a time $t = 2$ days has elapsed since injection. Constant parameters used in the evaluation of (15) and (16) are also detailed in the Figure.

From Fig. 1 we see that, for a given time, the rate displays a different behavior at observation points located close to or at some distance from the injection line. Close to the source, the effect of the type of injection, encapsulated by the second term in the parenthesis in (16), results first in the presence of a non-zero value of the rate for $K \to 0$. This is related to dominant diffusive effects in the presence of very low conductivities. This is then followed by a decreasing behavior of the rate with increasing K. This behavior persists until r vanishes and then starts increasing until it reaches a peak. The boundary effect is not felt at larger distances (i.e., at $x = 1\,\mathrm{m}$ the boundary effect is completely lost) so that (15) can be approximated by:

$$u_D = u_{D,0} + \frac{1}{2}\mathrm{erfc}\left[\left(\frac{Pe}{4t_D}\right)^{1/2}(1 - t_D)\right], \tag{17}$$

with $u_{D,0} = u^0/\Delta u^0$. We note that this is also an approximate solution for the transport problem (7) in the presence of different types of boundary conditions, including (*a*) the third-type boundary condition (Lindstrom et al., 1967), (*b*) the Kreft and Zuber (1978) condition, and (*c*) the constant point source condition (Sun, 1996). The total reaction rate then becomes

$$r = \underbrace{\frac{K_{eqD}}{\left(u_D^2 + 4\,K_{eqD}\right)^{3/2}}}_{A} \underbrace{\frac{\phi\,\Delta u^0}{2\pi t}}_{B} \underbrace{\left\{\exp\left[-\frac{1}{4D_L t}\left(x + \frac{JKt}{\phi}\right)^2\right]\right\}^2}_{C}. \tag{18}$$

In (18), we identify the product of three terms. The spatial distribution of r is mainly driven by term C. This term is of symmetric shape in x, and peaks at the value $K = -\phi\, x\, /\, (tJ)$, that is, it increases linearly with x. Term B provides the influence of the maximum with time. Basically, the maximum rate is proportional to t^{-1}. Term A provides a non-linear behavior that basically displaces the maximum towards smaller values of K (as evidenced by Fig. 1). Actually this term is driven by speciation, so that its variation with x is not that relevant. In any case, it is possible to write some bounds for the maximum reaction rate, as follows

$$\frac{K_{eqD}}{\left(\left(u_{D,0} + \frac{1}{2}\right)^2 + 4K_{eqD}\right)^{3/2}} \frac{\phi\,\Delta u^0}{2\pi t} \le r_{\max} \le \frac{K_{eqD}}{\left(u_{D,0}^2 + 4K_{eqD}\right)^{3/2}} \frac{\phi\,\Delta u^0}{2\pi t}. \tag{19}$$

3.2 *Statistical Moments of Reaction Rates*

Following Section 2.4, we are now in a position to write the following exact expressions for the mean and variance of r_D

$$\langle r_D(x,t)\rangle = \int_0^\infty f(K)\, p_K(K)\, dK, \tag{20}$$

$$\sigma_{rD}^2 = \int_0^\infty \left\{f(K) - \langle r_D\rangle\right\}^2 p_K(K)\, dK, \tag{21}$$

where the f function is provided in (16). The results in (17) and (18) are valid regardless of the functional format for the selected pdf. A common choice for the univariate distribution of K is the Log-Normal model, defined by:

$$p_K(K) = p(K) = \frac{1}{\sqrt{2\pi}\,\sigma_Y K} \exp\left\{-\frac{1}{2}\frac{(\ln K - \mu_Y)^2}{\sigma_Y^2}\right\}, \tag{22}$$

where μ_Y and σ_Y are, respectively, the mean and standard deviation of the natural logarithm of K, $Y = \ln K$. Thus, the mean and variance of the reaction rate can be obtained after a single quadrature, independently of the shape of the hydraulic conductivity variogram.

It is also important to note that the actual result for mean and variance is independent of the variogram model selected, since it involves only the univariate statistics of K. The main restriction to use the results of (20) and (21) in a single realization is that the parameter a in (14) a has to be much larger than λ_3, in order for ergodic conditions to hold.

4 Evaluation of the Reaction Rate Moments

Figure 2 depicts the dependence of the (ensemble) mean reaction rate, $\langle r \rangle$, on the distance from the line of injection and on time elapsed since injection when Y is Normal with $\mu_Y = 0$ and $\sigma_Y^2 = 1$. The constant input parameters are those of Fig. 1. The mean rate displays a non-monotonous behavior. It is characterized by a local maximum that (a) decreases in magnitude and (b) is displaced towards larger distances with time. The local maximum is then followed by a decreasing limb, whose rate of decay decreases with time.

The effect of the dispersion coefficient on $\langle r \rangle$ at a given time is illustrated in Fig. 3. Increasing the rate of mixing, as implied by large values of D_L, produces an increase of the local value of the mean reaction rate. The latter displays a larger spatial persistence for the largest D_L examined, thus evidencing the importance of this term in a geochemically active system. With the only exception of locations at

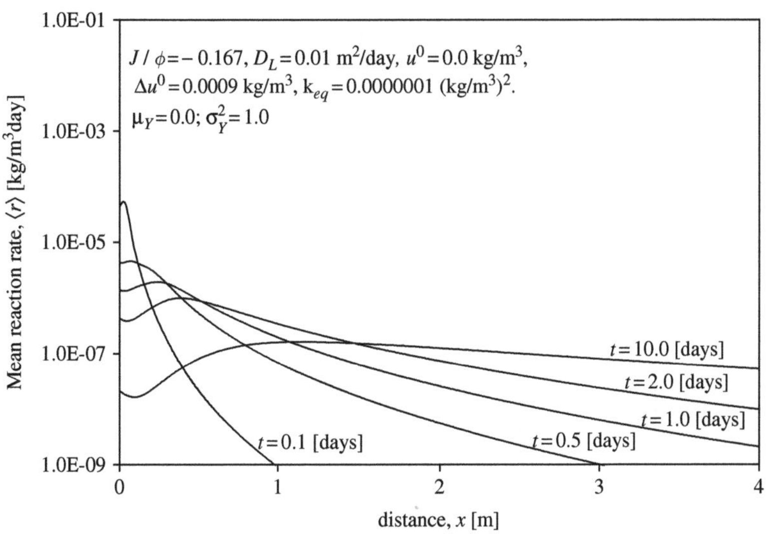

Fig. 2 Dependence of the (ensemble) mean reaction rate, $\langle r \rangle$, on the distance from the line of injection and on time elapsed since injection when Y is Normal with $\mu_Y = 0$ and $\sigma_Y^2 = 1$

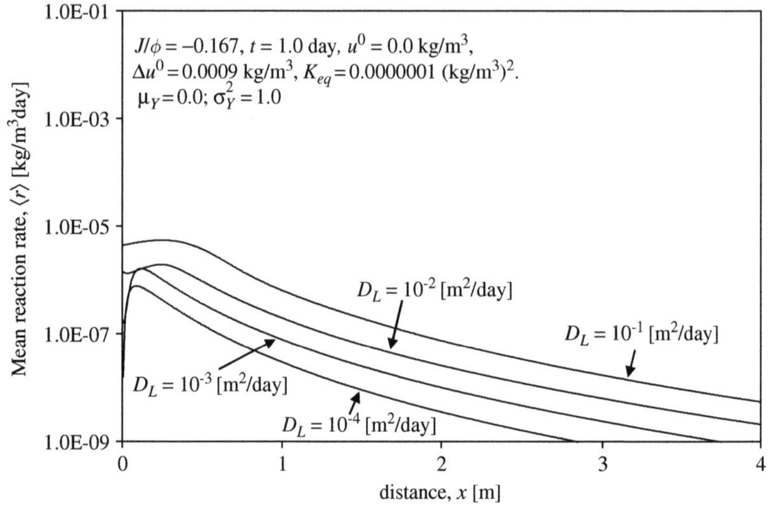

Fig. 3 Dependence of the (ensemble) mean reaction rate, $\langle r \rangle$, on the distance from the line of injection and on D_L for a fixed time when Y is Normal with $\mu_Y = 0$ and $\sigma_Y^2 = 1$

short distances from the source line, each increase of D_L of an order of magnitude produces a local increase of $\langle r \rangle$ of about half order of magnitude.

The standard deviation, σ_r, associated with the mean reaction rate is shown in Fig. 4 as a function of distance and time, when Y is Normal with $\mu_Y = 0$ and $\sigma_Y^2 = 1$. We start by noting that the σ_r is of the same order of magnitude as $\langle r \rangle$ for very short

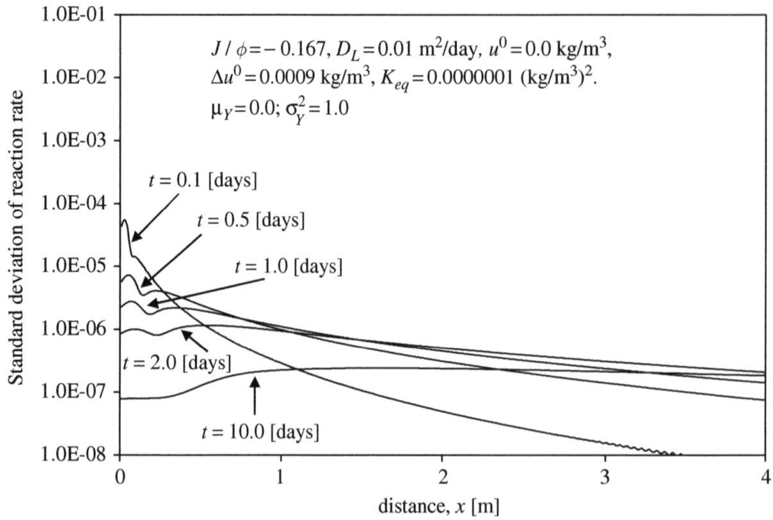

Fig. 4 Standard deviation of the reaction rate [in kg/m^3/day] versus distance from the line of injection and for different times. Y is Normal with $\mu_Y = 0$ and $\sigma_Y^2 = 1$

distances from the injection line. The rate of decay of σ_r with x is generally lower than that of $\langle r \rangle$. It follows that for the investigated time intervals σ_r is generally larger than $\langle r \rangle$ and much more so as x increases. This suggests that this type of structured heterogeneity results in coefficients of variations of the local rate generally larger than 100% for short times and/or distances. Since the system we analyze displays a high geochemical activity precisely in the range of short times and distances, we can conclude that uncertainty in the vertical distribution of hydraulic conductivity has a large negative influence on the predictability of the behavior of the system within regions where significant reaction rates occur.

As opposed to the mean rate, σ_r displays a local minimum that is located close to the source line and travels from the origin relatively slowly with time. In the example considered, σ_r displays a primary and a secondary peak, both of them decreasing with time. While the location of the primary peak is largely insensitive to the elapsed time (controlled by the boundary conditions), the position of the secondary peak travels along x as time increases.

The dependence of the spatial distribution of $\langle r \rangle$ on the variance of Y for times $t = 0.5$ and 2 days is depicted in Fig. 5. Corresponding depictions documenting the behavior of σ_r are offered in Fig. 6. It can be noted that, in general, both $\langle r \rangle$ and σ_r tend to increase with σ_Y^2. This tendency is less pronounced as time increases. It is also interesting to note that $\langle r \rangle$ and σ_r are relatively insensitive on the heterogeneity of the underlying log-conductivity field close to the line source. The distance within which this behavior persists increases with elapsed time.

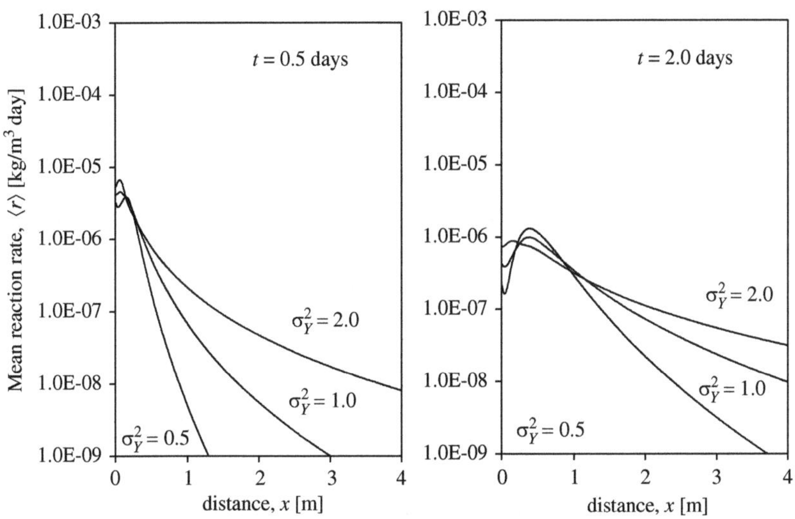

Fig. 5 Impact of the variance of Y on the spatial distribution of $\langle r \rangle$ versus travel distance, for times $t = 0.5$ and 2 days. Constant parameters used are: $J\ /\ \phi = -0.167$, $D_L = 10^{-2}\mathrm{m}^2/\mathrm{day}$, $u^0 = 0.0\,\mathrm{kg/m}^3$, $\Delta u^0 = 9 \times 10^{-4}\,\mathrm{kg/m}^3$, $K_{eq} = 10^{-7}(\mathrm{kg/m}^3)^2$, $\mu_Y = 0.0$

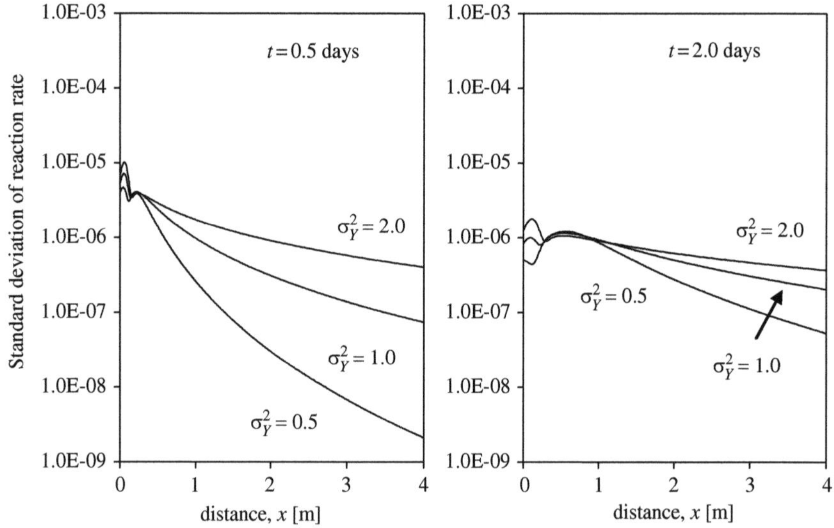

Fig. 6 Dependence of the variance of Y on the spatial distribution of σ_r versus distance for times $t = 0.5$ and 2 days. Constant parameters used are: $J\,/\,\phi = -0.167$, $D_L = 10^{-2}\,\mathrm{m^2/day}$, $u^0 = 0.0\,\mathrm{kg/m^3}$, $\Delta u^0 = 9\times10^{-4}\,\mathrm{kg/m^3}$, $K_{eq} = 10^{-7}(\mathrm{kg/m^3})^2$, $\mu_Y = 0.0$

5 Conclusions

We consider transport of reactive species in an anisotropic correlated random hydraulic conductivity field, with perfect correlation in the horizontal plane, while the vertical integral scale is finite. Flow is uniform and takes place in the x-direction. Uncertainty in the (vertical) distribution of the advective velocity causes the reaction rate to become a Spatial and Temporal Random Function. The low-order statistical moments (mean and variance) of the reaction rate for given space-time coordinates can be obtained in terms of a simple quadrature in probability space of hydraulic conductivity.

Our results highlight that, in general, both the mean and standard deviation of the reaction rate tend to increase with the level of heterogeneity of the hydraulic conductivity field. The coefficient of variation of the rate of reaction is always larger than 100%, thus evidencing the negative impact of uncertain hydraulic conductivity distribution on the predictability of such geo-chemical processes.

This work is a first stage toward advancing in the study of the evaluation of reaction rates in complex mutispecies transport of reactive species in heterogeneous media. This problem has profound implications in natural or enhanced attenuation systems. At larger distances it will be necessary to consider more realistic variogram models with finite correlation scales as well as to incorporate the effect of transverse dispersion.

References

Dentz M, Carrera J (2003) Effective dispersion in temporally fluctuating flow through a heterogeneous medium, Physical Review E, 68 (3): Art. No. 036310

De Simoni M, Carrera J, Sanchez-Vila X, Guadagnini A (2005) A procedure for the solution of multicomponent reactive transport problems, Water Resour Res, 41(11): Art. W11410

Gramling CM, Harvey CF, Meigs LC (2002) Reactive transport in porous media: A comparison of model prediction with laboratory visualization. Environ Sci Tech 36(11): 2508–2514

Kreft A, Zuber B (1978) On the physical meaning of the dispersion equation and its solutions for different initial and boundary conditions, Chem Eng Sci 33: 1471–1480

Lindstrom FT, Haque R, Freed VH, Boersma L (1967) Theory on the movement of some herbicides in soils: Linear diffusion and convection of chemicals in soils, J. Environ Sci Technol 1: 561–565

Matheron G, DeMarsily G (1980) Is Transport In Porous-Media Always Diffusive - A Counterexample, Water Resour Res 16 (5): 901–917

Ogata A, Banks RB (1961) A solution of the differential equation of longitudinal dispersion in porous media, U. S. Geol. Surv. Prof. Paper 411-A

Sun NZ (1996) Mathematical Modeling of Groundwater Pollution, Springer-Verlag, NY

Including Conceptual Model Information when Kriging Hydraulic Heads

M. Rivest, D. Marcotte and P. Pasquier

Abstract A reliable hydraulic head field is a key element to many hydrogeological, environmental or geotechnical studies. It enables quick identification of areas where high hydraulic gradients could threaten an earth dam's integrity. It also highlights probable contaminant flow paths and determines wells' influence areas. Furthermore, some inversion algorithms (direct methods) require an initial estimation of the entire head field to compute hydraulic conductivity. Interpolation techniques, such as kriging, have the advantage of reproducing the observed values. However, the shape of the interpolated head fields often lacks realism particularly near pumping wells, boundaries and lithological contrasts. In these cases, the flow equation is poorly reproduced by interpolation. On the other hand, numerical modeling can easily integrate the hydrogeological conceptual model and generate realistic head fields. Unfortunately, the numerical model is based on uncertain hard data which poorly reproduce the head observations.

We propose an approach based on kriging that uses the "shape" information present in a numerical conceptual model as an external drift. The performance of the method is first investigated using a 2D synthetic aquifer. In this case, several numerical head fields are used in the external drift to account separately for different aspects of the phenomenon (principle of superposition). A stepwise procedure is used to select the best set of numerical head fields. Kriging with external drift (KED) shows marked improvement over ordinary kriging and universal kriging with first order polynomials. The approach is also applied to the study of two large earth dams in which monitoring data is available. Cross-validation shows again the good performance of KED compared to ordinary kriging or universal kriging with first order polynomials. The approach can be used for 3D head field estimation.

M. Rivest
École Polytechnique de Montréal, Département des génies civil, géologique et des mines, C.P. 6079, Succ. Centre-ville, Montréal, Qc, Canada, H3C 3A7
e-mail: martine.rivest@polymtl.ca

A. Soares et al. (eds.), *geoENV VI – Geostatistics for Environmental Applications*, 141–151

1 Introduction

Many hydrogeological, environmental, and geotechnical studies require a reliable estimate of the hydraulic head field for design or monitoring purposes. A head map can be used directly to determine the zone of influence of a well or high gradients in an earth dam. It is also required as an initial solution for transient state problems, a first guess for non-linear numerical problems or a starting point in some inversion algorithms (Pasquier and Marcotte, 2005, 2006; Sagar et al., 1975) Head fields are typically obtained either by direct interpolation (e.g. kriging) or by numerical modelling. The latter requires knowledge of the geometry, the boundary conditions, and the hydrogeological parameters of the field under study. Heads obtained by a flow simulator respect the diffusion equation; however, uncertainties regarding the model's characteristics lead to important differences between computed and observed heads at measurement points. Model calibration helps to improve the head fit but can be time-consuming if done manually. Furthermore, since calibration is an ill-defined problem, many solutions are possible.

Direct head estimation, such as kriging, ensures exact interpolation at a data point, thus including measurement error. When measurement error is present, it is customary to filter out this component by avoiding estimating data points. Contrarily to numerical modelling, consistency of the resulting head field with the phenomenon is not guaranteed. In fact, kriging often fails to properly reproduce important features such as no-flow and constant head boundaries, as well as the influence of extracting wells (drawdowns and shape). Spurious head minimum or maximum could also be present on the kriged head field. Several techniques are helpful in improving the realism of kriged head fields. No-flow boundaries can be taken into account by the introduction of double points in the kriging system (Brochu and Marcotte, 2003; Delhomme, 1979). Analytical equations describing radial flow in a perfectly homogeneous and infinite medium define drift functions to further constrain the estimation and account for the presence of wells (Brochu and Marcotte, 2003; Tonkin and Larson, 2002). However, this approach remains limited to two-dimensional cases where analytical solutions are available.

In this article, we propose to combine the advantage of external drift kriging (KED) (Dehomme, 1979) to the capabilities of numerical flow simulation. The idea is to define the drift with numerical head fields, instead of using analytical equations. The performance of this approach is tested with two case studies. The first case consists of a synthetic two-dimensional aquifer composed of two contrasting homogeneous zones under the influence of pumping wells. In the second case, the method is applied to the mapping of hydraulic heads in two monitored earth dams. For each case, cross-validation is used to assess the precision of the method and to compare it with ordinary kriging (OK) and universal kriging with linear drift (UK). The realism of the kriged head fields is discussed. It is shown that KED with numerical drift significantly improves the head field estimation compared to OK and UK.

We emphasize that the proposed approach is not an inverse method. Its purpose is to obtain a realistic head map. This map, if desired, can then be used with inverse methods like the one presented in Pasquier and Marcotte (2006) to estimate the hydraulic conductivity field.

2 Methodology

In practice, site investigation provides the hydrogeologist with useful qualitative and quantitative information about the hydrogeological system. This information is synthesized into a conceptual model that is a simplified representation of the real field. Nevertheless, the conceptual model, when integrated into a flow simulator, yields valuable insight about the *shape* of the head field. This suggests the use of such a field as secondary variable in KED. The following sections present the theory pertaining to the method.

2.1 Kriging with an External Drift

Kriging with an external drift allows the use of secondary information to account for the spatial variation of the primary variable's local mean (non-stationarity of the mean). The secondary variable is chosen for its strong correlation with the variable of interest. Observed values of this variable should be available at every data point and every estimation point (Ahmed and de Marsily, 1987; Galli and Meunier, 1987). The secondary variable should vary more smoothly than the primary variable since it aims at representing the latter's expectation (Castelier, 1993). In KED, the secondary variable appears as additional unbiasedness constraints. For example, when two secondary variables are used, the unbiasedness constraints are:

$$\sum_{i=1}^{n} \lambda_i = 1 \quad ; \quad \sum_{i=1}^{n} \lambda_i f(x_i) = f(x_0) \quad \text{and} \quad \sum_{i=1}^{n} \lambda_i g(x_i) = g(x_0) \tag{2.1}$$

where λ_i are kriging weights, $f(x_i)$ and $g(x_i)$ are values of the secondary variables at data points, and $f(x_0)$ and $g(x_0)$ are values of the secondary variables at the estimation point. Note that the first equation is the usual ordinary kriging unbiasedness constraint. In matrix notation, the kriging system becomes:

$$\begin{bmatrix} \mathbf{K} & \mathbf{F} \\ \mathbf{F}' & \mathbf{0} \end{bmatrix} \begin{bmatrix} \boldsymbol{\lambda} \\ \boldsymbol{\mu} \end{bmatrix} = \begin{bmatrix} \mathbf{k} \\ \mathbf{f} \end{bmatrix} \tag{2.2}$$

where $\mathbf{K}$ is the $n \times n$ covariance matrix (observation-observation), $\mathbf{F}$ is the $n \times p$ drift matrix, λ is the $n \times 1$ kriging weights vector, μ is the $p \times 1$ vector of Lagrange multipliers corresponding to the unbiasedness conditions, $\mathbf{k}$ is the $n \times 1$ covariance vector (observation-estimation), and $\mathbf{f}$ is the $p \times 1$ drift functions evaluated at the estimation point. It is worth expressing KED in its dual form:

$$h^*(x_0) = \begin{bmatrix} \mathbf{b}' & \mathbf{c}' \end{bmatrix} \begin{bmatrix} \mathbf{k}' \\ \mathbf{f}' \end{bmatrix} \quad \text{where} \quad \begin{bmatrix} \mathbf{b}' \\ \mathbf{c}' \end{bmatrix} = \begin{bmatrix} \mathbf{K}' & \mathbf{F}' \\ \mathbf{F}' & \mathbf{0} \end{bmatrix}^{-1} \begin{bmatrix} \mathbf{h} \\ \mathbf{0} \end{bmatrix} \tag{2.3}$$

where **b** and **c** are the dual weights, **h** is the vector of observed heads, and $h^*(x_0)$ is the estimated head at x_0. The estimation of the hydraulic head h^* combines both a stochastic contribution ($\mathbf{b}'\mathbf{k}$) and a deterministic drift contribution ($\mathbf{c}'\mathbf{f}$). Note that the drift functions are individually weighted by vector **c**. This shows the advantage of splitting the conceptual model into simpler individual components to gain greater flexibility for the drift modelling.

2.2 Selecting Secondary Variable

As shown in Eq. 2.3, KED optimizes the weighting of each drift component.. For example, consider the aquifer presented in the synthetic case study (section 1, Fig. 1a). It is important to know the position of the contact between units and the transmissivity contrast in order to obtain the right shape of the drift with a single conceptual model. With two conceptual models, one corresponding to a homogeneous aquifer and the other to the two-unit aquifer, only the position of the contact is required. KED will adjust the weights to implicitly retrieve the transmissivity contrast compatible with the observed data.

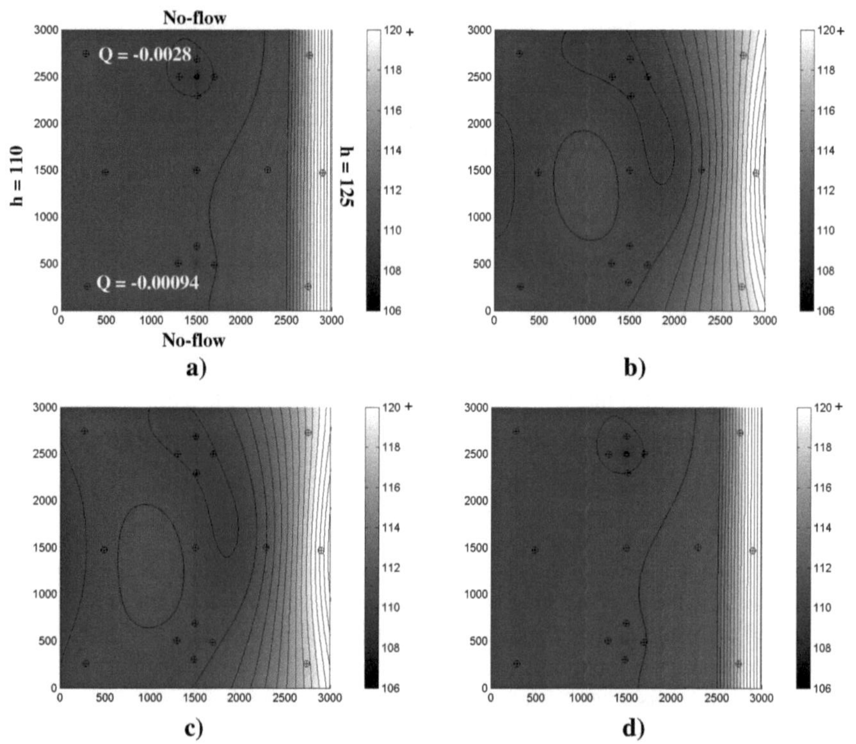

Fig. 1 Synthetic case study a) Reference head field, b) Head field obtained by OK, c) Head field obtained by UK, d) Head field obtained by KDE

It is important to limit the number of drift variables to avoid over-parameterization. Moreover, the p vectors of secondary constraints at data points must be linearly independent to ensure non-singularity of the kriging matrix (left hand side of Eq. 2.2). A stepwise selection procedure (Draper and Smith, 1981) of drift variables is used for this purpose. The approach proceeds sequentially by forward inclusion of drift variables. The candidate variable maximizes the coefficient of determination (R^2). It is retained if the variable yields a significant increase in R^2 (partial-F test). Then, the test determines whether or not the chosen drift variables can be removed (decrease in R^2 not significant). The procedure stops when no more variables can be included in or excluded from the selected set. The stepwise procedure ensures that the retained drift functions are linearly independent. Note that the partial F-test assumes statistical independence of the residuals. The test is only approximate since the residual head data is expected to be spatially correlated. The constant drift function is always included in the kriging.

2.3 Covariance Function

Hydraulic head is a non-stationary variable due to the physics of groundwater flow. The determination of its covariance function is difficult. The head covariance function must be differentiable at least four times because of the double differentiability of head implied by the diffusion equation. The Cauchy gravimetric covariance function has this property (Brochu and Marcotte, 2003). The range parameter and ratios C_0/C (ratio between nugget effect and sill) are chosen by cross-validation (Marcotte, 1995) using KED with the same drift functions used for the final head estimation.

Starting values for C_0/C cross-validation were chosen such as to provide reasonable values for the measurement error variance and head residual variance. Cross-validation parameters were checked by visual assessment of head maps produced by KED. Results were proven very robust to the choice of the model parameters.

For simplicity, we assume the covariance is isotropic. Alternatively, an anisotropic head covariance model could be used. However, our trials with anisotropic models have shown almost no difference with the isotropic model. This shows the strong conditioning imposed by the choice of the characteristics of the conceptual model (first order effect) compared to the fine tuning of the head residual covariance model (second-order effect).

2.4 Performance Criteria

For both synthetic and real case studies, cross-validation is used to quantify the precision of different types of kriging (OK, UK and KED). The statistic retained for comparison is the cross-validation mean absolute error (mae_{cv}), computed from the differences between the observations and the estimates. For the synthetic case, the entire reference head field of the aquifer is known. Therefore, direct comparison with the kriged field allows for the definition of a second statistic: the mean absolute error (mae_d), which is the difference between reference and kriged fields over the whole domain. In this case, hydraulic heads are interpolated at the center of each element and the size of the elements is used to compute the weighted average of the absolute differences between the reference and kriged values. Thus,

$$mae_d(h^K, h^R) = \sum_{j=1}^{n_{ele}} \left|h_j^K - h_j^R\right| \frac{A_j}{A_\Omega} \tag{2.4}$$

where h^K and h^R correspond respectively, to the kriged value and the reference value at the centre of jth element, A_j is the area of jth element, and A_Ω is the area of the entire field.

3 Results

3.1 Synthetic Aquifer

The proposed approach is tested with a synthetic aquifer obtained using the finite element solver Femlab 3.1 (Comsol, 2004). The study area is 3000 × 3000 [L^2]. A contact between two hydrogeological units is found at x = 2500 [L]. Transmissivity in the left-hand unit is 1×10^{-3} [L^2/T] and 4.5×10^{-5} [L^2/T] in the other unit. The right and left boundaries have constant heads (125 m and 110 m, respectively) while the upper and lower boundaries are impervious (no-flow). Two wells are present in the permeable unit yielding local radial flow ($Q_{upper} = -2.8 \times 10^{-3}$ [L^3/T], $Q_{lower} = -9.4 \times 10^{-4}$ [L^3/T]). The numerical solution for this aquifer yields the reference or "real" head field, which is then sampled at 16 locations (Fig. 1a).

When kriging, we use a ratio of $C_0/C=0.001$ and a range of 2090 m. We assume the following: the existence of the two pumping wells, the presence of two different hydrogeological units (the left one being more permeable than the right one), the location of the contact between units, and the presence of impervious boundaries along the upper and lower edges of the field. This information is split into four simple models:

1. A uniform aquifer with a homogeneous transmissivity (5×10^{-4} [L^2/T]) on the entire field;
2. A two-unit aquifer in which the position of the contact is known (offset = 0);
3. Same as 2) with an outflow in the upper well only;
4. Same as 2), with an outflow in the lower well only.

In all four models, heads imposed on the right and left boundaries are different from those of the reference model and produce stronger gradients. Similarly, the transmissivities of the two units are intentionally chosen to produce a contrast 100 times larger than that of the reference model. The above decomposition is far from unique. We have checked that other decompositions of the conceptual model provided equivalent results for this test case. The four head fields obtained are retained by the stepwise procedure (Partial F-test at $\alpha = 5\%$). Thus, KED is performed with

Table 1 Comparison of OK, UK and KED. The contact's location is specified in the conceptual model at the exact location minus offset

	OK	UK linear drift	KED offset 0	KED offset 25	KED offset 100
mae_{cv} (n = 16)	1.3	1.9	1.1×10^{-3}	7.1×10^{-2}	2.5×10^{-1}
mae_d	7.1×10^{-1}	6.9×10^{-1}	2×10^{-3}	6.3×10^{-2}	2.3×10^{-1}

the four head fields used as secondary variables. The resulting field is presented in Fig. 1d. The corresponding mae_{cv} and mae_d are respectively 1.1×10^{-3} and 2×10^{-3}. Such results are within numerical precision. Moreover, the estimates obtained by KED show major improvements over the estimates obtained by OK or UK (first 3 columns of Table 1, Fig. 1b and 1c). UK and OK both strongly underestimate the gradient in the right unit and create a large artificial maximum at coordinates (1000, 1250). In addition, only KED presents a head field compatible with the impervious boundaries (upper-lower) and the constant head boundaries (right-left).

The contact's position was shifted 25 units left on the external drift models (Table 1) to show the effect of uncertainty on this information. This was repeated for a 100 m shift. Even with an offset, the estimates obtained by KED are much more precise than those obtained by OK and UK.

Note that mae_{cv} and mae_d behave alike (Table 1). This suggests that the differences observed at data points by cross-validation (mae_{cv}) are a good estimate of the differences occurring over the whole field (mae_d).

3.2 Real Earth Dams

In this section, the method is applied on two real earth dams (location undisclosed for confidentiality reason). The dams were built using local borrow material: glacial till for the impervious core and various sands for the filters. Dam instrumentation consists of vibrating wire piezometers and of few observation wells. The Precision of measurements is reported to be of the order of magnitude of a meter.

3.2.1 Boundary Conditions

The boundary conditions used in the numerical model are based on a few assumptions: 1) the head loss between the reservoir and the upstream part of the core can be neglected; 2) the foundation is impervious due to grouting; 3) the hydraulic conductivity (K) ratio between the central core and the downstream filter is large enough to produce a seepage face at the interface. As a result, only the core needs to be considered in the model. The position of the seepage point is found numerically using an iterative method described by Chapuis and Aubertin (2001). No-flow conditions prevail above the seepage point, while the head below the seepage point is equal to the free surface elevation.

3.2.2 Kriging

Due to the simplicity of the problem, we use a single external drift model having homogeneous K and the boundary conditions described in 3.2.1.

Dummy values equal to the water level in the reservoir are included on the upstream side of the core and used in all krigings to enforce the constant head condition. These points were not considered in the computation of mae_{cv}. A value of 0.1 was chosen for C_0/C after a few cross-validation trials. This higher ratio compared to the one used with synthetic data reflects measurement error present in real data. Range parameters for Dam 1 and 2 are 23 m and 15 m, respectively.

Head fields for both dams are depicted in Fig. 2 and 3. All mae_{cv} values are compiled in Table 2. Note that kriging for Dam 1 was performed using 15 observations, whereas 12 observations were used for Dam 2.

The mae_{cv} shows strong improvement of KED and UK over OK for both dams. OK solutions (Fig. 2b and 3b) appear totally unrealistic. In both cases, the flow is converging toward a point in the lower part of the core. Head contours suggest flow through the foundation and also through the top of the core. Moreover, the head gradient at the toe is less with OK than with KED.

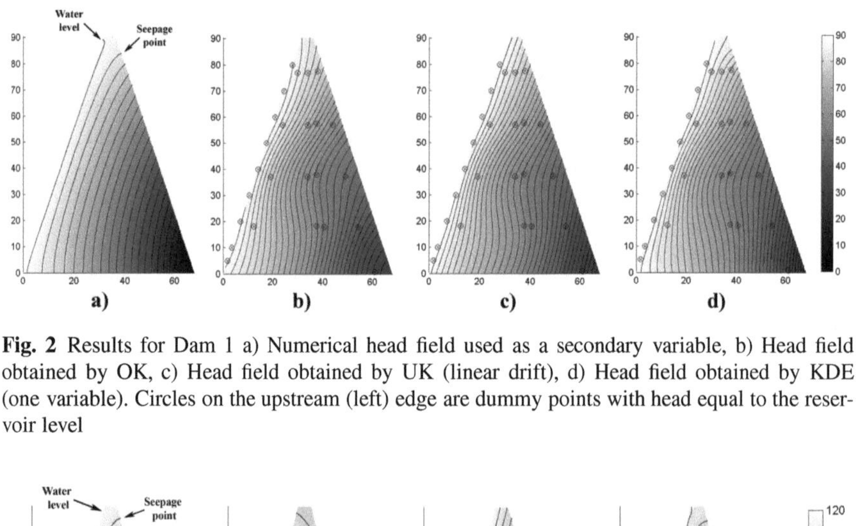

Fig. 2 Results for Dam 1 a) Numerical head field used as a secondary variable, b) Head field obtained by OK, c) Head field obtained by UK (linear drift), d) Head field obtained by KDE (one variable). Circles on the upstream (left) edge are dummy points with head equal to the reservoir level

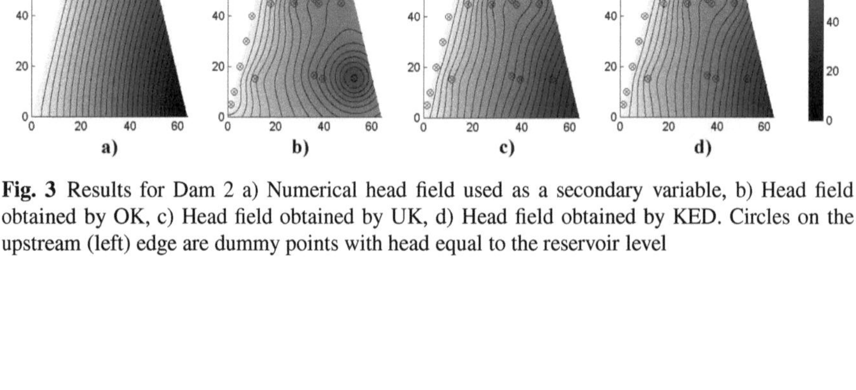

Fig. 3 Results for Dam 2 a) Numerical head field used as a secondary variable, b) Head field obtained by OK, c) Head field obtained by UK, d) Head field obtained by KED. Circles on the upstream (left) edge are dummy points with head equal to the reservoir level

Table 2 Cross-validation results for the dams

Method	Dam 1 - mae_{cv} (m)	Dam 2 - mae_{cv}(m)
OK	3.0	3.9
UK linear drift	2.0	2.2
KED	1.8	2.7

KED and UK provide comparable statistics, especially for the second dam, because the flow simulator solutions (Fig. 2a and 3a) are well approximated by a plane. Comparison of Fig. 2c and 3c to 2d and 3d respectively, also confirms the similarity between the two solutions. However, the no-flow condition is better reproduced by KED than by UK as KED head contours are more perpendicular to the boundaries.

KED does not exactly reproduce the seepage face condition. The head at the bottom downstream extremity of the dam is higher than zero. However, this estimate is consistent with positive pore pressures that are measured at the bottom of the filter, close to the seepage face. This shows that when data is incompatible with the specified conceptual model, KED gives priority to the data but still tries to match the constraints imposed by the model farther from the data (e.g. no-flow at the bottom boundary).

4 Discussion

In the aquifer synthetic case study, KED yielded a very small estimation error. The information required to obtain the numerical fields used in KED is realistic and limited to the main characteristics of the studied field. The values of the Dirichlet boundary conditions, the pumping rate of wells, or the transmissivity of formations need not be known with precision. Rough estimates used in the flow simulation will be implicitly adjusted by KED, as it is known that multiplying the drift function by a constant does not change kriging results. Hence, the choice for constant head values at the left and right boundaries has no impact on the results. In our case study, the transmissivity ratio between the different media could be changed by order of magnitudes without affecting the precision of KED results as long as the left unit is specified as more permeable. The position of the contact between units needs not to be known perfectly either. This demonstrates the great flexibility of the approach and its applicability in routine hydrogeological studies. With the earth dams' cases, we essentially reach the same conclusions as with the synthetic case. In the dams' examples, the UK solution was closer to the KED solution due to the fact that the studied problem possessed a head field close to a plane. As expected, the estimation errors are more important than in the ideal synthetic case. The errors could be due to: measurement errors; incompatibilities with a homogeneous numerical model; information loss in neglecting the effect of the downstream sand filter; assumption of impervious foundations not totally met; lack of information about what is really occurring at the downstream boundary; variations in the geometry of the dam along the third dimension (not modelled here).

Nonetheless, the results obtained by external drift kriging showed noticeable improvements over OK and, to a lesser extent, over UK. New geological information (e.g. foundation permeability; sand filter to consider) could improve the realism of the current numerical model. We stress that, although they are more realistic than OK and UK, the head fields obtained by KED do not totally comply with the conceptual model. As the earth dam case study showed, it depends on the compatibility of the data with the assumed conceptual model(s) used to define the numerical drift(s). In the dam case, data was not compatible with a seepage face along the downstream edge; therefore, the KED solution did not conform to it. However, KED was still able to reproduce faithfully the other boundary conditions. The results of KED could point out some weaknesses of the conceptual model that need improvement. Of course, after such a revision, drift functions could be recomputed and KED re-run.

Note that KED kriging variances provide a (non-local) measure of head uncertainty. Alternatively, head conditional simulations could be done as a way to assess hydraulic conductivity uncertainty in inverse methods, as well as to study the sensitivity to aspects of the conceptual model (e.g. geometry, boundaries) or to covariance parameters.

The approach was applied for the estimation of 3D head fields in the study of synthetic cases (Rivest, 2005). Results obtained (not shown) were in all points similar to the 2D case presented in this work. The proposed method enables fast estimation of entire hydraulic head fields showing a good degree of realism. The use of these fields in direct methods of inversion (e.g. Pasquier and Marcotte (2005, 2006)) is currently investigated.

5 Conclusion

Simple hydrogeological conceptual models, coupled with flow simulation, enable the definition of numerical drift functions to be used in kriging with external drift. The kriged head fields obtained by KED using this approach are more precise and more realistic than those obtained by ordinary kriging and universal kriging with a linear drift. It provides a simple way to transfer most valuable qualitative and semi-quantitative knowledge about the studied field directly in the kriged head maps without using calibration.

Acknowledgments Financial support for this work was provided by NSERC (Canada) research grant. Thanks to Jérémie Gaucher for editing the manuscript. Constructive comments from two anonymous referees helped to improve the manuscript.

References

Ahmed S, de Marsily G (1987) Comparison of geostatistical methods for estimating transmissivity and specific capacity. Water Resour Res 23 (no. 9):1717–1737

Brochu Y, Marcotte D (2003) A simple approach to account for radial flow and boundary conditions when kriging hydraulic head fields for confined aquifers. Math Geol 35 (no. 2): 111–136

Castelier E (1993) Dérive externe et régression linéaire: compte-rendu des journées de géostatistique, Fontainebleau: cahiers de géostatistique, Fascicule 3, pp 47–59

Chapuis RP, Aubertin M (2001) A simplified method to estimate saturated and unsaturated seepage through dikes under steady-state conditions, Can Geotech J 38 (no. 6):1321–1328

Comsol AB (2004) Femlab 3.0 User and reference manual, Stockholm, Sweden

Dehomme JP (1979) Étude de la géométrie du réservoir de Chemery. Internal report, Centre d'informatique géologique, École des Mines de Paris, Fontainebleau

Delhomme JP (1979) Kriging under boundary conditions, Presented at the American Geophysical Union fall meeting, San Francisco

Draper NR, Smith H (1981) Applied regression analysis (2nd edn). Wiley, New York

Galli A, Meunier G (1987) Study of a gas reservoir using the external drift method. In: Matheron G, Armstrong M (eds) Geostatistical case studies, Reidel D, Dordrecht pp 105–120

Marcotte D (1995) Generalized cross-validation for covariance model selection and parameter estimation. Math Geo 27 (no. 6):749–762

Pasquier P, Marcotte D, (2005) Solving the groundwater inverse problem by successive flux estimation. In: Renard et al (eds) GeoEnv 2004: Geostatistics for environmental applications. Springer, Dordrecht, pp 297–308

Pasquier P, Marcotte D, (2006) Steady- and transient-state inversion in hydrogeology by successive flux estimation. Adv Water Resour 29:1934–1952

Rivest M (2005) Rapport de projet de fin d'études, École Polytechnique de Montréal, p 41

Sagar B, Yakowitz S, Duckstein L (1975) A direct method for the identification of the parameters of dynamic nonhomogeneous aquifers, Water Resour Res 11 (no. 4):563–570

Tonkin MJ, Larson SP (2002) Kriging water levels with a regional-linear and point-logarithmic drift, Ground Water 33 (no. 1):185–193

Effect of Sorption Processes on Pump-and-Treat Remediation Practices Under Heterogeneous Conditions

M. Riva, A. Guadagnini and X. Sanchez-Vila

Abstract We analyze the impact of physical and chemical heterogeneity on solute travel time to a pumping well. Environmental applications related to our work include risk evaluation of a pump-and-treat aquifer remediation practice. We consider a non-conservative solute undergoing reversible linear instantaneous equilibrium sorption. Both the distribution coefficient, K_d, and the transmissivity field, T, are considered spatially variable, and are modeled as partially correlated spatial random functions. Groundwater flow and solute transport are then solved within the context of a Monte Carlo framework. Transport of the reactive solute is analyzed within a Lagrangian framework, upon neglecting the influence of local-scale dispersion. From a suite of scenarios, simple expressions of the first two statistical moments of particles travel time to the pumping well are derived as a function of: (i) physical and chemical degree of heterogeneity of the system, and (ii) level of correlation between physical and chemical properties. A key result is that the effects of the chemical and physical heterogeneities on the mean travel time can be decoupled. On the contrary, their relative role in governing travel time variance is more complex, and a separation of the two effects is not observed.

1 Introduction

A most typical aquifer remediation scheme is that of extracting the pollutants dissolved in groundwater through pumping. If the relative impact of the natural background flow is negligible, a fully convergent quasi-radial flow develops. Relevant aspects related to this remediation method include the estimation of the time required to reduce resident concentrations below a given threshold. For this purpose, a key concept is the evaluation of residence time, defined as the time that a particle injected in the system takes to reach the location of the pumping well. Demarcation of drinking well protection regions is also based on this concept.

M. Riva
Dipartimento Ingegneria Idraulica, Ambientale, Infrastrutture Viarie, Rilevamento (DIIAR), Politecnico di Milano, Piazza L. Da Vinci 32, 20133 Milano, Italy.
e-mail: monica.riva@polimi.it

A. Soares et al. (eds.), *geoENV VI – Geostatistics for Environmental Applications*, 153–164

A key issue in the analysis of this problem is the geochemical behavior of pollutants, since several processes lead effectively to an increase in solutes residence time. In particular we are interested in sorptive pollutants, and more specifically, in the impact of heterogeneity of the parameters describing sorption upon residence times. These parameters can be conveniently described by random fields, whose statistics are usually inferred from experimental data. This renders the corresponding flow and transport equations stochastic. As a consequence, solute particles residence time becomes a random variable.

The methodologies most commonly adopted to analyze the stochastic nature of solute residence times within extraction well fields are based either on numerical Monte Carlo simulations (either unconditional or conditional on a variety of information) or on the solution of approximate equations satisfied by the moments of travel times. The Monte Carlo method offers the appealing and convenient feature of allowing deriving information on the complete probability distribution of a given variable, as well as analyzing the response of the system in terms of specific target values of interest. Some target values include the time of first arrival and peak and late arrival times of contaminants to a pumping well.

General aspects for the evaluation of travel time moments of sorptive solutes under non-uniform flow conditions were presented by Cvetkovic et al. (1998) in an unconditional frame and by Sanchez-Vila and Rubin (2003) in a conditional framework.

The work presented here is an extension of Riva et al. (1999, 2006), who presented numerical (Monte Carlo-based) and analytical (based on recursive approximations of moment equations) results rendering the mean and variance of the residence time of a conservative solute in a two-dimensional field under convergent flow conditions. With the same flow configuration, we explore the low-order statistical moments of the residence time of reactive solute particles travelling in a physically and geochemically heterogeneous field. We consider that the solute undergoes reversible linear instantaneous equilibrium sorption, with spatially variable retardation factor, $R(\boldsymbol{x})$. This equilibrium model is based on the assumption that the typical time scale of the chemical reactions is small when compared to the typical time scales of advective-dispersive transport processes. This situation is frequently met in the presence of pumping well fields used in the context of remediation of contaminated aquifers (Valocchi, 1986).

2 Problem Statement

We consider incompressible steady state convergent flow created by a well located at the center of a circular randomly heterogeneous porous domain of radial extent L. The well pumps at a constant deterministic rate, Q. Hydraulic head, H_L, is deterministically prescribed along the outer circular boundary (Fig. 1). Here the quantities of interest are the ensemble mean and variance of the travel time of non-conservative solute particles released at time $t_0 = 0$ at a fixed point of polar coordinates $\mathbf{r}_0 \equiv (r_0, \theta_0)$, assuming the well located at $r = 0$. The location of the injection point is envisioned as a length characterizing the maximum extent of the initial polluted region.

Disregarding the effect of local dispersion, the residence time of a reactive solute particle is given by

$$t = \int_{r_0}^{0} R(\mathbf{r}) \frac{dr}{V_r\,(r, \varphi\,(r, \mathbf{r_0}))} \tag{1}$$

Here, V_r is the radial component of the seepage velocity [LT^{-1}] evaluated along the trajectory, $\varphi\,(r, \mathbf{r}_0)$, of a particle which originates from $\mathbf{r}_0$ (see Figure 1), and R is the retardation factor [−], given by

$$R = 1 + \frac{\rho_b\, K_d}{n}, \tag{2}$$

where K_d is the distribution coefficient [L^3M^{-1}], and n and ρ_b respectively are advective porosity [−] and the bulk density [ML^{-3}] of the solid matrix.

Assuming that transmissivity, T, and K_d are Spatial Random Functions, the residence time becomes a random variable. There are experimental evidences of a certain degree of correlation between T and K_d (Roberts et al., 1986; Robin et al., 1991; Allen-King et al., 1998, 2002). The assumption of an imperfect correlation between these two variables relies on the observation that, while T depends, for a given fluid, only on the characteristics of the solid matrix of the medium, which are in turn related to grain size and orientation, K_d is also a function of the type of contaminant and of the chemical properties of the medium (including pH and organic content). We adopt the following model, postulated by Robin et al. (1991) to describe the relationship between T and K_d:

$$Z = ln K_d = ln K_{dG} + \beta Y' + W. \tag{3}$$

Here, $Y' = Y - \langle Y \rangle$, $\langle \cdot \rangle$ indicates ensemble averaging, $Y = ln\,T$, K_{dG} is the geometric mean of K_d and β is a dimensionless coefficient reflecting the degree of linear correlation between the two variables (Y and Z). The actual value of β

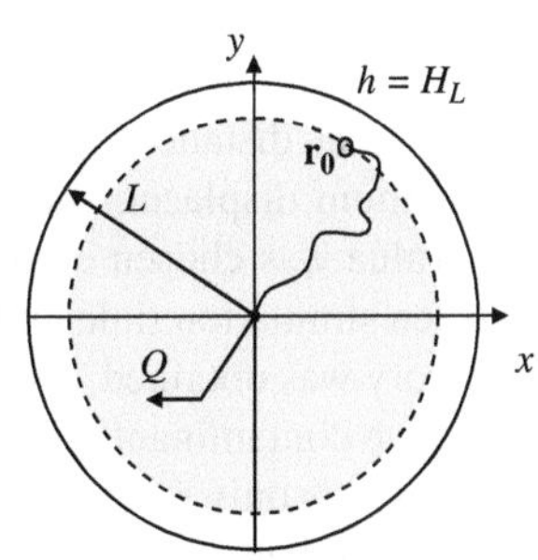

Fig. 1 Sketch of the computational domain

depends on the measurement scale, the solute and the medium mineralogy. It can be either negative (Robin et al., 1991) or positive (Mackay et al., 1986; Allen-King et al., 1998). The quantity W is modeled as a Gaussian process independent of Y and accounts for imperfect correlation between Z and Y. In this sense W includes the effects of factors which are independent from grain size distribution. It also comprises measurement uncertainties.

Without any loss of generality, we model Y and W as statistically independent stationary random fields, with zero mean, variance σ_Y^2 and σ_W^2, respectively, and spherical isotropic covariance structures

$$C_i\left(h\right) = \sigma_i^2 \begin{cases} 1 - 1.5\left(\frac{h}{\lambda_i}\right) + 0.5\left(\frac{h}{\lambda_i}\right)^3 & h \leq \lambda_i \\ 0 & h > \lambda_i \end{cases} \qquad i = Y,\ W, \tag{4}$$

where λ_Y and λ_W are the corresponding correlation scales. On these premises, the autocovariance of Z is provided by the following expression:

$$C_Z(h) = \beta^2 C_Y(h) + C_W(h). \tag{5}$$

Our analysis considers the impact on the statistics of particles travel times of the following quantities: (a) degree of correlation between T and K_d; (b) the degree of heterogeneity of the aquifer; and (c) the values of integral distances I_Y and I_W.

3 Numerical Simulations

Numerical Monte Carlo simulation of flow and particle movement are performed within a square domain of 100 × 100 elements of uniform size ($\Delta x = \Delta y = \delta = 0.2$). A pumping well discharging at a constant rate $Q = 100$ is located at the central node ($x_w = 10$; $y_w = 10$). A circular boundary of radius $L = 50\delta$ was defined around the well by designating all external cells as inactive. Head was prescribed in the outer boundary. The following parameters where fixed during the simulations: $n = 0.3$; $\rho_b = 1.6$; $K_{dG} = 0.28$; and $\sigma_Y^2 = 1.0$. All the above quantities are defined in consistent units.

Flow is solved by Galerkin finite elements using bilinear shape functions. Solute movement in each realization is modeled by Particle Tracking, upon injecting a particle at distance $|\mathbf{r}_0| = 9.0$ from the well. Advection is modeled by using a maximum displacement equal to δ / 1000 between two subsequent time steps. This value was chosen empirically by adjusting it until an acceptable compromise between simulation time and numerical accuracy in the reproduction of the particle trajectory was obtained. We note that, in the absence of dispersion, the trajectory of a reactive contaminant coincides with that of a conservative solute (the retardation factor affects only residence times, not trajectories). Tracking was stopped when the particle reached one of the cells sharing the well node. The remaining travel time was then computed analytically by means of the solution for homogenous media, $t' = \pi n r^2/Q$, r being the remaining distance to the well.

The combined impact of physical and chemical heterogeneity of the aquifer is studied by means of unconditional Monte Carlo simulations for a number of scenarios. We set $\sigma_Y^2 = 1$ and $\lambda_Y = 2.67$, and adopt various values of β ($\beta = -1.0, -0.5, 0.0, 0.5, 1.0$), λ_w ($\lambda_w = 0, \lambda_Y, 10\,\lambda_Y$) and σ_W^2. The 39 unconditional scenarios explored are detailed in Table 1. The values of σ_W^2 were chosen so as to obtain variances of Z, σ_z^2, equal to 0.5, 1.25, 2. With these combinations of parameters, the mean retardation factors range between 2.9 and 5.1 depending on the simulations considered. These values are typical of organic contaminants (where $R \approx (2 \ldots 10)$). Additional runs were performed with $\sigma_Y^2 = \sigma_W^2 = 1$, $I_Y = 10.0$ and two values of λ_w ($\lambda_w = \lambda_Y, 0.1\,\lambda_Y$) and with $\sigma_Y^2 = 0.5, 1.5$ (by keeping $I_Y = I_w = 1.0$).

The unconditional realizations of the independent random variables Y and W were obtained using the Gaussian sequential co-simulator code GCOSIM3D (Gómez-Hernández and Journel, 1993). Each realization constitutes, then, a sample from a multivariate Gaussian, statistically homogeneous field, with ensemble mean $\langle Y \rangle = 0$, $\langle W \rangle = 0$, given variances, and isotropic spherical covariance functions given by (4). The random realizations of Z are obtained from (3).

From preliminary runs, the number of Monte Carlo simulations needed to obtain stable values of the travel time moments increases with (1) the order of the statistical moment considered, (2) increasing σ_Z^2 and I_Z, and (3) decreasing β. We observe that negative values of β have a negative impact on the rate of stability of mean and variance of particles travel time for a given σ_Z^2/σ_Y^2 ratio. This behavior can be explained since the $\beta < 0$ regions imply that large T values are associated with small retardation factors, thus, originating small local residence times for a particle. Contrariwise, low transmissivity regions, associated with large R values, imply large particle residence times. As a consequence, the variance of residence times

Table 1 Main controlling parameters adopted in the Monte Carlo simulations

$I_Y = 1.0$; $\sigma_Y^2 = 1.0$

Run No.	I_W/I_Y	β	σ_W^2	Run No.	I_W/I_Y	β	σ_W^2	Run No.	I_W/I_Y	β	σ_W^2
1		−1	0.25	14		−1	0.25	27		−1	0.25
2			1	15			1	28			1
3		−0.5	0.25	16		−0.5	0.25	29		−0.5	0.25
4			1	17			1	30			1
5			1.75	18			1.75	31			1.75
6	0	0.0	0.5	19	1	0.0	0.5	32	10	0.0	0.5
7			1.25	20			1.25	33			1.25
8			2	21			2	34			2
9		0.5	0.25	22		0.5	0.25	35		0.5	0.25
10			1	23			1	36			1
11			1.75	24			1.75	37			1.75
12		1	0.25	25		1	0.25	38		1	0.25
13			1	26			1	39			1

increases when $\beta < 0$, and a large number of Monte Carlo iterations is needed to attain stability. The opposite behavior is observed for positive β values. We found that convergence is obtained with 5000 Monte Carlo iterations for each unconditional case, with the only exception of the scenario characterized by $\sigma_z^2 = 2.0$ and $I_W/I_Y = 10$, where stability is not reached after 10000 simulations.

4 Dimensional Analysis

Results from Monte Carlo simulations were treated in the framework of dimensional analysis. In this context, the (random) residence time of a non-conservative solute particle can be expressed by the following functional format

$$t = f(Q, L, r_0, H_L, T, K_d, \beta, n, \rho_b). \tag{6}$$

Here, the two random fields T and K_d are fully defined in terms of their geometric mean, variance, shape of covariance function, correlation scale and, eventually, measured values at conditioning points. In the absence of conditioning, the functional form rendering the *i-th* order statistical moment of travel time, $E^i[t]$, becomes,

$$E^i\,[t] = f(Q, L, r_0, H_L, n, \rho_b, T_G, \sigma_Y^2, \lambda_Y, \alpha_Y, K_{dG}, \sigma_Z^2, \lambda_Z, \alpha_Z, \beta). \tag{7}$$

Here α_j (with $j = Y, Z$) is a dimensionless coefficient which accounts for the model adopted for the auto-correlation function, and $T_G = \exp(\langle Y \rangle)$.

We start by noting that the deterministic hydraulic head H_L at the boundary does not affect the contaminant travel time. We then select L, K_{dG} and Q as fundamental quantities and rewrite Eq. (7) in dimensionless format as

$$E^i\left[t\frac{Q}{L^2}\right] = f\left[\frac{r_0}{L}, n, \rho_b K_{dG}, \frac{T_G L}{Q}, \frac{\lambda_Y}{L}, \frac{\lambda_Z}{\lambda_Y}, \sigma_Y^2, \sigma_Z^2, \alpha_Y, \alpha_Z, \beta\right]. \tag{8}$$

In this study we do not explore the dependence of travel time moments on the injection location and consider fixed values for T_G, n and ρ_b. Furthermore, as already indicated, we adopt a spherical covariance function for Y. This implies that we are concerned with the following functional dependences:

$$E^i\left[t\frac{Q}{L^2}\right] = E^i\,[\bar{t}] = f\left[\frac{\lambda_Z}{\lambda_Y}, \frac{\lambda_Y}{L}, \sigma_Z^2, \sigma_Y^2, \alpha_Z, \beta\right]. \tag{9}$$

5 Mean Travel Time

Figure 2 depicts the dependence of mean dimensionless travel time, $\langle \bar{t} \rangle$, on β and σ_Z^2 for $I_W/I_Y = 0$, $I_Y = 1.0$ and $\sigma_Y^2 = 1.0$. The mean value for a conservative solute, $\langle t_{NR} \rangle$, and that corresponding to a solute subject to a constant retardation factor, with $K_d = K_{dG}$ (that is $\langle t_{HR} \rangle = R_{HR} \quad \langle t_{NR} \rangle$, where $R_{HR} = 1 + \rho_b K_{dG}/n$), in a randomly heterogeneous Y field with the same values of I_Y and σ_Y^2, are also reported for comparison.

We note that (i) when Z is heterogeneous the mean residence time is always larger than that obtained when the reactive process is modeled by means of a constant $K_d = K_{dG}$; (ii) $\langle \bar{t} \rangle$ shows a supra-linear increase with the degree of chemical heterogeneity, as expressed by σ_z^2; and (iii) $\langle \bar{t} \rangle$ decreases as β increases, and this effect is more visible for large σ_z^2 values. The latter observation arises from the fact that regions with low transmissivity and large retardation factors (this combination occurs for negative β) dominate the mean residence time. Additional scenarios (not reported) show a negligible effect of I_W/I_Y in mean travel time.

On the basis of the complete set of Monte Carlo-based results displayed in Fig. 3, we propose the following empirical expression, in terms of dimensionless groups, relating the difference between $\langle \bar{t} \rangle$ and $\langle \bar{t}_{HR} \rangle$:

$$\langle \bar{t} \rangle - \langle \bar{t}_{HR} \rangle = \sigma_Z \left[\sigma_Z^2 - \beta \right]. \tag{10}$$

The key result encapsulated in Fig. 3 is that the effects of the geological and chemical heterogeneity on the mean residence time can be separated in the range of parameters studied. While $\langle t_{HR} \rangle$ depends on geological heterogeneity (i.e., σ_Y^2 and L/I_Y) and can be interpreted, for instance, by the analytical solution of Riva et al. (2006), the quantity $\langle \bar{t} \rangle - \langle t_{HR} \rangle$ depends on the degree of geochemical heterogeneity and it appears to be practically insensitive to variations in I_W/I_Y, σ_Y^2 and L/I_Y.

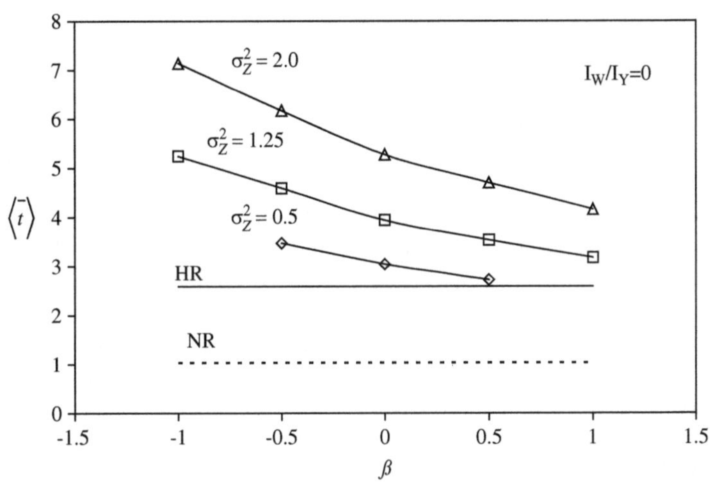

Fig. 2 Dependence of mean dimensionless travel time, $\langle \bar{t} \rangle$, on β and σ_Z^2 for $I_W/I_Y = 0$. The values corresponding to a conservative contaminant (dotted line, labeled NR) and that obtained for constant retardation factor, with $K_d = K_{dG}$, in a randomly heterogeneous Y field (continuous line, labeled HR) are also reported

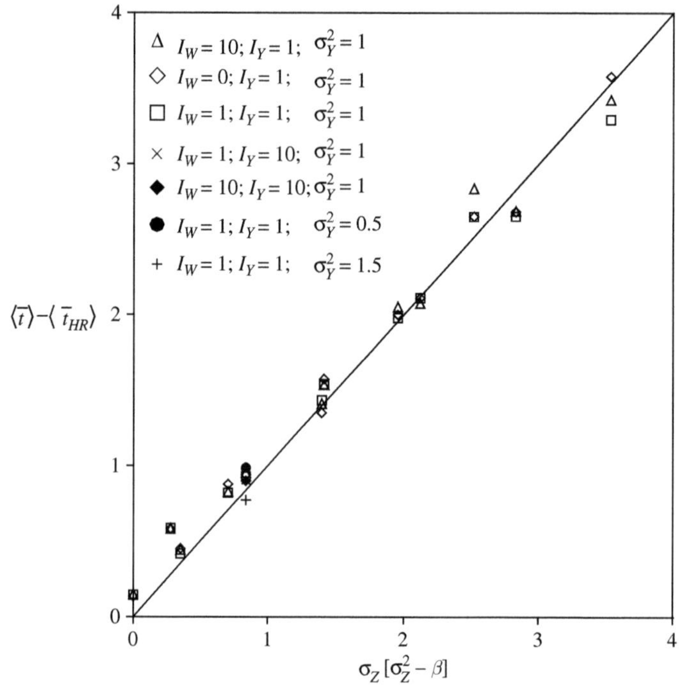

Fig. 3 Mean travel time as a function of $\sigma_z [\sigma_z^2 - \beta]$. The mean values for different scenarios and the 1:1 line are displayed

6 Variance of Travel Time

Figure 4 depicts the variance of the dimensionless log-travel time, $\sigma^2_{ln\bar{t}}$, as a function of β and σ_z^2 for $I_W/I_Y = 0$, 1.0. We present the logarithm of travel time since the travel time distribution in radial flows is more closely related to a log-Normal than to a Normal distribution. The result corresponding to conservative solute (equal to that corresponding to any constant retardation factor) is also displayed.

We note that the variance of log-travel time decreases as β increases. This is consistent with previous observations about the interplay between physical and chemical heterogeneity of the system, as expressed by β. This decreasing in variance is causing the decrease in mean travel time with β, since the travel time distribution is positively skewed. We also note that for positive β values, modeling the subsurface as a chemically homogeneous system can lead to significant overestimation of the uncertainty associated with the process.

The variance of the logarithm of travel time increases with σ_Z^2 and is also strongly affected by the ratio I_W/I_Y. As opposed to what is observed for mean travel time, the variance significantly increases with I_W/I_Y. This observation is consistent with the fact that increasing I_W/I_Y leads to fields with larger spatial persistence of chemical properties. This implies a large variability in probability space. The rate of increase of $\sigma^2_{ln\bar{t}}$ with I_W/I_Y is amplified for large σ_Z^2 values.

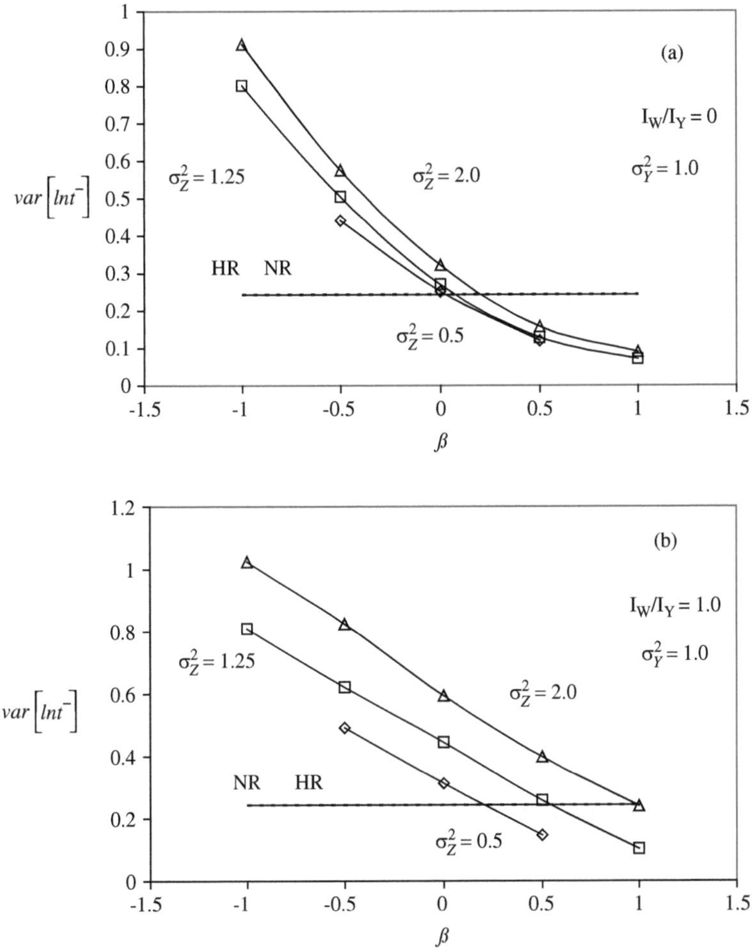

Fig. 4 Variance of natural logarithm of normalized travel time as a function of β and σ_Z^2, with $\sigma_Y^2 = 1$, $I_Y = 1$ and (a) $I_W/I_Y = 0$; (b) $I_W/I_Y = 1$. The values corresponding to a conservative contaminant (dotted line, labeled NR) and that obtained for constant retardation factor, with $K_d = K_{dG}$, in a randomly heterogeneous Y field (continuous line, labeled HR) are also reported

On the basis of the complete set of Monte Carlo simulations we propose the following empirical expression, relating the difference between the variance of the travel times obtained in a geochemically heterogeneous system, $\sigma^2_{ln\bar{t}}$, and that typical of a chemically homogeneous (with $K_d = K_{dG}$) but geologically heterogeneous system, $\sigma^2_{ln\bar{t},NR}$, to the dimensionless groups appearing in (9):

$$\sigma^2_{ln\bar{t}} - \sigma^2_{ln\bar{t},NR} = \beta\sigma_z^{2\varepsilon} + \alpha\,\sigma_z^2\,(1-\upsilon\beta) - \upsilon\,\beta\sigma^2_{ln\bar{t},NR}. \tag{11}$$

Here α, ν, and ε are positive coefficients, which depend only on I_W/I_Y:

$$\alpha = 0.2\, e^{-2\frac{I_W}{I_Y}} - 0.4\, e^{-\frac{I_W}{I_Y}} + 0.3; \qquad 0.1 \le \alpha \le 0.3 \tag{12}$$

$$v = -e^{-2\frac{I_W}{I_Y}} + 2.9\, e^{-\frac{I_W}{I_Y}} + 2.4; \qquad 2.4 \le v \le 4.3 \tag{13}$$

$$\varepsilon = -1.7\, e^{-2\frac{I_W}{I_Y}} + 1.7\, e^{-\frac{I_W}{I_Y}} + 0.3; \qquad 0.3 \le \varepsilon \le 0.7 \tag{14}$$

The values for $\sigma^2_{ln\bar{t},\,NR}$ can be obtained, for instance, by the Monte Carlo-based results of Riva et al. (1999) or by the analytical solution of Riva et al. (2006). The expressions (12)–(14) were derived on the basis of the tests reported in Table 1 (with fixed $I_Y = 1$ and $\sigma_Y^2 = 1$). The synthesis of the results corresponding to all the scenarios analyzed is reported in Fig. 5.

In order to analyze the robustness and reliability of Eq. (11)–(14) for varying degrees of geological heterogeneity of the system, we also juxtaposed to the results of Fig. 5 the outcome of the tests performed by increasing I_Y to 10 (while keeping unit variance of Y) and setting $\sigma_Y^2 = 0.5, 1.5$, while keeping $I_Y = 1$.

The points corresponding to the scenario characterized by $\sigma_z^2 = 2.0$ and $I_W/I_Y = 10$ are not presented, since we did not obtain stable results after 10000

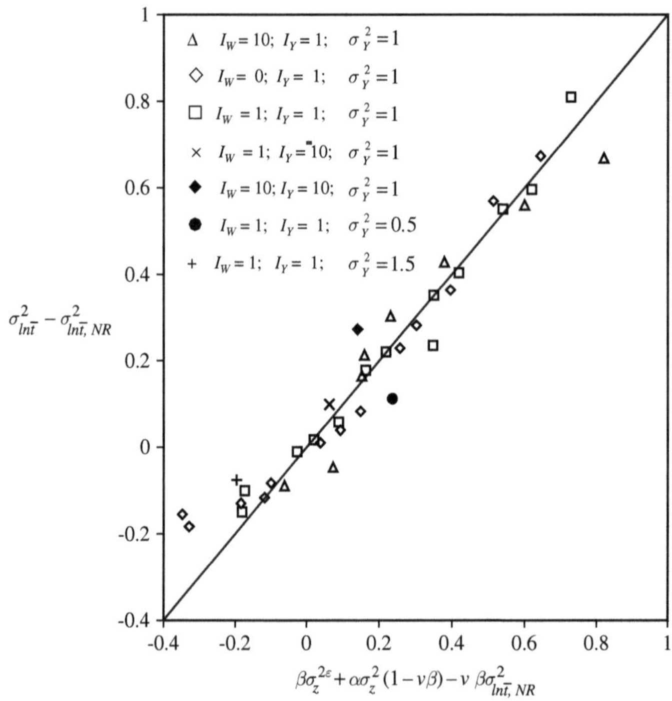

Fig. 5 Variance of log-travel time as a function of the empirical expression provided in (11) with the coefficients obtained from (12) to (14). The values obtained for the different scenarios analyzed through Monte Carlo simulations are presented to show the goodness of the fit

simulations. With reference to the structure of the relationship provided by (11), the following observations can be made:

1. under chemically homogeneous conditions ($\sigma_z^2 = 0$ and $\beta = 0$), the trivial solution $\sigma^2_{ln\bar{t}} = \sigma^2_{ln\bar{t},\,NR}$ can be retrieved;
2. $\sigma^2_{ln\bar{t}}$ display a (generally) nonlinear dependence on σ_z^2;
3. when $\beta = 0$ (i.e., chemical and physical heterogeneities lack correlation) the difference $\sigma^2_{ln\bar{t}} - \sigma^2_{ln\bar{t},\,NR}$ increases linearly with the degree of chemical heterogeneity (σ_z^2) and it is not influenced significantly by the physical heterogeneity;
4. when $\beta < 0$ the effect of the physical heterogeneity is amplified (see the last term appearing in (11)) and $\sigma^2_{ln\bar{t}}$ increases; the opposite holds for $\beta > 0$;
5. the degree of chemical heterogeneity, i.e., σ_z^2, causes the difference $\sigma^2_{ln\bar{t}} - \sigma^2_{ln\bar{t},\,NR}$ to increase if $\beta\sigma_z^{\varepsilon} > \alpha\,(\upsilon\beta - 1)$.

7 Conclusions

The residence time of reactive solute particles displaced in a convergent flow which takes place in a geochemically heterogeneous aquifer is analyzed in the context of a numerical unconditional Monte Carlo framework. In our approach we disregard the effects of local scale diffusion and dispersion processes and model randomness of solute particle trajectories as a process which depends solely on the random physical heterogeneity of the system. On the basis of several synthetic scenarios accounting for the influence of variations in the statistical parameters characterizing $Y = lnT$ and $Z = lnK_d$ (variance, covariance, integral scale and degree of correlation) we derive simple empirical expressions of the first two statistical moments of the travel time of contaminant particles. These expressions relate ensemble mean and variance of particles residence time to (*i*) the physical and chemical degree of heterogeneity of the system, and (*ii*) the level of correlation between physical and chemical properties. A key result is that the effects of the chemical and physical heterogeneities on the mean travel time can be clearly separated. On the other hand, a clear separation of these same two effects on the residence time variance could not be identified. The results of the study are relevant in the context of *a priori* analyses of the efficiency of a pump and treat method in that they are conducive to an estimate of the residual contaminant mass in the groundwater after a given operational time. More generally, they are aimed at supporting the evaluation of the time required for groundwater remediation actions and at providing a measure of the associated risk.

On these basis, the empirical expressions which are derived can be used (in principle) only within the range of parameters examined, for convergent flow conditions and for linear instantaneous reversible sorption. Furthermore, our results do not incorporate the effect of conditioning on different types of information. As such, they can be useful in providing preliminary analyses of the risk involved in pump-and-treat scenarios. More sophisticated design protocols capable of assimilating different types of data, information and/or geochemical processes are then required in the final design stage of the remediation practice.

References

Allen-King RM, Halket RM, Gaylord DR, Robin MJL (1998) Characterizing the heterogeneity and correlation of perchloroethene sorption and hydraulic conductivity using a facies-based approach, Water Resour Res, 34 (3): 385–396

Allen-King RM, Grathwohl P, Ball WP (2002) New modeling paradigms for the sorption of hydrofobic organic chemicals to heterogeneous carbonaceous matter in soils, sediments, and rocks, Adv Water Resour, 25: 985–1016

Cvetkovic V, Dagan G, Cheng H (1998) Contaminant transport in aquifers with spatially variable hydraulic and sorption parameters, Proceedings Royal Soc. London A, 454: 2173–2207

Gómez-Hernández JJ, Journel AG (1993) Joint sequential simulation of multi-Gaussian field. In: Geostatitics Troia'92, vol 1. pp 85–94

Mackay, DM, Freyberg DL, Roberts PV, Cherry JA (1986) A natural gradient experiment on solute transport in a sand aquifer 1. Approach and overview of plume movement, Water Resources Research, 22 (13): 2017–2029

Riva M, Guadagnini A, Ballio F (1999) Time related capture zones for radial flow in two-dimensional randomly heterogeneous media, Stoch. Environ. Res. Risk Assess. 13 (3): 217–230

Riva M, Sánchez-Vila X, Guadagnini A, De Simoni M, Willmann M (2006) Travel time and trajectory moments of conservative solutes in two-dimensional convergent flows, J. Contam. Hydrol., 82: 23–43

Roberts PV, Goltz MN, Mackay DM (1986) A natural gradient experiment on solute transport in a sand aquifer 3. Retardation estimates and mass balances for organic solutes, Water Resour Res, 22 (13): 2047–2058

Robin MJL, Sudicky EA, Gillham RW, Kachanoski RG (1991) Spatial Variability of Strontium Distribution Coefficients and Their Correlation With Hydraulic Conductivity in the Canadian Forces Base Borden Aquifer, Water Resour Res, 27 (10): 2619–2632

Sanchez-Vila X, Rubin Y (2003) Travel time moments for sorbing solutes in heterogeneous domains under nonuniform flow conditions, Water Resour Res, 39(4): 1086, doi: 10.1029/2002WR001399

Valocchi J (1986) Effect of radial flow on deviation from local equilibrium during sorbing solute transport through homogeneous soils, Water Resour Res, 22 (12): 1693–1701

A Stochastic Approach to Estimate Block Dispersivities that Includes the Effect of Mass Transfer Between Grid Blocks

D. Fernàndez-Garcia and J. J. Gómez-Hernández

Abstract Efficiency constraints force the use of a coarse discretization of the numerical transport model compared with the detailed scale required for the most adequate description of the physical properties. Upscaling encompasses the methods that transfer small-scale information to the computational scale. The loss of small-scale information of aquifer properties to construct a numerical model by upscaling largely modifies the true heterogeneous structure of the aquifer compromising the final predictions of solute transport. Within this context, we present extensive Monte Carlo solute transport simulations in heterogeneous porous media to investigate the impact of upscaling on the evolution of solute plumes, and we analyzed the benefits of using enhanced block dispersion tensors in the advection-dispersion equation to compensate for the loss of information. In doing this, we show that when enhancing the block dispersion tensor to compensate for the loss of small-scale information, mass transfer between grid blocks is in turn amplified largely reducing macrodispersion in the upscaled model. We conclude that block dispersivities should consider not only the fluctuation of aquifer properties inside the block but also the simultaneous effect of enhanced mass transfer between all blocks of the numerical model. Then, using a stochastic approach, we present a new concept of block dispersivity that accounts for both effects: block heterogeneity and mass transfer between grid blocks. As a result, we quantified the amount of contribution that mass transfer effects has on block dispersivity.

1 Introduction

In order to efficiently make solute transport predictions in real field settings, complex transport models cannot afford to describe heterogeneity at the necessary detail scale required for an adequate description of the underlying processes. As a result, models are often used with a coarse grid discretization of the media. This implies a

D. Fernàndez-Garcia
Polytechnic University of Valencia, Ingeniería Hidráulica y Medio Ambiente, Camino de Vera s/n., 46022 Valencia, Spain
e-mail: dafernan@dihma.upv.es

A. Soares et al. (eds.), *geoENV VI – Geostatistics for Environmental Applications*, 165–173

simplification of the physical problem, since not all the subgrid information on the spatial variability of the parameters is transferred to the numerical grid. In this context, upscaling is used to transfer small-scale information to the computational scale.

We present Monte Carlo solute transport simulations in heterogeneous porous media to investigate the impact of upscaling on the evolution of solute plumes. We show that usual upscaled transport models can largely underestimate the spreading of solute plumes even if block dispersivities are calculated as being representative of within-block heterogeneity. Two major effects were identified that can restrict the growth of the solute plume in the numerical upscaled model: (i) tensorial nature of hydraulic conductivity; and (ii) mass transfer effects between blocks of the numerical model. This paper focuses on the latter effect. In particular, we investigate the concept of block dispersivity and its relation with mass transfer between grid blocks.

2 Computational Investigations

2.1 Design of Solute Transport Monte Carlo Simulations

Transport simulations consider a square bidimensional confined aquifer with uniform mean flow in the x-direction. The domain extends 240 units in the x and y directions. Boundary conditions were no-flux for boundaries parallel to the mean flow and constant-head otherwise (mean hydraulic gradient J equal to 0.01). At the small scale, the hydraulic conductivity tensor is isotropic. The aquifer is heterogeneous and described by a spatially varying hydraulic conductivity such that the lnK(**x**) follows a multi-Gaussian random function. The geometric mean of the lnK(**x**) field is $K_G = 1$. The random function model is described by an isotropic exponential covariance function with the correlation scale (λ) set to 4 units. A very fine grid is used to generate a reference lnK(**x**) field through the GCOSIM3D code (Gómez-Hernández and Journel 1993) representing the real aquifer. The resolution of the fine-scale model is 4 grid-cells per correlation scale. The Monte Carlo transport simulation scheme consists in 50 realizations for each $\sigma^2_{\ln K}$ that ranged from 0.06 to 4. Each realization of the lnK(**x**) field is upscaled to a resolution referred to as 30-by-30, which correspond to the upscaling process of transferring the small-scale information (240×240 cells) to a regular computational grid of 30×30 blocks. For simplicity, at the fine-scale the transport model is purely advective. After upscaling, at the coarse-scale, solute transport is governed by the advection-dispersion equation that is used with an equivalent block hydraulic conductivity tensor and an equivalent block dispersivity tensor.

The impact of upscaling was then evaluated by comparing Monte Carlo solute transport simulations of a large plume moving through the reference lnK(**x**) fields with their corresponding upscaled model. A seven-point finite difference groundwater flow model, MODFLOW2000 (Harbaugh et al. 2000) was used to solve the flow problem and a transport code based on the Random Walk Particle Method (Fernàndez-Garcia et al. 2005) was used to simulate solute transport. Transport simulations start by injecting a large number of particles (5,000) equidistantly distributed in a line transverse to the mean flow direction. This line is 35λ long and is centered with respect to the transverse dimension (Fig. 1). The first arrival time and the position of particles passing through 20 control planes transverse to the mean flow direction and located at several distances away from the source were tracked until particles exited the lower constant head boundary. This allowed measuring longitudinal macrodispersivity at control planes. We calculate macrodispersivities from Monte Carlo simulations by using the method of temporal moments as applied to particle tracking transport codes in Fernàndez-Garcia et al. (2005).

We note that the objective of this work is not to examine the performance of solute transport under different choices of boundary and initial conditions, but to evaluate transport behavior with a change of support scale under the same conditions. Thus, we consistently estimate block properties (upscaling rule) and solute transport behavior always using a slug injection to ultimately estimate transport behavior/properties through flux-averaged concentrations.

2.2 Upscaling Methodology

The selected method for the calculation of block hydraulic conductivities $\mathbf{K}_b$ is known as the Simple Laplacian method with skin (Wen and Gómez-Hernández 1996). For a given realization of the lnK(**x**) field, the region being upscaled is isolated from the rest

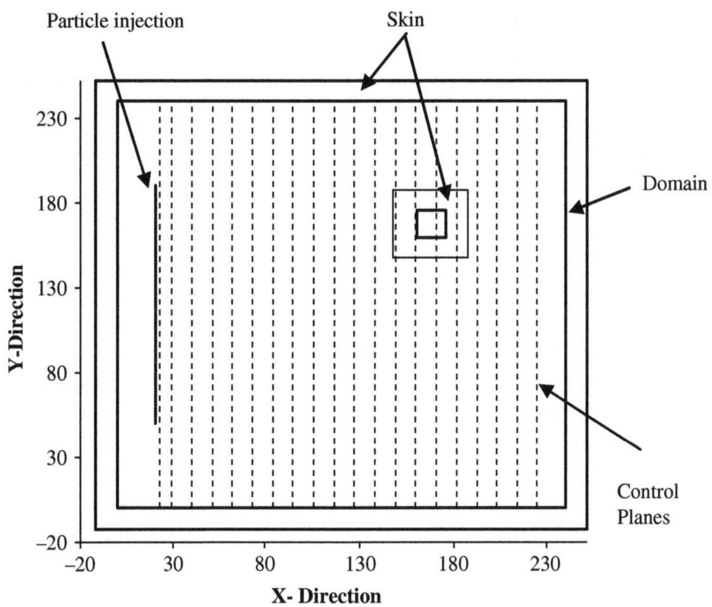

Fig. 1 Sketch of transport simulations showing the initial location of particles, the control planes where mass fluxes are measured

of the system. This region not only comprises the portion of the heterogeneous aquifer delineated by the grid-block, but also includes a small portion of the heterogeneous aquifer adjacent to the grid-block referred to as the skin (Fig. 1). The skin is designed to approximately emulate the original water head boundary conditions on the grid-block without having to solve the flow problem for the entire domain. In this work, the skin spans over 3λ to minimize boundary effects and block hydraulic conductivity $\mathbf{K}_b$ is assumed to be a diagonal second-order tensor with principal directions parallel to the block sides. The principal components were calculated as

$$K_{b,ii}(\mathbf{x}) = \frac{\int\limits_{V(\mathbf{x})} q_i(\mathbf{u})d\mathbf{u}}{\int\limits_{V(\mathbf{x})} -\partial h/\partial x_i(\mathbf{u})d\mathbf{u}} \quad (1)$$

where V(x) denotes the volume of the grid-block the centroids of which is at **x**, **q** is the darcy velocity, and h is the piezometric head. We considered a diagonal $\mathbf{K}_b$ tensor for being the usual assumption underlying most benchmark groundwater flow models. We proposed a numerical method that evaluates block dispersivities by simulating a natural-gradient tracer test inside the isolated block region, so that the solute tracer only samples the block heterogeneity. This choice stems from the fact that field tracer tests are often attempted as a means of estimating input dispersivities for transport models. For each isolated block with skin, steady-state flow is achieved using the needed boundary conditions (i.e. linearly varying pressure head at the block boundaries) to originate a mean flux equal to the block averaged one. Then, a Dirac-input tracer line source is injected in a line transverse to the block averaged flux and situated at the upgradient limit of the block. Block dispersivity values, A_L and A_T, are estimated from the mass flux breakthrough curve by the method of temporal moments as,

$$A_L = \frac{L}{2}\frac{\sigma_t^2}{T_a^2} - \alpha_L \quad (2)$$

$$A_T = \frac{\sigma_y^2}{2L} - \alpha_T \quad (3)$$

where L is the size of the block in the mean flow direction, α_L and α_T are respectively the longitudinal and transverse local dispersivity, $\sigma^2{}_t$ is the variance of travel times of particles exiting the block, $\sigma^2{}_y$ is the variance of transverse displacements of particles exiting the block, and T_a is the mean arrival time.

2.3 Simulation Results

Figure 2 shows the scale-dependence of longitudinal dispersivity as a function of travel distances for different $\sigma^2{}_{\ln K}$ and transport models. We distinguish two important features. At early times, when particles have still not travelled through various

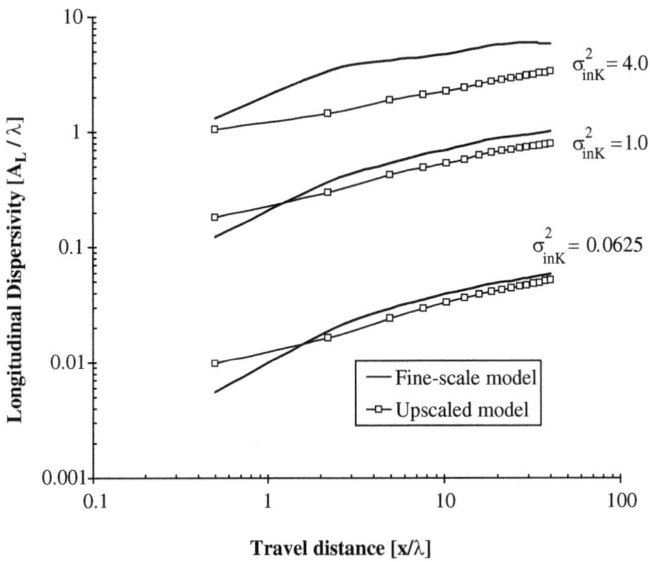

Fig. 2 Comparison of the scale-dependence of longitudinal macrodispersivity obtained from fine-scale simulations with those for the upscaled model

grid-blocks, dispersivities are larger than those corresponding to the fine-scale model mainly because block dispersivity was not considered a time-dependent parameter but represents the time-average behavior within a block. This effect rapidly vanishes when particles passes through few blocks.

At late times, block dispersivity approaches an asymptotic dispersivity value that is significantly smaller than those for the fine-scale model. Asymptotic behavior is seen in all cases yet not clearly appreciated in Fig. 2 because we used logarithmic scales. This can be attributed to several effects (Fernàndez-Garcia and Gómez-Hernández 2006). Among them, we suspect that mass transfer interaction between grid-blocks of the numerical model and the tensorial nature of hydraulic conductivity can largely affect the behavior of solute transport in the upscaled model. For instance, it is suspected that when enhancing the block dispersion tensor to compensate for the loss of velocity variability through upscaling, transverse mass transfer between grid blocks is in turn amplified reducing macrodispersion. This issue is analyzed in the following sections based on the small perturbation stochastic approach.

3 A Stochastic Approach to Estimate Block Dispersivities

3.1 General Formulation

This section presents a novel stochastic approach to the problem of upscaling dispersivity in heterogeneous formations. Considering a nonreactive solute plume moving through a stationary lnK random field under steady-state mean uniform flow parallel to the x_1 coordinate, the effective spatial moments of a solute at large travel distances are written as (e.g., Gelhar and Axness 1983)

$$\lim_{x_1 \to \infty} \frac{1}{2} \frac{M_{ij}(x_{m,1})}{x_{m,1}} = \left(\alpha_i + \frac{\tau D_d \phi}{q_m, 1} \right) \delta_{ij} + \sigma_f^2 \lambda_1 B_{ij}(\alpha_i, \lambda_i) \quad (4)$$

where $x_{m,1}$ is the mean travel distance, M_{ij} is the effective spatial moment tensor, α_i are the local dispersivity coefficients, ϕ is the porosity, τ is the tortuosity, $\sigma^2{}_{\ln K}$ is the variance of the natural log of K, λ_i are the correlation scales in the ith-direction, and B_{ij} is a real function expressed as

$$B_{ij} = \frac{1}{\sigma_{\ln K^2 \lambda_1}} \int_{-\infty}^{+\infty} \frac{1}{q_{m,1}^2 \beta(\mathbf{k})} S_{q_i q_j}(\mathbf{k}) d\,\mathbf{k} \quad (5)$$

$$\beta(\mathbf{k}) = ik_1 + \alpha_i k_i^2 \quad (6)$$

where $S_{\mathbf{qq}}$ is the spectrum of the darcy velocity field. Einstein's convention is used. Following Dagan (1994), the basic requirement for upscaling is that the statistics of the spatial moments at the fine-scale should be the same as those obtained in the upscaled model,

$$\frac{1}{2} \frac{M_{ij}(x_{m,1})}{x_{m,1}} = \frac{1}{2} \frac{M_{ij}^m(x_{m,1})}{x_{m,1}} \quad (7)$$

where the superscript m denotes that the quantity is related to the simulated values given by the numerical model. The problem is reduced to resolve M^m. To achieve this, we view the process of upscaling the hydraulic conductivity field as a filtering process. The filter is such that suppresses the high frequency lnK fluctuations that cannot be represented by the numerical model. We used a low-pass filter function denoted as $F_g(\mathbf{k})$. In commercial groundwater systems based on the classical advection-dispersion equation, the increase in block dispersivity to account for a coarse discretization is directly translated in the large time growth of spatial moments as

$$\lim_{x_1 \to \infty} \frac{1}{2} \frac{M_{ij}(x_{m,1})}{x_{m,1}} = \left(\alpha_i + \frac{\tau D_d \phi}{q_{m,1}} \right) \delta_{ij} + A_i^b \delta_{ij} + \sigma_f^2 \lambda_1 B_{ij}^m(A_i^b, \alpha_i, \lambda_i) \quad (8)$$

$$B_{ij}^m = \frac{1}{\sigma_{\ln K}^2 \lambda_1} \int_{-\infty}^{+\infty} \frac{1}{q_{m,1}^2} \beta^m(\mathbf{k}) F_g(\mathbf{k}) S_{q_i q_j}(\mathbf{k}) d\,\mathbf{k} \quad (9)$$

$$\beta^m(\mathbf{k}) = ik_1 + \alpha_i k_i^2 + A_i^b k_i^2 \quad (10)$$

where A^b denotes the increase in block dispersivity due to block heterogeneity. Since the terms (real part) multiplying the velocity spectrum in the integration of B_{ij} and $B^m{}_{ij}$ are all even functions, and knowing that $S_{\mathbf{qq}}$ is odd if $i \neq j$, in this case, the requirement of upscaling (at large travel distances) is fulfilled and reduced to the following nonlinear system of n equations with n unknowns ($A^b{}_i$), being n the dimension of the problem,

$$\sigma^2_{\ln K}\lambda_1 B_{ii}(\alpha_i, \lambda_i) = A^b_i + \sigma^2_{\ln K}\lambda_1 B^m_{ii}(A^b_i, \alpha_i, \lambda_i, F_g) \quad i = 1, \ldots, n \tag{11}$$

3.2 Evaluating Mass Transfer Effects on Block Dispersivity

In this section we discuss the influence of mass transfer effects on block-effective dispersivities for the case of a two-dimensional aquifer with an isotropic exponential covariance function. We only focus on the underestimation of the longitudinal spatial moment due to mass transfer effects in the upscaled model. Thus, the objective is not to exactly solve the coupled system of equations but to understand and quantified mass transfer effects in modeling solute transport with a numerical code. Assuming an isotropic dispersivity, i.e., $A^b{}_i = A^b$ and $\alpha_i = \alpha$, the problem of upscaling dispersivity is simplified to find the root of the following equation,

$$\sigma^2_{\ln K}\lambda_1 B_{11}(\alpha, \lambda) - A^b - \sigma^2_{\ln K}\lambda_1 B^m_{11}(A^b, \alpha, \lambda, F_g) = 0 \tag{12}$$

To obtain simple analytical expressions of B^m, we employed a low-pass filter function similar to Rubin's Nyquist model (Rubin et al. 1999), defined as a function that takes the value of unity if $|\mathbf{k}| \leq \pi/\Delta$, where Δ is the domain discretization, assumed constant for all directions. It can be shown that this definition, which is mathematically convenient, yields analytical solutions which effectively behave as Rubin's Nyquist model for negligible mass transfer (A^b approaching zero in $B^m{}_{11}$). Defining $\varepsilon = (\alpha + A^b)/\lambda$ and $\xi = \pi/(\Delta/\lambda)$, using the relationship between the velocity spectrum and the lnK spectrum (Gelhar and Axness 1983), and expressing B_{11} in polar coordinates, we obtain after integration,

$$B^m_{11}(\varepsilon, \xi) = \frac{1}{2}\int_0^{\xi} (-3\varepsilon z - 2\varepsilon^3 z^3 + 2(1 + \varepsilon^2 z^2)^{3/2})(1 + z^2)^{-3/2} dz \tag{13}$$

Using (13) in (12), we solve for A^b by means of finding the root of equation (12). Knowing that $\alpha << A^b$ for general aquifer conditions, we consider α negligible in the analysis. We quantified the contribution of mass transfer to block dispersivity using the relative increase in A^b due to mass transfer defined as,

$$\varepsilon_r = \frac{A^b(\varepsilon, \Delta) - A^b(\varepsilon = 0, \Delta)}{A^b(\varepsilon = 0, \Delta)} \times 100 \tag{14}$$

Figure 3 shows the relative increase in A^b due to mass transfer effects as a function of the size of grid-block, Δ/λ, and degree of heterogeneity, $\sigma^2_{\ln K}$, for the case of

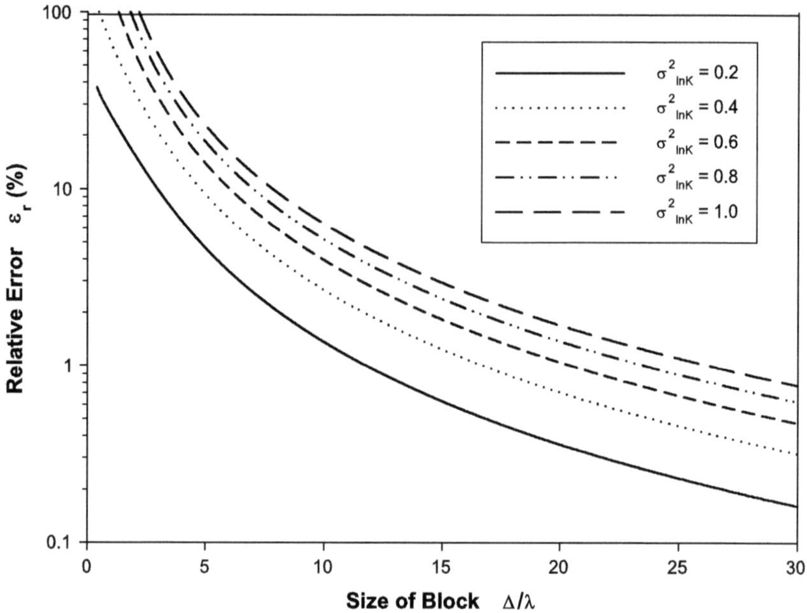

Fig. 3 Relative increase in block dispersivity due to mass transfer as a function of size of grid-block and degree of heterogeneity for the case of a two-dimensional isotropic exponential covariance function

a two-dimensional isotropic exponential covariance function. The relative contribution of mass transfer decays with block size and $\sigma^2_{\ln K}$. Note that as Δ/λ approaches zero from the right ε_r tends to infinity because block dispersivity with negligible mass transfer $A^b(\varepsilon{=}0,\Delta)$ approaches zero faster than $A^b(\varepsilon,\Delta)$. For the usual case of block sizes of few correlation scales, we see that the contribution of mass transfer is very important, for instance, ε_r is about 57 per cent for the case of $\Delta/\lambda{=}3$ and $\sigma^2_{\ln K}{=}1$.

We note that in realistic modeling applications the form of the dispersivity tensor is usually anisotropic with $A^b_1/A^b_2 \approx 10$. In this case, mass transfer effects can be significantly decreased, being less dramatic. Nonetheless, our combined numerical-analytical approach suggests that mass transfer interaction between blocks of the numerical model should be taken into consideration when quantifying block dispersivity values.

4 Conclusions

We used an analytical stochastic approach combined with Monte Carlo simulations to study the meaning of block dispersivity as frequently utilize in modeling contaminant transport. Specifically, we focus on the relationship between mass transfer between blocks of the numerical model and block dispersivities. We found that block dispersivities used as input in transport models should not only reflect the underlying heterogeneous structure filtered out by the model but also consider the effect of mass transfer and the interaction between grid-blocks of the numerical model. These effects are estimated to be significant with decreasing block size and increasing degree of heterogeneity.

References

Dagan G (1994) Upscaling of dispersion coefficients in transport through heterogeneous porous formations, In: Peters A et al (eds) Computational methods in water resources X, Kluwer Acad, Norwell, Mass pp 431–439

Fernàndez-Garcia D, Gómez-Hernández JJ (2006) Impact of upscaling on solute transport: travel times, scale-dependence of dispersivity and propagation of uncertainty. Water Resour Res doi: 10.1029, In review

Fernàndez-Garcia D, Illangasekare TH, Rajaram H (2005) Differences in the scale-dependence of dispersivity and retardation factors estimated from forced-gradient and uniform flow tracer tests in three-dimensional physically and chemically heterogeneous porous media, Water Resour Res 41, W03012, doi:10.1029/2004WR003125

Gelhar LW, Axness CL (1983) Three-dimensional stochastic anlysis of macro dispersion in aquifers. Water Resour Res 19(1):161–180

Gómez-Hernández JJ, and Journel AG (1993) Joint simulation of multi-Gaussian random variables. In: Soares A (ed) Geostatistics Tróia'92, vol 1. Dordrecht, Kluwer, pp 85–94

Harbaugh AW, Banta ER, Hill MC, and McDonald MG, MODFLOW-2000, The US Geological Survey Modular Ground-Water Model–user guide to modularization concepts and the ground-water flow process, Open-file Report 00-92

Rubin Y, Sun A, Maxwell R, and Bellin A (1999) The concept of block effective macrodispersivity and a unified approach for grid-scale- and plume-scale-dependent transport. J Fluid Mech 395:161–180

Wen X-H, and Gómez-Hernández JJ (1996) Upscaling hydraulic conductivities in heterogeneous media: An overview. J of Hydrology, 183 (1–2):ix–xxxii

Fracture Analysis and Flow Simulations in the Roselend Fractured Granite

D. Patriarche, E. Pili, P. M. Adler and J.-F. Thovert

Abstract Understanding of flow and transport in fractured media requires a good knowledge of fractures and fracture networks, which are privileged pathways for water and solutes. The Roselend underground laboratory (French Alps) gives the opportunity to fully investigate flow and solute transport through such a medium. Fracture traces and water fluxes have been determined along the Roselend tunnel.

The major objectives of this work are to derive a three dimensional fracture network consistent with the observations to calculate its percolation properties, and the macroscopic permeability of the medium.

In the tunnel, fractures can be classified into two families of large and small fractures. While large fractures intersect entirely the tunnel, small fractures partially intersect it. Variograms of both trace length and fracture orientation do not show any significant correlation with distance along the tunnel axis or with distance between fractures.

A stereological analysis of the trace length probability densities of the small fractures provides the fracture diameter probability density distribution which is best described by a power law. The large fractures are assumed monodisperse, with a radius equal to 5 m. Numerical simulations show that the networks obtained by combining large and small fractures do percolate while networks constituted of small fractures only do not percolate.

For three different sections along the gallery reflecting the major contrasts in dripping water fluxes, fracture networks are repeatedly generated according to the observed fracture densities. The permeability of these networks is systematically calculated. Results compare well to conductivity properties of similar media and show good consistency with observed water fluxes.

D. Patriarche
Commissariat à l'Energie Atomique, Département Analyse Surveillance Environnement, BP 12, 91680 Bruyères-le-Châtel, France. Now at Gaz de France, Saint-Denis La Plaine Cedex, France
e-mail: delphine.patriarche@gazdefrance.com

A. Soares et al. (eds.), *geoENV VI – Geostatistics for Environmental Applications*, 175–190

1 Introduction

Fractures determine the macroscopic behavior of many natural rocks by impacting their mechanical, hydraulic, and transport properties. In hard rock environments such as gneissic or granitic media, the role of fractures in flow and transport is enhanced since fractures are privileged pathways to water flow and in a first approach, matrix can be considered impervious. The main objectives of this work is to show that field measurements of fractures associated to water fluxes records in an underground cavity can be used for characterizing the hydraulic behavior of a fractured medium. This is illustrated through the example of the Roselend site (French Alps) where the fractured granite is investigated thanks to the presence of a dead-end tunnel (Pili et al. 2004, Provost et al. 2004).

As shown in many studies (e.g., Berkowitz 1994, Koudina et al. 1998), the crucial feature for fluid flow and solute transport is the connectivity of fractures. Characterization of fracture networks is challenging since full three-dimensional representation of fractures is generally impossible to obtain from direct in-situ measurements, which most of the time are scarce.

A useful approach to obtain such a three-dimensional representation is to extrapolate one- or two-dimensional measurements of fractures through stereological analysis. Usually, one-dimensional and two-dimensional field data are obtained from intersections of fractures with a borehole (Sisavath et al. 2004), and with a ground surface or an accessible fault plane (e.g., Warburton 1980a, b, Piggott 1997, Berkowitz and Adler 1998). Additionally, pseudo tri-dimensional field information may also be obtained from intersections of fractures with an underground tunnel (Mauldon and Mauldon 1997, Peacock et al. 2003, Gupta and Adler 2006).

Percolation which occurs above a certain fracture density, is an essential property since a fracture network can be permeable only if it percolates. Bogdanov et al. (2003), Huseby et al. (1997), and Koudina et al. (1998) have shown that a dimensionless fracture density can be defined and that it yields a density threshold independent of the fracture shape.

The Roselend site is described in Section 2. In Section 3, fracture orientations and traces in the tunnel are analyzed. Fractured networks for three zones in the tunnel are generated and their percolation probability is calculated through Monte-Carlo simulations. In Section 4, the macroscopic permeability of the reconstructed tri-dimensional fracture networks is assessed through flow simulations and compared with the observed water fluxes along the tunnel.

2 The Roselend Site

The Roselend tunnel is located in the French Alps, 25 km southwest from Mont Blanc. The tunnel entrance is at an altitude of 1576 m, close to the west shore of the artificial Roselend Lake, and 19 m above the lake at its highest water level. This dead-end tunnel was drilled in the granites, gneisses and micaschists of the Méraillet massif. The thickness of rock overburden increases from 7 m at the tunnel entrance

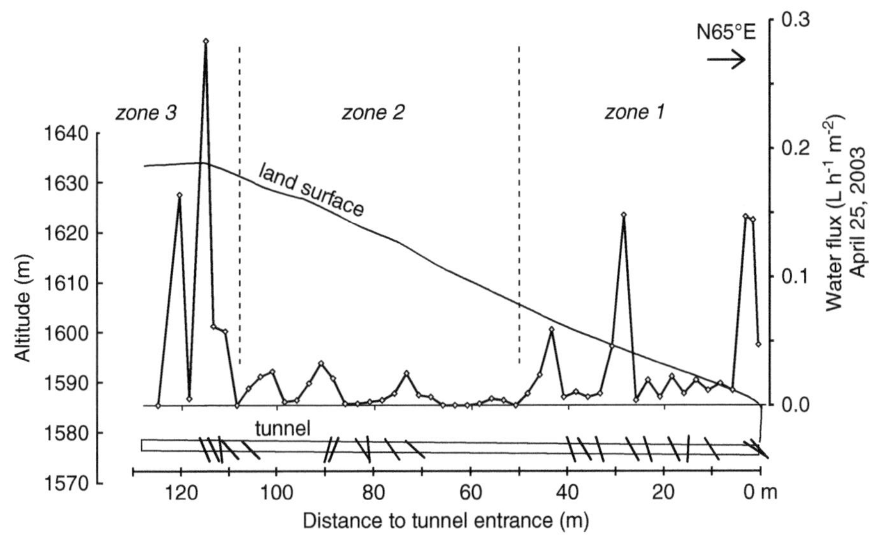

Fig. 1 Vertical cross-section of the granitic massif along the tunnel. Water fluxes measured in the tunnel are reported. Zones 1, 2, 3 for which flow is qualitatively considered as mild, low and high, respectively, are indicated. Black bars in the tunnel represent fractures fully intersecting it

to 55 m at the tunnel closed end (Fig. 1). The 128 meter-long tunnel has a roughly cylindrical shape with a 2.4 m diameter ϕ_t. The average air temperature in the tunnel is stable at 6.8 ± 0.2°C over the whole year. The air relative humidity is close to saturation.

Fractures traces along the tunnel walls larger than 20 cm and with an observable face were digitized and georeferenced (Dezayes and Villemin 2002). Along the gallery cylinder, the metric fracture density varies from 1.8 to 0.6 per meter. Fracturation along the tunnel is heterogeneous, with 21 large fractures entirely intersecting the tunnel (Fig. 1), and 172 small ones only partially intersecting it.

The high-range and steep mountainous environment is characterized by contrasted precipitation regimes with alternating snow, rain, and drought periods. Seasonal flow dynamics arise from dominant rain and melted snow infiltrations from late summer to mid-spring, increasing water content in the medium.

A profile of water fluxes measurements along the tunnel (Fig. 1) was performed. Three zones with contrasted hydrological behavior can be identified: from 0 to 50 m, 50 to 108 m, and 108 to 128 m from the tunnel entrance.

3 Trace Length Analysis and Percolation of Fracture Network

3.1 Fracture Trace Lengths and Orientations

Small fracture trace length is determined from starting and ending point coordinates of the considered trace and from the tunnel shape. Large fracture trace length is determined from the fracture normal dip angle (i.e. the angle in the vertical plane between the normal to the fracture, and its horizontal projection) and from the tunnel shape.

The histogram of the 193 fracture trace lengths clearly shows two distinct populations (Fig. 2a). 172 fractures present trace lengths much lower than $\pi\phi_t$ the tunnel circumference, and the 21 fractures recognized on the field since large fractures, show trace lengths larger than $\pi\phi_t$. The lack of trace lengths between 3.2 and 7.5 m strongly suggests that the fracture network consists of two families of fractures.

Lengths can be made dimensionless by dividing by R_t, the cylinder radius, and are denoted with a prime. For instance metric trace lengths c can be written as $c' = \frac{c}{R_t}$.

The orientational distributions of small and large fractures are represented in Fig. 3. Three or even four subfamilies could be distinguished for the small fractures. However, in some recent studies (Gonzalez-Garcia et al. 2000, Sisavath et al. 2004, Thovert and Adler 2004), it has been shown that the properties of networks composed of a few subfamilies were very close to the properties of networks with an isotropic distribution. Therefore, the orientations of the small fractures will be assumed to be isotropic in most of this paper.

This is not true for the large fractures which seem to be composed of two main subfamilies. Three variants can be proposed to model these orientations. A first one which is not realistic in view of Fig. 3, consists in assuming an isotropic distribution. A second one consists in assuming that all the large fractures are vertical and perpendicular to the tunnel axis. A third one consists in choosing at random the orientation of each large fracture among the 21 orientations displayed in Fig. 3. These three variants will be called isotropic, vertical and discrete, respectively.

3.2 Partial Intersections

To reconstruct in three dimensions the small fracture population, correlations considering distance between fractures along the tunnel, and considering fracture event

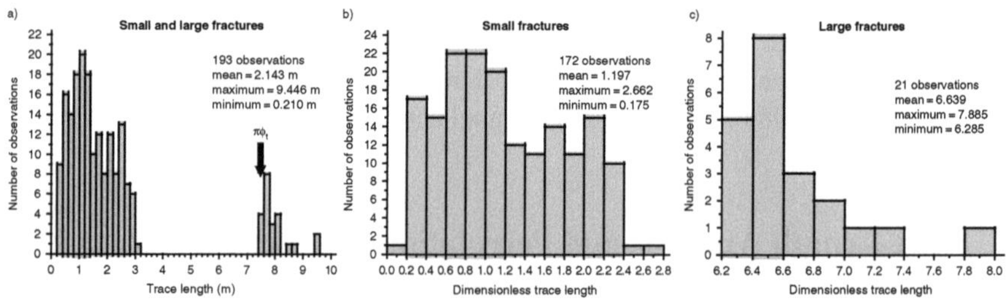

Fig. 2 (a) Histograms of metric fracture lengths observed in the tunnel. The arrow indicates the minimal trace length of a fracture fully intersecting the tunnel. Histograms of dimensionless fracture lengths for (b) small fractures, partially intersecting the tunnel, and for (c) large fractures fully intersecting the tunnel

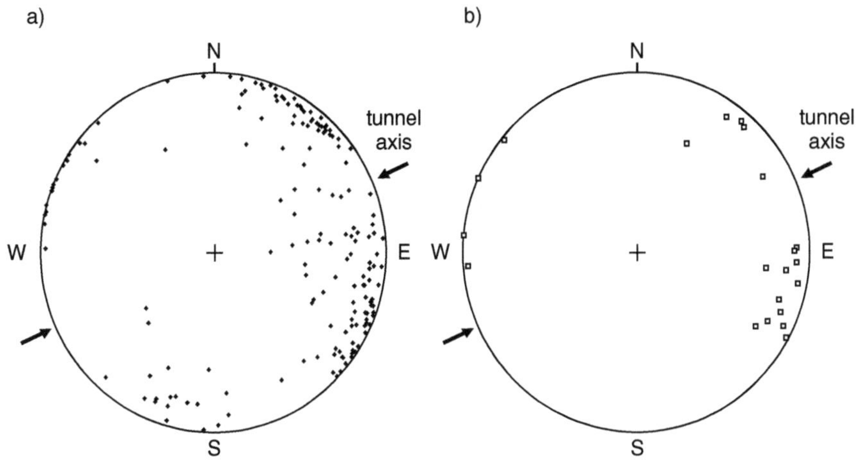

Fig. 3 Orientation of the fracture normals. Symbols correspond to the projection on the equatorial plane of the normal vector pointing onto the upper unit hemisphere for (a) small and (b) large fractures. The tunnel axis direction is indicated

lag were investigated. Variograms of dimensionless trace lengths and fracture orientations, according to the metric distance to the tunnel or to the number of separating fractures show only pure nugget effect. Therefore, since no spatial structure could be recognized, results from a classical stereological approach where fractures are statistically independent can be used.

Gupta and Adler (2006) propose an approximate solution, verified by Monte-Carlo simulations, of the mean trace length of a disk-fracture population (monodisperse and isotropic) intersecting a cylinder. The solution is identical to the one proposed by Berkowitz and Adler (1998) for the intersection of a disk-fracture population with a plane, which is

$$\langle c' \rangle = r'_d \frac{\pi}{2}, \tag{1}$$

where r'_d is the dimensionless radius of the disk. Equation (1) is valid up to $r'_d = 3$ (Gupta and Adler 2006).

3.2.1 Data Inversion

The left part of the histogram (Fig. 2b) provides the distribution of c' for the small fractures. Since the maximal dimensionless trace length of the small fracture population is 2.662, which implies $r'_d = 1.7$ using equation (1), one can investigate this population by the plane-disk intersection analysis made by Berkowitz and Adler (1998). Therefore, the trace length probability density $g(c')$ is related to the fracture diameter probability density $h(\phi')$ by

$$g(c') = \begin{cases} \frac{1}{\langle\phi'\rangle} \int_{c'}^{\phi'_M} \frac{c'}{(\phi'^2 - c'^2)^{1/2}} h(\phi') d\phi', \phi'_m \le c' \le \phi'_M \\ \frac{1}{\langle\phi'\rangle} \int_{\phi'_m}^{\phi'_M} \frac{c'}{(\phi'^2 - c'^2)^{1/2}} h(\phi') d\phi', c' \le \phi'_m \end{cases}, \tag{2}$$

where ϕ'_m and ϕ'_M are the minimum and maximum dimensionless disk diameters of the disk population. This formula can be inverted numerically (Berkowitz and Adler 1998), and results for the small fracture family at Roselend are shown in Fig. 4.

3.2.2 Fit of the Inverted Data by Classical Laws

$h(\phi')$ is tentatively fitted by the lognormal, exponential, and power law functions (cf. Berkowitz and Adler (1998)). The quality of the fit is evaluated by the criterion q_1, the mean over N-1 observations (not considering the first data, for valid comparison between the three functions) of the squared relative error between computed and observed density probabilities

$$q_1 = \frac{1}{N-1} \sum_{1}^{N-1} \left(\frac{h_{computed} - h_{observed}}{h_{observed}} \right)^2. \tag{3}$$

The best results for lognormal, exponential, and power law functions are plotted in Figs. 4a, b, and c, respectively. The best fit is obtained for the following power law

$$h(\phi') = \alpha \phi'^{-a}, \tag{4}$$

with $\phi'_m = 0.175$, $\phi'_M = 2.662$, and $a = 0.7$. The calculated mean diameter of fractures $\langle\phi'\rangle$ is equal to 0.9188.

3.2.3 Data Simulations

$g(c')$ derived by equation (2) is compared to simulations based on the classical functions studied in the previous section (Fig. 5). Simulations are performed by

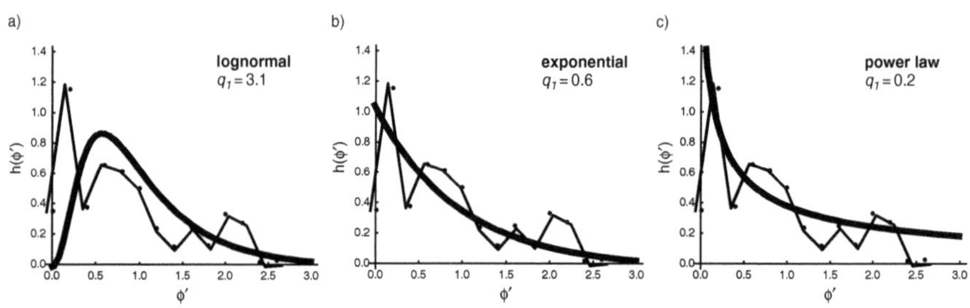

Fig. 4 Diameter probability density of small fractures observed in the Roselend tunnel obtained through inversion of equation (2), using the probability density of small fracture traces (thin lines). Best fits of inverted data (bold lines) through the (a) lognormal, (b) exponential, and (c) power law distributions

generating 10^6 disks. Results are compared by means of q_2, the mean over N-2 observations (not considering the first and last data, due to artificial cut-offs at ϕ'_m and ϕ'_M values) of the squared relative error between the computed and observed density probabilities. As expected, the simulation based on the inverted data shows a very good fit. Among the three classical functions, the best fit is obtained for the power law simulation with $q_2 = 0.05$.

To investigate the result dispersion, disks are generated until 172 intersections with a plane are obtained. The corresponding probability density of trace lengths is then calculated (stars; Fig. 5). This is repeated ten times, for ten various random number seeds, and for the four investigated cases. The envelope of these data (grey areas; Fig. 5) gives a rough estimate of the dispersion of the results for the intersections obtained from 10^6 disks. Dispersions for the lognormal, exponential and power laws are quite large (Figs. 5a, b, and c, respectively). Since the verification

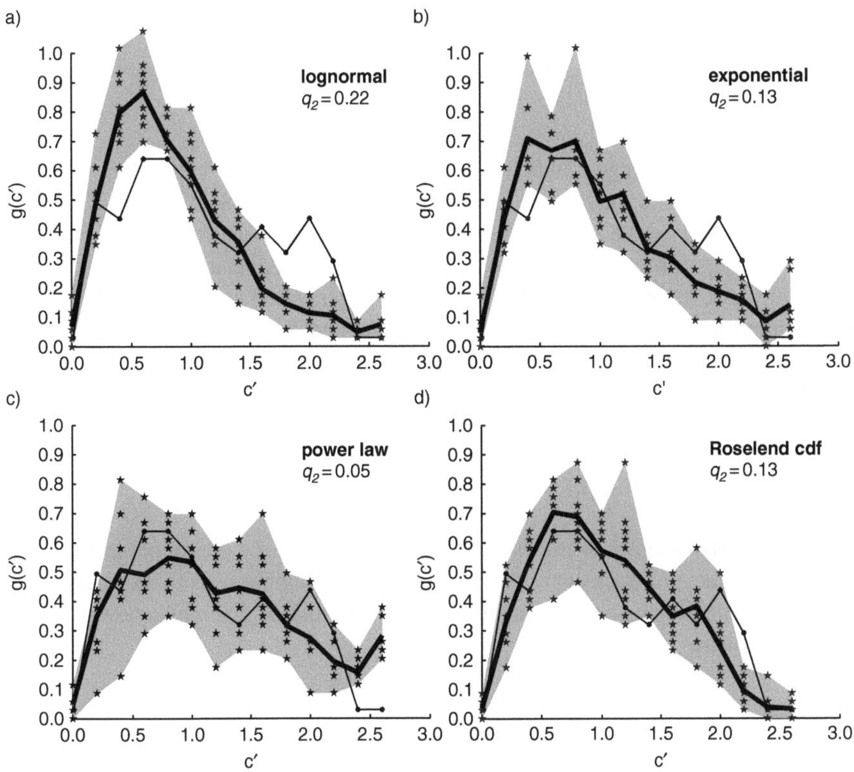

Fig. 5 Observed (thin lines) and calculated (bold lines) trace lengths probability densities for small fractures observed in the Roselend tunnel. Calculations are performed using the (a) lognormal, (b) exponential, and (c) power law distributions, given in Fig. 4, and (d) using the inverted data. Stars represent trace lengths probability densities calculated with 10 various random number seeds until 172 intersections were obtained. The shaded area illustrates the dispersion that is obtained using the various functions

test based on the cumulative distribution function of observed traces presents a large dispersion (Fig. 5d), it is concluded that dispersion intrinsically resides in the data set of observed trace lengths, and particularly in the fact that only 172 observed traces are available.

Despite the minimization of dispersion by the exponential law, the best description of the trace length density probabilities is obtained for the power law distribution of fracture diameters with $a = 0.7$.

3.2.4 Percolation of Small Fracture Networks

The whole network results from the periodic juxtaposition in space of infinitely many replicas of a unit cell τ, on which periodic boundary conditions are applied (Adler 1992). The percolation of the fracture network is checked along the three directions of space, by using a pseudodiffusion algorithm (Thovert et al. 1993, Huseby et al. 1997).

The average number of fractures ρ in a reference volume can be made dimensionless by the relation

$$\rho_s' = \rho_s V_{ex}. \tag{5}$$

ρ_s' represents the average number of intersections per fracture with other fractures in the network, i.e. a mean coordination number. The excluded volume V_{ex} of an object is defined as the volume into which the center of another object may not enter if overlap of the two objects is to be avoided (Balberg et al. 1984).

In our case, the network of small fractures is generated according to the power law defined in equation (4); 20-gones are generated, i.e. polygons with 20 equal sides that are approximately circular. The number of generated fractures is determined according to the volumetric small fracture density (ρ_s).

$$\rho_s = \frac{4N_s}{\pi^2 L_t \phi_t \langle\phi\rangle}, \tag{6}$$

where N_s is the number of traces of small fractures, and L_t the tunnel length (m). Considering the 172 traces over the 128 meter-long tunnel, $\rho_s = 0.206$ small fractures m^{-3}. The dimensionless density for disks distribution given by a power law, is calculated using

$$\rho_s' = \rho_s \; v_{ex} < R^3 >, \tag{7}$$

where v_{ex} is a factor shape associated to the geometrical object used for representing a fracture (for 20-gones, $v_{ex} = 8.895887$) and $< R^3 >$ is the third moment of the disk radius (Mourzenko et al. 2005), and is equal to 0.545. In our case, $\rho_s' = 0.999$, which is much lower than polydisperse networks percolation thresholds (always superior to 2.3) determined by Mourzenko et al. (2005). Therefore, the small fracture network is not expected to percolate. This is verified by Monte-Carlo simulations.

Percolation probabilities of the small fracture network P_s are calculated for a cubic cell τ of size L equal to 12, 16, and 20 meters, over 200 realizations.

Here, independently of the cell size, percolation probabilities are null ($P_{s12} = P_{s16} = P_{s20} = 0.$), showing that water dripping in the Roselend tunnel is unlikely to occur through small fractures only. Thus, the role of large fractures as water pathways has to be investigated. This is done in the following section through the analysis of full intersections.

3.3 Full Intersections

3.3.1 Data

Large fracture trace lengths C correspond to the ellipse circumference of a plane entirely intersecting the tunnel cylinder. They were calculated for each of the 21 large fractures according to their dip normal angle. The average of these dimensionless trace lengths is equal to 6.6 (Fig. 2c).

The estimation of the lateral extension of these large fractures is difficult. If these fractures are assumed to be isotropic disks of radius R_L, equation (1) gives $r'_d = 4.2$ for $<C'> = 6.6$. Therefore, the large fracture disk radius R_L is estimated to be 5.0 m. The numerical solution for full intersections with a cylinder would imply an even greater R_L (Fig. 11 in Gupta and Adler (2006)). Since relatively few observations of large fractures ($N_L = 21$) are available, the minimum $R_L = 5.0$ m is considered hereafter.

Due to the lack of knowledge on the distribution of large fractures, several hypotheses are tested. In a first approach, the large fracture density ρ_L is calculated according to Thovert and Adler (2004) for isotropic fractures ($\rho_L = 0.0042$ fractures m^{-3}), and for subvertical anisotropic fractures ($\rho_L = 0.0021$ fractures m^{-3}).

3.3.2 Percolation of the Networks Made of the Small and Large Fractures

Knowing ρ_L and ρ_s, percolation tests are performed, over 200 realizations, for networks composed of small and large fractures, and for 21 meter-long cubic cells. For the isotropic and the subvertical anisotropic cases, the percolation probabilities P_Z of small plus large fracture networks along the vertical direction are equal to 1.000 and 0.800, respectively. Therefore, percolation tests for networks made of both small and large fracture families show that large fractures are essential for the networks to percolate, in agreement with observations.

4 Permeability Assessments

4.1 Observed Water Fluxes and Zones

Three major zones (Fig. 1) can be distinguished along the tunnel, reflecting major contrasts in dripping water fluxes. They correspond to the zones 0–50, 50–108, and 108–128 m. Water fluxes were determined by collecting water drips during several minutes in a plastic sheet of 3.8 m^2. This operation was performed on April 25, 2003, which corresponds to snowmelt end; therefore the system can be considered as nearly saturated. Water fluxes range from 0 to $0.283 \times 10^{-3}\,m\,h^{-1}$ (Fig. 1). For comparison purpose with permeability results, water fluxes are normalized to their maximum value (Fig. 6).

For each zone, three cases with different large fracture orientations are investigated; 1) isotropic, 2) vertical and perpendicular to the tunnel, and 3) discrete (i.e. orientations of fractures are randomly chosen among the observations). The studied zones differ by their density of small and large fractures. The latter are calculated (Table 1) according to formulas presented by Thovert and Adler (2004).

Percolation analysis, over 200 realizations, is performed for each of the three cases and for the three zones, for cubic cells of length equal to 21 m. All zones, for all cases, present percolation probabilities (Table 2) above 0.5 except when zone 3 presents a percolation probability of 0.135 in the vertical direction, which correspond to vertical large fractures which are perpendicular to the tunnel. Since zone

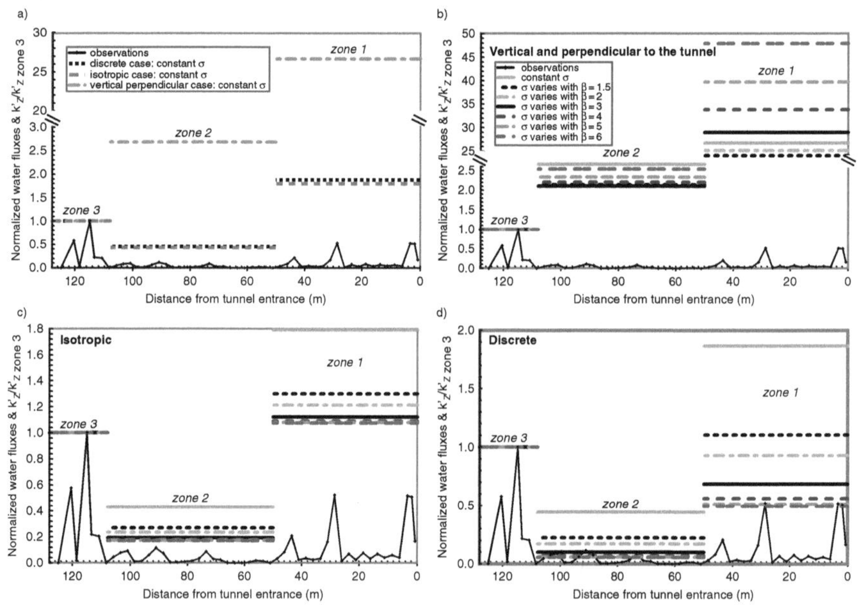

Fig. 6 Normalized water flux along the tunnel and normalized vertical permeabilities in the three zones in the tunnel for fracture networks composed of small and large fractures, where small fracture distribution is isotropic and polydisperse described by the power law function (4), and where large fractures are considered monodisperse with a radius $R_L = 5$ m. (a) Results are presented for the three various cases of large fracture orientations: vertical perpendicular to the tunnel, isotropic and discrete (i.e. randomly chosen among the observations). Fracture permeabilities are constant. Results are presented for the case where large fracture orientations are (b) vertical and perpendicular to the tunnel, (c) isotropic and (d) discrete, and for fractures with constant permeability or varying permeability according to equation (14) with $\beta = 1.5, 2, 3, 4, 5, 6$. Conventions in (c) and (d) are the same as in (b)

Table 1 Number of small and large fractures for the three zones of the tunnel. Small fracture volumetric density is given for an isotropic network. Large fracture volumetric densities are given for the three investigated cases; vertical perpendicular to the tunnel axis, isotropic, and discrete (where fracture orientations are randomly chosen among measured orientations)

	Small fractures		Large fractures			
		All cases Isotropic		Vertical perpendicular	Isotropic	Discrete
	Number of observations	ρ_s fractures m^{-3}	Number of observations	ρ_L fractures m^{-3}	ρ_L fractures m^{-3}	ρ_L fractures m^{-3}
zone 1 (0–50 m)	82	0.252	10	2.546E-3	5.093E-3	4.000E-3
zone 2 (50–108 m)	82	0.217	7	1.537E-3	3.074E-3	2.414E-3
zone 3 (108–128 m)	8	0.061	4	2.546E-3	5.093E-3	4.000E-3

Table 2 Absolute percolation probability P (overall) as well as both horizontal (P_X and P_Y) and vertical (P_Z) percolation probabilities, for reconstructed networks considering isotropic small fractures for the three different cases of large fracture orientations.

Case	Vertical perpendicular to the tunnel				Isotrope				Discrete			
Percolation probability	P	P_X	P_Y	P_Z	P	P_X	P_Y	P_Z	P	P_X	P_Y	P_Z
zone 1 (0–50 m)	0.86	0.61	0.99	0.98	1.00	1.00	1.00	1.00	1.00	1.00	1.00	1.00
zone 2 (50–108 m)	0.36	0.03	0.49	0.58	1.00	1.00	1.00	1.00	0.93	0.88	0.96	0.96
zone 3 (108–128 m)	0.08	0.00	0.10	0.14	1.00	1.00	1.00	1.00	0.98	0.96	0.97	1.00

3 is the zone where the largest fluxes are observed, this model seems inadequate, and flow simulations are expected to present more interest for isotropic and discrete orientations for large fractures.

4.2 Flow Simulations

For each case and each zone, flow simulations are conducted for cells on which percolation tests were performed. Flow simulations consist in solving flow equations according to the chosen geometry for the fracture networks. Even networks presenting low percolation probabilities are examined.

The solid matrix containing the fractures is assumed to be impervious. The flow of a Newtonian fluid at low Reynolds number is governed by the Stokes equations within a fracture, i.e. at a local scale characterized by a typical fracture aperture b_0 which is assumed to be much smaller than the typical lateral extent ϕ of the fracture (i.e. its diameter, if represented as a disk). Because of the classical Poiseuille law, the permeability σ_0 of a fracture is expected to be to the order of

$$\sigma_0 = \frac{b_0^3}{12}. \tag{8}$$

At a scale between b_0 and ϕ, the saturated flow is governed by the Darcy equation

$$\mathbf{q} = -\frac{\boldsymbol{\sigma}}{\mu} \cdot \overline{\nabla p}. \tag{9}$$

$\mathbf{q}$ and $\overline{\nabla p}$ are the locally averaged flow rate per unit width [$L^2\ T^{-1}$] and the pressure gradient. μ is the fluid dynamic viscosity [$M\ L^{-1}\ T^{-1}$], and $\boldsymbol{\sigma}$ is the fracture permeability tensor [L^3]. The mass conservation equation becomes

$$\nabla_s \cdot \mathbf{q} = 0, \tag{10}$$

where ∇_s is the two-dimensional gradient operator in the mean fracture plane.

These equations supplemented by adequate boundary conditions are systematically solved. $\overline{\overline{\nabla p}}$ is the macroscopic pressure gradient applied upon the medium ($\overline{\overline{\nabla p}} = 1$, in our case). The flux at the medium outlet is related to the pressure gradient by Darcy's law:

$$\overline{\mathbf{U}} = -\frac{\mathbf{k}}{\mu} \cdot \overline{\overline{\nabla p}}. \tag{11}$$

$\mathbf{k}$ is the macroscopic permeability tensor [L^2], to be determined from equation (11), once the problem from equations (9) and (10) has been solved.

σ_0 together with a characteristic radius R_c (equal to 1 m in our case), and a reference pressure p_0 are used to recast the equations in a dimensionless form. Flow equations are indeed solved in their dimensionless form since it is more convenient. Dimensionless parameters (with primes) are defined for instance by

$$\mathbf{q} = \frac{\sigma_0 p_0}{\mu R_c}\mathbf{q}', \quad \sigma = \sigma_0 \sigma', \quad \mathbf{k} = k_0 \mathbf{k}' = \frac{\sigma_0}{R_c}\mathbf{k}', \quad R_m = R_c R'_m. \tag{12}$$

The numerical method applied to solve the flow problem was described by Koudina et al. (1998). Macroscopic permeabilities given in the following are always averaged over 100 realizations, for each set of model parameters.

4.3 Permeability Assessment Compared to Percolation Observations

4.3.1 Networks with Constant Fracture Permeability

Dimensionless permeabilities k' for the three zones and for each of the three cases of large fracture orientations are obtained considering a constant fracture permeability σ'. According to equation (11), the water flux at the medium outlet should be proportional to the vertical macroscopic permeability. In order to assess if geometrical settings and fracture features chosen in the various cases are appropriate, vertical permeabilities may be compared to water fluxes recorded along the tunnel. Such comparison is performed with water fluxes and vertical permeabilities normalized according to zone 3 in the tunnel, where highest water fluxes occur. Figs. 6a show that none of the three cases is able to reproduce the general pattern of water fluxes observed in the tunnel (i.e. mild, low and high flux in zones 1, 2, and 3, respectively).

4.3.2 Networks with Variable Fracture Permeability

Available data (Vermilye and Scholz 1995, Johnston and McCaffrey 1996, Walmann et al. 1996, Gudmundsson et al. 2001) show that the fracture average aperture b_0 follows a power law of the form

$$b_0 = FR^{\kappa}, \tag{13}$$

where κ varies between 0.5 and 2, F is a multiplicative factor and R is the metric radius of the considered fracture. In order to obtain the distribution of fracture conductivities, it is assumed in this study that the fracture hydraulic conductivity σ_0 is related to its mean geometrical aperture b_0 via equation (8). In many cases, this relation is not exactly satisfied for rough walled fractures, but it can be used for a first estimate of a fracture hydraulic aperture. The scaling relationship (13) and the cubic law (8) imply

$$\sigma' = R'^{\beta}, \tag{14}$$

where $\beta = 3\kappa$, with a possible range of variation $1.5 < \beta < 6$, and where R' is the dimensionless radius of the considered fracture. The model (14) is used here in a straightforward manner with $\beta = 1.5$, 2, 3, 4, 5, and 6. Comparison of normalized permeabilities to normalized water fluxes is performed for the three cases (Figs. 6b, c, d), with fracture permeabilities varying according to equation (14).

A detailed study of Figs. 6b to 6d shows that the geometrical model which reproduces best the general patterns of water fluxes, is the discrete case with fracture permeabilities given by equation (14) and $\beta = 3$. Note that the normalized permeability still overestimates the normalized water flux in zone 1. This overestimation remains moderate and was somehow expected due to flow occurring along the tunnel walls near the tunnel entrance, which artificially leads to a slight underestimation of the measured flow. This model is therefore used in the following to assess the dimensional permeabilities of the medium.

4.3.3 Macroscopic Permeability and Hydraulic Conductivity

k, the macroscopic permeability [L^2], is derived from $k = \frac{\sigma_0}{R_c} k'$ and $\sigma_0 = \frac{b_0^3}{12}$ with b_0 equal to 10^{-4} m, the approximate value for fracture aperture with R_c radius in the Roselend tunnel. K_Z the vertical hydraulic conductivity of the medium [$L\,T^{-1}$] is obtained with $K_Z = \frac{\rho_f g}{\mu} k_Z$, where ρ_f and μ are the density [$M\,L^{-3}$] and the dynamic viscosity of the fluid [$M\,L^{-1}\,T^{-1}$] respectively, and g is the gravitational acceleration [$L\,T^{-2}$].

K_Z were calculated considering ρ_f and μ equal to $1000\,kg\,m^{-3}$ and 1.518×10^{-3} Pa s (values for water at 5°C), and $g = 9.81\,m\,s^{-2}$. Permeability values k_Z with $\beta = 3$ are estimated to 2.1×10^{-13}, 3.1×10^{-14}, and $3.0 \times 10^{-13}\,m^2$ for zones 1, 2, and 3, respectively. The corresponding hydraulic conductivities K_Z are equal to 1.3×10^{-6}, 2.0×10^{-7}, and $1.9 \times 10^{-6}\,m\,s^{-1}$. Such values compare very well with fractured hard rocks hydraulic features. Indeed, classical hydraulic conductivity values usually range between 8×10^{-9} and $3 \times 10^{-4}\,m\,s^{-1}$, with a narrower range of $3.3 \times 10^{-6} - 5.2 \times 10^{-5}\,m\,s^{-1}$ for weathered granites (Domenico and Schwartz 1998). These values are also in a very good agreement with the hydraulic conductivity ($K = 2.9 \times 10^{-7}\,m\,s^{-1}$) determined by a pumping test (W. Epting, pers. written comm., November 2003) performed at the vicinity of the tunnel entrance.

5 Perspectives

This work constitutes a small step toward the quantification of macroscopic properties of fractured media. Through the example of the Roselend tunnel, developments of many contributors are applied in order to determine geometrical properties of real fracture networks, their percolation character and the macroscopic properties of the fractured medium, such as its permeability. Determining this property constitutes a real challenge since it varies over $\sim$ 13 orders of magnitude. It is even more challenging for a fractured medium where strong heterogeneity is intrinsic, the medium being made of permeable fractures and an impermeable matrix.

Further work should be conducted to investigate flow in double permeability medium and transport. Simulations in two phase flow, and transport simulations would be interesting to performed, considering long-term time series of water flux and chemical species in few specific points in the tunnel. Hydromechanical coupling is also a promising field of investigations for characterizing varying fracture permeability over time, and its impact on solute transport.

Acknowledgments The authors thank Electricité de France and the city of Beaufort. Investigations of the Roselend site could be performed thanks to the support from D. Calmet, J. Bouchez, and R. Chiappini from CEA. We also wish to thank V. Mourzenko for his assistance in computing management. We are deeply grateful to S. Bureau, P. Richon, F. Perrier and C. Dezayes for their technical support, field investigations, data collection, and subsequent fruitful discussions, and to an anonymous reviewer for his thorough review.

References

Adler PM (1992) Porous media: Geometry and Transports. Butterworth-Heinemann, Boston

Balberg I, Anderson CH, Alexander S, Wagner N (1984) Excluded volume and its relation to the onset of percolation. Phys Rev B 30(7):3933–3943

Berkowitz B (1994) Modelling flow and contaminant transport in fractured media. Adv Porous Media 2:397–451

Berkowitz B, Adler PM (1998) Stereological analysis of fracture network structure in geological formations. J Geophys Res-Solid Earth 103(B7):15339–15360

Bogdanov II, Mourzenko VV, Thovert JF Adler PM (2003) Effective permeability of fractured porous media in steady state flow. Water Resour Res 39(1):1023 DOI: 10.1029/2001WR000756

Dezayes C, Villemin T (2002) Etat de la fracturation dans la galerie CEA de Roselend et analyse de la déformation cassante dans le massif du Méraillet, Technical report CEA contract n 46 000 32745: Université de Savoie LGCA

Domenico PA, Schwartz FW (1998) Physical and chemical hydrology. John Wiley and sons, New York

Gonzalez-Garcia R, Huseby O, Thovert JF, Ledesert B, Adler PM (2000) Three-dimensional characterization of a fractured granite and transport properties. J Geophys Res 105(B9):21387–21401

Gudmundsson A, Berg SS, Lyslo KB Skurtveit E (2001) Fracture networks and fluid transport in active fault zones. J Struct Geol 23(2–3):343–353

Gupta AK, Adler PM (2006) Stereological analysis of fracture networks along cylindrical galleries. Math Geol 38(3) DOI: 10.1007/s11004-005-9018-4

Huseby O, Thovert JF Adler PM (1997) Geometry and topology of fracture systems. J Phys A-Math Gen 30(5):1415–1444
Johnston JD, McCaffrey KJW (1996) Fractal geometries of vein systems and the variation of scaling relationships with mechanism. J Struct Geol 18(2-3):349–358
Koudina N, Garcia RG, Thovert JF, Adler PM (1998) Permeability of three-dimensional fracture networks. Phys Rev E 57(4):4466–4479
Mauldon M, Mauldon JG (1997) Fracture sampling on a cylinder: From scanlines to boreholes and tunnels. Rock Mech Rock Eng 30(3):129–144
Mourzenko VV, Thovert JF, Adler PM (2005) Percolation of three-dimensional fracture networks with power-law size distribution. Phys Rev E 72:036103 DOI: 10.1103/PhysRevE.72.036103
Peacock DCP, Harris SD, Mauldon M (2003) Use of curved scanlines and boreholes to predict fracture frequencies. J Struct Geol 25(1):109–119
Piggott AR (1997) Fractal relations for the diameter and trace length of disc-shaped fractures. J Geophys Res-Solid Earth 102(B8):18121–18125
Pili E, Perrier F, Richon P (2004) Dual porosity mechanism for transient groundwater and gas anomalies induced by external forcing. Earth Planet Sci Lett 227(3-4):473-480 DOI: 10.1016/j.epsl.2004.07.043
Provost A-S, Richon P, Pili E, Perrier F, Bureau S (2004) Fractured porous media under influence: the Roselend experiment. Eos Trans AGU 85:113
Sisavath S, Mourzenko V, Genthon P, Thovert JF, Adler PM (2004) Geometry, percolation and transport properties of fracture networks derived from line data. Geophys J Int 157(2):917–934 DOI: 10.1111/j.1365-246X.2004.02185.x
Thovert JF, Adler PM (2004) Trace analysis for fracture networks of any convex shape. Geophys Res Lett 31(22):L22502 DOI: 10.1029/2004GL021317
Thovert JF, Salles J, Adler PM (1993) Computerized characterization of the geometry of real porous-media - Their discretization, analysis and interpretation. J Microsc-Oxf 170:65–79
Vermilye JM, Scholz CH (1995) Relation between vein length and aperture. J Struct Geol 17(3):423–434
Walmann T, Malthe-Sorenssen A, Feder J, Jossang T, Meakin P, Hardy HH (1996) Scaling relations for the lengths and widths of fractures. Phys Rev Lett 77(27):5393–5396
Warburton PM (1980a) A stereological interpretation of joint trace data. Int J Rock Mech Min Sci 17(4):181–190
Warburton PM (1980b) Stereological interpretation of joint trace data - Influence of joint shape and implications for geological surveys. Int J Rock Mech Min Sci 17(6):305–316

Assessment of Groundwater Salinisation Risk Using Multivariate Geostatistics

A. Castrignanò, G. Buttafuoco and C. Giasi

Abstract The risk assessment at regional scale requires modelling spatial variability of environmental variables. Traditional approach, based on estimating point environmental indicators, cannot be considered satisfactory for this purpose, because it does not take into account spatial dependence between variables. We propose the application of an approach to the problem of groundwater salinisation, in which multivariate geostatistics and GIS are combined to integrate primary information with exhaustive secondary information. The dataset consisted of 454 private wells used for irrigation and located in Apulia region (south Italy). Three variables were processed: concentration of chlorides and nitrates, as primary variables, and the distance from the coast, as auxiliary variable. The approach highlighted the widespread degradation of water resources in the Apulian groundwater. The maps of the global indicator allowed us to delineate the zones at high risk of groundwater contamination and also to identify those parameters most responsible for water degradation, so that a wiser management of water resources could be planned. This approach can be used as operational support to a wide range of activities and in decision making among several remediation alternatives.

1 Introduction

Mediterranean is one of the most beautiful and richest ecosystems of the world, but it is also one of the most fragile and vulnerable ones, owing to prolonged drought periods, high soil erodibility, high frequency of forest fires, abandonment of vast rural areas that become marginal, excessive exploitation of water resources, massive concentration of economic and touristic activities along coastlines, soil impermeabilization and salinisation. Above all the rapid socio-economic growth over the last decades has caused severe stress to the Apulian hydrogeological system in southern Italy, because large quantities of water have been drawn from groundwater for domestic, irrigation and industrial uses. Apulia is affected by two major types of

A. Castrignanò
CRA - Agronomic Research Institute, Bari, Italy
e-mail: annamaria.castrignano@entecra.it

A. Soares et al. (eds.), *geoENV VI – Geostatistics for Environmental Applications*, 191–202

anthropic pollution: salt contamination, spread over large portions of the land, and chemical-physical and biological pollution, mainly confined to urban and suburban areas (Cotecchia and Polemio, 1997). Keeping apart northern Apulia, the remaining hydrogeological units share some common features: the predominant rock material of the aquifers is either limestone or limestone-dolomite, affected by karst and fracturing phenomena occurring below the sea level. Therefore, in the vicinity of the coast the aquifer may be deep enough to allow seawater intrusion, which underlies fresh groundwater owing to differences in density. Salt contamination of the Apulian groundwater is a well known and thoroughly investigated phenomenon (Cotecchia, 1977): it is quite evident a strong relationship between the increase of salt contamination and the lowering of piezometric levels, due to groundwater overdraft and/or natural decrease in groundwater recharge. The severity of the present situation urges to define some indicators in order not only to assess groundwater quality but also to monitor impact of prevention and mitigation programmes. The assessment of groundwater salinisation risk at regional scale requires modelling spatial variability of environmental variables. Traditional approach, based on estimating point environmental indicators, cannot be considered satisfactory for this purpose, because it does not take into account spatial dependence between variables. Moreover, it should also be pointed out that the reliability of a spatially distributed indicator is affected by both density of the sampling and specific technique used for space/time interpolation of point data. Data from different locations are often interpolated without any critical use of the adopted method, spanning from Thiessen (1911) approach to advanced geostatistical algorithms, whereas the choice of a suitable method is a quite important aspect. Geostatistics, which is based on the theory of regionalized variables (Goovaerts, 1997), is generally preferred, because it allows to take into account spatial correlation between neighbouring observations to predict attribute values at unsampled locations. One of the major advantages of interpolation technique of kriging over simpler methods is that sparsely observations of the primary attributes can be complemented by secondary attributes that are more densely sampled. In the 1990s, with the diffusion of geographical information systems (GIS) and remote sensing techniques, exhaustively mapped secondary variables were used to directly estimate environmental variables. Distance from the sea coast is one of the main landscape features which can be used as quantitative information to model seawater intrusion into Apulian aquifers. Digital terrain models (DTM) are quantitative representations of variables describing topographic surface, and digital modelling can offer a set of quantitative methods to analyse the landsurface and the relationships between topography and groundwater quality. However, defining specific individual indicators of groundwater quality is not sufficient to assess an integrated feature like salinisation, which should include information from all the identified critical factors. The approach in defining quality should then be holistic and not reductionistic, with the diverse data integrated in a fashion that each individual indicator is combined and weighted appropriately (Rodale Institute, 1991). This integration needs to be flexible enough to be applicable to all types of information and be able to incorporate all types of groundwater degradation information, both numerical (hard) and nominal (soft) information. The integration method should be regionalized across larger areas, with each indicator receiving importance or weighting as a function of geographic region, and give some provision for estimating groundwater quality whereas data are lacking. Finally, the integration method should be able to evaluate groundwater quality at a variety of spatial scales, in order to propose effective specific management plans or water use policies. Therefore, a common characteristic of these indicators is that spatial variability of each of them has to be mapped at a scale compatible with the one chosen for the final report, so that all the layers of information can be inserted into a Geographical Information System (GIS) to carry the mathematical and geostatistical procedures needed.

We propose the application of an approach to the problem of groundwater salinisation, in which multivariate geostatistics and GIS are combined to integrate primary information, as a global index of salinisation risk of groundwater, with exhaustive secondary information as distance from coast.

2 Methodology

2.1 Study Site

The hydrogeological units of the study area located in the southern part of Apulia region (South Italy), excluded the Tavoliere and Gargano units, (Fig. 1) consist of large and deep Mesozoic carbonate aquifers. In the low Murge Plateau (Murgia) aquifers are under pressure except on a restricted coastline strip, whereas in the Salentine Peninsula (Salento) subsurface water flow under phreatic conditions is prevailing. The Salentine hydro-geological unit is the only one which is lapped by the sea on both sides. The porous and fractured nature of aquifer allows seawater intrusion where aquifer is deep enough (Cotecchia and Polemio, 1997).

The dataset consisted of 454 private wells (Fig. 2), used by farms for field irrigation, and sampling was performed on several times. Chemical analyses were carried out for the major ions and anions concentrations of the water samples using a spectrophotometer (APHA, AWWA, WPCF, 1985). In this study we used only the concentrations (mg l^{-1}) of nitrate (NO_3^-) and chloride (Cl^-) ions.

2.2 Geostatistical Approach

2.2.1 The Indicator Kriging Approach

The method is based on a set of applied statistics techniques known as geostatistics (Goovaerts, 1997). Geostatistical methods can be used to detect, model, estimate and simulate spatial patterns of different kinds of data (Castrignanò et al. 1998). The geostatistical approach consists of different steps: it begins with traditional univariate and bivariate analysis, followed by descriptive and diagnostic variography,

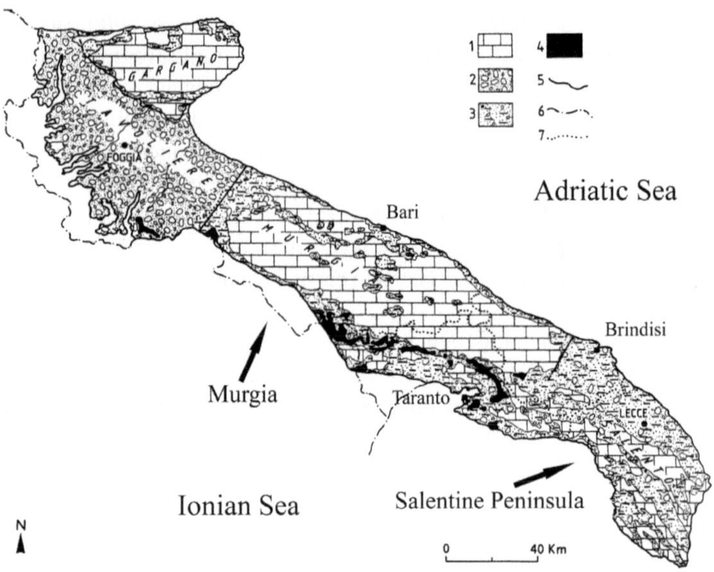

Fig. 1 Apulian hydrogeological units. 1) Carbonate rock outcrops of Gargano, Murgia and Salento units; 2) Tavoliere unit, mainly conglomerate and sands; 3) shallow aquifers and permeable lithotypes, calcarenites, clayey sands, sands, gravel, or conglomerates; 4) low permeable lithotypes, blue marly clays; 5) hydrogeological unit boundary, dashed where uncertain; 6) regional boundary; 7) provincial boundary.(Modified after Cotecchia et al., 1999)

a process whereby the similarity between samples is determined as a function of their separation distance. Then, this spatial dependence is modelled and used in an interpolation procedure, such as kriging. Detailed description of the theory can be found in Isaaks and Srivastava (1989) and Goovaerts (1997). In the present work the Indicator Kriging was used, which is a non-parametric type of ordinary kriging

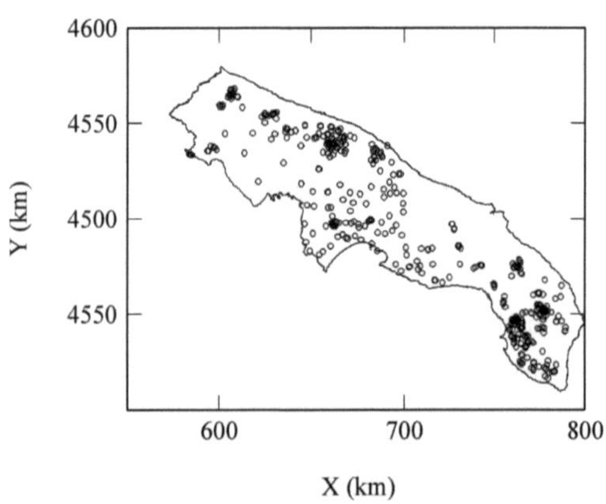

Fig. 2 Sample locations

(Journel, 1983; Isaaks and Srivastava, 1989). It has the advantage of being resistant to the effects of outlier values and is useful for analysing skewed data sets.

The proposed approach is based on a simple binary transformation whereby each datum is transformed into an indicator, before variography and kriging. By convention, data are coded as 0s or 1s, if they lie below or above given critical threshold values, in conformity with a definition of salinisation risk. Each indicator is then mapped as the probability that the corresponding variable is beyond the defined threshold (Journel, 1988), i.e. the probability that the area may be affected by the risk of groundwater salinisation in relation to that parameter.

2.2.2 FKA Analysis

Matheron (1982) developed a geostatistical method called factorial kriging analysis (FKA) for multiple-variable analysis. FKA is based on the three following steps:

1. modelling the coregionalization of the individual indicators, using the so called linear model of coregionalization (LMC);
2. analysing the correlation structure among the individual indicators by applying principal component analysis (PCA);
3. estimating the regionalised factors by cokriging at each spatial scale and mapping.

The application of FKA allows to define a restricted number of regionalized factors, summarising the effects of several individual indicators on groundwater salinisation at each given spatial scale.

Co-kriged estimates of the global indicators are summarised in choropleth maps. These values, standardised to the interval [0, 1], can be interpreted as the probability that groundwater quality, in a particular location, does not jointly fulfil the threshold criteria for the individual parameters considered.

In order to assess groundwater salinisation risk, three variables were selected: two water parameters, chlorine concentration (mg l^{-1}) and nitrate concentration (mg l^{-1}) as primary variables, and one topographic parameter, distance from the coast (m) as auxiliary variable. The critical threshold values for the two primary variables were: 250 mg l^{-1} for chloride concentration and 25 mg l^{-1} for nitrate concentration, corresponding to the maximum acceptable concentrations in groundwater.

The technique used to integrate secondary information in primary variable modelling was "Multi-Collocated Cokriging" (Rivoirard, 2001), where the influence of the secondary variable on the primary variables is explicitly taken into account through the estimation of cross-variograms. The approach is quite similar to ordinary cokriging with the only difference in the neighbourhood search. As using all secondary information contained within the neighbourhood may lead to an intractable solution because of too many information, the secondary variable is used at the target location and also at all the locations where the primary variable is defined within the neighbourhood. The modified version is less precise than full cokriging, not using all the auxiliary information contained within the neighbourhood.

However, because the collocated secondary datum tends to screen the influence of further away secondary data, actually there is little loss of information.

2.3 Decision Making in the Presence of Uncertainty

Many surveys are aimed at making important decisions, such as declaring some area of groundwater polluted. Decisions are very often made in the presence of uncertainty, because the estimates are always affected by errors, whichever the interpolation technique used. It is then quite critical to assess uncertainty of estimation. One approach consists in delineating all locations where groundwater contamination is beyond a maximum value tolerable for human consumption. This approach requires the kriging estimation of chloride and nitrate concentrations at the unsampled locations but, because of estimation error, there is a risk of declaring 'polluted' a save location and, conversely, 'save' a polluted location. These two misclassification risks can be assessed from the probability that the value of the variable z for the pollute concentration at any unsampled site $\mathbf{u}$ ($\mathbf{u}$ is the location coordinates vector) is not greater than a given threshold z_k, (Buttafuoco et al., 2000; Castrignanò and Buttafuoco, 2004). Indicating by F the conditional cumulative distribution function of probability, it follows that:

(1) the risk $\alpha(\mathbf{u})$ (false positive), *i.e.* the probability of wrongly declaring a location $\mathbf{u}$ 'polluted', is given by

$$\alpha(\mathbf{u}) = \text{Prob}\left\{Z(\mathbf{u}) \leq z_k \left| z^*(\mathbf{u}) > z_k \; (n)\right.\right\} = F(\mathbf{u}; z_k \left|(n)\right.) \tag{1}$$

for all locations $\mathbf{u}$ such that the kriging estimate $z^*(\mathbf{u}) > z_k$. In other words, $\alpha(\mathbf{u})$ measures the probability that the actual value is less than the critical threshold, whereas the estimated value $z^*(\mathbf{u})$ by kriging is greater than the threshold;

(2) the risk $\beta(\mathbf{u})$ (false negative), i.e. the probability of wrongly declaring a location $\mathbf{u}$ as 'save', is given by

$$\beta(\mathbf{u}) = \text{Prob}\left\{Z(\mathbf{u}) > z_k \left| z^*(\mathbf{u}) \leq z_k \; (n)\right.\right\} = 1 - F(\mathbf{u}; z_k \left|(n)\right.) \tag{2}$$

for all locations $\mathbf{u}$ such that the kriging estimate $z^*(\mathbf{u}) \leq z_k$. The symbol (n) means: conditional to the n sample data. More explicitly, $\beta(\mathbf{u})$ measures the probability that the actual value is greater than the critical threshold, whereas the kriged estimate is less than the threshold.

The risk $\beta(\mathbf{u})$ is not defined where the risk $\alpha(\mathbf{u})$ is present and conversely. The main working difficulty met in this approach consists in appropriately choosing a probability threshold for each type of risk prompting to some mediation action (Goovaerts, 1997).

All statistical and geostatistical analyses were done by using the software package ISATIS, release 5.1.8 (Geovariances, 2006).

3 Results

The binary transformation has shown that the proportion of samples above the thresholds was 0.35 for chloride concentration (Cl^-) and 0.40 for nitrate concentration (NO_3^-). These figures underline that less of 50% of the surveyed area may be affected by groundwater pollution. However, the previous analysis gives us no information about the location of the polluted sites; therefore we preferred to apply a geostatistical approach. The correlation matrix evidenced a significative correlation only between the distance from the coast and $Cl^-(-0.40)$ but not with NO_3^-. Nevertheless, as the seawater intrusion into the Mesozoic limestones of the Apulian groundwater is a very long observed and well studied phenomenon, we decided to treat the two indicator variables chloride (INDCL) and nitrate (INDNO3) together with the distance as auxiliary variable according to a multivariate approach. Due to the different sizes of the variables, distance was standardized to the interval [0.1] before fitting an isotropic Linear Model of Coregionalization (LMC), no anisotropy being observed in the variogram maps (not shown). The LMC (Table 1) includes three basic structures: 1) a nugget effect; 2) a spherical model with range = 8000 m and 3) a spherical model with range = 25522 m.

From the values of the eigenvalues it results that most variation occurs at 1) micro-scale, within a distance less than lag = 6000 m, and is erratic, largely affected by measurement error and 2) longer scale within the distance of about 25000 m, which is half of the average distance between the coasts of the two seas (Adriatic sea and Ionian sea) lapping Apulia region. On the contrary, the contribution of the spatially short-range component (range = 8000 m) to the total variance is the least, which induces us to make some considerations about the probable causes of groundwater pollution: one is quite site-specific, related to land use and local management practices; the other one, acting at longer scales, is more affected by the predominant rock material of aquifers and by the land shape.

The predictions of the contents of chloride and nitrate, obtained with multi-collocated cokriging and using a 10 m by 10 m grid, are shown in Fig. 3, where

Table 1 Decomposition in the regionalized factors

a) Nugget effect					
	Norm. dist.	INDNO3	INDCL	Eigenvalue	Percentage
Factor 1	−0.0012	0.9824	0.1869	0.1252	61.07
Factor 2	0.0385	−0.1867	0.9817	0.0798	38.93
Factor 3	0.9993	0.0083	−0.0376	0.0000	0.00
b) Spherical model (Range = 8000.00 m)					
Factor 1	0.0279	−0.9652	−0.2601	0.0923	100.00
Factor 2	0.0000	−0.2602	0.9656	0.0000	0.00
Factor 3	0.9996	0.0270	0.0073	0.0000	0.00
c) Spherical model (Range = 25522.04 m)					
Factor 1	0.2253	0.2138	−0.9505	0.1567	90.45
Factor 2	0.9122	−0.3890	0.1287	0.0165	9.55
Factor 3	0.3422	0.8961	0.2827	0.0000	0.00

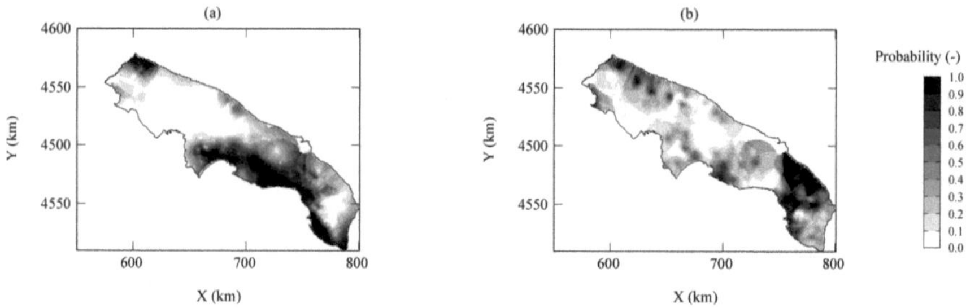

Fig. 3 Maps of chloride (a) and nitrates (b) indicator variables

it is possible to delineate the areas characterised by high probability that the criteria of good quality of groundwater cannot be fulfilled.

Figure 3a shows that water of acceptable quality may be found in the Murge Plateau, where seawater intrusion has an impact but only on a restricted length of coastline. On the contrary the problems related to salt contamination of groundwater become more severe on the Ionian side of the Murge plateau and in the Salentine hydrogeological unit. The risk of groundwater salinisation is generally less important along the Adriatic coast, with the exception of some restricted areas around Brindisi and to the North of Bari. The Fig. 3b shows a different pattern of the areas at hazard, quite likely due to the human-related pollution. Most of the Salentine Peninsula is at high risk of nitrate contamination and also the Adriatic coast is severely interested owing to the several tourist villages.

To have a synthesis picture of the results and spot the areas jointly characterised by a groundwater with poor hydrochemical features and affected by a relevant anthropic impact, we performed a factorial kriging analysis. Table 1 reports the decomposition into regionalised factors for each basic structure, showing that factor 1 accounts for about 61%, 100% and more than 90% of the total variation, respectively at micro-, short- and long range. Factor 1 is particularly related to INDNO3, positively at nugget effect and negatively at shorter range; on the contrary INDCL seems to impact more (negatively) on the regionalised factor at longer scales. Figure 4 shows the maps of the factor 1 at short and long range; the one corresponding to nugget effect was omitted, because mostly affected by measurement errors.

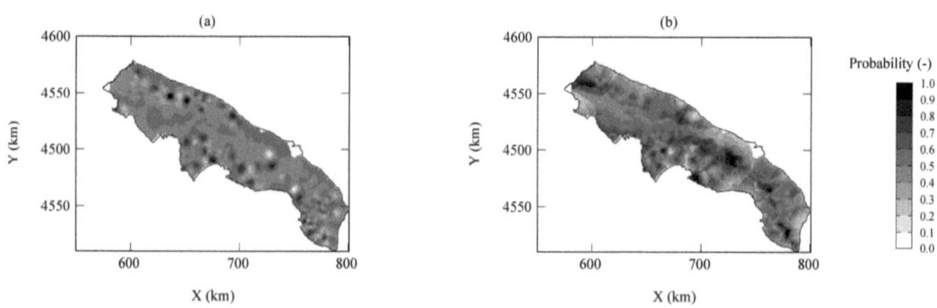

Fig. 4 Maps of normalized short (a) and long-range (b) factors

To make easier reading the maps, we have normalised the values of the spatial components of factor 1 to the interval [0, 1], taking into account that the correlation of factor 1 is negative with INDNO3 at short-range and with INDCL at long-range: the scores increase as a function of the impact of the most influential indicator. The pattern of spatial distribution shown in Fig. 4a is quite erratic, with a probability of 50–60% on average and few spots characterised by high probability values of groundwater pollution induced by agricultural and domestic uses. The map of Fig. 4b looks more spatially structured, with generally low risk of pollution along the coast of the Adriatic Sea and in some spots along the Ionian coast. The inner areas at high hazard reveal that salt contamination due to seawater intrusion is enhanced by human-related pollution. At this scale the two types of contamination risk are directly correlated (Table 1).

Finally, to assess the risks of a misclassification based only on estimate, we produced the kriging maps of the raw values of chloride and nitrate concentrations (Fig. 5). As the two water parameters showed highly positively skewed distributions with long tails, we preferred to normalize the data by using an expansion in terms of 100 Hermite polynomials so that the histogram shape of the transformed data matches quite well with the one of the standardized Gaussian distribution. Taking into account the significative correlation only between distance from the coast and chloride concentration, we applied a multivariate approach (multi-collocated cokriging) for chlorine, whereas a univariate approach (ordinary kriging) for nitrate. The LMC for the gaussian variable cor-responding to chloride includes three basic structures: 1) a nugget effect; 2) a cubic model with range = 30000 m and 3) a spherical model with range = 67000 m. These features of the model mean that seawater intrusion extends to large distances; however most of variation occurs within 30000 m (first eigenvalue = 1.36) where the impact of the coast distance is greater: the influence of distance on groundwater salinisation actually decreases at longer distances (first eigenvalue = 0.61 for range = 67000 m).

The variogram model for the gaussian variable of nitrate includes three basic structures: 1) nugget effect; 2) a cubic model with range = 8000 m and 3) a spherical model with range = 68000 m. Even if nitrate variation is more erratic, nevertheless spatial dependence extends within large distances (68000 m). The gaussian variables were interpolated on the same grid used for the indicator variables and the estimates were back-transformed to the original data.

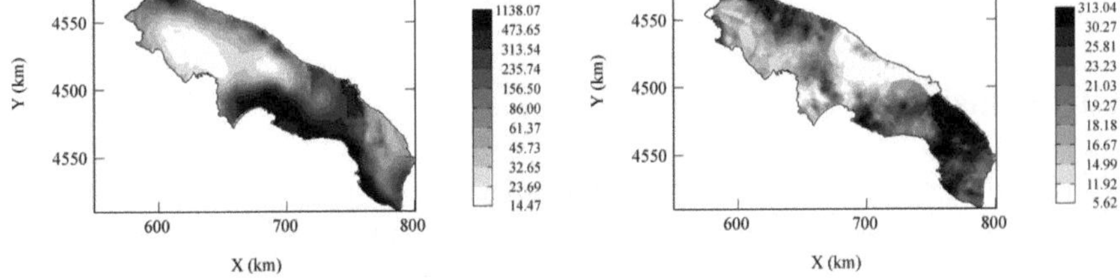

Fig. 5 Maps of chloride (a) and nitrate (b) concentrations

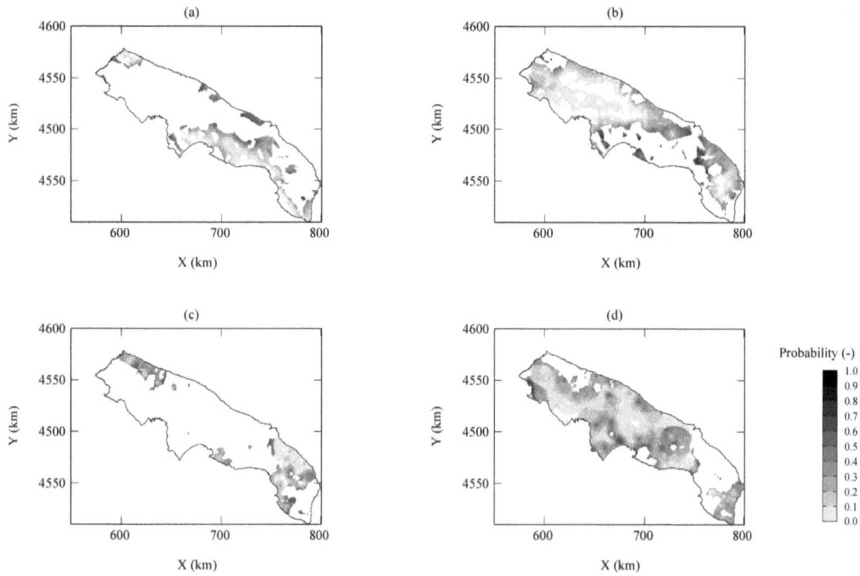

Fig. 6 Maps of: error α for chloride (a) and nitrate (c); error β for chloride (b) and nitrate (d)

In Fig. 5 the maps of the estimates of chloride and nitrate concentrations are shown which allow to delineate the areas at hazard of groundwater contamination. They are located in most of the Salentine Peninsula, along the Ionian coast and towards the northern Adriatic coast.

In Fig. 6 the maps for error α and β of chloride and nitrate are reported.

Even if such errors cannot be disregarded, they are generally low, less than 10% on average, with the exception of few restricted areas with error about 40–60%. The latter highlight where further sampling should be performed so to increase precision of pollution prediction, because the high error may be due to either large variability or sparse sampling or to both causes. Figure 6 reveals also that the estimate for both parameters is more biased towards underestimation, which might cause severe effects for environment deriving from not treating a high contaminated site. On the contrary acting based on a false alarm might result in unwarranted high investments in remediation procedures. A decision maker has to balance the consequences from correct or false decisions and then the information about the probabilities of occurrence of these errors might give him a useful support.

4 Conclusions

A probabilistic approach, based on multivariate geostatistics, is proposed to assess the contamination degree of the Apulian groundwater, which highlights the widespread degradation of water resources. The power of this tool is that it is flexible and could be used to make direct comparison of groundwater quality among different regions, by using a common list of indicator parameters with their corresponding critical thresholds or allowable ranges of values. The maps of the global indicator allow us to delineate the zones at high risk of groundwater contamination and also to identify those parameters most responsible for water degradation, so that a wiser management of water resources can be planned. The method is actually of great practical value, because, through its use, it is possible to identify and manage areas of poor groundwater quality and monitor the progress and the effectiveness of management treatments. Moreover, the application of GIS techniques allows us to include impacts associated with spatial variation of aquifers and then optimize the hydro-geological monitoring network. However, the goodness of the procedure will depend essentially on how individual indicators are related to degradation process of groundwater.

Finally, this approach could be used as operational support to a wide range of activities, not only for estimating, assessing and mapping the extent of groundwater salinisation, but also for determining the causes, quantifying the impacts, justifying the reclamation costs, monitoring the efficiency of the measures taken and making decisions among several remediation alternatives.

References

APHA, AWWA, WPCF (1985) Standard methods for the examination of water and wastewater (16th edn), Washington, DC, American Public Health Association p 905

Buttafuoco G, Castrignanò A, Stelluti M (2000) "Accounting for local un-certainty in agricultural management decision making". 7th ICCTA – International congress for computer technology in agriculture. Florence, 15th – 18th November 1998, pp 510–517

Castrignanò A, Buttafuoco G (2004) Geostatistical Stochastic Simulation of Soil Water Content in a Forested Area of South Italy. Biosystems Engineering 87:257–266

Castrignanò A, Mazzoncini M, Giugliarini L (1998) Spatial characterization of soil properties. Adv GeoEcology 31:105–111

Cotecchia V (1977) Studi e Ricerche sulle acque sotterranee e sull'intrusione marina in Puglia (Penisola Salentina). (Studies and Research on underground waters and on seawater intrusion in Apulia (Salentine Peninsula)) (in Italian) Quad 1st Ric Acque, Rome, XX. Water Research Institute

Cotecchia V, Limoni PP, M Polemio (1999) Identification of typical chemical and physical conditions in apulian groundwater (southern Italy) through well multi-parameter logs. XXXIX IAH Congress "Hydrogeology and land use management", Bratislava, 353–358

Cotecchia V, Polemio M (1997) Salinization and pollution of main Apulian aquifers (Southern Italy). Proceedings International Conference: Water management, salinity and pollution control towards sustainable irrigation in the Mediterranean region. CIHEAM, Ist. Agronomico Mediterraneo, September 22–26, 1997. Bari, Italy, 2:211–214

Géovariances (2006) Isatis Technical References, version 6.0.0. Geovariances & école Des Mines De Paris, Avon Cedex, France

Goovaerts P (1997) Geostatistics for natural resources evaluation. New York: Oxford University Press

Isaaks EH, Srivastava RM (1989) An introduction to applied geostatistics. New York: Oxford University Press

Journel AG (1983) Non-parametric estimation of spatial distributions. Math Geol 15:445–468

Journel AG (1988) Non-parametric geostatistics for risk and additional sampling assessment. In L. Kieth (Ed.), Principles of Environmental Sampling, Amer Chem Soc Publ, Washington, DC, pp 45–72

Matheron G (1982) Pour une analyse krigeante des données régionalisées. Rapport N-732, Centro de Géostatistiques, École des Mines de Paris, Fontainebleau

Rivoirard J (2001) Which Models for Collocated Cokriging?: Math Geol 33(2):117–131

Rodale Institute (1991) Conference Report and Abstracts, International conference on the assessment and monitoring of soil quality. Rodale Press, Emmaus, PA

Thiessen AH (1911) Precipitation average for large areas. Monthly Weather Review, 39:1082–1084

A MultiGaussian Kriging Application to the Environmental Impact Assessment of a New Industrial Site in Alcoy (Spain)

J. R. Ilarri and J. J. Gómez-Hernández

Abstract In this chapter an application to a real case of the multiGaussian kriging technique to assess the environmental impact of a new industrial site in the area of the La Canal Aquifer is presented. The La Canal Aquifer area in Alcoy (Valencia – Spain) has been intensively studied during the past few years as a potential location for a new industrial site. The high environmental value of this area, which is located near the Font Roja Natural Park, is because of its biotic reserve value (flora and fauna) and also because the aquifer formation that underlines the area is the main water-supplying source for the city of Alcoy.

The main objective of the hydrogeological study was to analyse the potential risk of contamination of the aquifer as a result of a failure in the industrial waste management system. The aquifer had previously been characterized on a regional level. Afterwards, a field data sampling campaign was designed to obtain local conductivity data and in situ soil characteristics. The set of conductivity data was obtained from Lefranc tests.

The multiGaussian kriging technique was used to determine the risk maps associated with the different thicknesses of the clay layer located over the limestone aquifer. This clay layer is the natural formation that prevents the limestone aquifer from being contaminated by providing an additional security to the artificial systems in case of a local or permanent spill of a contaminant in the surface. A simple one-dimensional contamination model in the non-saturated zone was developed based on the previous characterization of the aquifer system to obtain the concentrations of a contaminant in the soil and groundwater considering both a permanent spill coming from an improperly preserved pipeline and an accidental local spill of contaminant to the ground.

J. R. Ilarri
Universidad Politécnica de Valencia, Instituto de Ingeniería del Agua y Medio Ambiente, Camino de Vera s/n., 46022 Valencia, Spain
e-mail: jrodrigo@upv.es

A. Soares et al. (eds.), *geoENV VI – Geostatistics for Environmental Applications*, 203–210

1 Introduction

The municipality of Alcoy is located in the southern part of the Valencian Community and has historically been one of the most important industrial towns of this Mediterranean region, especially in the 1900–1930s. During the past few decades a high percentage of the municipality land has been protected thorough environmental legislation particularly after the creation of the Font Roja Natural Park.

Almost all of the municipality land is either devoted to urban areas or affected by the high level of environmental protection afforded because of the natural park. The lack of industrial land in the municipality implied that new industries were located in other municipalities. This fact has caused an important loss for the municipality both in economical and social aspects.

In recent years the municipality of Alcoy has been interested in developing studies to analyse the possibility of locating a new industrial site in their land. These studies proposed nine different locations as potentially available for locating new industries. One of these locations is the La Canal zone, in the southern part of the municipality as shown in Fig. 1.

This preliminary study (Municipality of Alcoy (1999)) suggests not locating the industrial site in the La Canal area due to the potential risk of contamination of the underlying aquifer, which is one of the main water supplies of Alcoy. In order to develop a complete analysis of the potential risk of contamination of the aquifer, a geological sampling campaign was designed to obtain information about the geology distribution and local permeability data. Previous results showed that there

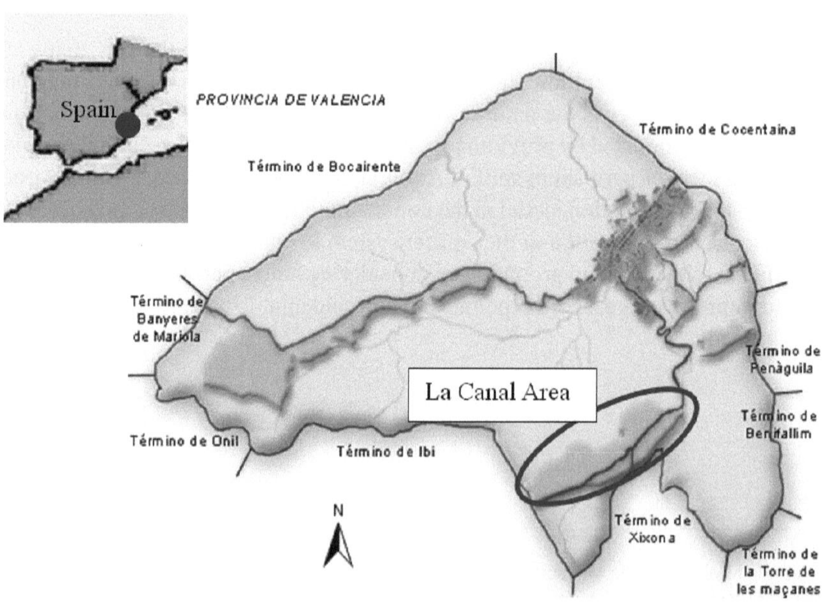

Fig. 1 Location of the study area inside the municipality of Alcoy. Southern part of the Valencian Community (Spain)

exists a low-conductivity layer over the higher-permeability formations that works as a geological barrier, preventing the aquifer from being contaminated (Instituto Técnológico y Minero de España (2000)).

Within such a framework, the objectives of the study were to analyse the spatial distribution of the thickness of the low-conductivity layer and its effectiveness as a barrier to the underlying aquifer formation.

2 The Study Area

The study area is located south of Alcoy city and has an irregular polygonal shape divided by National Road N-340 as shown in Fig. 2. The La Canal area is located over a variable-thickness low-permeability layer of loam and clay (Instituto Geológico y Minero de España (1978)). The high-permeability limestone aquifer is located under this clay layer. This protection does not appear some hundreds of metres down the valley in the Barranc de la Batalla area, in which limestone formations outcrop.

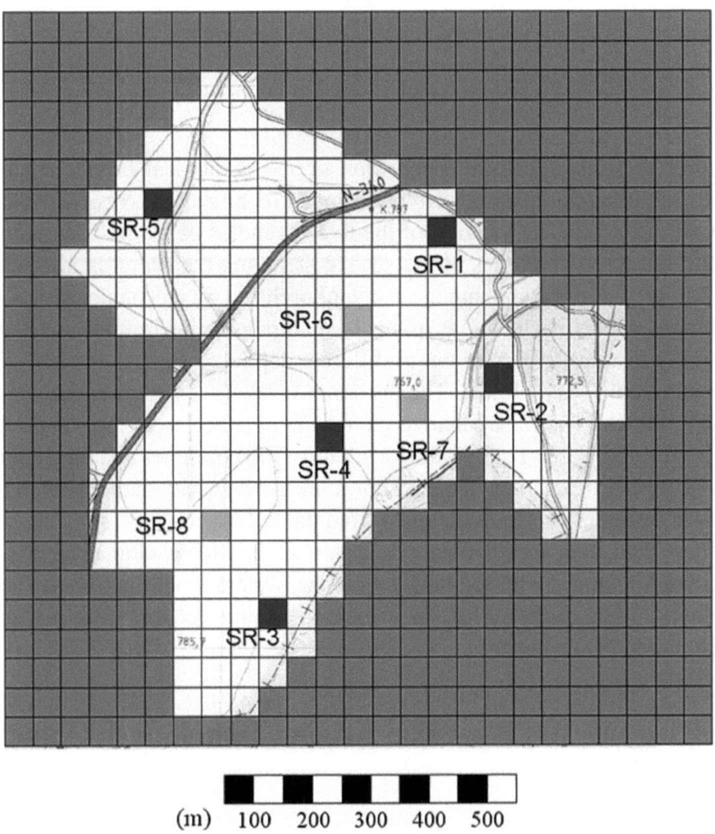

Fig. 2 Study area and location of the boreholes on the first and second sampling campaigns (first dark grey, second light grey)

Table 1 Location of boreholes and thickness of the low-K layer

Borehole	X local	Y local	Thickness (m)
SR-1	769	869	11,80
SR-2	869	619	2,40
SR-3	469	219	6,00
SR-4	569	519	5,50
SR-5	269	919	18,80
SR-6	619	719	8,10
SR-7	719	569	5,40
SR-8	369	369	4,70

Table 2 Conductivity data obtained from the Lefranc tests

Borehole	Depth (m)	K (m/s)
SR-1	−5.00	$8{,}9 \times 10^{-5}$
SR-2	−4.00	$6{,}07 \times 10^{-4}$
SR-2	−11.00	$2{,}63 \times 10^{-8}$
SR-3	−5.10	$8{,}2 \times 10^{-5}$
SR-3	−11.50	$1{,}15 \times 10^{-4}$
SR-4	−5.50	$1{,}02 \times 10^{-5}$
SR-4	−13.00	$1{,}29 \times 10^{-5}$
SR-5	−6.50	$1{,}94 \times 10^{-5}$
SR-5	−13.00	$8{,}96 \times 10^{-4}$

In order to study the local disposition of the geological layers in the study area, two sampling campaigns were performed (Technical University of Valencia–Incivsa S.L. (2002)). The location of the five boreholes of the first sampling campaign and the three boreholes of the second sampling campaign is also shown in Fig. 2. Samples were taken from these eight boreholes by mechanical extraction of soil columns. At the five locations of the first campaign, nine Lefranc constant-head permeability tests were carried out at different depths to analyse the behaviour of all the stratigraphic units found. The piezometric level was not found in any of the boreholes. Tables 1 and 2 summarize the results obtained from the sampling campaigns performed. Table 1 shows the locations of the boreholes and the thickness of the low-K layer in each one. Table 2 shows the conductivity data obtained in each borehole at different depths as obtained from the Lefranc test.

All these data were used to obtain the risk maps and to develop a simple groundwater contamination model to analyse how the system would behave if there is a spill of toxic waste at the surface or if there is a failure of the waste management system in the industrial site.

3 Applying the MultiGaussian Kriging Technique to Obtaining the Risk Maps

Using the `gamv2` routine from the Geostatistical Software Library (GSLIB) (Deutsch and Journel, 1992), the experimental variogram of the variable thickness shown in

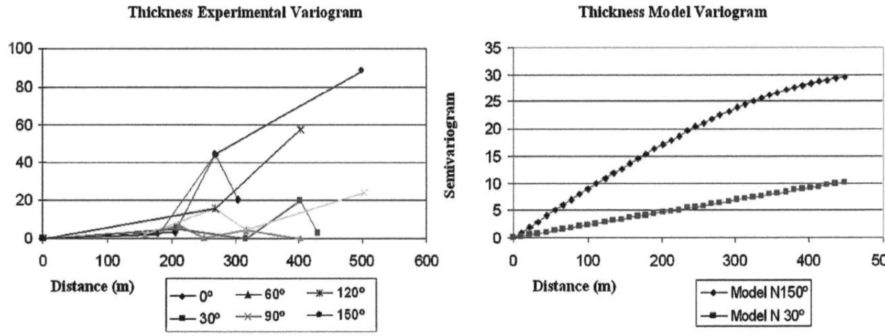

Fig. 3 Experimental and model variogram of thickness

Table 1 has been obtained for different directions. Figure 3 shows this experimental variogram and the model variogram used for kriging.

The multiGaussian kriging technique was developed as follows. First of all, an ordinary kriging of the available data (Table 1) was performed. According to Deutsch and Journel (1992) the ordinary kriging estimator is:

$$Z^*_{OK}(u) = \sum_{\alpha=1}^{n} \nu_\alpha(u)\, Z(u_\alpha)$$

and the stationary OK system:

$$\begin{cases} \sum_{\beta=1}^{n} \nu_\beta(u)\, C(u_\beta - u_\alpha) + \mu_u \\ \sum_{\beta=1}^{n} \nu_\beta(u) = 1 \end{cases}$$

where the $\nu_\alpha(u)$ are the OK weights and μ_u is the Lagrange parameter associated with the constraint $\sum_{\beta=1}^{n} \nu_\beta(u) = 1$. This process has been performed using the GSLIB routine `okb2d`.

Afterwards, if we assume in each cell a local Gaussian distribution with mean value and variance obtained by ordinary kriging, the probability maps associated with different values of the thickness of the low-K layer can then be obtained by reading the probability distribution functions on each cell. To do so, it is necessary to perform a standardization of the variable.

$$\frac{Z - \mu}{\sigma} \approx N\,[0, 1]$$

where Z is the thickness of the low-K layer, μ is the ordinary kriging estimator, and σ is the ordinary kriging variance.

The `gauinv` routine (Kennedy and Gentle, 1980) has been used to compute the inverse of the standard normal cumulative distribution function with a numerical approximation. By doing so, probability distribution functions have been estimated in each cell. These functions allow associating a probability value to each value of the variable thickness. The security level is higher as the thickness of the low-K layer increases. The risk maps are shown in Fig. 4 and have been obtained by plotting the probability values associated with the following security levels:

Level A: thickness = 2 m
Level B: thickness = 3 m
Level C: thickness = 4 m
Level D: thickness = 5 m

Given these results, the safety characterization of the aquifer is intrinsically related to assuming a certain risk level, which is associated with a value of the probability for a specific thickness. Once the minimum thickness to ensure aquifer security has been defined, the decision-maker can use the correspondent risk map to define the area that verifies a security level with a high probability value.

The USEPA VLEACH model (a one-dimensional finite difference vadose zone leaching model) (US Environmental Protection Agency (1996)) was used in order to determine the minimum thickness of low-K formation acceptable. The results of this model showed that 2-m thickness of low-K formation under the foundations

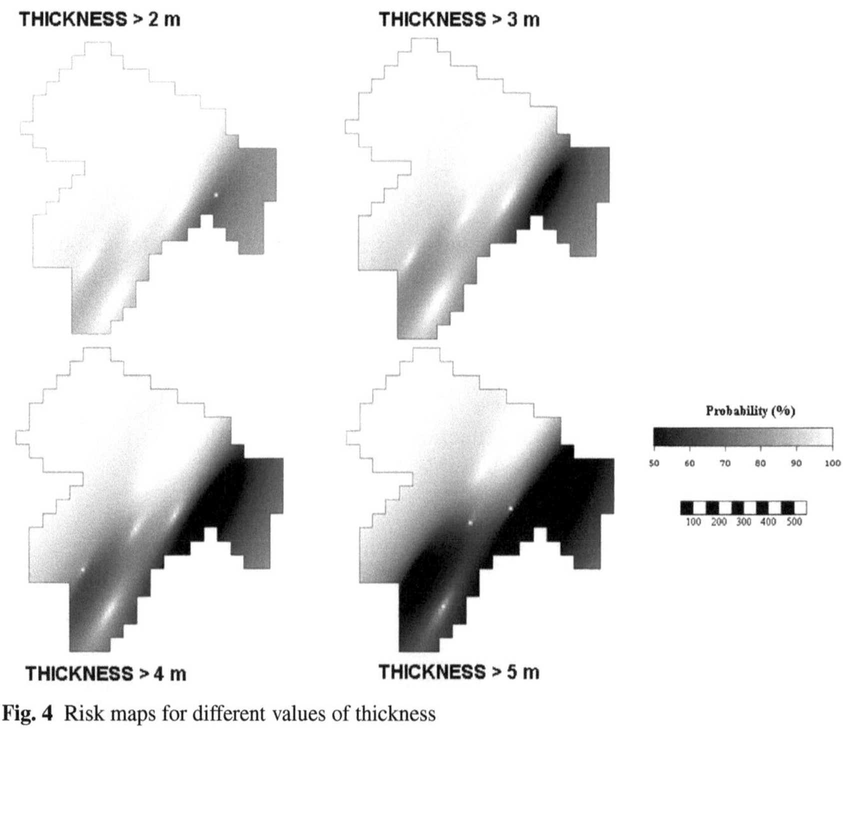

Fig. 4 Risk maps for different values of thickness

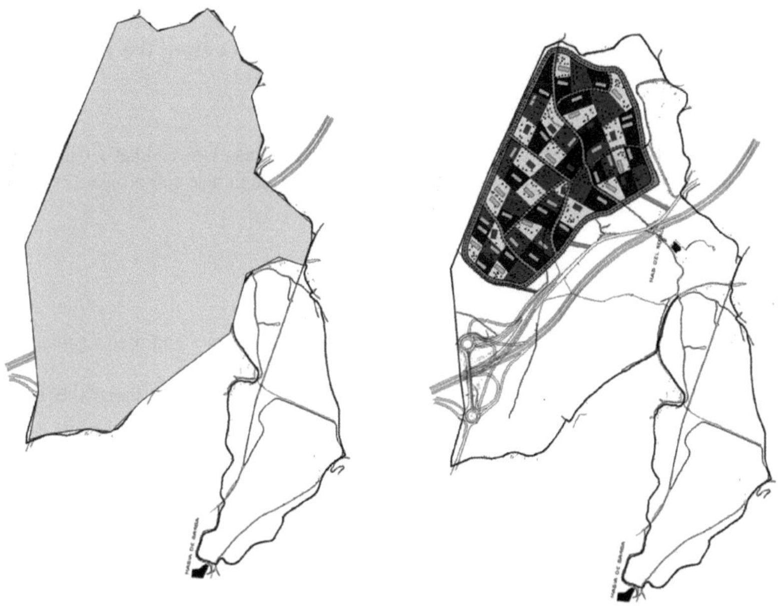

Fig. 5 Proposed area accounting only aquifer-protection matters (left) and final proposed area including every environmental issue into consideration (right)

of the buildings of the industrial area give a security level that is high enough. Using this conclusion, the decision-makers decided to use the Level C risk map (4-m thickness) and the 90% probability contour line to define the area in which to build the industrial area.

Afterwards, a complete environmental impact assessment (EIA) report was done according to the European Union environmental legislation. This EIA report took into account every topic of interest that was related to the environmental protection of the area such as flora, fauna, landscape, hydrology and interactions with other specific legislation (roads, surface water, urbanism).

Figure 5 shows the area that was selected on the basis of aquifer protection and the comparison with the final area, taking every environmental issue into consideration.

4 Conclusions

- An application of the multiGaussian kriging technique to the EIA of a new industrial site in Alcoy (Spain) has been presented.
- The methodology was used for obtaining risk maps with a small number of field data.
- Its simplicity allowed its use in a real-world EIA report, which could easily be reviewed by technicians who were not familiar with geostatistical methods.

- In combination with a simple contamination model, the risk maps allowed the decision-makers to determine the suitable area for locating the new industrial site.

Acknowledgments Thanks are due to the Conselleria de Empresa Universidad y Ciencia of the Generalitat Valenciana, which has funded this investigation through research project GV06/250.

References

Deutsch CV, Journel AG (1992) GSLIB: Geostatistical software library and user's guide. Oxford University Press

Instituto Geológico y Minero de España (1978), Mapas geológicos de España E:1:50.000. Villajoyosa, Alcoy, Castalla

Instituto Técnológico y Minero de España (2000) Unidades hidrogeológicas de España. Mapa y datos básicos

Kennedy WJ Jr, Gentle JE (1980) Statistical Computing p 95

Municipality of Alcoy (1999) Inventario de Suelos Aptos para el Uso Industrial en el término municipal de Alcoi

Technical University of Valencia – Incivsa S.L. (2002) Informe de reconocimiento del terreno de un solar situado en la zona denominada "La Canal" de Alcoi (Alicante)

US Environmental Protection Agency (1996). CSMOS. VLEACH Model version 2.2

Hydrogeological Modeling of Radionuclide Transport in Heterogeneous Low-Permeability Media: A Comparison Between Boom Clay and Ieper Clay

M. Huysmans and A. Dassargues

Abstract Deep low-permeability clay layers are considered as possible suitable environments for disposal of high-level radioactive waste. In Belgium, the Boom Clay is the reference host formation and the Ieper Clay an alternative host formation for research and safety and feasibility assessment of deep disposal of nuclear waste. In this study, two hydrogeological models are built to calculate the radionuclide fluxes that would migrate from a potential repository through these two clay formations. Transport parameters' heterogeneity is incorporated in the models using stochastic sequential simulation of hydraulic conductivity, diffusion coefficient and diffusion accessible porosity, using primary information and several types of secondary information, i.e. resistivity, gamma ray and grain size. The calculated radionuclide fluxes in the two clay formations are compared. Results show that in the Ieper Clay larger differences between the fluxes through the lower and the upper clay boundary occur than in the Boom Clay, larger total output radionuclide amounts are calculated than in the Boom Clay, and a larger effect of parameter heterogeneity on the calculated fluxes is observed, compared to the Boom Clay.

1 Introduction

Safe disposal of nuclear waste is an important environmental challenge. Several countries are investigating deep geological disposal as a long-term solution for their high-level waste. In Belgium, the Oligocene Boom Clay is the reference host formation for research purposes and for the safety and feasibility assessment of the deep disposal of high-level and/or long-lived radioactive waste. The clay layers of the Eocene Ieper Group (the Kortrijk Formation and Kortemark Member) are an alternative host formation for the research and assessment of a deep disposal solution for high-level and/or long-lived radioactive waste in Belgium (ONDRAF/NIRAS, 2002).

M. Huysmans
Applied Geology and Mineralogy, Department of Geology-Geography, Katholieke Universiteit Leuven, Belgium
e-mail: marijke.huysmans@geo.kuleuven.be

A. Soares et al. (eds.), *geoENV VI – Geostatistics for Environmental Applications*, 211–219

The aim of this study is to calculate and compare the radionuclide fluxes that would migrate from a potential repository through the clays into the surrounding aquifers. Radionuclide transport through the clays into the surrounding aquifers is calculated by means of a hydrogeological model of both clay formations. The model results for both potential host formations are analyzed and compared. Since the previous studies of the Boom Clay (Huysmans and Dassargues, 2005a; Huysmans and Dassargues, 2005b) showed that spatial variability of the transport parameters may have an effect on the calculated radionuclide fluxes, the hydrogeological models in this study incorporate parameter heterogeneity. Hydraulic conductivity, diffusion coefficient and diffusion accessible porosity heterogeneity was included in the hydrogeological models using geostatistical simulation.

2 Method

2.1 Study Sites

The Mol-Dessel zone (province of Antwerp) is the reference site for research, development and demonstration studies on the Oligocene Boom Clay. In this zone, an underground experimental facility (HADES-URF) was built in the Boom Clay at 223 m depth. In this area, the Boom Clay has a thickness of about 100 m and is overlain by 180 m of water bearing sand formations. The Doel nuclear zone (province of Antwerp) is an alternative reference site for methodological studies regarding the Eocene Ieper Clay. In this zone, the clay layers of the Ieper Group (the Kortrijk Formation and Kortemark Member) are situated at a depth of approximately 340 m and have a thickness of about 100 m.

2.2 Data Analysis

Two deep boreholes on the Mol/Dessel site and the Doel nuclear zone respectively provide the data for this study. On the Mol/Dessel site, a 570 m deep borehole (Mol-1 borehole) was drilled. Several transport and geological parameters (hydraulic conductivity K, diffusion coefficient D_e, diffusion accessible porosity η and grain size) have been intensively measured in the laboratory on cores taken at the Mol-1 borehole. To complement the knowledge about the primary variables of interest, measurements of secondary variables were also collected. Geophysical logging was performed in the Mol-1 borehole to obtain logs of gamma ray, resistivity and nuclear magnetic resonance. The resulting data set for the Boom Clay comprises 52 hydraulic conductivity values, 41 diffusion coefficient and diffusion accessible porosity measurements, a gamma ray log, an electrical resistivity log, 71 grain size measurements and a porosity log estimated from the nuclear magnetic resonance log. On the Doel nuclear zone, a series of boreholes was drilled near the Doel nuclear power station (Van Marcke and Laenen, 2005). The deepest borehole reaches a

Table 1 Average and variance of Boom Clay and Ieper Clay parameters (Huysmans and Dassargues, 2006)

	Boom Clay	Ieper Clays
Vertical hydraulic conductivity average (m/s)	7.03e-12	5.84e-12
Vertical hydraulic conductivity variance (m^2/s^2)	3.42e-22	1.29e-22
Iodide diffusion coefficient average (m^2/s)	1.62e-10	2.30e-10
Iodide diffusion coefficient variance (m^4/s^2)	8.16e-21	9.65e-21
Iodide diffusion accessible porosity average (-)	0.16	0.23
Iodide diffusion accessible porosity variance (-)	0.00037	0.0012
Grain size (d_{40}) average$^{(1)}$ (μm)	3.79	7.43
Grain size (d_{40}) variance$^{(1)}$ (μm^2)	33.93	20.83
Gamma ray average (gAPI)	84.55	78.40
Gamma ray variance ($gAPI^2$)	104.41	116.65
Resistivity average (ohm m)	7.01	1.76
Resistivity variance ($ohm^2\ m^2$)	5.83	0.05

$^{(1)}$ Grain size is expressed by the parameter d_{40}, the grain size for which 40% of the total sample has a smaller grain size.

depth of 688 m. Laboratory experiments on cores from the Doel boreholes provided 25 hydraulic conductivity values, 25 diffusion coefficient and diffusion accessible porosity measurements and 49 grain size measurements of the Ieper Clay. Geophysical logging provided logs of gamma ray and resistivity.

Comparison of the statistics of the parameters of the Boom Clay and Ieper Clay (Table 1) shows that the transport parameters have similar values for both clays. Comparison of the correlation coefficients between the parameters (Table 2) shows that hydraulic conductivity and diffusion coefficient are better correlated with the secondary variables in the Boom Clay than in the Ieper Clay.

Geostatistical estimators, i.e. variograms and cross-variograms, were calculated and modeled for all primary and secondary measurements. Variograms and cross-variograms of variables are modeled as the sum of a nugget model and a spherical model with a range of 35 m for the Boom Clay and 24 m for the Ieper Clay.

Table 2 Correlation coefficients between the parameters of the Boom Clay and the Ieper Clays (Huysmans and Dassargues, 2006)

	Boom Clay	Ieper Clays
$\log_{10} K_v$ - D	0.97	0.88
$\log_{10} K_v$ - η	0.44	0.81
$\log_{10} K_v$ - GR	–0.65	–0.53
$\log_{10} K_v$ - RES	0.73	0.41
$\log_{10} K_v$ - grain size	0.95	0.78
D - η	0.36	0.80
D - GR	–0.63	–0.38
D - RES	0.66	0.53
D - grain size	0.93	0.92
η - GR	–0.20	–0.49
η - RES	0.20	0.36
η - grain size	0.28	0.03

The sills are fitted by the optimization program LCMFIT2 (Pardo-Iguzquiza and Dowd 2002).

2.3 Stochastic Sequential Simulation of the Transport Parameters

The Boom Clay and the Ieper Clay shows a lateral continuity that largely exceeds the extent of the local scale model. Therefore it is assumed that the properties of the clays do not vary in the horizontal direction and one-dimensional vertical realizations of hydraulic conductivity, diffusion accessible porosity and diffusion coefficient are generated. The simulations are performed using direct sequential simulation with histogram reproduction (Oz et al. 2003). Figures 1 and 2 show examples of simulated fields of hydraulic conductivity, diffusion coefficient and diffusion accessible porosity in the Boom Clay and the Ieper Clay.

2.4 Hydrogeological Models

A local 3D hydrogeological model of the Boom Clay and a model of the Ieper Clay are constructed. Both models have the same size ($20\,\mathrm{m} \times 15\,\mathrm{m} \times 102\,\mathrm{m}/104\,\mathrm{m}$) and grid spacing (between 0.2 m and 1 m). The vertical boundary conditions for groundwater flow are zero flux boundary conditions since the hydraulic gradient is vertical. The horizontal boundary conditions for groundwater flow are Dirichlet conditions. The vertical hydraulic gradient is approximately 0.02 in the Boom Clay (Wemaere and Marivoet, 1995) and 0.25 in the Ieper Clay (ONDRAF/NIRAS 2002). The vertical hydraulic gradient in the Ieper Clay is more than twelve times larger

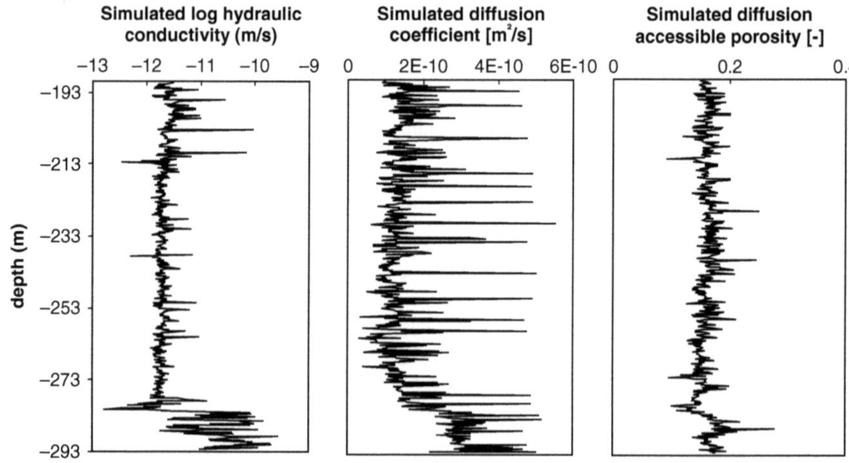

Fig. 1 Simulated hydraulic conductivity (m/s), diffusion coefficient (m^2/s) and diffusion accessible porosity (-) of the Boom Clay in the Mol-1 borehole (Huysmans and Dassargues, 2006)

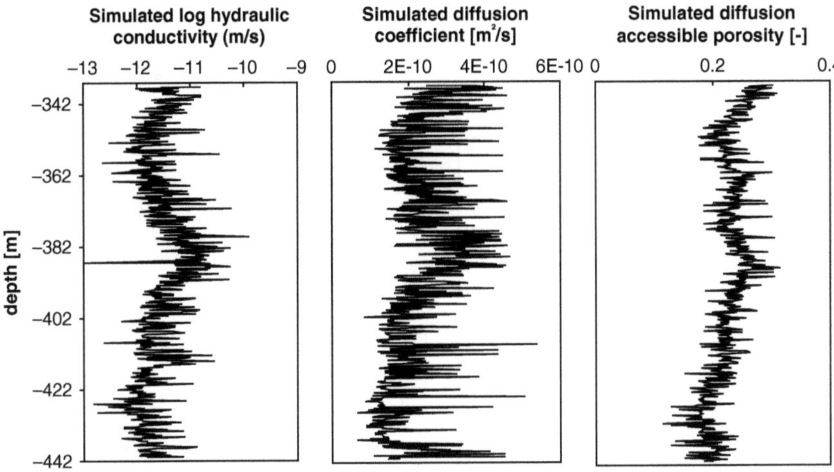

Fig. 2 Simulated hydraulic conductivity (m/s), diffusion coefficient (m^2/s) and diffusion accessible porosity (-) of the Ieper Clay in the Doel borehole (Huysmans and Dassargues, 2006)

than in the Boom Clay and oriented in the opposite direction. Although it is likely that these gradients vary over the long time periods considered, they are assumed to be constant in this study.

Transport by advection, dispersion, molecular diffusion and radioactive decay is calculated for 3 radionuclides: ^{79}Se, ^{129}I and ^{99}Tc. The boundary conditions for transport at the upper and lower boundaries are zero concentration boundary conditions (Mallants et al., 1999) since the hydraulic conductivity contrast between the clay and the aquifer is so large that solutes reaching the boundaries are assumed to be flushed away by advection in the aquifer. In both models, the same source term is inserted: an applied flux source or an applied concentration source depending on the effect of the solubility limit (Mallants et al., 1999). The initial transport condition is a zero concentration condition.

The 2 local 3D hydrogeological models are run with FRAC3DVS, a simulator for three-dimensional groundwater flow and solute transport in porous, discretely-fractured porous or dual-porosity formations (Therrien and Sudicky, 1996; Therrien et al., 2003). The models are run for 10 different random combinations of simulations of hydraulic conductivity, diffusion coefficient and diffusion accessible porosity.

3 Results

Figure 3 shows the computed total radionuclide fluxes through the lower and upper clay boundaries of the Boom Clay and the Ieper Clay for 10 different equally probably simulations. The total amount of radionuclides leaving the clay M (Bq) was calculated as flux integrated over time for each simulation and is also indicated

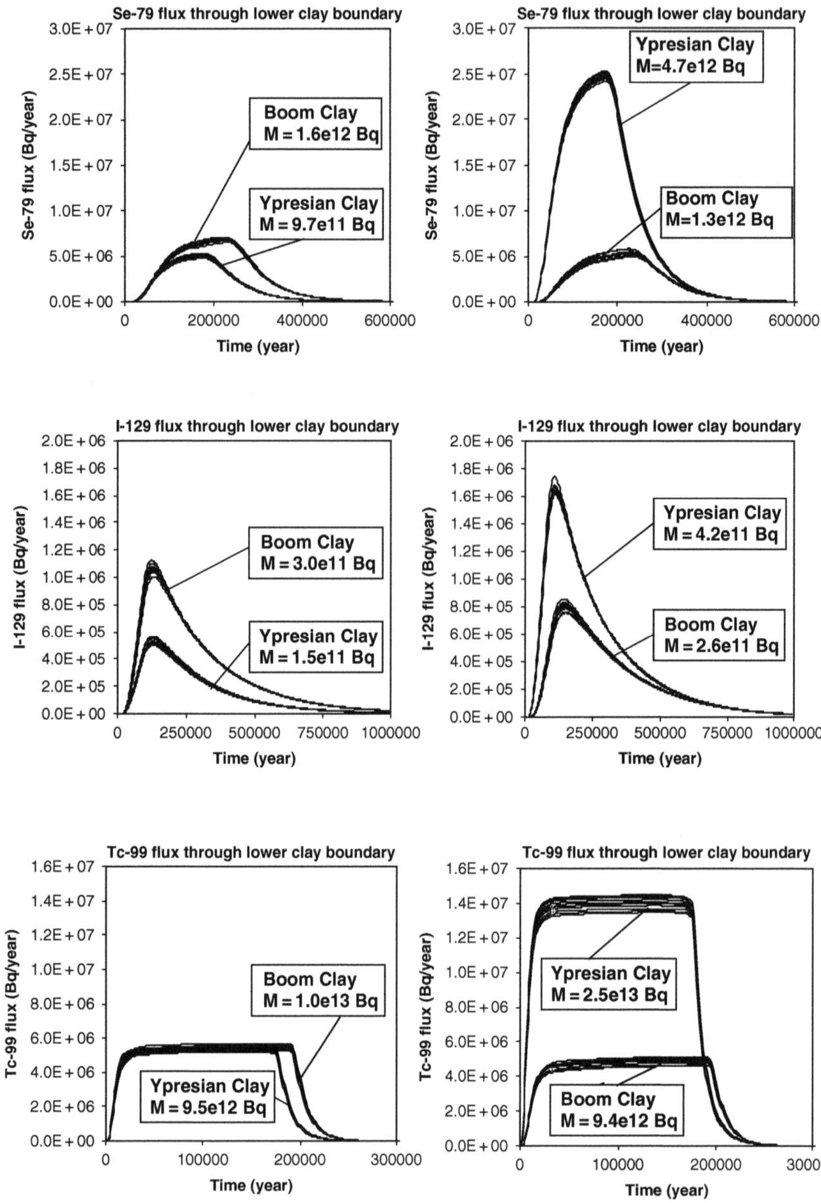

Fig. 3 Computed total radionuclide fluxes (Bq/year) versus time (year) through the lower and upper clay boundaries of the Boom Clay and the Ieper Clay for 10 different realizations of random fields of the transport parameters (Huysmans and Dassargues, 2006)

on Fig. 3. For the Boom Clay, the difference between the total radionuclide amount leaving through the lower and upper clay boundary is between 6% (^{99}Tc) and 23% (^{79}Se). For the Ieper Clay, the total radionuclide amount leaving through the upper clay boundary is between 2.6 (^{99}Tc) and 4.8 (^{79}Se) times larger than the total radionuclide amount leaving through the lower clay boundary. Comparison of the total radionuclide amounts leaving the Boom Clay and the Ieper Clay also shows that approximately twice as much radionuclides leave the Ieper Clay compared to the Boom Clay.

A comparison is made between the radionuclide amounts leaving the clays calculated with heterogeneous simulations and homogeneous models with a homogeneous hydraulic conductivity, diffusion coefficient and diffusion accessible porosity equal to the arithmetic averages of the measurements. Arithmetic averages instead of effective parameters were chosen to compare the heterogeneous models of this study with earlier homogeneous models made by other agencies that used the arithmetic average. For the Boom Clay, there is a maximum difference of 27% between the radionuclide amounts calculated by the homogeneous and heterogeneous models. For the Ieper Clay, there is a maximum difference of 59% between the radionuclide amounts calculated by the homogeneous and heterogeneous models. These values show that incorporating parameter heterogeneity has a larger effect in the Ieper Clay than in the Boom Clay.

4 Discussion

In the Ieper Clay, larger differences between the fluxes through the lower and the upper clay boundary and larger total output radionuclide amounts are calculated than in the Boom Clay. Differences between the fluxes through the lower and the upper clay boundaries can only be attributed to transport by advection since in a pure diffusion model with a source in the middle of the clay the output fluxes through the lower and the upper clay boundary would be identical. These results show that the effect of upward advective transport in the Ieper Clay is much larger than the effect of downward advective transport in the Boom Clay. Since all flow and transport parameters have similar values in both formations, this difference in results is probably due to the difference in hydraulic gradient. The gradient in the Ieper Clay is more than twelve times larger than in the Boom Clay and oriented in the opposite direction. This results in a larger contribution of transport by advection in the Ieper Clay than in the Boom Clay, in larger differences between the fluxes through the lower and the upper clay boundary and larger total output radionuclide amounts in the Ieper Clay than in the Boom Clay.

The larger effect of parameter heterogeneity in the Ieper Clay compared to the Boom Clay can also not be completely explained by differences in parameter variability. All transport parameters have mean values and variances in the same order of magnitude, as demonstrated by the statistics in Table 1. Detailed examination of the effect of heterogeneity in the Ieper Clay shows that the heterogeneity of hydraulic conductivity has a larger effect than the heterogeneity of the diffusion parameters in this clay. The larger effect of parameter heterogeneity in the Ieper Clay is therefore mainly a larger effect of hydraulic conductivity heterogeneity in the Ieper Clay compared to the Boom Clay. Since the hydraulic conductivity variation is not significantly larger in the Ieper Clay compared to the Boom Clay, the higher effect of K heterogeneity is probably also caused by the larger gradient. Since the gradient is larger, tranport by advection is a more important process in the Ieper Clay. Therefore, the results are more sensitive to K heterogeneity.

5 Conclusions

In this study, the radionuclide fluxes that would migrate from a potential nuclear waste repository through the Boom Clay and the Ieper Clay were modeled and compared. Two hydrogeological models were built to calculate the radionuclide fluxes through these two clay formations. Transport parameter heterogeneity was incorporated in the models using geostatistical co-simulations of hydraulic conductivity, diffusion coefficient and diffusion accessible porosity. The calculated radionuclide fluxes in the two clay formations were compared with the results from homogeneous models and with the results of the other clay formation.

A first conclusion of this study is that differences of up to 59% of the calculated output radionuclide amounts between heterogeneous and homogeneous models are observed. This study thus demonstrates that parameter heterogeneity can have an important effect on the results and should be incorporated in transport studies in low permeability media.

Comparison of the results of the Boom Clay and the Ieper Clay show that in the Ieper Clay (1) larger differences between the fluxes through the lower and the upper clay boundary occur, (2) larger total output radionuclide amounts are calculated and (3) a larger effect of parameter heterogeneity on the calculated fluxes is observed, compared to the Boom Clay. These results are explained by the larger and inversely oriented hydraulic gradient in the Ieper Clay that results in a larger importance of transport by advection in this clay. Since both the radionuclides fluxes and the effect of heterogeneity on these fluxes are largely affected by the direction and magnitude of the hydraulic gradient and since the gradient in nuclear waste disposal studies is subject to large uncertainty due to the large time periods considered, this study illustrates the importance of using a range of possible hydraulic gradients as input for safety studies.

Acknowledgments The authors wish to acknowledge the Fund for Scientific Research – Flanders for providing a Research Assistant scholarship to the first author. We also wish to thank ONDRAF/NIRAS (Belgium agency for radioactive waste and enriched fissile materials) and SCK-CEN (Belgian Nuclear Research Centre) for providing the necessary data for this study. We also thank René Therrien and Rob McLaren for providing Frac3dvs and for their assistance.

References

Huysmans M, Dassargues A (2005a) Stochastic analysis of the effect of heterogeneity and fractures on radionuclide transport in a low permeability clay layer. Enviromental Geology 48(7): 920–930

Huysmans M, Dassargues A (2005b) Stochastic analysis of the effect of spatial variability of diffusion parameters on radionuclide transport in a low permeability clay layer. Proceedings of ModelCARE2005, the fifth international conference on calibration and reliability in groundwater modelling: From uncertainty to decision making, The Hague (Scheveningen), The Netherlands, 6–9

Huysmans M, Dassargues A (2006) Hydrogeological modeling of radionuclide transport in low permeability media: a comparison between Boom Clay and Ypresian Clay. Environ Geo 50 (1): 122–131

Mallants D, Sillen X, Marivoet J (1999) Geological disposal of conditioned high-level and long lived radioactive waste: Consequence analysis of the disposal of vitrified high-level waste in the case of the normal evolution scenario. ONDRAF/NIRAS report R-3383, Brussel, Belgium

ONDRAF/NIRAS (2002) Safety Assessment and Feasibility Interim Report 2 - SAFIR 2. NIROND 2001 - 06 E, Brussel, Belgium

Oz B, Deutsch CV, Tran TT, Xie Y (2003) DSSIM-HR: A FORTRAN 90 program for direct sequential simulation with histogram reproduction. Comput & Geosci 29 (1): 39–51

Pardo-Iguzquiza E, Dowd PA (2002) FACTOR2D: a computer program for factorial cokriging. Comput and Geosci 28(8): 857–875

Therrien R, Sudicky EA (1996) Three-dimensional analysis of variably-saturated flow and solute transport in discretely-fractured porous media. J Contam Hydro 23 (1–2): 1–44

Therrien R, Sudicky EA, McLaren RG (2003) FRAC3DVS: An efficient simulator for three-dimensional, saturated-unsaturated groundwater flow and density dependent, chain-decay solute transport in porous, discretely-fractured porous or dual-porosity formations User's guide

Van Marcke Ph, Laenen B (2005) The Ieper clays as possible host rock for radioactive waste disposal: an evaluation. ONDRAF/NIRAS

Wemaere I, Marivoet J (1995) Geological disposal of conditioned high-level and long lived radioactive waste: updated regional hydrogeological model for the Mol site (The north-eastern Belgium model). ONDRAF/NIRAS Report R-3060, Brussel, Belgium

Topological Kriging of Runoff

J. O. Skøien and G. Blöschl

Abstract In this paper we spatially interpolate hourly runoff data by a Top-kriging (Skøien 2006) approach and compare the results with ordinary kriging and a deterministic rainfall-runoff model. Cross-validation indicates that the Top-kriging approach performs better than both ordinary kriging and the deterministic model for a large number of catchments in Austria. We suggest that the Top-kriging approach can be used for filling in temporal gaps in observed runoff time series and for real time spatial mapping of the flow situation.

1 Introduction

Stream flow related variables are in many ways different from other variables that are generally interpolated using geostatistical interpolation methods. Although variables are usually measured at a single location (e.g. runoff gauge, temperature measurement, concentration of a pollutant), the measured variable is usually an integrated value, a value that has been filtered by the catchment both in space and in time. The variable can be seen to have a finite support. Additionally, there is a connection between catchments in the sense that water (and physical characteristics or what is dissolved in the water) from an upstream catchment will also be a part of the measurement of a downstream catchment.

This is in contrast to traditional kriging, where the measurements are either point variables or have a non-overlapping support (e.g. pixels in a map). Because of this, geostatistics have rarely been used for direct interpolation of stream flow characteristics along streams. It has been more common to regionalize flow characteristics by regionalizing parameters of deterministic rainfall-runoff models (Bárdossy 2006; Merz and Blöschl 2005; Parajka et al. 2005). Some authors have developed alternative stochastic models for random variables defined on trees (Monestiez et al., 2005, Bailly et al., 2006) but these methods are not for conservative variables as stream

J. O. Skøien
Department of Physical Geography, Utrecht University, P.O. box 80115, 3508 TC Utrecht, The Netherlands
e-mail: j.skoien@geo.uu.nl

A. Soares et al. (eds.), *geoENV VI – Geostatistics for Environmental Applications*, 221–231

flow. Notable exceptions are Gottschalk and co-workers (Gottschalk 1993a; Gottschalk 1993b; Gottschalk et al. 2006; Sauquet et al. 2000). Skøien et al. (2006) worked with similar methods and presented Top-Kriging for interpolation along stream networks. Skøien and Blöschl (2006) used geostatistics to analyze the filtering effect of catchments on runoff. In this paper, we will present the first results from combining the methods above, to interpolate runoff time series.

One particular reason for considering geostatistics, is that geostatistical interpolation in many cases can be more accurate than regionalization of parameters of a rainfall-runoff model. Whereas runoff can be measured with relatively high accuracy, regionalization of a deterministic rainfall-runoff models is complicated by a range of uncertainties including the identification of parameters, uncertainties regarding possible pedo-transfer functions between catchment characteristics and model parameters, model uncertainties and input data uncertainties (precipitation as the most important). In areas with a relatively dense net of runoff gauges, we argue that a direct interpolation of runoff in many cases will be equally or more accurate than regionalization of the parameters of a deterministic rainfall-runoff model.

2 Data

The data used in this paper stem from a comprehensive hydrographic data set of Austria. Austria has a varied climate with mean annual precipitation ranging from 500 mm in the eastern lowland regions up to about 3000 mm in the western alpine regions. Runoff depths range from less than 50 mm per year in the eastern part of the country to about 2000 mm per year in the Alps. Potential evapotranspiration is in the order of 600–900 mm per year.

Altogether, we have runoff time series from 591 catchments, with areas ranging from 3 to 130 000 km^2. The series have been recorded with breakpoints (high temporal resolution when runoff changes fast, lower temporal resolution when runoff does not change), and interpolated into time series with 60 min temporal resolution. Catchments smaller than 10 km^2, stream gauges with short records and catchments with significant anthropogenic effects and lake effects were excluded from the data set, which gave a total of 488 stream gauges available for the analysis. Figure 1 shows the centroids of the catchments.

3 Methods

3.1 Concept

Skøien and Blöschl (2006) refer to catchments as spatio-temporal filters. Assuming that runoff is generated at the point scale, they suggest that the runoff of a catchment i, measured at a certain time t is:

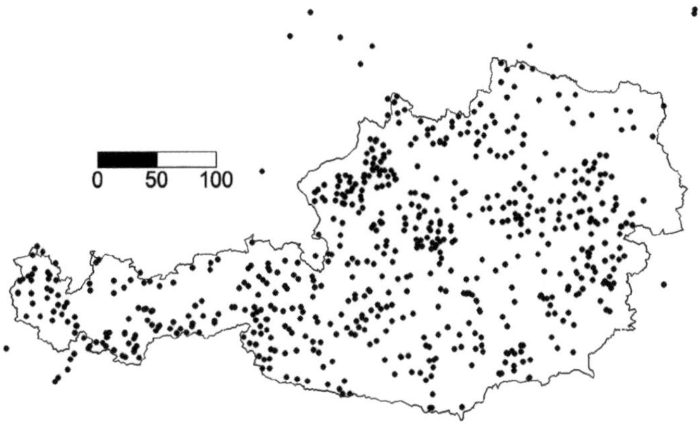

Fig. 1 Centroids of gauged catchments

$$Q_i(t) = \iint\limits_{A_i} \int\limits_{t-T_i}^{t} W(x, y, \tau)u(x, y, \tau)d\tau dx dy \tag{1}$$

where $W(x, y, t)$ is a function describing the runoff generated at point (x, y) and time t, while the unit hydrograph $u(x, y, t)$ can be seen as local unit hydrograph or a weighting function for each point in space and time. A_i refers to the catchment area and T_i is the time interval that influences the output, with τ as the temporal integration variable. The equation describes that the instantaneous runoff at a runoff gauge consists of runoff generated within the catchment during a time period T_i. The weight function $u(x, y, t)$ is dependent on the temporal distribution function of water from a point within the catchment to the outlet. This weight function will then take into account that runoff generated in areas close to the outlet or close to a stream inside the catchment will reach the outlet faster than runoff generated further away from streams, and that the water generated at a certain point and time step will experience dispersion before reaching the outlet. If we use a constant T_i for the catchment, it is likely that $u(x, y, t) = 0$ for τ close to the upper and lower integration limits for most parts of the catchments. $u(x, y, t)$ for a certain point will also change with time, taking into account the effect of changed velocities in the catchment due to changes in catchment state variables.

The function $u(x, y, t)$ is therefore likely to be too complicated in space and time for modeling purposes. As a first simplification, Skøien and Blöschl (2006) suggested to set

$$\begin{aligned} u(x, y, t) &= 1/T_i, \qquad & t - T_i < \tau < t \\ u(x, y, t) &= 0 & \tau < t - T_i \end{aligned} \tag{2}$$

This means that runoff generated at a certain time step will be distributed over a time T_i at the outlet, or considered in the opposite way, that the runoff at the outlet at a certain time step is equal to the average of the runoff generated within the catchment and within the previous period T_i. Skøien and Blöschl (2006) suggested that T_i is related to the catchment area A_i as:

$$T_i = \mu A_i^{\kappa} \tag{3}$$

where μ and κ are parameters to be fitted.

3.2 Interpolation of Time Series

The method presented in this paper is based on the Top-kriging approach by Skøien et al. (2006). In a review paper, Kyriakidis and Journal (1999) suggests viewing spatio-temporal random fields either as vectors of temporally correlated spatial random fields, or as vectors of spatially correlated time series. As time series of runoff have recordings with a relatively high resolution in time and lower in space, we are more interested in interpolation in space, in addition to filling in temporal gaps of existing stations. It is therefore sensible to model runoff as a set of spatially correlated time series.

The simplest way to perform interpolation of time series is to consider each time step as a separate spatial field, and then carry out the interpolation for each time step t_ω separately:

$$\hat{q}(\mathbf{x}_i, t_\omega) = \sum_{j=1}^{n} \lambda_j q(\mathbf{x}_j, t_\omega) \tag{4}$$

where n is the number of stations used for the interpolation, λ_j is the interpolation weight of the measurements at position $\mathbf{x}_j$, and q refers to the specific runoff ($q(\mathbf{x}_i, t_\omega) = Q(\mathbf{x}_i, t_\omega)/A_i$). If we assume that we can use the same weights for all time steps, we find these weights from solving the kriging system:

$$\begin{aligned} \sum_{k=1}^{n} \lambda_k \gamma_{jk} + \mu = \gamma_{ij} \quad j = 1, \ldots, n \\ \sum_{j=1}^{n} \lambda_j = 1 \end{aligned} \tag{5}$$

γ_{jk} refers to the gamma value or the expected semivariance between two measurements j and k, as found from a theoretical semivariogram model. μ is the Lagrange parameter. For point kriging γ_{jk} can be found directly from a theoretical variogram model fitted to a sample variogram (Matheron 1965).

Although it is likely that the correlation structure will change with different event types, we will estimate the weights only once for each catchment to be interpolated. Due to the squared differences, the variograms are likely to be more affected by the larger events. These are also the most difficult to estimate properly. Hence, we find this a reasonable assumption.

Top-kriging does not take precipitation into account, and does hence not necessarily fulfill the mass balance (precipitation, evapotranspiration, runoff, storage), as it only attempts to estimate the runoff part of the mass balance. This is not necessarily bad, as a deterministic rainfall-runoff model will fulfill the mass balance given input variables (precipitation, evapotranspiration) with large uncertainties.

3.3 Estimation of Regularized Gamma Values

As described in 3.1, we see runoff measurements as variables with a spatio-temporal support. It is therefore necessary to modify the kriging system from Eq. 5. The important difference is that we need to take the support of the measurements into account when estimating the gamma values. Following Cressie (1991, p. 66) and Skøien and Blöschl (2006), the gamma value γ_{ij} in Eq. 5 for a pair of catchments i and j is found by the help of a point variogram $\gamma_p(h_s, h_t)$:

$$\begin{aligned}
&\gamma_{ij}(h_s \left|(A_i, A_j)\right.) = \\
&\frac{1}{A_i A_j T_i T_j} \int_{A_i}\int_{A_j}\int_0^{T_i}\int_0^{T_j} \gamma_p(|\,\mathbf{r}_1 - \mathbf{r}_2|\,, |\tau_1 - \tau_2|) d\tau_1 d\tau_2 d\mathbf{r}_1 d\mathbf{r}_2 \\
&-0.5 * \left[\frac{1}{A_i^2 T_i^2} \int_{A_i}\int_{A_i}\int_0^{T_i}\int_0^{T_i} \gamma_p(|\,\mathbf{r}_1 - \mathbf{r}_2|\,, |\tau_1 - \tau_2|) d\tau_1 d\tau_2 d\mathbf{r}_1 d\mathbf{r}_2 \right. \\
&\left. + \frac{1}{A_j^2 T_j^2} \int_{A_j}\int_{A_j}\int_0^{T_j}\int_0^{T_j} \gamma_p(|\,\mathbf{r}_1 - \mathbf{r}_2|\,, |\tau_1 - \tau_2|) d\tau_1 d\tau_2 d\mathbf{r}_1 d\mathbf{r}_2 \right]
\end{aligned} \tag{6}$$

where $h_s = |\mathbf{h}_s|$ is the spatial lag, h_t refer to temporal lags, $\mathbf{r}_1$ and $\mathbf{r}_2$ are integration vectors in space within the two catchments while τ_1 and τ_2 are integration variables in time within the temporal supports of the two catchments. The theoretical point variogram $\gamma_p(h_s, h_t)$ has to be inferred from a sample variogram.

We back-calculated a spatio-temporal point variogram in a similar way as Skøien and Blöschl (2006). They divided the catchments into three size classes, and inferred spatio-temporal sample variograms separately for each of the three classes based on runoff time series of 15 min temporal resolution. We did also infer the cross-variograms (also referred to as pseudo-cross variograms, Myers, 1991) between different catchments size classes S_i and S_j:

$$\begin{aligned}
&\hat{\gamma}_{st}(S_i, S_j, h_s, h_t) = \\
&\frac{1}{2 \sum\limits_{k=1}^{m(S_i,S_j,h_s)} n_k(h_t)} \sum_{k=1}^{m(S_i,S_j,h_s)} \sum_{l=1}^{n_k(h_t)} (q(\mathbf{x}_k + \mathbf{h}_s, t_l + h_t) - q(\mathbf{x}_k, t_l))^2
\end{aligned} \tag{7}$$

where $m(S_i,S_j,h_s)$ is the number of pairs of stations with distance h_s, and $n_k(h_t)$ is the number of pairs of points in time with time lag h_t. h_s was in Eq. 7 taken as the distance between the centers of gravity of the catchments. Skøien and Blöschl (2006) then assumed that all catchments of a certain class had the same size. It was then possible to back-calculate a point variogram by fitting regularized gamma values (similar to Eq. 6) to the observed gamma values. We used the exponential variogram also tested by Skøien and Blöschl (2006):

$$\gamma_{st}(h_s, h_t) = a(1 - \exp(-((ch_t + h_s)/d)^b)) + a_s h_s^{b_s} + a_t h_t^{b_t} \tag{8}$$

The first term of this variogram represents the stationary part. a gives the variance of the (stationary) process, c relates space and time, d is the combined correlation length, and b gives the slope of the variogram. The second and the third terms give the non-stationary parts of the variogram in space and time, respectively.

3.4 Cross Validation

To test the method, we performed a cross validation of the measurements in the data set. Using the years 1997–1999 for testing, we estimated the hourly time series of runoff for each catchment based on the runoff measurements of neighboring catchments. For comparison, we also estimated runoff by ordinary kriging. From the 488 stations used for calculating the variogram, 421 stations had measurements from this three year period. We used 8 neighbors in all cases for simplicity.

For the estimated time series from each catchment i, we calculated the model efficiency (ME_i) according to Nash and Sutcliffe (1970):

$$ME_i = 1 - \frac{\frac{1}{\Omega}\sum_{\omega=1}^{\Omega}(q_i(\omega) - \hat{q}_i(\omega))^2}{\sigma_{q_i}^2} \tag{9}$$

where Ω is the number of time steps interpolated and $\sigma_{q_i}^2$ is the variance of the observations in the same time period. This gives $ME \leq 1$, where $ME = 1$ indicates perfect interpolation and $ME = 0$ means that the efficiency of the interpolation method performs no better than the mean of the observations. Negative values indicate yet worse performances.

We first compared the Top-kriging approach with ordinary kriging. We computed spatial variograms averaged over the time series as in Eq. 7, but with $h_t = 0$ and without separating the catchments according to catchment size. To this sample variogram, we fitted the spatial version of Eq. 8:

$$\gamma_s(h_s) = a(1 - \exp(-(h_s/d)^b)) + a_s h_s^{b_s} \tag{10}$$

The same parameter symbols are used to show the similarity between the equations.

In addition to the comparison between the two different kriging methods, we also compared our results with results from simulations with a deterministic rainfall-runoff model from a study of Parajka et al. (2005). The runoff model is a conceptual soil moisture accounting scheme that uses precipitation and air temperature data as inputs and runs on a daily time step. It consists of a snow routine, a soil moisture routine and a flow routing routine and involves 14 model parameters. Three of the parameters were preset in their study, leaving 11 parameters to be found by model calibration. Parajka et al. (2005) first calibrated the model to 320 catchments in Austria. They then regionalized the calibrated model parameters by different methods and examined the model performance for the ungauged catchment case by cross-validation. The comparison includes 207 stations that are common for both studies. As Parajka et al. (2005) used daily runoff values, we averaged the hourly estimates of the Top-kriging approach and the observations of these stations to daily values. The model efficiencies found here were then compared to the model efficiencies obtained by the deterministic rainfall-runoff model using two different parameter sets. In the first case, parameters have been calibrated to the runoff time series for each of the catchments, i.e. at site calibration. In the second case the parameters have been inferred from neighboring catchments using kriging, i.e. regionalized calibration.

4 Results

The estimated parameters of the point variogram Eq. 8 and the spatial variogram for ordinary kriging Eq. 10 are shown in Table 1.

Figure 2 shows the cumulative distribution functions (cdf) of the model efficiencies, both from Top-kriging and from ordinary kriging. There is a small, but consistent increase in the model efficiency from ordinary kriging to Top-kriging. The Top-kriging approach is able to model three quarter of the catchments with $ME>0.5$, and that the median is 0.79, whereas the median of the ordinary kriging approach is 0.76. A comparison of individual estimates indicates that Top-kriging outperforms ordinary kriging for 266 of the 421 catchments.

There are some catchments where the method does not perform well. About 10 percent of the catchments have $ME<0$, which is too low for any application purposes. We have not yet examined the reasons for these results and how we can identify poor interpolation results without cross-validation, which is important for estimation of

Table 1 Parameters of the point variogram Eq. 12 and Eq. 10 estimated for the Innviertel region

Parameter	Spatio-temporal point variogram	Spatial variogram	Units
A	0.0017	0.0019	$m^6s^{-2}km^4$
B	0.32	0.53	
C	0.36		$km\,hr^{-1}$
D	3.5	490.	km
a_s	0.012	0.00012	$m^6s^{-2}km^4$
a_t	0.0049		$m^6s^{-2}km^4$
b_s	0.00025	0.0010	–
b_t	0.015		–
μ	1.4		hrs
κ	0.13		–

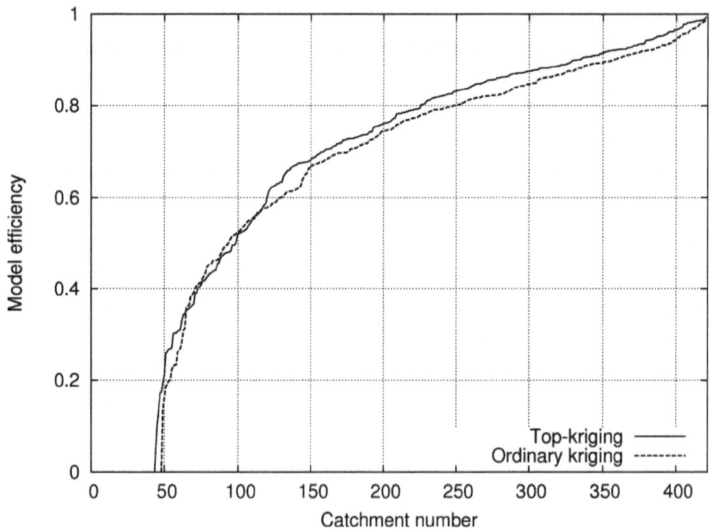

Fig. 2 Comparison of cumulative distribution functions of model efficiencies from estimation of hourly runoff from Top-kriging (solid) and ordinary kriging (dashed)

ungauged catchments. But as a starting point, we can assume that if one catchment is an outlier (it has a time series that does not behave according to the assumed homogeneity of the process), this time series will also contaminate the interpolated time series of its neighbors, giving poor results for a complete region. A closer analysis of spatial pattern of catchments with poor interpolation performance did indeed indicate that most of them were concentrated in smaller regions. We therefore think a thorough examination of regions with poor results will substantially increase the level of the left part of the cumulative distribution function.

Figure 3 shows the cumulative distribution functions of model efficiencies from daily runoff, together with the model efficiencies from the deterministic rainfall-runoff model from Parajka et al. (2005), both the results from model runs with regionalized parameters, and from model runs with the parameters calibrated at site. This indicates that the Top-kriging approach is much better than the deterministic model for most of the catchments. The cdf of Top-kriging is always larger than that of the deterministic model with regionalized parameters. However, if the parameters of the deterministic model are fitted individually for each of the catchments, the deterministic model performs better for some catchments. The model efficiencies of Top-kriging has a median of 0.88, it is only 0.75 for the deterministic model calibrated at site and 0.67 for the regionalized deterministic model. A comparison of individual estimates indicates that Top-kriging outperforms the deterministic model calibrated at site for 154 of the 207 catchments. Top-kriging outperforms the regionalized deterministic model for 175 of the 207 catchments.

If we compare Figs. 2 and 3, it is interesting to note that the model efficiencies increase when we interpolate daily runoff instead of hourly runoff. The most

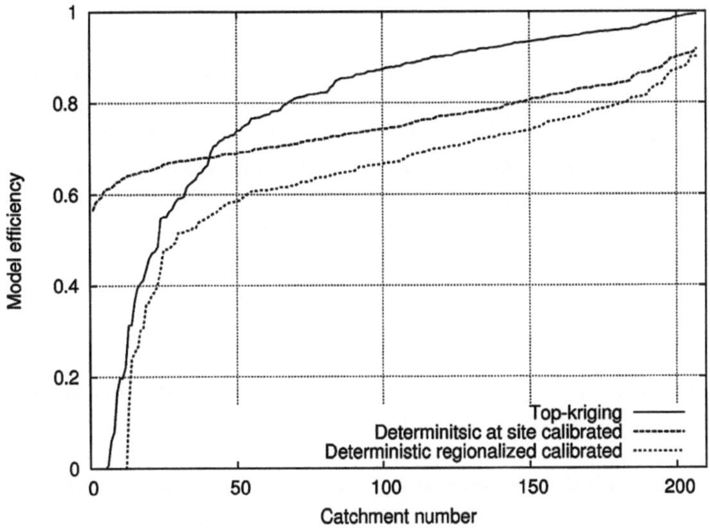

Fig. 3 Comparison of cumulative distribution functions of model efficiencies from estimation of daily runoff from Top-kriging (solid) and from the deterministic rainfall runoff model (from regionalized parameters dotted, from at site calibrated parameters dashed)

obvious reason is that temporal shifts of an estimated flow peak in comparison to the observed flow peak will have larger impact for hourly runoff than for daily runoff.

As an example, Fig. 4 shows hydrographs from a two month period in 1998 for the station Riedau in Innviertel. The upper part of the figure shows the estimated and the observed runoff time series for this period. The lower part of the figure shows

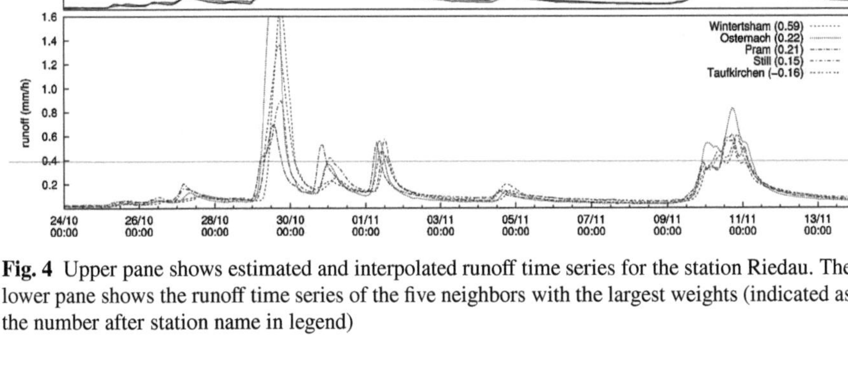

Fig. 4 Upper pane shows estimated and interpolated runoff time series for the station Riedau. The lower pane shows the runoff time series of the five neighbors with the largest weights (indicated as the number after station name in legend)

the runoff time series for the five neighbors with the largest weights. This region is a homogenous region, where floods occur practically at the same time, which explains that this station could be estimated with a model efficiency of 0.88. One of the neighboring stations (Taufkirchen) has been allocated a negative weight. The reason is that both Taufkirchen and Winertsham are downstream neighbors of Riedau, with Taufkirchen being the downstream station of both. This is thus an effect of the ability of kriging to reduce the weights for clustered observations. Pram is the upstream neighbor of Riedau, while the last two stations do not share catchment area.

5 Discussion

We have tested the use of a modified Top-kriging approach for interpolating runoff time series in space. The cross validation shows that there is an improvement in comparison to ordinary kriging. As important as the improvement is that the weights are more reliable in a hydrological context, in the sense that upstream and downstream gauges, everything else being equal, get more weight than non-nested catchments, and that measurements from large catchments are seen as regionally more representative than measurements from smaller catchments (see Skøien et al., 2006 for spatial examples).

The comparison with results from a deterministic rainfall-runoff model showed that the Top-kriging approach performed better than the deterministic model. The reason is partly that many regions in Austria are densely gauged, there are strong correlations between measurements, and in particular between upstream and downstream neighbors. Secondly, while regionalization of a rainfall-runoff model introduces a range of uncertainties that will accumulate, the uncertainties of geostatistical interpolation is limited to the assumption of spatial homogeneity of runoff and the spatial distribution of runoff gauges.

Although the Top-kriging approach performed well, the method yielded poor results for some stations. As the poorer results were mostly concentrated in small regions, we think it is possible to reduce the number of stations that are not modeled satisfactorily by a more thorough consideration of the hydrology and the physical properties of the catchments, and by examining the input and the output data closer. Another possibility for identifying stations were the Top-kriging approach performs poorly is through analyses of the kriging uncertainty. Skøien et al. (2006) used estimates of kriging uncertainty to identify catchments that were difficult to model due to the spatial configuration of runoff gauges. This can help to identify catchments that are likely to be poorly estimated due to the configuration of gauges. However, these analyses will not help us for catchments that do not correspond to the assumption of homogeneity.

It is an underlying assumption that the point process is constant, or at least intrinsic. Although this might contradict with the intuitive idea that there are more and quicker runoff from high altitudes and steep slopes, many of these catchments are small and will also of that reason have a larger variance if compared to the neighbour. If we assume that the distribution of slopes, rocks etc. are also random variables that fulfills the intrinsic hypothesis, then we can also assume that the point runoff is an intrinsic random variable.

There are a number of applications for the method as presented in this paper. It can be used for filling in temporal gaps in runoff time series. In that case it will be possible to do a cross-validation with the existing time series, to quantify the interpolation errors. Another application of the method is real-time mapping of the flow situation in a region.

Acknowledgments Our research work has been supported financially by the Austrian Academy of Sciences project HOE18. We would like to thank the Hydrographic Office at the Federal Ministry of Agriculture, Forestry, Environment and Water Management in Austria for providing the Austrian data.

References

Bailly JS, Monestiez P, Lagacherie P (2006) Modelling spatial variability along drainage networks with geostatistics. Math Geol 38:515–539

Bárdossy A (2006) Calibration of hydrological model parameters for ungauged catchments. Hydr Earth Syst Sci Discuss 3:1105–1124

Cressie N (1991) Statistics for spatial data, Wiley, New York, NY

Gottschalk L (1993a) Correlation and covariance of runoff. Stochastic Hydr Hydraulics, 7:85–101

Gottschalk L (1993b) Interpolation of runoff applying objective methods. Stochastic Hydr Hydraulics, 7:269–281

Gottschalk L, Krasovskaia I, Leblois E, Sauquet E (2006) Mapping mean and variance of runoff in a river basin. Hydr Earth Syst Sci Discuss 3:299–333

Kyriakidis PC, Journel AG (1999) Geostatistical space-time models: A review. Math Geol 31: 651–684

Matheron G (1965) Les variables regionalisées et leur estimation, Masson, Paris, France

Merz R, Blöschl G (2005) Flood frequency regionalisation – Spatial proximity vs. catchment attributes. J Hydrol 302:283–306

Monestiez P, Bailly JS, Lagacherie P, Voltz M (2005) Geostatistical modelling of spatial processes on directed trees: Application to fluvisol extent. Geoderma 128:179–191

Myers DE (1991) Pseudo-cross variograms, positive definiteness, and cokriging. Math Geol 23:805–816

Parajka J, Merz R, Blöschl G (2005) A comparison of regionalisation methods for catchment model parameters. Hydr Earth Syst Sci 9:157–171

Sauquet E, Gottschalk L, Leblois E (2000) Mapping average annual runoff: a hierarchical approach applying a stochastic interpolation scheme. Hydr Sci J 45:799–815

Skøien JO, Blöschl G (2006) Catchments as space-time filters - a joint spatio-temporal geostatistical analyses of runoff and precipitation. Hydr Earth Syst Sci Discuss 3:941–985

Skøien JO, Merz R, Blöschl G (2006) Top-kriging - geostatistics on stream networks. Hydr Earth Syst Sci 10:277–287

Part III
Meteorology

Quantifying the Impact of the North Atlantic Oscillation on Western Iberia

R. M. Trigo

Abstract The main objective of this work is to evaluate the influence, both physical and socio-economical, of the most important large-scale phenomenon of the Northern Hemisphere, the North Atlantic Oscillation (hereafter NAO), on the climate of the western Iberian region. Using high and low NAO index composites, statistically significant anomaly fields of climate variables are then interpreted based on physical mechanisms associated with anomalous large-scale circulation.

The Iberian Peninsula precipitation and river flow regimes are characterized by large values of inter-annual variability, with large disparities between wet and dry years. This is a major problem for water resource management, in general, and for the production of hydroelectricity, in particular. We have assessed the impact of the NAO on Iberian winter precipitation and river flow regimes for the three main international Iberian river basins, namely the Duero (north), the Tagus (centre) and the Guadiana (south). Results show that the large inter-annual variability of these three river flows is mostly modulated by the NAO phenomenon. Throughout most of the 20th century, the January-to-March river flow is better correlated with the 1-month-lagged (December-to-February, DJF) NAO index than is the simultaneous (DJF) river flow. Correlation values for the period 1973–1998 are highly significant, –0.76 for Duero, –0.77 for Tagus and –0.79 for Guadiana, being consistently of higher magnitude than those obtained in previous decades.

The majority of landslide episodes in the area north of Lisbon are associated with rainfall events of short (less than 3 days) or long duration (more than 20 days). Results for the low NAO class are crucial because these months are more likely associated with long-lasting rainfall episodes responsible for large landslide events. This is confirmed by the application of a 3-month moving average to both the NAO index and the precipitation time series. This procedure allows the identification of virtually all months with landslide activity as being characterized by negative average values of the NAO index and high values of average precipitation (above 100 mm/month).

R. M. Trigo
CGUL, Departamento de Física, Faculdade de Ciências, Universidade de Lisboa, Portugal and Departamento de Engenharias, Universidade Lusófona, Lisboa, Portugal
e-mail: rmtrigo@fc.ul.pt

A. Soares et al. (eds.), *geoENV VI – Geostatistics for Environmental Applications*, 235–246

1 Introduction

The North Atlantic Oscillation (NAO) has been regarded for more than 80 years as one of the major patterns of atmospheric teleconnection across all seasons (Walker, 1924). This important circulation mode has been characterized more comprehensively over the last two decades (e.g. Barnston and Livezey, 1987; Hurrell, 1995; Trigo et al., 2002, 2004). In simple terms, the NAO corresponds to a large-scale meridional oscillation of atmospheric mass between the subtropical anticyclone near the Azores and the subpolar low-pressure system near Iceland (Trigo et al., 2002). Recently, a number of different studies have shown the relevance of the NAO to the winter surface climate of the Northern Hemisphere in general and over the Atlantic/European sector in particular (Hurrell, 1995; Trigo et al., 2002). Additional studies have established links between different NAO modes and changes in the associated activity of North Atlantic storm tracks (Serreze et al., 1997; Osborn et al., 1999). Two contemporaneous and possibly related winter-time trends between the 1960s and 1990s are a trend towards the positive phase of the NAO and a trend towards warmer northern Eurasian land temperatures, which has now been established (Hurrell and van Loon, 1997).

In this work we will summarize the different impacts of the NAO as investigated by the author in recent years, namely on the European climate (Trigo et al., 2002), on the Iberian water resources (Trigo et al., 2004) and on Portuguese landslide activity (Trigo et al., 2005). Climatologists generally evaluate the impact of major teleconnections such as the NAO or El Niño using low-resolution surface climate data sets. That is also the case with this work and the vast majority of references given. This low resolution does not account for local effects, such as mountains, lakes, estuaries. Therefore the intense use of post-processing tools to downscale to higher-resolution fields is required. This is where geographical information systems (GIS) and environmental geostatistics are bound to play a major role, in order to improve the spatial resolution that is currently used by the climatological and geostatistics communities.

2 NAO and Large-Scale Data

The NAO index used in this study was developed by the Climatic Research Unit (University of East Anglia, UK) and is defined, on a monthly basis, as the difference between the normalized surface pressure at Gibraltar (southern tip of the Iberian Peninsula) and Stykkisholmur in Iceland (Jones et al., 1997). The NAO index for winter months presents a positive trend over the last 30 years; as a consequence its distribution is dominated by positive values, with monthly averages above zero (Jones et al., 1997). Therefore we decided to normalize the entire winter NAO index (average of NDJFM values) so that it has zero mean and a standard deviation of one. Finally, we defined the seasonal high NAO composite (low NAO composite) to be a combination of all winters with an NAO index greater than 0.5 (less than –0.5). Between 1923 and 1998 (76 winters), the number of winter seasons with high

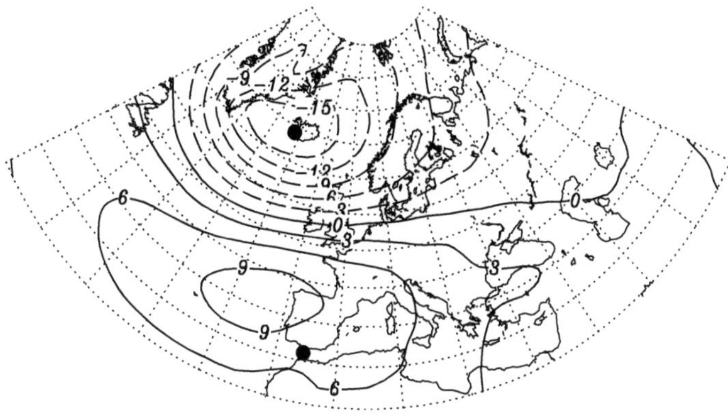

Fig. 1 Difference in sea level pressure SLP (hPa) between winter months (NDJFM) with an NAO index > 0.5 and with an NAO index < −0.5 between 1958 and 1997. Black circles show the location of station observations from Gibraltar and Iceland. (adapted from Trigo et al., 2005)

NAO index (24) is equal to the number characterized by a low NAO index (24). The remaining winters (28) are characterized by "near normal" values of the NAO index.

The large-scale gridded data used in this study were retrieved from the National Centers for Environmental Prediction/National Center for Atmospheric Research (NCEP/NCAR) reanalysis data set (Kalnay et al., 1996). Daily values of sea level pressure (SLP) and precipitation rate were extracted for the NCEP 2.5° latitude by 2.5° longitude grid. The gridded data cover an area from 30° N to 80° N and 60° W to 70° E, for the period 1958–1997. The advantages and the precautions on the use of this data set for Iberia are beyond the scope of this work and have been addressed previously in greater detail by Trigo et al. (2002, 2004).

The spatial signature of the NAO is represented by the difference in SLP between composites of winter months (DJFM) with (normalized) NAO index >0.5 and NAO index <–0.5 from 1958 to 1997 (Fig. 1). This pattern shows the expected dipole between the Iceland and the Azores regions. The southern centre of action is not centred over the Gibraltar station used in the NAO index, because the variance of SLP is lower there than over the Azores, and thus the magnitude of the composite anomaly is greater over the Azores (Trigo et al., 2002). It should be stressed that the NCEP reanalyses does not cover the entire 76-year period (available for the NAO index and precipitation in Portugal). Therefore the spatial analysis presented in Fig. 1 as well as the discussion on large-scale precipitation impact in the following section is performed for the shorter 40-year-long period spanning between 1958 and 1997.

3 Precipitation

SLP and precipitation rate anomaly fields for winter months characterized by high and low NAO index values were computed and are shown in Figs. 2a and 2b, respectively. Differences of the SLP between winter months (NDJFM) with high and low

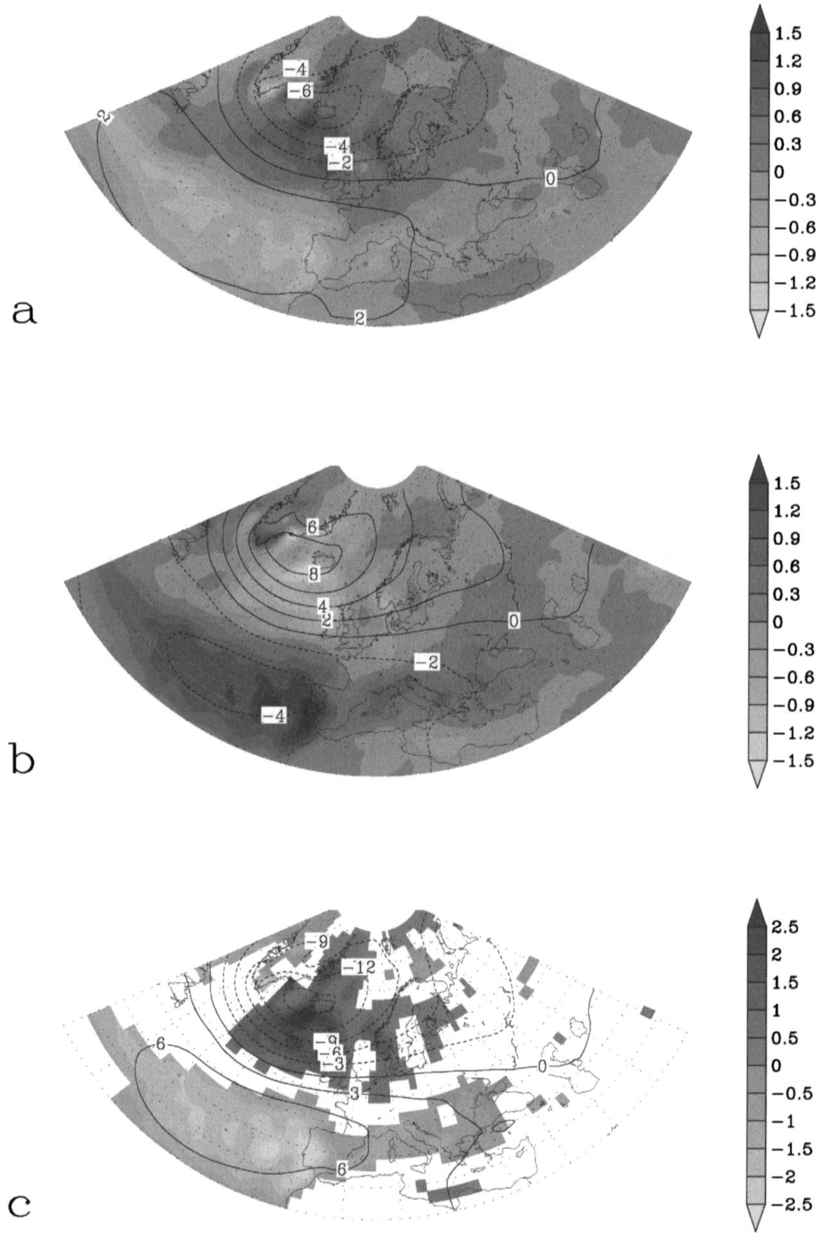

Fig. 2 Precipitation rate anomaly fields (mm/day) from the National Centers for Atmospheric Research NCEP/NCAR reanalysis for winter months with (a) high NAO index >0.5, (b) low NAO index <–0.5 and, and, c) their difference (represented only if significant at the 5% level). Positive (solid) and negative (dashed) isolines of the sea level pressure anomaly field (hPa) are also represented

NAO index (solid contour lines) are shown in Fig. 2c. Corresponding differences in precipitation rate, between high and low NAO composites, are also represented in Fig. 2c, wherever those differences are statistically significant at the 5% level (grey scale).

Several conclusions can be drawn from this figure:

i. the impact of the NAO on the Northern Hemisphere winter precipitation field, for both phases of the NAO index, is not restricted to the European continent but extends over large sectors of the North Atlantic, confirming results from previous works (Hurrell, 1995; Osborn et al., 1999; Trigo et al., 2004);
ii. Figure 3 shows quasi-zonal bands of opposite anomaly signs, with positive differences extending from eastern Greenland to Finland, with maximum values south of Iceland. At lower latitudes, a strong band of negative differences extends from the Azores archipelago in the mid-Atlantic Ocean to the Balkan Peninsula, with larger differences located west of Iberia, and particularly over Portugal.

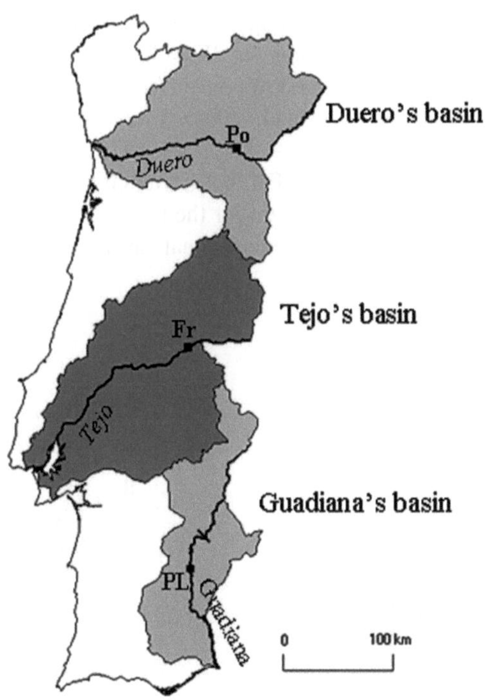

Fig. 3 Location of the Portuguese section of the three river basins considered. Small black dots show the location of river flow gauges used in Duero (Pocinho), Tagus (Fratel) and Guadiana (Pulo do Lobo)

4 NAO Impacts on Iberian Water Resources

It is natural to expect that the strong impact of the NAO on precipitation over the central and western sectors of Iberia should extend to the flow of rivers located within these sectors. River flow reflects precipitation integrated both spatially (over the catchment) and temporally, and therefore a seasonal (rather than monthly) timescale is considered here, taking the December through February (DJF) average of the NAO index. Monthly river flow data were provided by the Portuguese Institute of Water (INAG). The location of the river gauges within the Portuguese section of the rivers is depicted in Fig. 3.

The impact on the hydrological cycle (from October to September) for years characterized by winters with large positive and negative NAO index anomalies is shown in Figs. 4a, 5a and 6a for the Duero, Tagus and Guadiana rivers, respectively. In all the three cases there is a clear partition of mean river flow in the winter and spring months between the high and low NAO composites. For river Duero these differences are significant (at the 5% significance level) only between January and April while for rivers Tagus and Guadiana they are consistently significant between January and September.

Correlation coefficients between winter river flow (DJF) and contemporaneous winter NAO index (lag 0) were computed, using the entire period of data available for each river. We have also computed the 1-month-lagged correlation between the NAO index for DJF and the river flow for JFM. Recent studies have highlighted the possibility that the impact of the NAO on surface climate changes slowly with time (e.g. Osborn et al., 1999; Trigo et al., 2004). Therefore, correlation coefficients for three successive sub-periods, 1923–1947, 1948–1972 and 1973–1998, were also computed for Tagus and Duero rivers. For the shorter Guadiana time series we have only computed the correlation coefficient for the two most recent sub-periods. Results are summarized in Table 1. The inter-annual variability of winter (JFM) river

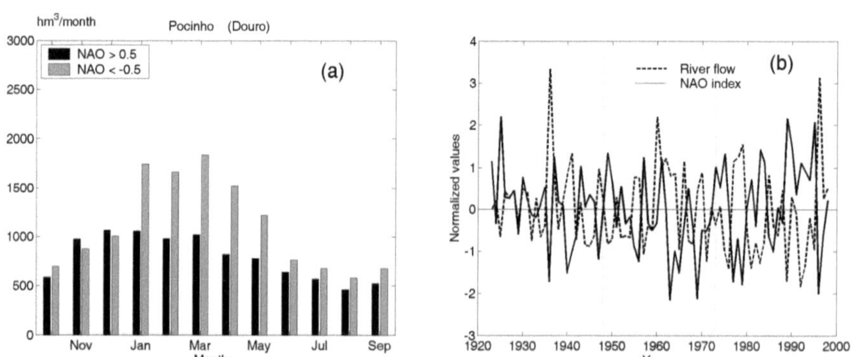

Fig. 4 (a) Monthly river flow of river Duero at Pocinho during and following winters with high NAO index (black bars) and winters with low NAO index (grey bars), (b) Inter-annual variability of the mean winter (JFM) river flow (solid curve), for river Duero at Pocinho, and the lagged winter (DJF) NAO index Both curves have been normalised

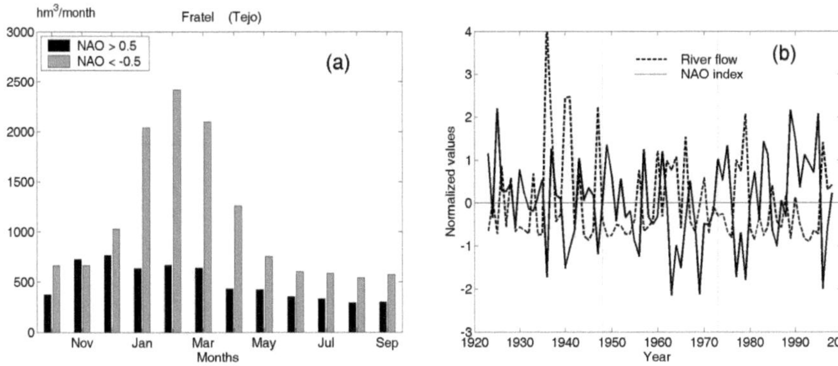

Fig. 5 The same as in fig. 4 but for river Tagus at Fratel

flow against the lagged winter NAO index (DJF) can be observed in Figs. 4b, 5b and 6b for the Duero, Tagus and Guadiana rivers, respectively. Sub-periods are delimited by a vertical dotted line while both curves (river flow and NAO index) were normalized and represented between 1920 and 2000 to facilitate comparisons between all three rivers. The simultaneous analysis of these figures and Table 1 suggests the following:

1. The magnitude of all 1-month-lagged correlation coefficients is consistently higher than the corresponding non-lagged values. In particular, for the entire period (1923–1998) the correlation coefficient increases, for all three rivers, when the NAO index is lagged by 1 month. This is relevant because it emphasizes the potential use of these relationships for forecasting purposes.
2. For both Duero and Tagus rivers there is a decrease in the magnitude of correlation coefficient values between the first and the second sub-periods followed by a major increase between the second and third. The change in coherence between the time series is obvious in Figs. 4b, 5b and 6b. One-month-lagged correlation values for the most recent sub-period (1973–1998) are the strongest overall: –0.76 for Duero, –0.77 for Tagus and –0.79 for Guadiana.

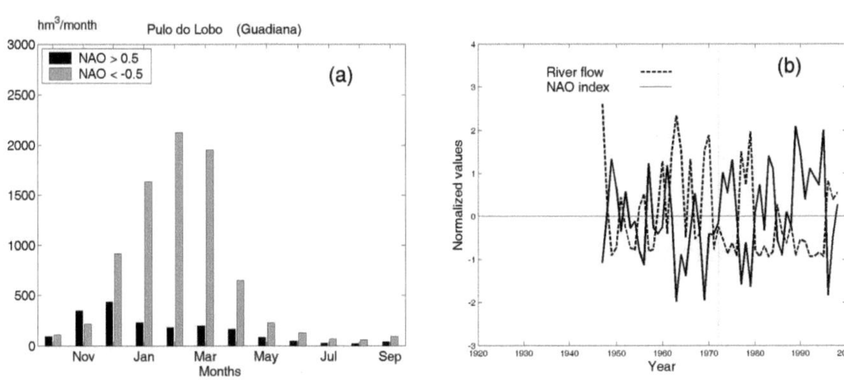

Fig. 6 The same as in fig. 4 but for river Guadiana at Pulo do Lobo

Table 1 Seasonal one-month lagged (lag 1) and simultaneous (lag 0) correlation coefficients between winter NAO index and river flow in Pulo do Lobo (Guadiana), Fratel (Tagus) and Pocinho (Duero). Significance level greater than 95% is highlighted in bold

		River flow Lag 1 NAO (DJF), flow (JFM)	River flow Lag 0 NAO (DJF), flow (DJF)
Guadiana	1948–1972	**−0.60**	**−0.45**
	1973–1998	**−0.79**	**−0.67**
	1948–1998	**−0.69**	**−0.57**
Tagus	1923–1947	**−0.56**	**−0.54**
	1948–1972	−0.37	−0.25
	1973–1998	**−0.77**	**−0.61**
	1923–1998	**−0.52**	**−0.45**
Duero	1923–1947	**−0.48**	−0.27
	1948–1972	−0.29	−0.10
	1973–1998	**−0.76**	**−0.58**
	1923–1998	**−0.55**	−0.35

3. Despite these consistently high values of anti-correlation, it is evident that the river flow of both the Guadiana and the Tagus did not reduce to the low levels expected from the high NAO index values in the 1980s and 1990s.

Therefore the impact of NAO on precipitation, river flow and consequently on hydroelectric production presents high values of inter-decadal variability (Trigo et al., 2004). Recent studies have shown that the northern centre of the NAO dipole has moved closer to Scandinavia (Jung and Hilmer, 2001). This shift has major implications for the Northern Hemisphere climate, in general (Lu and Greatbatch, 2002), and for the precipitation field over Iberia, in particular (Goodess and Jones, 2002). It is not obvious if this variability is natural or induced by climate change. Using a multi-century control run Osborn et al. (1999) have shown a remarkable range of inter-decadal variability on the magnitude of the association between NAO and Northern Hemisphere temperatures.

An objective comparison with similar studies for other river basins can be hampered by the use of different periods, seasonal aggregations, etc. Nevertheless, it should be emphasized that neither rivers Tigris and Euphrates (Cullen and deMenocal, 2000) nor the Danube (Rîmbu et al., 2002) nor the rivers in England and Wales (Wedgbrow et al., 2002) demonstrate such strong correlation with the NAO index as found here for the Iberian rivers. The author is not aware of any European river presenting a seasonal correlation coefficient value higher than 0.75 with any large-scale climate index, such as the NAO and the El Niño Southern Oscillation (ENSO).

5 The Impact of NAO on Landslide Activity

In the previous section it was shown that the phase of the NAO circulation mode is the most important to model the temporal precipitation distribution over Iberia. However, it is not straightforward to correlate the NAO index with landslide activity, and this is partly due to both the small number of landslide episodes and the low number of recognized slope movements for some of these episodes.

The landslide data set was obtained by detailed geomorphological mapping carried out in five sample areas in the Lisbon area (Trigo et al., 2005). The reconstruction of past landslide activity was supported by field work, archive investigation and local interviews (Zêzere and Rodrigues, 2002). The most recent slope instability events (after 1978) are better documented than the older ones. Table 2 shows that for the 45-year period that spans between 1956 and 2001, only 19 landslide-triggering events were reported in the study area. The closest rain gauge station is São João da Talha (hereafter SJT), located about 20 km north of Lisbon (Trigo et al., 2005). Table 2 summarizes the most relevant features of each event, including date, return period and critical rainfall amount/duration.

Most landslides recognized for the Lisbon area are shallow movements with slip surface depth less than 10 m and almost always induced by rainfall. Shallow translational slides are usually triggered by the water infiltration in unconsolidated slope deposits, which cover impermeable rocks. The soil saturation is responsible for the reduction of the shear strength of the soil, by the temporary rise in pore water pressure and by the loss of the apparent soil cohesion (Gostelow, 1991). These landslides are mostly activated in the study area by intense and concentrated rainfall 1 to 15 days in duration (Zêzere, 2000; Zêzere and Rodrigues, 2002). Translational slides, rotational slides and complex and composite slope movements are triggered by a rise in the groundwater level and the shear strength reduction (Gostelow, 1991). Such hydrological conditions occur as a consequence of rainfall periods 40 to 90 days long in the Lisbon area (Zêzere, 2000).

Table 2 Temporal occurrence and major characteristics of rainfall-triggered landslides in the Lisbon area from 1956 to 2001

Episode	Date (yy/mm/dd)	Critical rainfall amount/duration mm (dd)	Return period (years)
1	1958/12/19	149 (10)	2.5
2	1959/03/09	175 (10)	4
3	1967/11/25	137 (1)	60
4	1968/11/15	350 (30)	6.5
5	1978/03/04	204 (15)	3.5
6	1979/02/10	694 (75)	20
7	1981/12/30	174 (5)	13
8	1983/11/18	164 (1)	200
9	1987/02/25	52 (1)	2
10	1989/11/22	164 (15)	2
11	1989/11/25	217 (15)	4.5
12	1989/12/05	333 (30)	5.5
13	1989/12/21	495 (40)	20
14	1996/01/09	544 (60)	10
15	1996/01/23	686 (75)	18
16	1996/01/28	495 (40)	20
17	1996/02/01	793 (90)	24
18	2001/01/06	447 (60)	5
19	2001/01/09	467 (60)	5.5

A 3-month moving average was applied to filter the original NAO index (non-normalized) and SJT precipitation time series, restricting the analysis to monthly data from the wet season (NDJFM). October values are used to compute November and December averages. Thus, values for February of year n correspond to the rainfall and NAO index averages computed between December of year $n-1$ and February of year n, while values for November of any year are restricted to the October and November values for that same year (Trigo et al., 2005). For landslides episodes that occurred in the first 5 days of the month we considered only the NAO index and precipitation values from the previous 2 months (episodes 5, 12 and 17). Figure 7 represents the scatter plot between both the filtered time series, and it is limited to the months between November and March. Small open diamonds represent months where landslide episodes did not occur (or were not reported) and black circles correspond to months with only one recognized landslide episode (e.g. February 1979, episode 6), while black triangles represent those months with more than one slope instability episode (e.g. January 1996, episodes 14, 15 and 16). Numbers close to black symbols represent the landslide episode number reported in Table 2.

The following important conclusions can be drawn after analysing Fig. 7:

1. The linear correlation between both the time series is –0.63, a value statistically significant at the 1% level. Nevertheless, this relationship increases to –0.73 if we consider only those months affected by landslide activity and the regression line is also represented.
2. Most of the months for which landslide events were reported were above the horizontal line indicating the threshold of 101 mm/month (mean value of 3-month

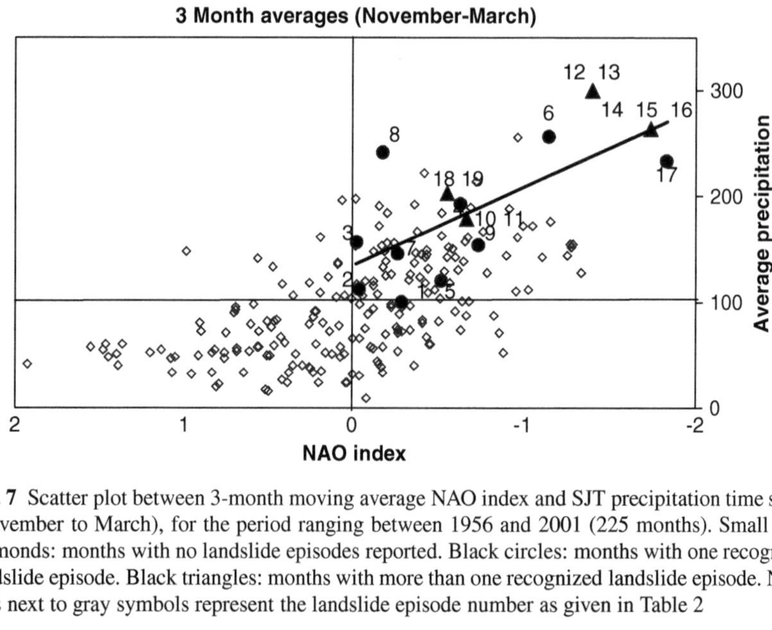

Fig. 7 Scatter plot between 3-month moving average NAO index and SJT precipitation time series (November to March), for the period ranging between 1956 and 2001 (225 months). Small open diamonds: months with no landslide episodes reported. Black circles: months with one recognized landslide episode. Black triangles: months with more than one recognized landslide episode. Numbers next to gray symbols represent the landslide episode number as given in Table 2

average precipitation). Moreover, all cases are characterized by negative values of the averaged NAO index.

3. Six of the landslide episodes (12 to 17 that have occurred in 1989 and 1996, Table 2 are characterized by extremely negative average values of the NAO index (lower than –1) and extremely high values of average precipitation (clearly above 200 mm/month).
4. The majority of landslide episodes located at the lower end of the regression line (episodes 1, 2, 5, 9, 10 and 11) are characterized by the lowest critical return period values (<5 years, Table 2).

6 Conclusions

We evaluated the magnitude and the spatial extent of statistically significant impact of the NAO mode on the precipitation field over the entire European continent. The relevance of this large-scale atmospheric circulation mode was then evaluated for the winter precipitation and river flow in three important Iberian river basins. The January-to-March mean river flow was shown to be better associated with the 1-month-lagged (December-to-February) NAO index than the simultaneous (DJF) river flow. To study temporal changes in the NAO–river flow relationship, the correlation coefficient values were compared for three non-overlapping sub-periods: 1923 to 1947, 1948 to 1972 and 1973 to 1998. The highest correlation values were always obtained for the most recent period (1973–1998) and, with magnitudes in the range –0.75 to –0.80, can be regarded as highly significant (–0.76 for Duero, –0.77 for Tagus and –0.79 for Guadiana).

Despite the non-stationary nature of the NAO impact I believe that the potential predictability of the NAO index and, consequently, of the precipitation field and river flow regimes over the Iberian Peninsula and the associated hydroelectric production provides large potential economic advantages. In fact, different researchers have started developing statistical (e.g. Gámiz-Fortis et al., 2002) as well as dynamical models to predict precipitation over Europe several months in advance. We believe that, based on NAO–rainfall and NAO–flow relationships, there is a large scope for further development of useful statistical and dynamical models. Such models should be developed with the double purpose of providing water resource managers in Iberia with seasonal forecasting tools and for assessing changes in river flow regime under climate change scenarios.

Finally we confirmed the relevance of this large-scale atmospheric circulation mode to landslide events. This was done through the application of a 3-month moving average to both NAO index and precipitation time series. It allowed the identification of months with landslide activity as being characterized by negative average values of the NAO index and high values of average precipitation (above 100 mm/month).

Acknowledgments This work was supported by the Portuguese Science Foundation (FCT) through project PREDATOR (Seasonal Predictability and Downscaling over the Atlantic European Region), Contract POCI/CTE-ATM/62475/2003. Landslide data were provided by J.L. Zêzere from CEG.

References

Barnston AG, Livezey RE (1987) Classification, seasonality and persistence of low-frequency atmospheric circulation patterns. Mon Weather Rev 115:1083–1127

Cullen HM, deMenocal PB. (2000) North Atlantic influence on Tigris–Euphrates streamflow. Int J Climatol 20:853–863

Gámiz-Fortis S, Pozo-Vázquez D, Esteban-Parra MJ, Castro-Díez Y. (2002) Spectral characteristics and predictability of the NAO assessed through singular spectral analysis. J Geophys Res 107 (D23), 4685. Doi:10.1029/2001JD001436

Goodess CM, Jones PD (2002) Links between circulation and changes in the characteristics of Iberian rainfall. Int J Climatol 22:1593–1615

Gostelow P (1991) Rainfall and landslides. In : Almeida-Teixeira M et al (eds) Prevention and control of landslides and other mass movements. CEC, Bruxels, pp 139–161

Hurrell JW (1995) Decadal trends in the North Atlantic oscillation: regional temperatures and precipitation. Sci 269:676–679

Hurrell JW, van Loon H (1997) Decadal variations in climate associated with the North Atlantic Oscillation. Climatic Change 36:301–326

Jones PD, Johnson T, Wheeler D (1997) Extension to the North Atlantic Oscillation using instrumental pressure observations from Gibraltar and south-west Iceland. Int J Climatol 17:1433–1450 greenhouse gas. Clim Dyn 15:551–559

Jung T, Hilmer M (2001) On the link between the North Atlantic Oscillation and Arctic sea ice export through Fram Strait. J Clim 14:3932–3943

Kalnay E, Kanamitsu M, Kistler R, Colins W, Deaven D, Gandin L, Iredell M, Saha S, White G, Wollen J, Zhu Y, Leetmaa A, Reynolds R, Chelliah M, Ebisuzaki W, Higgins W, Janowiak J, Mo KC, Ropelewski C, Wang J, Jenne R, Joseph D (1996) The NCEP/NCAR 40-years reanalyses project. Bull Am Meteorol Soc 77:437–471

Lu J, Greatbatch RJ (2002) The changing relationship between the NAO and the northern hemisphere climate variability. Geophys Res Lett vol 29. (7) 10.1029/2001GLO14052

Osborn TJ, Briffa KR, Tett SFB, Jones PD, Trigo RM (1999) Evaluation of the North Atlantic Oscillation as simulated by a climate model. Clim Dyn 15:685–702

Rîmbu N, Boroneanþ C, Carmen B, Mihai D (2002) Decadal variability of the Danube river flow in the lower basin and its relation with the North Atlantic Oscillation. Int J Climatol 22:1169–1179

Serreze MC, Carse F, Barry RG, Rogers JC (1997) Icelandic Low cyclone activity: climatological features, linkages with the NAO, and relationships with recent changes in the Northern Hemisphere circulation. J Climatol 10:453–464

Trigo RM, Osborn TJ, Corte-Real JM (2002) The North Atlantic Oscillation influence on Europe: climate impacts and associated physical mechanisms. Climate Res 20:9–17

Trigo RM, Pozo-Vazquez D, Osborn TJ, Castro-Diez Y, Gámis-Fortis S, Esteban-Parra MJ (2004) North Atlantic Oscillation influence on precipitation, river flow and water resources in the Iberian Peninsula. Int J Climatol 24:925–944

Trigo RM, Zêzere JL, Rodrigues ML, Trigo IF (2005) The influence of the North Atlantic Oscillation on rainfall triggering of landslides near Lisbon. Nat Hazards 36:331–354, DOI 10.1007/s11069-005-1709-0

Walker GT, 1924, Correlations in seasonal variations of weather. IX Mem Ind Meteorol Dept 24:275–332

Wedgbrow CS, Wilby RL, Fox HR, O'Hare G (2002) Prospect for seasonal forecasting of summer drought and low river flow anomalies in England and Wales. Int J Climatol 22:219–236

Zêzere, JL (2000) Rainfall triggering of landslides in the area North of Lisbon. In Bromhead E et al (eds) Landslides in Research, Theory and Practice, vol 3. London, Thomas Telford, pp 1629–1634

Zêzere JL, Rodrigues ML (2002) Rainfall Thresholds for Landsliding in Lisbon Area (Portugal). In: Rybar et al (eds) Landslides, Balkema AA, Lisse, pp 333–338

Monthly Average Temperature Modelling

M. Andrade-Bejarano

Abstract This research is associated with the goal of the horticultural sector of the Colombian southwest, which is to obtain climatic information, specifically, to predict the monthly average temperature in sites where it has not been measured. The data correspond to monthly average temperature, and were recorded in meteorological stations at Valle del Cauca, Colombia, South America. Two components are identified in the data of this research: (1) a component due to the temporal aspects, determined by characteristics of the time series, distribution of the monthly average temperature through the months and the temporal phenomena, which increased (El Niño) and decreased (La Niña) the temperature values, and (2) a component due to the sites, which is determined for the clear differentiation of two populations, the valley and the mountains, which are associated with the pattern of monthly average temperature and with the altitude. Finally, due to the closeness between meteorological stations it is possible to find spatial correlation between data from nearby sites. In the first instance a random coefficient model without spatial covariance structure in the errors is obtained by month and geographical location (mountains and valley, respectively). Models for wet periods in mountains show a normal distribution in the errors; models for the valley and dry periods in mountains do not exhibit a normal pattern in the errors. In models of mountains and wet periods, omni-directional weighted variograms for residuals show spatial continuity. The random coefficient model without spatial covariance structure in the errors and the random coefficient model with spatial covariance structure in the errors are capturing the influence of the El Niño and La Niña phenomena, which indicates that the inclusion of the random part in the model is appropriate. The altitude variable contributes significantly in the models for mountains. In general, the cross-validation process indicates that the random coefficient model with spatial spherical and the random coefficient model with spatial Gaussian are the best

M. Andrade-Bejarano
School of Biological Sciences, Statistics Section, Harry Pitt Building, The University of Reading, RG6 6FN, Reading, United Kingdom and Universidad del Valle, Cali, Colombia, South America.
e-mail: snr02ma@reading.ac.uk

A. Soares et al. (eds.), *geoENV VI – Geostatistics for Environmental Applications*, 247–262

models for the wet periods in mountains, and the worst model is the model used by the Colombian Institute for Meteorology, Hydrology and Environmental Studies (IDEAM) to predict temperature.

1 Introduction

The county of Valle del Cauca (Colombia, South America; study zone) covers an area of 2,214,000 ha, of which 350,000 ha (15.8%) is used for agriculture (SAG, 2001). Agriculture is the main sector of the regional economy. Specifically, vegetables represent 1.5% of the agricultural area and the tomato crop represents 45% of this area. Tomato production is realised in farms with extension of less than 5 ha (the majority being between 0.5 and 2 ha), especially concentrated in hillside areas. Three kind of sowing areas are found: (1) low tropic (flat zone of Valle del Cauca), (2) middle tropic (low hillside), and (3) high tropic (high hillside). The tomato crop grows in low tropic and middle tropic areas, at altitudes between 0 and 2000 m.

The vegetable crop zones and specifically the tomato crop zones have technological problems, which have affected social and economical aspects in the region, generating low life levels and high inversion risk in this sector of the Colombian economy. The problems identified are concerned with (1) adaptation problems of the seeds, (2) the tomato varieties cultivated being susceptible to a minimum of four fungus pathogens, two bacterial and two viral, and (3) farmers using pesticides as a predominant tool to eliminate phytosanitary problems. The Colombian Institute for Agricultural and Farm Research (CORPOICA) has identified three research lines with the objective of obtaining knowledge to provide proposals for solving the technological problems identified with the tomato crop: (1) estimation of climatic and edaphic variables in sites where it has not been measured, (2) determination of adaptation ranges of the varieties and species and their relation with climatic and environmental factors, (3) ecology and biology of plagues, fungus and virus and its relation with climatic and environmental factors. This research is associated with the first objective of CORPOICA and is focused on the analysis of climatic variables, specifically for the prediction of monthly average temperature in sites where it has not been measured (Osorio, 1999).

2 Data

The data correspond to monthly average temperature and were recorded in meteorological stations at Valle del Cauca, Colombia, South America. In this project all meteorological stations in the study zone, with records of monthly average temperature and altitudes less than 2000 m.a.s.l., were included in the analysis. This corresponds to 28 meteorological stations, located within latitudes from 3°19′N to 4°44′N, within longitudes from 75°49′W to 76°45′W and within altitudes from 920

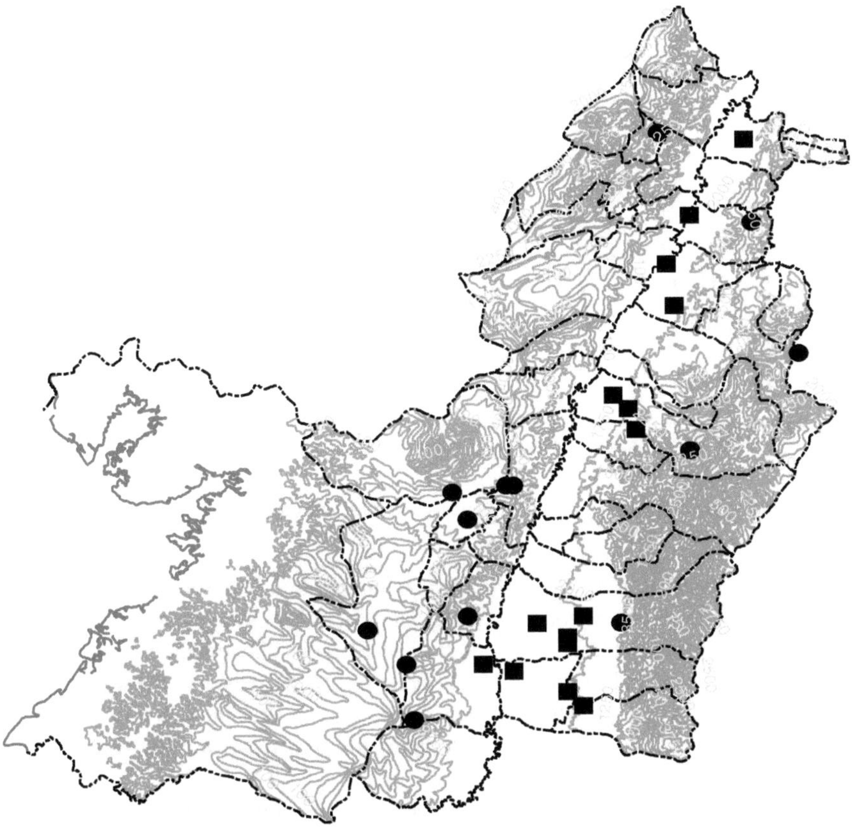

Fig. 1 Location of meteorological stations on map of Valle del Cauca, included in monthly average temperature modelling. ●: Meteorological station located in mountains, ■: Meteorological station located in the valley

to 1950 m (Fig. 1). Fifteen meteorological stations are located in the valley, at altitudes between 950 and 1100 m.a.s.l., and 13 in the mountains, at altitudes from 1233 to 1950 m.a.s.l.

3 Exploratory Data Analysis and The Research Problem

Two components that characterise this research are identified: (1) A component due to the temporal aspects, corresponding to the data recorded during the period 1971 to 2002. (2) A component due to the sites (meteorological stations) and their local geographic characteristics that determine it. Several aspects of these two components are explored in the temporal component: Some of the time series from meteorological stations show a trend to increase (Fig. 2), indicating that these trends should be modelled or time series should be de-trended (Box et al., 1994; Vandaele, 1983).

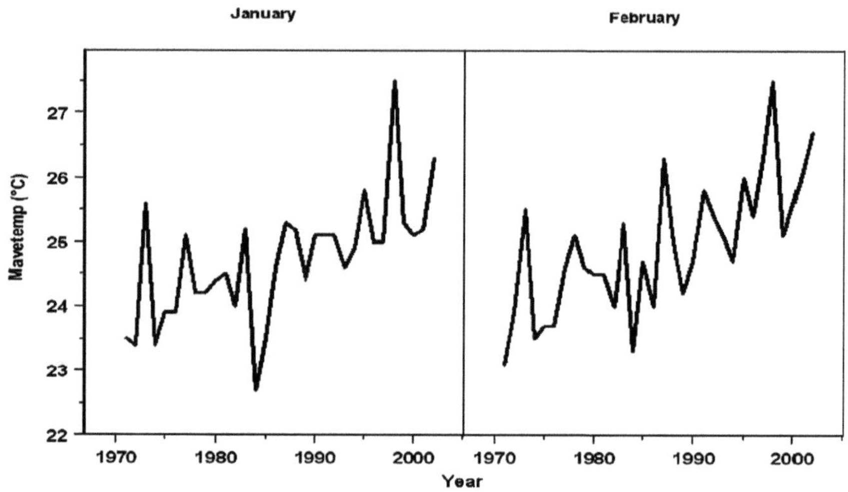

Fig. 2 Time series graphics from meteorological station localised in Valley at 1040 m, for the driest period January – February

A seasonal variation in the monthly average temperature is observed (Fig. 3). Two dry periods and two wet periods are clearly identified. Temporal phenomena affected the monthly average temperature values. The El Niño phenomenon increased the temperature values and in the La Niña phenomenon decreases in the values were observed (Figs. 4 and 5), which indicates that year to year there are variations in the mean values of monthly average temperature and these variations should be modelled.

With respect to the component due to the sites, two populations are clearly identified (Fig. 6); the valley and mountains behave differently with respect to altitude: there is a linear relation between monthly average temperature and altitude in the data belonging to mountains (Fig. 7(b)) and in the valley the pattern of monthly average temperature is constant (Fig. 7(a)). Therefore the modelling of monthly average temperature may be considered separately for the valley and mountains.

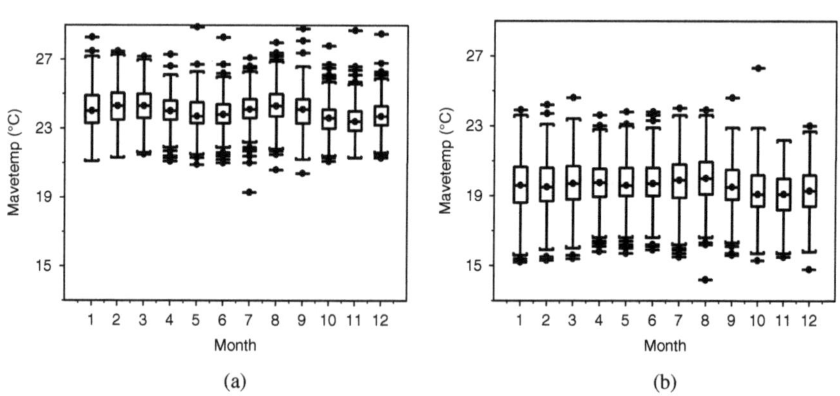

Fig. 3 Box plots for monthly average temperature for (a) The Valley (b) Mountains

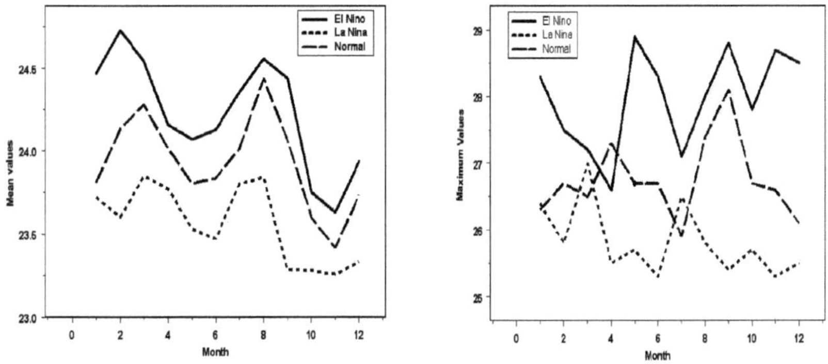

(a) (b)

Fig. 4 Graphs of the three phenomena, for the valley (a) mean values of monthly average temperature versus months, and (b) maximum values of monthly average temperature versus months

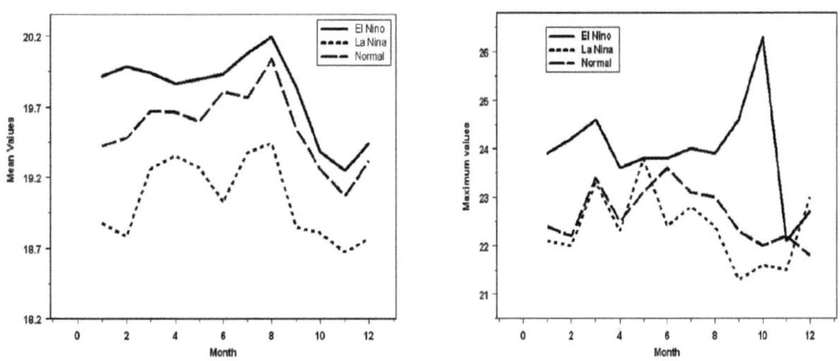

(a) (b)

Fig. 5 Graphs of the three phenomena, for mountains (a) mean values of monthly average temperature versus months (b) maximum values of monthly average temperature versus months

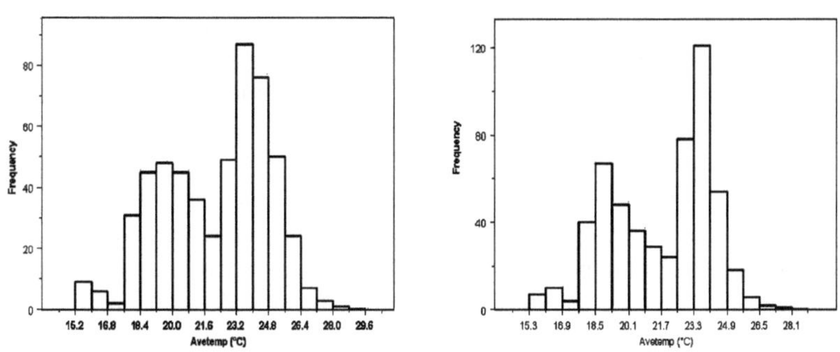

(a) (b)

Fig. 6 Monthly average temperature histograms, including data of the Valley and Mountains for (a) the dry month January, (b) the wet month October

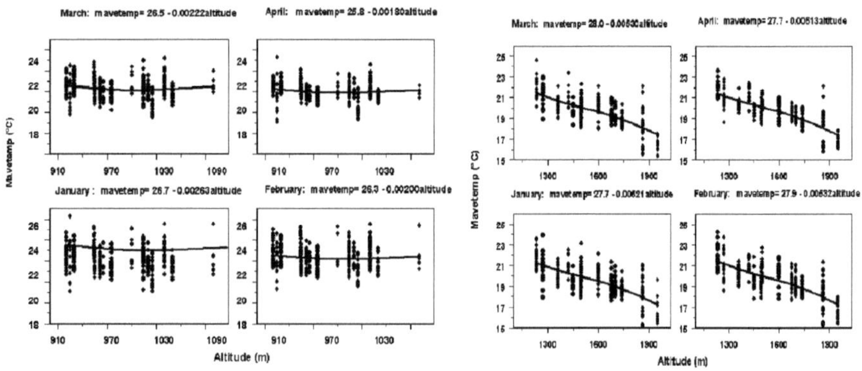

(a) (b)

Fig. 7 Monthly average temperature vs. altitude by month for (a) meteorological stations located in the valley, from January to April (b) meteorological stations located in mountains, from January to April

Finally, due to the closeness between meteorological stations it is possible to have spatial correlation between data from nearby meteorological stations, and if this kind of correlation is found, it should be modelled.

4 The Model

The following are in agreement with the results obtained in the exploratory analysis of data:

1. Regularly, the El Niño phenomenon increases the temperature between 1°C and 2°C. Due to the influence of the El Niño phenomenon, the variability from year to year should be modelled through the inclusion of a random effect common to all sites.
2. With monthly models the variation due to seasonal changes can be eliminated from the model containing the climatic variables. However, in this case some time series show a trend. These time series are de-trended before being included in the modelling, using the method of least squares (Vandaele, 1983; Brockwell and Davis, 1996; Chan et al., 1977).
3. Covariates associated with the sites are included in the modelling. Altitude is included for the modelling of monthly average temperature in mountains, due to its influence on the temperature in this geographical place. In the Andean zone (study zone), the air temperature is determined by altitude; as the altitude above sea level increases, the average temperature decreases (Pabón et al., 2002).
4. If in the modelling, near sites in coordinates are included, it will be possible to have spatial correlation between errors. The spatial correlation in the errors is modelled through isotropic models (Cressie, 1993; Isaaks and Srivastava, 1989; Chilès and Delfiner, 1999).

The model formulated for monthly average temperature corresponding to the jth month, separately for mountains and the valley, is

$$y_{ik} = \alpha + a_k^* + \beta x_i + \varepsilon_{ik} \tag{1}$$

$i = 1, \ldots.., 13$; $k = 1, \ldots., 32$ for mountains,

$$y_{ik} = \alpha + a_k^* + \varepsilon_{ik}$$

$i = 1, \ldots.., 15$; $k = 1, \ldots., 32$ for the valley, where i denotes the sub index for the site and k the sub index of years,

$$a_k^* = a_k - \alpha$$

where y_{ik} is the monthly average temperature at the site i, in the year k; α denotes the expectation value of the intercept; β is the slope of the model for the altitude variable; a_k^* is the deviation of the regression coefficient a_k from its expectation α; a_k is the intercept in the year k; x_i is the altitude value in the site i; and ε_{ik} denotes the spatial correlated error in the site i and the year k.

$$Var(\varepsilon_{ik}) = \sigma^2{}_{ik}; Cov(\varepsilon_{ik}, \varepsilon_{i'k'}) = \sigma^2[f(d_{ii'})]$$

where $d_{ii'}$ is the distance between the site s_i and $s_{i'}$. $f(d_{ii'})$ is a stationary and isotropic model.

$a_k^* \sim iddN(0, \sigma^2{}_a)$; a_k^* and ε_{ik} are uncorrelated; $\alpha + \beta x_i$ is the fixed part of the model; and $a_k^* + \varepsilon_{ik}$ is the random part of the model.

4.1 Results for the Random Coefficient Model Without Spatial Covariance Structure in the Errors

In first instance, results for the random coefficient model (1) without spatial structure in the errors are obtained.

4.1.1 Results for the Valley

For all the months, the covariance parameter estimate values for residual are higher than for years. The null model likelihood ratio test (Little et al., 1996) indicates that the inclusion of the random intercept in the model is correct. The model captures the effect of the El Niño and La Niña phenomena. For all the months, the highest estimated values of the random intercept are recorded in 1972, 1973, 1976, 1983, 1987, 1988, 1991, 1994, 1997 and 1998. All of these years correspond to the El Niño phenomenon. The lowest estimated values are recorded in 1974, 1975, 1984, 1999 and 2000. With the exception of 1984 all of these years belong to the La Niña phenomenon. For all the months there is no obvious trend in the random part of the model.

In the analysis of the fixed part of the model, t-test $p$values indicate that the intercept contributes significantly at $\alpha = 0.05$ in all the months, and the altitude variable contributes in the months January, February, March, April, August and November. For all the months, histograms, Q-Q normal plots and the Anderson–Darling test (A^2) (Anderson and Darling, 1954) indicate that errors do not have a normal pattern in the valley. Graphics of residuals versus predicted values display a uniform spread about the zero error line; this indicates the homogeneity of variances in the errors.

4.1.2 Results for Mountains

For all the months the covariance parameters estimates values for residuals are higher than for years. The null model likelihood ratio test values are significant at $\alpha = 0.05$. As in the valley, the inclusion of the random intercept is correct and the model captures the effect of the El Niño and La Niña phenomena. For all the months, the highest random coefficient estimated values were recorded in 1982, 1983, 1987, 1988, 1990, 1992, 1994, 1997, 1998 and 2001. With the exception of 1988, 1990 and 2001, all these years correspond to the El Niño phenomenon. The lowest estimated values were recorded in 1971, 1973, 1974, 1975, 1976, 1984 and 1996. With the exception of 1984 and 1996, these years belong to the La Niña and El Niño phenomena. There is no obvious trend in the random part of the model.

In all the months, t-test p values indicate that the intercept and the variable altitude contribute significantly at $\alpha = 0.05$. For all the months the relation between monthly average temperature, and altitude and monthly average temperature and year have been well specified in the models' histograms, and Q-Q normal plots and the Anderson–Darling test (A^2) show that errors do not have a normal pattern in the dry periods January–February and July–August, as well as in September. In the other months the pattern of the errors is normal. Graphics of residuals versus predicted values show the homogeneity of variances in the errors.

5 Analysis of the Spatial Correlation in the Errors for Mountains

Variograms of the residuals are obtained by year, for data of mountains in wet periods, because in agreement with the results discussed in Sects. 4.1.1 and 4.1.2, in this geographical location and season, the residuals show a normal pattern. Thirty-two years are included in the analysis. Finally, a weighted variogram of residuals, including all variograms by year, is obtained.

The variogram of residuals is

$$\gamma(h) = (1/2N(h)) \sum_{i=1}^{N} [r(s_i) - r(s_{i+h})]^2$$

where $N(h)$ is the count of the pairs separated (approximately) by the lag h, $r(s_i)$ and $r(s_{i+h})$ are the residual values of the variable in the points s_i and s_j, respectively.

In this research there are monthly average temperature data from 13 meteorological stations, located in mountains. The data belong to 32 years; however, in some meteorological stations and years there are missing data. Because of this situation a weighted variogram of the errors is calculated by taking into account the number of pairs of points to calculate the variogram, per year:

$$\gamma(h) = \sum_{i=1}^{n} N_i(h)\gamma_i(h) / \left(\sum_{i=1}^{n} N_i(h) \right),$$

where $\gamma_1(h),\ldots,\gamma_i(h)$ are the variograms from n years and $N_1(h),\ldots,N_i(h)$ are the counts of pairs of points separated (approximately) by the lag $h$from each year.

5.1 Omni-directional Variograms

In agreement with the location of the meteorological stations in the mountains, the lag distance used is 12 km and the maximum lag distance is 4 km. The results shows that with the exception of the lag 0, the number of the pairs by lag is higher than 30. All the variograms show a spatial continuity up to 23.4 km; at this distance there is a spatial discontinuity in all the months evaluated. This is because between 23.4 km and 35.6 km there are no meteorological stations in the study zone and the variograms show the physical spatial discontinuity. In Fig. 8, graphics of variograms for the months April and October can be seen. In Fig. 9, the physical spatial discontinuity between meteorological stations can be seen.

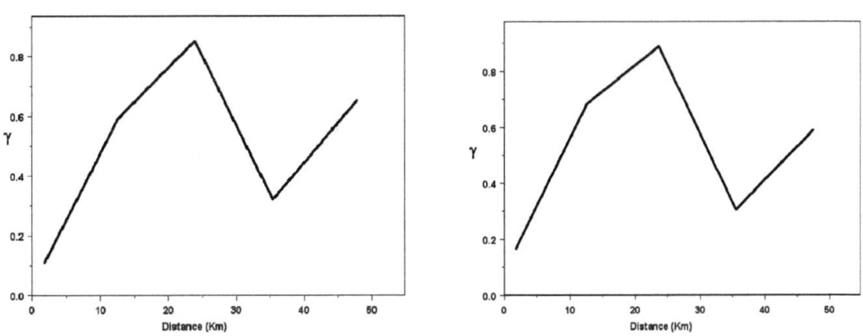

(a) (b)

Fig. 8 Omni-directional Variograms for (a) April (b) October

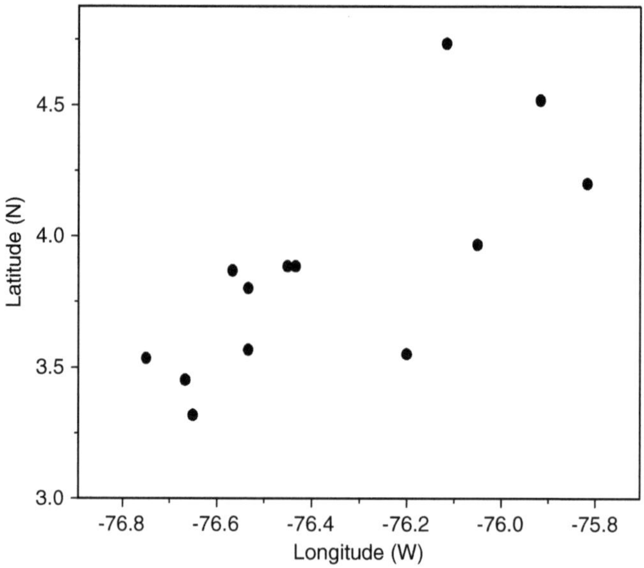

Fig. 9 Meteorological Stations Located in Mountains. Coordinates in Decimals

5.2 Evaluation of Anisotropy

Due to the small number of pairs in the evaluation of the variograms in different directions – 0°, 45°, 90°, 135° and 30°, 60° – an isotropic pattern in monthly average temperature is assumed, and isotropic models are used in the estimation of the stationary covariance functions.

5.3 Results for the Random Coefficient Model With Spatial Covariance Structure in the Errors

5.3.1 Results for the Wet Months in Mountains

Different models are fitted with the objective to select the model with the lowest fit values. The estimation is obtained by the restricted maximum likelihood (REML) method (SAS, 1999).

Table 1 shows that for all the months, with the exception of June, the random coefficient model with spatial spherical has the lowest values of Akaike's information criterion (AIC) and Schwarz's Bayesian criterion (BIC). The random coefficient model with spatial Gaussian shows the lowest values in June. As a result, a random coefficient model with spatial Gaussian is chosen for the modelling of June and a random coefficient model with spatial spherical is chosen for the rest of the months.

Table 1 Fit statistics by month for different models of spatial covariance structure in the errors, variable monthly average temperature

Month	n	Random Coefficient Model without Spatial Covariance Structure in the Errors			Spatial Spherical			Spatial Gaussian			Spatial Exponential		
		-2Rll	*AIC*	*BIC*	*-2Rll*	*AIC*	*BIC*	*-2Rll*	*AIC*	*BIC*	*-2Rll*	*AIC*	*BIC*
March	238	615.1	619.1	622.1	607.4	607.4	607.4	616.6	616.6	616.6	613.3	613.3	613.3
April	233	553.3	557.5	560.4	548.5	548.5	548.5	562.4	562.4	563.4	551.5	551.5	551.5
May	236	553.1	557.1	560.1	546.9	546.9	546.9	552.3	552.3	552.3	555.3	555.3	555.3
June	242	634.2	638.2	641.1	622.8	622.8	622.8	620.4	620.4	620.4	628.7	628.7	628.7
Oct.	240	591.3	595.3	598.2	589.2	589.2	589.2	607.3	607.3	607.3	597.0	597.0	597.0
Nov.	241	588.1	592.1	595.1	585.2	585.2	585.2	586.0	586.0	586.0	587.2	587.2	587.2
Dec.	240	623.7	627.7	627.7	616.0	616.0	616.0	624.3	624.3	624.3	618.8	618.8	618.8

-2RLL: -2 Restricted Log Likelihood; *AIC:* Akaike's Information Criterion; *BIC:* Schwarz's Bayesian Criterion

Table 2 Estimated covariance functions by month $Cov(\varepsilon_{ik}, \varepsilon_{i'k'}) = \sigma^2[f(d_{ii'})]$

Month	Model	Covariance Function
March	SP(SPH)	$C(d_{ij}) = 0.60[1 - 1.5(d_{ij}/19.4) + 0.5(d_{ij}/19.4)^3]1(d_{ij} < \rho)$
April	SP(SPH)	$C(d_{ij}) = 0.50[1 - 1.5(d_{ij}/15.6) + 0.5(d_{ij}/15.6)^3]1(d_{ij} < \rho)$
May	SP(SPH)	$C(d_{ij}) = 0.50[1 - 1.5(d_{ij}/16.1) + 0.5(d_{ij}/16.1)^3]1(d_{ij} < \rho)$
June	SP(GAU)	$C(d_{ij}) = 0.70[exp(d_{ij}/10.2)^2]$
October	SP(SPH)	$C(d_{ij}) = 0.60[1 - 1.5(d_{ij}/14.4) + 0.5(d_{ij}/14.4)^3]1(d_{ij} < \rho)$
November	SP(SPH)	$C(d_{ij}) = 0.60[1 - 1.5(d_{ij}/18.0) + 0.5(d_{ij}/18.0)^3]1(d_{ij} < \rho)$
December	SP(SPH)	$C(d_{ij}) = 0.60[1 - 1.5(d_{ij}/21.2) + 0.5(d_{ij}/21.2)^3]1(d_{ij} < \rho)$

SP(SPH): Spatial Spherical; SP(GAU): Spatial Gaussian

For all the months, there is not any trend in the random part of the model and the models capture the effect of the El Niño and La Niña phenomena. This can be appreciated in the random part of the model: the lowest values of random intercept estimates belong to 1971, 1973, 1974, 1975, 1976, 1977, 1984 and 1999, and with the exception of 1973 and 1984, all these years correspond to the La Niña phenomenon; the highest values belong to 1983, 1987, 1988, 1990, 1992, 1997, 1998 and 2001, and with the exception of 1988 (the La Niña phenomenon) and 1990, 2001 (normal conditions), the rest of the years belong to the El Niño phenomenon. Table 2 displays the covariance functions by month.

A likelihood ratio test is performed to evaluate the significance of the spatial variability in the data. It is estimated as the difference between ($-$2Rll value of the random coefficient model without spatial covariance structure in the errors) and ($-$2Rll of the random coefficient model with spatial covariance structure in the errors of the best model selected). The values are compared with χ^2 (1). With the exception of October and November, there is significant spatial variability in the data at $\alpha = 0.05$. Table 3 displays the results obtained.

Table 4 displays the results of estimates by month for the fixed part of the models. The t-test p values indicate that the intercept and the altitude variable contribute significantly at $\alpha = 0.05$ in the monthly average temperature modelling.

Model Checking

For all the months, the relation between monthly average temperature and altitude, and monthly average temperature and year, have been well specified in the models.

Table 3 Likelihood Ratio Test values by month

Month	Likelihood Ratio Test
March	7.7
April	4.8
May	6.2
June	13.8
October	2.1
November	2.9
December	7.7

Table 4 Estimates of fixed effects for the model with better fit, variable monthly average temperature

Month	Intercept			Altitude		
	Estimate	Standard Error	t-test p value	Estimate	Standard Error	t-test p value
March	28.5308	0.3980	< 0.0001	-0.00567	0.000245	< 0.0001
April	28.4522	0.3717	< 0.0001	-0.00563	0.000232	< 0.0001
May	28.0447	0.3705	< 0.0001	-0.00539	0.000230	< 0.0001
June	28.2044	0.3750	< 0.0001	-0.00542	0.000232	< 0.0001
October	27.4659	0.3988	< 0.0001	-0.00528	0.000248	< 0.0001
November	27.3613	0.3983	< 0.0001	-0.00527	0.000247	< 0.0001
December	27.3810	0.3953	< 0.0001	-0.00513	0.000242	< 0.0001

Histograms and Q-Q normal plots and the Anderson–Darling test show that sufficient evidence to reject the normality of the errors at $\alpha = 0.05$ is not there. Graphics of residuals versus predicted values display a uniform spread about the zero error line, indicating the homogeneity of variances in the errors for all the months.

Cross-validation

Three models are cross-validated: (1) random coefficient model without spatial structure in the errors (for the months October and November), (2) random coefficient model with spatial covariance structure in the errors, and (3) the IDEAM model, which is used to predict in Colombia a temperature value in a site where it has been not measured (information supplied by the meteorology section of IDEAM).

The IDEAM model is based on the theoretical knowledge of the temperature pattern with altitude: the altitude reduces average temperature by about 0.625°C per 100 m (McGregor and Nieuwolt, 1998). IDEAM fitted a simple regression model $y = 29.36872–0.00618x$, where y is the average temperature value and x is the altitude variable.

Table 5 Mean square error values for the three models

Month	Random coefficient model without spatial covariance structure in the errors		Random coefficient model with spatial covariance structure in the errors		IDEAM Model	
	Bias	MSE	Bias	MSE	Bias	MSE
March			–0.01218	0.81753	-0.07259	1.05726
April			–0.07710	0.60518	0.01722	0.68054
May			–01.0221	0.65890	-0.10595	0.72279
June			–0.03119	0.76049	0.03899	0.78885
October	0.08875	0.54472	0.07761	0.53384	–0.44548	0.94416
November	0.07350	0.62506	0.05467	0.61510	–0.39685	0.80702
December			–0.03753	0.86850	–0.38558	1.06190

The cross-validation results show that the random coefficient model with spatial covariance structure in the errors is the model with the lowest values of mean square error (Table 5) and the IDEAM model displays the highest values. In general the IDEAM model shows the highest residual values in comparison with the other models.

6 Discussion

Two models are evaluated for the monthly average temperature modelling: the random coefficient model without spatial covariance structure in the errors and the random coefficient model with spatial covariance structure in the errors (spherical and Gaussian). These models are compared with the IDEAM model. The random coefficient model without spatial covariance structure in the errors and the random coefficient model with spatial covariance structure in the errors capture the influence of the El Niño phenomenon. This indicates that the inclusion of the random part in the model is appropriate.

In the modelling of spatial covariance, variograms of the residuals by year are obtained, using 13 meteorological stations for the mountains. The numbers vary by year, because of the missing values of monthly average temperature in some meteorological stations. The area studied has only these meteorological stations. There are not many meteorological stations; however, the variogram used to model the spatial covariance in the errors is the weighted residuals variogram, obtained from variograms over 32 years. The objective is to estimate the spatial covariance; the prediction with kriging models is not the objective of this study. Carroll et al. (1997) in the article "Ozone Exposure and Population Density in Harris County, Texas, USA" used between 9 and 11 meteorological stations, the number varying by year, and hourly ozone measurements between 1980 and 1993. The model fitted by Carroll et al. (1997) consists of two components: a deterministic function of month, hour, temperature and other meteorological data and a random process, which contains a spatial and temporal variation. Bel (2004) indicates that environmental processes are rarely stationary and isotropic and in some cases few measures are available, in which case classical kriging performs badly. He proposes a non-parametric estimator of the variogram to compare the estimation with the parametric estimation. The non-parametric estimation of the variogram gives better results than the parametric estimation.

The results of the cross-validation process for months belonging to the wet period show that the random coefficient model with spatial covariance structure in the errors is the model with the lowest values for mean square error, in spite of the not significant spatial variability in October and November. The IDEAM model displays the highest values for mean square error. In general the IDEAM model shows the highest residual values in comparison with the other two models. It means that the two models fitted contribute to the estimation of monthly average temperature, because they improve the common way of estimating temperature with the

IDEAM model, based on a simple regression model, whose slope value of altitude (–0.00618) is very close to the theoretical value (–0.00625). In this research the slope value of altitude varies between –0.00567 (for March) and –0.00513 (for December).

The normality assumption on the errors is rejected for dry periods in the mountains and for all the months in the valley. Non-normality in the data caused extreme values to be recorded during the El Niño years. A classical model in geostatistics assumes that the data $Y=[y(x_i); i=1, ...n]$ are a realisation of a Gaussian intrinsic random function (Bradley and Haslett, 1992; Diggle et al., 1998). Alternative solutions proposed and used to solve this problem are (1) the transformation of the original variables (Nunes et al., 2004; Diggle et al., 1998) and (2) the application of generalised linear models to model variables with non-Gaussian distribution and a Bayesian inferential framework to estimate the spatial covariance structure.

Acknowledgments This research has been funded by Universidad del Valle, Cali, Colombia, and COLFUTURO (Fundación para el Futuro de Colombia).

References

Anderson T, Darling D (1954) A Test of Goodness of Fit. J Am Stat Assoc 49(268):716–723

Bel, L (2004). Nonparametric variogram estimator. Application to air pollution data. In: Sanchez-Vila X, Carrera J, Gómez Hernández JJ (eds) "Geostatistics for environmental applications, geoENV IV. Proceedings of the fourth European conference on geostatistics for environmental applications. Barcelona, Spain, November 27–29, 2002". Kluwer Academic Publisher, pp 29–40

Box GEP, Jenkins GM, Reinsel GC (1994) Time series analysis, forecasting and control. Englewood Cliffs, NJ: Prentice-Hall

Bradley R, Haslett J (1992) High-interaction diagnostics for geostatistical models of spatially referenced data. The Statistician 41(3):371–380

Brockwell PJ, Davis RA (1996) Introduction to time series and forecasting. Springer-Verlag, New York Inc

Carroll RJ, Chen EI, George TH Li, Newton J, Schmiediche H, Wang N (1997) Ozone exposure and population density in Harris county, Texas. J Am Stat Assoc 92(438):392–404

Chan KH, Hayya JC, Ord K (1977) A Note of trend removal methods: the case of polynomial regression versus variate differencing. Econometrica 45(3):737–744

Chilès JP, Delfiner P (1999) Geostatistics. Modelling spatial uncertainty. John Wiley and Sons, Inc, United States

Cressie, N.A.C. (1993). Cressie NAC (1993) Statistics for spatial data. John Wiley and Sons, Inc, United States

Diggle PJ, Tawn JA, Moyeed RA (1998) Model-based geostatistics. Appl Stat 47(3):299–350

Isaaks EH, Srivastava RM (1989) An introduction to applied geostatistics. Oxford University Press. United States

Little R, Milliken G, Stroup W, Russell D (1996) SAS System for Mixed Models, Cary, NC, SAS Institute Inc, United States

McGregor G, Nieuwolt S (1998) Tropical climatology. An introduction to the climates of the low altitudes. John Wiley and Sons, England

Nunes C, Soares A (2004) Geostatistical space–time simulation model for characterization of air quality. In:Sanchez-Vila X, Carrera J, Gómez Hernández JJ (eds) "Geostatistics for environmental applications, geoENV IV. Proceedings of the fourth European conference on geostatistics for environmental applications. Barcelona, Spain, November 27–29, 2002". Kluwer Academic Publisher, pp 103–113

Osorio J (1999) Osorio J (1999) Plan de Investigacion y Transferencia de Tecnologia para la Competitividad y el Desarrollo Sostenible de la Producción de Hortalizas en el Suroccidente Colombiano. Corporación Colombiana para la Investigación Agropecuaria, CORPOICA

Pabón JD, Zea J, LeóN G, Hurtado G, González OC, Montealegre JE (2002) La Atmósfera, el Tiempo y el Clima. Instituto de Hidrología, Meteorología y Estudios Ambientales IDEAM. Colombia

SAS Institute Inc (1999) SAS/STAT. User's Guide. Version 8. SAS, Institute Inc

Vandaele W (1983) Vandaele W (1983) Applied time series and Box–Jenkins models. Academic Press, Inc

Improving the Areal Estimation of Rainfall in Galicia (NW Spain) Using Digital Elevation Information

J. M. M. Avalos and A. P. González

Abstract Rainfall is an intermittent phenomenon in both space and time and it displays large spatio-temporal variability. The most commonly used interpolation methods provide good estimates of the total amount of rainfall but they do not model accurately its complex spatio-temporal structure. Better descriptions of rainfall spatial variability should be obtained from a digital elevation model. The application of a geostatistical technique that should improve rainfall estimation by integrating elevation data in Galicia (NW Spain) is discussed. The algorithm used is kriging with an external drift. The results are compared with methods that do not account for the elevation information data, such as ordinary kriging, conditional simulation and the inverse squared-distance weighting. The data set used in this exercise consists of monthly rainfall data from a maximum of 121 pluviographs corresponding to a period of 48 months (from January 1998 to December 2001) and a digital elevation model with cells of 500 m by 500 m size covering an area of 29750 km^2. For 15 out of 48 monthly semivariograms a pure nugget effect was observed and during 33 months spatial dependence was modelled by a nugget effect component plus a spherical, exponential or Gaussian component. Gaussian conditional simulation gave higher rainfall mean values than ordinary kriging and kriging with external drift. However, kriging with external drift accounting for elevation gave results that were thought to be the best descriptor of the effect of topography on the rainfall. The results, while in the line of similar applications in other fields, favor the geostatistical methods including the secondary information; however, the scores of the different methods are very similar, making it difficult to justify complex geostatistical analysis in this specific case study. Reasons for this performance should be found in the weak spatial correlation of the rainfall and between the rainfall and elevation.

1 Introduction

Rainfall is one of the main characteristics that define the climate of a region so its spatial characterization is highly relevant. The highest rainfall values observed in a

J. M. M. Avalos
Faculty of Sciences. University of Coruña, A Zapateira 15071, A Coruña, Spain
e-mail: jmirasa@udc.es

A. Soares et al. (eds.), *geoENV VI – Geostatistics for Environmental Applications*, 263–273

map of Galicia with annual isohyets obtained by classic interpolation methodologies and for an average of 30 years (De Uña Álvarez, 2001) were observed on the eastern mountain ranges and on the precoastal elevations and the lowest ones were found in the interior fluvial valleys.

Both facts that rainfall is a spatio-temporal intermittent phenomenon displaying large spatio-temporal variability and that rain gauge networks only collect point estimates of rainfall make obtaining an estimate of rainfall spatial distribution within a catchment area to be a problem of interpolation.

Several studies have demonstrated the convenience of performing geostatistical analyses for mapping rainfall in different geographical locations (Abtew *et al.*, 1993; Goovaerts, 2000; Gómez-Hernández *et al.*, 2001; Militino *et al.*, 2001). However, the use of a complex technique is not a guarantee of a better performance (Gómez-Hernández *et al.*, 2001).

Similarly, this study compares the efficiency of several interpolation techniques used to map rainfall in Galicia (NW Spain). The main objective of this study was to compare four different interpolation techniques, a deterministic one (inverse distances) and three geostatistical methods (ordinary kriging, kriging with external drift and Gaussian conditional simulation) and to improve the quality of the estimates by including topographical information into the method used.

2 Material and Methods

The data sets used for this work corresponded to total monthly rainfall (in millimeters) during the period January 1998 to December 2001 from 121 climatological stations distributed irregularly over the surface of Galicia. To conduct a spatial interpolation using geostatistical techniques, an exhaustive analysis of the spatial structure of this data set was performed using the software GSTAT (Pebesma, 2000) integrated in a GIS called PCRaster (Van Deursen and Wesseling, 1992).

A digital elevation model of Galicia (Thonon and Paz González, 2004) was used to perform the interpolations, by both geostatistical and deterministic techniques. This model consists of regular cells of 500 by 500 meters covering an area of 29750 km^2 (Fig. 1). The rain gauge network distribution is shown in Fig. 2.

The data sets were statistically characterized; this description included the calculation of mean, median, mode, minimum, maximum, variation coefficient and standard deviation. Once the monthly data sets were statistically characterized, the absence of outliers was verified. Then, stationarity was analyzed, and when observed, any drift was filtered. Next, correlations between rainfall and altitude and between rainfall and distance to the coast were checked. This analysis was performed by linear regression. The information obtained was used to define external trends by universal kriging.

Inverse distances method was used as a reference for mapping monthly rainfall data. This method takes into account both proximity and gradual variation associated to trend surfaces.

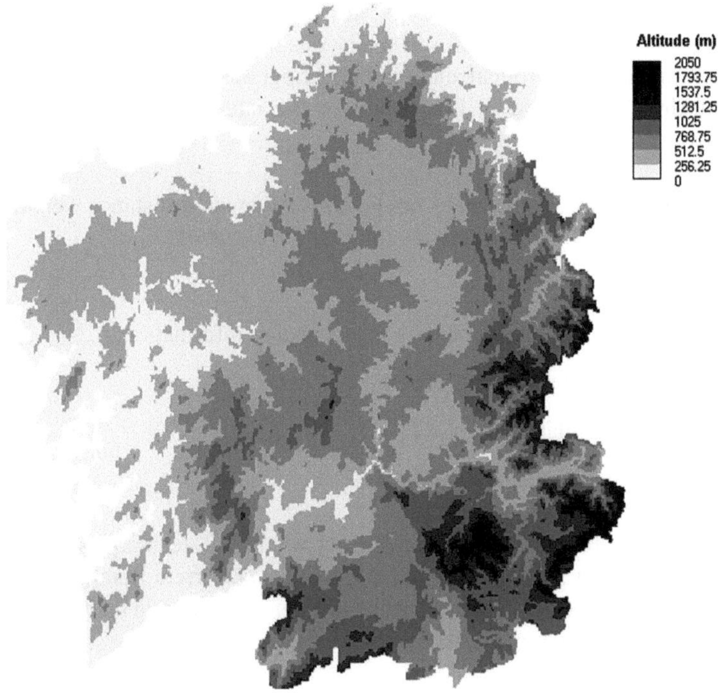

Fig. 1 Digital elevation model of Galicia

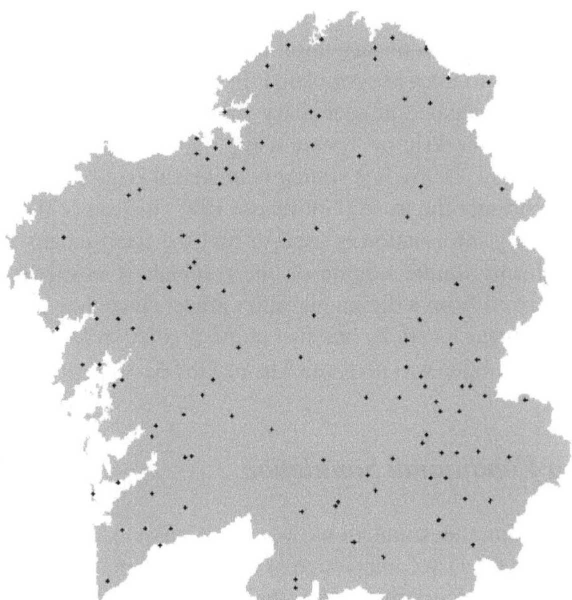

Fig. 2 Distribution of rain gauges

Because of the lack of data only the omnidirectional semivariogram was computed, and hence, the spatial variability was assumed to be identical in all directions. Classic criteria for calculating semivariograms were taken into account.

The techniques applied in this paper for the interpolation of rainfall are as follows:

- Traditional non-geostatistical technique: Inverse squared-distance weighting;
- Geostatistical techniques not accounting for the elevation:
  - Ordinary kriging (OK);
- Geostatistical technique accounting for secondary information:
  - Kriging with external drift (KED);
- Geostatistical simulation:
  - Gaussian conditional simulation (GS).

These methods are considered standard geostatistical techniques (Chilés and Delfiner, 1999; Goovaerts, 1997; Samper Calvete and Carrera Ramírez, 1996). Ordinary kriging is not described here, while kriging with external drift and Gaussian conditional simulation are described only briefly.

2.1 Kriging with External Drift

This method incorporates secondary information into the kriging system when the main and secondary variables are correlated. In this case both altitude and distance to the shoreline are evaluated as secondary variables. KED integrates the universality conditions into the kriging system using one or more of the ancillary drift variables (Wackernagel, 1995). It is similar to universal kriging, but it uses an ancillary variable to represent the trend (Goovaerts, 1997) instead of the spatial coordinates. The secondary information is used to find the local means of the primary variable and performs simple kriging on the residuals (Goovaerts, 2000). These variables were derived from a digital elevation model since they need to be known not only at the sampled locations but also at the prediction locations (McBratney et al., 2000).This technique was performed using GSTAT software (Pebesma, 2000).

2.2 Gaussian Conditional Simulation

The method used for the simulations was the Gaussian conditional simulation through LU (lower-upper) decomposition of the variance matrix. This method is the preferred Gaussian-based algorithm when the total number of conditioning data plus the number of nodes to be simulated is small and many realizations are requested (Goovaerts, 1997).

For conditional simulation analysis, 100 different realizations were run. From these data, a mean value was computed. When a sufficient realization number is used, mean values obtained by simulation are expected to be similar to those obtained by kriging.

3 Results and Discussion

Preliminary statistical analysis showed the spatial heterogeneity of the rainfall in Galicia; this variability was higher in the dry season. From the values of skewness and kurtosis coefficients, it was assumed that monthly rainfall data follows a Gaussian distribution (data not shown, Mirás Avalos, 2004).

Spatial dependence was described in 33 out of the 48 months analyzed. When no spatial dependence was found, rainfall distribution was approached by the inverse distance technique.

The correlation between rainfall and elevation ranged from 0.04 to 0.55 and the correlation between rainfall and distance to the shoreline oscillated from 0.004 to 0.71. Sometimes this information is not worth considering but it was used when the correlation coefficient was higher than 0.2 ($n = 100$) by performing KED.

Inverse distance method was suitable for a quick estimation of rainfall at the study level. Output maps showed, in general, a discontinuous appearance.

From an analysis of the semivariograms (Table 1) and their fitted parameters we gather that the monthly rainfall showed a variable nugget effect ranging from 0 to 64.02% of the sill value, which is a dependence ratio (Cambardella et al., 1994), and a rather short range of spatial dependence oscillating from 2.99 and 59.87 Km.

Selected cross-validation parameters (r^2, MSE and NMSE) of the fitted semivariograms are shown in Table 1 as well.

A summary of the variogram models fitted to the experimental data is shown in Table 1. The low magnitudes of the nugget effects during 1998 might reflect that the network would be enough for detecting the rainfall spatial variability structure. However, the fact that no spatial dependence was observed in six months of 1998 suggested that the available network is not enough for a proven analysis of the precipitation spatial structure. In summary, theoretical models fitted quite well to experimental semivariograms, except in the cases of August and November. The parameters provided by cross-validation were appropriate. Fitted model ranges were lower than 20 Km, as a consequence, kriging and simulation interpolations would be oriented to local environments.

In 1999, 11 data sets showed spatial autocorrelation; the exception was June. Theoretical models for 1999 showed cross-validation parameter values close to the ideal values. Nugget effects were high in most cases. Most spatial dependence ranges were lower than 20 Km except in February, April, May, July and September (Table 1).

In 2000, only January, February and June did not show a spatial dependence pattern. Nugget effects were low or moderate. Theoretical models fitted well to

Table 1 Theoretical model parameters for experimental variograms (C_0 = nugget effect; C_0+C_1 = sill; $\%C_0$ = percentage of the nugget effect; a = range; r^2 = correlation coefficient; MSE = mean square error; NMSE = non-dimensional mean square error)

Month	Trend	Model	C_0	C_0+C_1	$\%C_0$	a (Km)	r^2	MSE	NMSE
January 1998	Linear	Exponential	0	2445.2	0	3.0	0.66	0.0157	0.966
June 1998	Linear	Spherical	10	145.47	6.43	14.2	0.45	−0.0042	1.138
July 1998	–	Exponential	0	396.18	0	8.8	0.66	0.0072	1.032
August 1998	Quadratic	Gaussian	3	56	5.08	10.0	0.27	0.086	1.621
November 1998	Quadratic	Spherical	0	946.61	0	18.6	0.25	0.0334	1.226
December 1998	–	Exponential	69.1	2702.5	2.49	9.7	0.51	0.025	0.887
January 1999	–	Exponential	589.16	1235.36	32.29	11.1	0.68	−0.013	1.102
February 1999	–	Spherical	145.96	301.02	32.65	44.8	0.72	0.004	0.984
March 1999	Linear	Exponential	327.5	2164	13.16	15.9	0.89	0.01	0.902
April 1999	–	Spherical	929.31	1727.7	34.98	23.9	0.63	0.013	0.970
May 1999	–	Spherical	548.4	858.76	38.97	37.7	0.58	0.008	1.047
July 1999	–	Spherical	28.51	102.73	21.72	33.0	0.49	0.04	1.065
August 1999	–	Exponential	25	712.71	3.39	8.3	0.59	−0.002	0.992
September 1999	–	Spherical	1796.1	4084.9	30.54	28.4	0.67	0.011	0.953
October 1999	–	Exponential	0	5430.82	0	4.1	0.61	0.0001	1.042
November 1999	Linear	Exponential	0	813.85	0	3.6	0.79	−0.01	1.005
December 1999	Linear	Exponential	2619.3	5321	32.99	18.1	0.65	0.01	0.927
March 2000	–	Exponential	37.07	104.7	26.15	14.6	0.54	−0.0031	0.986
April 2000	–	Exponential	100	9413.5	1.05	16.8	0.79	0.0195	1.007
May 2000	–	Exponential	213.85	1338.7	13.77	16.8	0.66	0.0205	0.984
July 2000	–	Exponential	0	555.56	0.00	26.4	0.72	−0.015	1.060
August 2000	–	Spherical	33.54	492.31	3.38	59.9	0.87	0.0066	1.205
September 2000	–	Exponential	90	372.47	19.46	10.6	0.68	0.0056	1.055
October 2000	–	Spherical	883.95	1515.4	36.84	32.4	0.66	0.002	1.046
November 2000	–	Spherical	4262.43	12126.6	26.01	35.6	0.66	0.0255	0.991
December 2000	–	Exponential	6424.94	21501.8	23.01	33.3	0.77	0.0172	0.925
January 2001	–	Exponential	0	21146.1	0	11.7	0.67	0.01	0.992
March 2001	Cubic	Exponential	100	20196.1	0.49	9.9	0.71	−0.075	1.060
April 2001	Linear	Spherical	742.81	417.43	64.02	34.64	0.49	0.019	1.040
July 2001	Linear	Spherical	100	700	12.5	30	0.67	0.008	0.964
August 2001	Linear	Exponential	20	423.01	4.51	10.36	0.49	0.016	0.877
October 2001	Cubic	Spherical	909.1	37263.6	17.71	37.26	0.87	−0.007	0.987

experimental semivariograms, as the cross-validation showed. Spatial dependence ranges were higher than 20 Km in more than half the cases (Table 1).

Rainfall spatial dependence was observed in six months of 2001. Theoretical models showed variable nugget effects (Table 1). Theoretical models fitted well to experimental semivariograms, the cross-validation parameters being close to their ideal values. The ranges of spatial dependence were higher than 20 Km on three occasions.

Taking into account the entire data set, values of NMSE were close to 1 in most cases. This was the main criterion for deciding which fitted model was the best one for each monthly data set. Other parameters, such as correlation coefficient (r^2) and MSE were taken into account as well.

Figure 3 shows an example of the estimate maps using the four different techniques applied in this study.

Ordinary kriging offered smoothed maps and high errors. The distribution patterns of maps obtained by this technique were similar to those of maps obtained by inverse distances; however, ordinary kriging, reproduced extreme values and smoothed estimate map boundaries. Kriging results depend on a number of factors

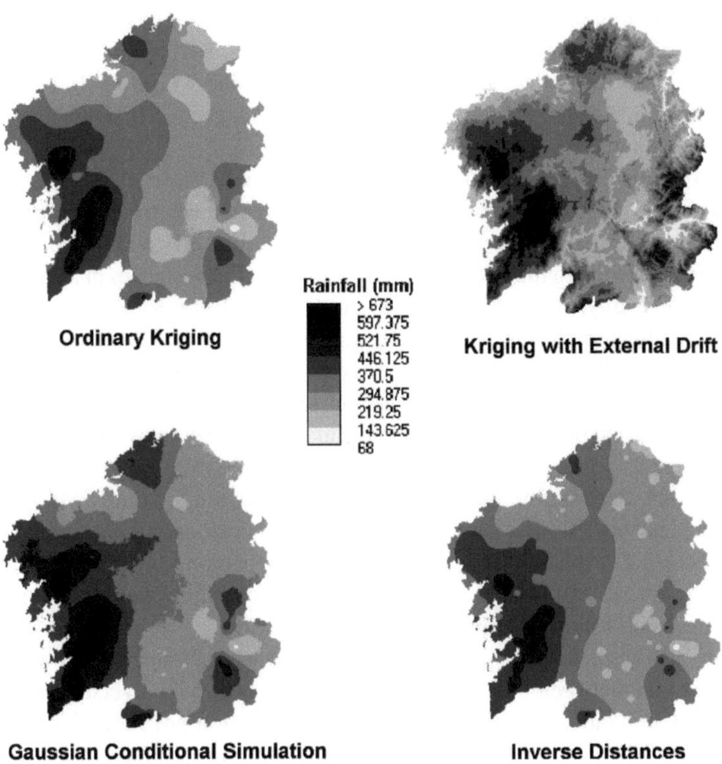

Fig. 3 Rainfall distribution for April 2000 interpolated by different techniques (KED accounting for altitude)

which include the fitted model, rain gauge network characteristics, trend, etc. Estimated maps were too smooth to reproduce maximum and minimum values which appeared in the original data sets and kriging error maps tended to show a high and uniform uncertainty pattern.

Soft information (topography, distance to the shoreline) was used to enhance the previous estimation by using the KED technique. A noticeable influence of topography over the rainfall estimated values was observed. KED results seemed to be reasonable and they were similar to the descriptions about topography influence on rainfall found in the literature. Although the KED results showed effects which ordinary kriging did not show, the drift degree approach and semivariogram model were quite arbitrary.

Finally, the Gaussian conditional simulation method was used and it gave maps that presented a realistic variability, reproducing, simultaneously, measured values. This methodology is able to reproduce lower amplitude fluctuations than ordinary kriging and it is not so sensitive to the presence of extreme values. Nugget effect amplitude influences the extreme values generated by conditional simulation.

Mean monthly rainfall values interpolated using the methodology described in this paper are shown in Table 2.

KED results using altitude as soft information showed positive and negative differences regarding those of ordinary kriging and KED using distance to the shoreline as soft information. For example, in January 1998, KED estimated mean value (122.4 mm) was lower than the value obtained by ordinary kriging (144.2 mm). On the contrary, in March 2001, KED estimated mean value was 476.8 mm, higher than the values obtained by ordinary kriging (459.7 mm).

When comparing the two different approaches of KED, the results showed a certain dispersion; thus, sometimes higher values were obtained when altitude was used as a secondary information source and other times the higher values corresponded to KED using distance to the coast as soft information. For instance, in January 1998, or April, May, July, August, September and October 1999, KED using altitude as soft information gave lower mean rainfall values; in March and August 2000 or 2001 this modality of KED gave higher mean rainfall values than the other KED approach.

Inverse distance technique results were close to the registered values, their estimation ratio varied between 4% of underestimation and 3% of overestimation of the rainfall values.

Taking into account the mean of the results from the 33 months which showed spatial dependence (Table 3), we observed that Gaussian conditional simulation was the method that provided the highest mean value for rainfall (162 mm) and KED accounting for distance to the shoreline as soft information the one that provided the lowest mean value (145 mm).

When all the methodologies used in this work were compared, more similarities were found between the rainfall distributions obtained by inverse distances and conditional simulation. Moreover, conditional simulation gave the highest rainfall mean values while the mean values obtained by ordinary kriging were lower than those obtained by inverse distances, generally. KED results showed positive and negative differences when comparing with ordinary kriging and inverse distance

Table 2 Mean rainfall values estimated in Galicia by the different methodologies used in this study [KED using as soft information (a) altitude and (b) distance to the coast]. Data displayed in mm

Month	Sample Mean	Inverse distances	OK	KED (a)	KED (b)	CS
January 1998	143.2	142.8	144.2	122.4	138.9	158.9
February 1998	46.9	45.6	–	–	-	–
March 1998	71.7	71.1	–	–	-	–
April 1998	367.2	366.4	–	–	–	–
May 1998	84.9	85.3	–	–	–	–
June 1998	31.4	31.1	30.8	32.2	31.0	33.3
July 1998	43.8	44.9	45.8	43.5	45.7	50.6
August 1998	6.6	6.9	7.3	7.8	7.3	8.9
September 1998	165.2	164.8	–	–	–	–
October 1998	69.8	69.8	–	–	–	–
November 1998	82.1	82.0	81.5	87.1	82.2	86.6
December 1998	121.4	120.8	120.2	112.8	118.2	132.6
January 1999	121.2	121.2	120.7	116.1	117.8	133.1
February 1999	60.5	60.2	60.1	60.7	59.1	66.1
March 1999	164.2	166.3	163.3	172.4	166.0	183.6
April 1999	159.0	160.3	158.6	146.0	156.6	177.4
May 1999	123.0	123.3	124.1	121.1	123.3	136.4
June 1999	31.0	31.0	–	–	–	–
July 1999	19.8	20.2	20.0	19.8	20.1	22.6
August 1999	70.9	70.5	70.6	66.8	70.0	78.4
September 1999	232.3	233.3	230.6	221.5	229.1	260.8
October 1999	222.9	222.8	225.2	203.9	223.6	248.1
November 1999	81.8	82.3	84.0	86.8	83.5	90.1
December 1999	210.3	206.9	202.3	213.8	204.2	226.7
January 2000	42.2	40.6	–	–	–	–
February 2000	56.9	66.8	–	–	–	–
March 2000	29.7	30.4	30.8	31.0	30.8	34.5
April 2000	314.8	312.3	306.9	313.9	306.2	342.3
May 2000	96.0	95.9	95.2	96.7	95.3	106.5
June 2000	16.3	16.6	17.0	18.1	16.8	18.6
July 2000	60.9	60.2	62.3	62.4	61.6	68.7
August 2000	47.3	46.0	43.7	44.7	43.7	48.2
September 2000	79.7	79.2	79.1	73.1	77.0	87.0
October 2000	173.2	172.2	171.5	161.4	168.5	189.1
November 2000	410.3	404.7	398.4	397.3	394.7	439.9
December 2000	445.4	451.2	458.9	461.1	451.9	508.1
January 2001	368.1	370.2	365.0	373.9	365.4	404.4
February 2001	131.0	141.4	–	–	–	–
March 2001	467.4	465.1	459.7	476.8	460.4	511.6
April 2001	82.5	81.5	81.7	84.7	80.9	88.3
May 2001	96.6	94.3	–	–	-	–
June 2001	16.1	16.2	–	–	-	–
July 2001	82.1	82.8	82.8	86.4	83.4	92.0
August 2001	56.6	57.0	56.0	59.3	56.2	60.8
September 2001	48.9	49.3	–	–	–	–
October 2001	226.9	224.1	216.5	224.7	216.4	251.2
November 2001	18.0	18.5	–	–	–	–
December 2001	40.6	40.5	–	–	–	–

Table 3 Mean rainfall values for the 33 months which showed a spatial dependence in rainfall. [KED using as soft information (a) altitude and (b) distance to the coast]

	Sample Mean	Inverse Distances	OK	KED (a)	KED (b)	GS
Mean rainfall value (mm)	147.02	146.82	145.90	145.46	145.02	161.98

results. Inverse distances interpolation is considered an appropriate technique for rapid estimation of the rainfall spatial distribution at the scale studied.

4 Conclusions

Differences between mean monthly values measured at the stations and those obtained by inverse distances technique were lower than ± 10 mm. This result might reflect the fact that the network is representative despite the irregular distribution of the rain gauges. Maps obtained by inverse distances showed a division in large areas, and within them, smaller areas were observed; usually, punctual areas; corresponding to high or low rainfall values giving a discontinuous appearance to the maps.

Ordinary kriging gave values similar to those obtained by inverse distances but it provided estimation error maps as well. Resulting maps showed similar patterns to those obtained by inverse distances. These maps were too smooth for reproducing original maximum and minimum data; moreover, error maps tended to show a high and uniform uncertainty pattern.

KED offered reasonable results but drift approach seemed to be arbitrary. Rainfall spatial distributions and monthly rainfall mean values obtained by the two different KED approaches, usually, differed notably. Generally, when distance to the coast was used as soft information, isohyets showed a notorious smoothing, thus, the maps were similar to those obtained by ordinary kriging. Taking into account the topographic characteristics, interpolation performed using altitude as external drift was more plausible.

Finally, conditional simulation gave maps that presented a variability which made them seem real, reproducing, simultaneously, measured values. Rainfall average map from 100 simulated maps obtained by conditional simulation seemed to fit very well to what occurred in reality.

Acknowledgments This study was financed by Xunta de Galicia (Spain), project PGIDIT02REM1 6201PR. We thank the *Centro de Investigaciones Forestales de Lourizán (Spain)* and the *Centro Meteorológico Territorial de Galicia* of the Spanish Ministry of Environment and Energy for providing us with data for this study.

References

Abtew W, Obeysekera J, Shih G (1993) Spatial analysis for monthly rainfall in South Florida. Water Res Bull 29(2):179–18

Cambardella CA, Moorman TB, Novak JM, Parkin TB, Karlen DL, Turco RF, Konopka AE (1994) Field-scale variability of soil properties in Central Iowa soil. Soil Sci Soc Am J 58:1501–1508
Chilés JP, Delfiner P (1999) Geostatistics. Modeling spatial uncertainty. Wiley Series in Probability and Statistics. John Wiley and Sons, Inc p 695
De Uña Álvarez E (2001) El Clima. In: Precedo Ledo A, Sancho Comíns J (eds) Atlas de Galicia. Tomo 1: Medio Natural. Sociedade para o Desenvolvemento Comarcal de Galicia. Xunta de Galicia. pp 137–156
Gómez-Hernández JJ, Cassiraga EF, Guardiola-Albert C, Álvarez Rodríguez J (2001) Incorporating information from a digital elevation model for improving the areal estimation of rainfall. In: Monestiez P, Allard D, Kluwer RF (eds) geoENV III – Geostatistics for Environmental Applications. Quantitative Geology and Geostatistics. Academy Publishers, pp 67–78
Goovaerts P (1997) Geostatistics for natural resources evaluation. Applied Geostatistics Series. New York, p 483
Goovaerts P (2000) Geostatistical approaches for incorporating elevation into the spatial interpolation of rainfall. J Hydrol 228:113–129
McBratney AB, Odeh IOA, Bishop TFA, Dunbar MS, Shatar TM (2000) An overview of pedometric techniques for use in soil survey. Geoderma, 97:293–327
Militino AF, Palacios MB, Ugarte MD (2001) Robust predictions of rainfall in Navarre, Spain. In: Monestiez P, Allard D, Kluwer RF (eds) geoENV III – Geostatistics for Environmental Applications. Quantitative Geology and Geostatistics. Academy Publishers, pp 79–90
Mirás Avalos JM (2004) Estimación y Simulación de la Precipitación en Galicia a Escala Mensual. Doctoral Thesis, University of A Coruña, p 264
Pebesma EJ (2000) Gstat User's Manual. Department of Physical Geography. Utrecht University, p 100
Samper Calvete FJ, Carrera Ramírez J (1996) Geoestadística. Aplicaciones a la Hidrología Subterránea (2ª edición). Centro Internacional de Métodos Numéricos en Ingeniería. Barcelona, p 484
Thonon I, Paz González A (2004) A geostatistically interpolated digital elevation model of galicia (NorthWest Spain). In: Sánchez Vila X, Carrera J, Gómez Hernández JJ (eds) GeoENV 2002. Fourth European conference on geostatistics for environmental applications. Kluwer Academy Publishers, pp 532–533
Van Deursen WPA, Wesseling CG (1992) The PCRaster Package. Vakgroep Fysische Geografie. Faculteit Ruimtelijke Wetenschappen. Universiteit Utrecht, Utrecht (The Netherlands), p 192
Wackernagel H (1995) Multivariate Geostatistics. Springer, Berlin, p 256

Identification of Inhomogeneities in Precipitation Time Series Using Stochastic Simulation

A. C. M. Costa, J. Negreiros and A. Soares

Abstract Accurate quantification of observed precipitation variability is required for a number of purposes. However, high quality data seldom exist because in reality many types of non-climatic factors can cause time series discontinuities which may hide the true climatic signal and patterns, and thus potentially bias the conclusions of climate and hydrological studies. We propose the direct sequential simulation (DSS) approach for inhomogeneities detection in precipitation time series. Local probability density functions, calculated at known monitoring stations locations, by using spatial and temporal neighbourhood observations, are used for detection and classification of inhomogeneities. This stochastic approach was applied to four precipitation series using data from 62 surrounding stations located in the southern region of Portugal (1980–2001). Among other tests, three well established statistical tests were also applied: the Standard normal homogeneity test (SNHT) for a single break, the Buishand range test and the Pettit test. The inhomogeneities detection methodology is detailed, and the results from the testing procedures are compared and discussed.

Introduction

Precipitation is one of the most important climate variables. Accurate quantification of its observed variability is required for a number of purposes. Long series of reliable precipitation records are essential for climate changes monitoring, general circulation models and regional climate models, modelling of erosion, runoff and pollutant transport, among other applications for ecosystem and hydrological impact modelling. However, high quality data seldom exist because in reality many types of non-climatic factors (e.g. monitoring stations relocations, changes of the surroundings, different observational and calculation procedures, etc.) can cause time series discontinuities which may hide the true climatic signal and patterns, and thus potentially bias the conclusions of climate and hydrological studies. Therefore,

A. C. M. Costa
ISEGI, Universidade Nova de Lisboa, Campus de Campolide, 1070-312 Lisbon, Portugal
e-mail: ccosta@isegi.unl.pt

A. Soares et al. (eds.), *geoENV VI – Geostatistics for Environmental Applications*, 275–282

it is recommended that, besides routine quality control, the homogeneity testing of data to be evaluated before performing those studies (Aguilar et al. 2003).

Several techniques have been developed for detecting inhomogeneities in time series of weather elements. The approaches underlying the homogenization techniques are quite different and typically depend on the type of element (temperature, precipitation, pressure, evaporation, etc.), the temporal resolution of the observations (annual, seasonal, monthly or sub-monthly), the availability of metadata (station history information) and the monitoring stations network density (spatial resolution). A review of different statistical methods is presented in (Peterson et al. 1998) and guidelines on homogenization from the World Meteorological Organization are provided by (Aguilar et al. 2003).

We propose the direct sequential simulation (DSS) approach, introduced by (Soares 2001), for inhomogeneities detection in precipitation time series. Unlike most of the traditional testing procedures described in the literature (e.g. Peterson et al. 1998, Aguilar et al. 2003), the proposed technique accounts for the joint spatial and temporal dependence between observations, and enhances the pre-eminence of the closer stations, both in spatial and correlation terms.

The inhomogeneities detection procedures applied in this study used an annual resolution testing variable derived from the daily precipitation data, namely the wet day count with 1mm as threshold, which is expected to be representative of important characteristics of variation at the daily scale (Wijngaard et al. 2003).

The stochastic simulation approach was applied to the testing variable data from four ("candidate") stations using data from 62 surrounding stations ("reference" stations, presumed homogeneous) located in the southern region of Portugal (1980–2001). As with other relative homogeneity testing approaches, reference stations data are used to account for regional climate changes and to isolate the effects of station irregularities (Peterson et al. 1998).

Among other tests, three well established statistical tests were also applied to the candidate stations time series: the Standard normal homogeneity test (SNHT) for a single break (Alexandersson 1986), the Buishand range test (Buishand 1982) and the Pettit test (Pettit 1979).

The inhomogeneities detection methodology is detailed, the results from the testing procedures are compared and discussed, and finally some conclusions of this study are drawn.

Precipitation Data and Previous Homogeneity Testing

The daily precipitation series used in this study were compiled from the European Climate Assessment (ECA) dataset and the National System of Water Resources Information (SNIRH – Sistema Nacional de Informação de Recursos Hídricos, managed by the Portuguese Institute for Water) database, and are available through free downloads from the ECA website (http://eca.knmi.nl) and the SNIRH website (http://snirh.inag.pt), respectively. The analysed precipitation series, from monitoring stations located in the southern region of Portugal with records within the period

1980–2001, were downloaded during the first semester of 2004. All stations with at least 30 years with less than 5% of observations missing were selected. In order to select a larger set of reference series, shorter series with at least 10 years lacking a maximum of 5% of data were also chosen, and hence the series with too many gaps were discarded. Using those criterions, long-term series of daily precipitation from 42 weather stations were selected, and 54 shorter series were accepted as eligible reference series.

Before being compiled for this study, the daily series of the ECA dataset had already been subject to four statistical homogeneity tests which were applied to each station data separately (absolute tests were applied rather than relative tests because of the sparse density of the ECA station network). For precipitation, the testing variable used was the annual wet day count using 1 mm as threshold. The precipitation series compiled from the ECA dataset for this study were all marked as "useful", as the four homogeneity tests did not reject the homogeneity hypothesis, at the 1% significance level. For further details see (Wijngaard 2003, Wijngaard et al. 2003), and the ECA project website (http://eca.knmi.nl).

Regarding the series from the SNIRH database, some homogeneity testing of the annual precipitation amounts has already been carried out by (Nicolau 1999), for the period 1959/60–1990/91. This author used three absolute homogeneity tests and one *subjective* relative test and found no inhomogeneities in the annual precipitation series of the stations considered here.

As recommended by (Auer et al. 2005), we assumed that the 96 daily precipitation series could contain potential breaks, and thus several homogenization procedures were applied to all of them in order to select a subset of series with quality data. The homogeneity testing followed the hybrid approach proposed by (Wijngaard et al. 2003) for the ECA dataset, and used the same annual resolution testing variable. The absolute approach used comprises the application of six statistical tests to the testing variable at all locations and using the full length of the series: the Mann-Kendall test (Mann 1945, Kendall 1975), the Wald-Wolfowitz runs test (Wald and Wolfowitz 1943), the Von Neumann ratio test (Von Neumann 1941), the Standard normal homogeneity test (SNHT) for a single break, the Pettit test, and the Buishand range test.

Series for which two or more absolute tests rejected the homogeneity hypothesis, at a 5% significance level, were excluded from this study. Consequently, a set of 66 series (Fig. 1) was selected to be used in the stochastic simulation approach: 14 long-term and 42 short-term series were considered as homogeneous by all tests, and for 10 long-term series only one of the six tests rejected the homogeneity hypothesis. To illustrate the proposed methodology, four candidate stations were chosen: Beja (ECA 666), Aljezur (SNIRH 30E.01), Alferce (SNIRH 30G.01), and Santiago do Escoural (SNIRH 22H.02).

Note that if an absolute test detects a break in a station's time series it may indicate an inhomogeneity or it may simply indicate an abrupt change in the regional climate. Historic metadata support is then essential for evaluating the breaks detected through absolute testing. Therefore, relative approaches are usually preferred, as they intend to isolate the non-climatic influences (Peterson et al. 1998).

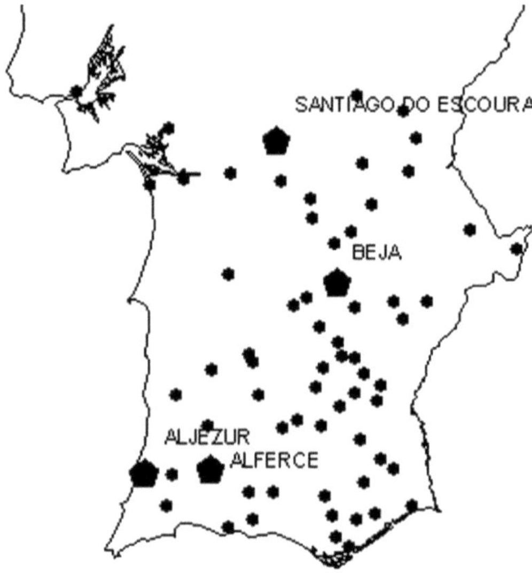

Fig. 1 Locations of the 66 monitoring stations. Candidate stations are marked with pentagons

Methodological Framework

For the detection of inhomogeneities, we propose the DSS algorithm (Soares 2001) to calculate the local probability density functions (pdfs) at candidate stations locations, by using spatial and temporal observations from nearby reference stations and *without* taking into account the candidate data. The local pdfs from each year can then be used to verify the existence of irregularities: a break year is identified whenever the interval of a specified probability p (e.g. 0.95) centred in the local pdf does not contain the observed (real) value of the candidate station. In practice, the local pdfs are provided by the histograms of simulated maps, thus this rule implies that if the observed (real) value lies below or above the predefined percentiles of the histogram of a given year then it is not considered as homogeneous.

The inhomogeneities detection procedures used in this study followed the hybrid approach proposed by (Wijngaard et al. 2003) for the ECA dataset, and used as testing variable the annual wet day count with 1 mm as threshold. For illustration purposes, the stochastic simulation approach was applied to the testing variable data from four candidate stations.

Techniques that use series from surrounding stations, some times run the test once, relying the reference to be homogeneous, or engage in an iterative procedure in which all the stations in the data set are seen consecutively as candidates and references (Aguilar et al. 2003). Following this methodology, the local pdfs of each year of the candidate series, derived from 50 simulated maps, were computed using data not only from the 62 references but also from the other 3 candidate stations. The analysed period was 1980–2001.

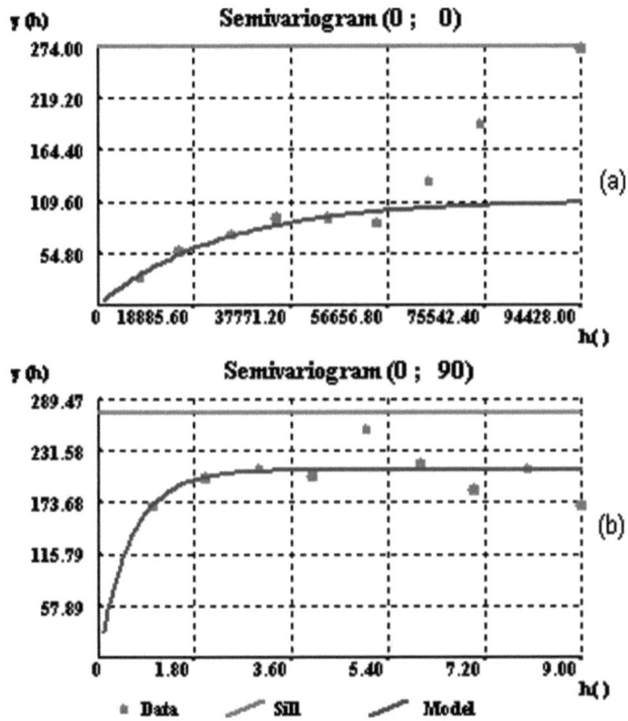

Fig. 2 Spatial and temporal experimental semivariograms of the testing variable data with the model fitted: (a) spatial isotropic semivariogram, (b) temporal semivariogram

The stochastic simulations used a spherical semivariogram model fitted to the testing variable data from the complete set of 66 monitoring stations (Fig. 2): the spatial dimension was modelled using an isotropic semivariogram with a range of 72 km, and the temporal one with the range equal to 1.8 years.

The results from the proposed technique are compared with the results from three well established homogeneity tests that used two reference series for each candidate and the full length of the series. The SNHT, Pettit and Buishand range tests were applied to composite (ratio) reference series (Alexandersson and Moberg 1997), which were derived from the testing variable. Further methodological details and additional results from this approach are described by (Costa and Soares 2006).

Results and Discussion

The proposed approach allowed us to identify several inhomogeneities by comparing the observed (real) values of the candidate series, for each year, with the 2.5% and the 97.5% percentiles of the corresponding histograms of 50 simulated maps. This methodology identified not only the same break years (or within one-year range) as the other three testing procedures, but also revealed inhomogeneities

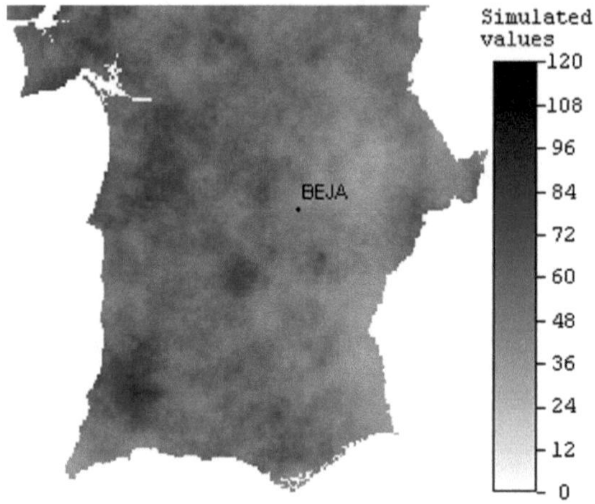

Fig. 3 One simulated realization of the annual wet day count in 1991, computed without data from Beja, at the nodes of a 1 km × 1 km grid

in other years that were not detected by any of the three statistical tests, at a 5% significance level.

For station Aljezur, the four approaches considered the series as homogeneous. The series from Beja was considered as homogeneous by the three statistical tests, whereas the stochastic approach identified a break in 1991 (Fig. 3 and Fig. 4).

For station Alferce, the SNHT concluded the series as homogeneous, the Buishand and Pettit tests detected a break in 1984, and the stochastic approach identified a break in 1983.

The candidate series from Santiago do Escoural was considered as inhomogeneous by all techniques: the SNHT detected a break in 1989, the Buishand and Pettit tests identified a breakpoint in 1988, and the proposed procedure detected breaks in 1987, 1988 and 1996.

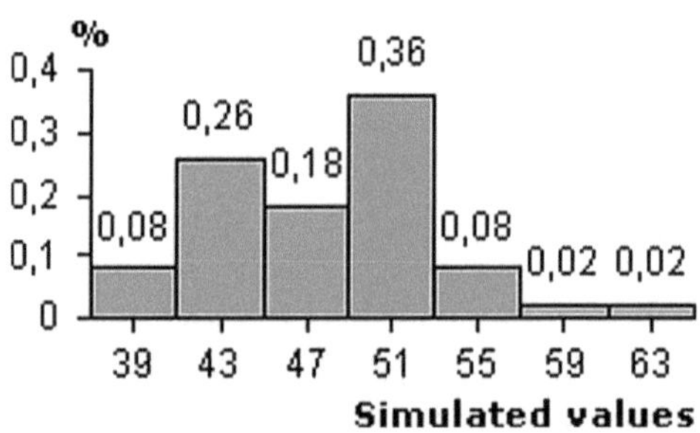

Fig. 4 Histogram of the 50 simulated realizations, computed without data from Beja, of the annual wet day count in 1991 at Beja location (the real value at Beja is 60 days)

Final Remarks

The promising results from this case study indicate the stochastic approach as a valuable tool for inhomogeneities detection in climate time series. All break years identified by the three well established statistical tests considered were also detected by the proposed technique. Moreover, the stochastic simulations approach allowed for the identification of breaks near the end of the series that were not detected by the other methods. In fact, this is one of the advantages of the proposed methodology relatively to other testing procedures commonly used which have less power in detecting breakpoints near the start and end of a series (Aguilar et al. 2003).

The evidences provided by the case study results indicate that the potential advantages of the proposed methodology are that it allows to:

- account for the detection of multiple breaks simultaneously
- identify breakpoints near the start and end of a series
- use different sets of neighbouring stations at different years, including shorter and non-complete records
- enhance the pre-eminence of the closer stations, both in spatial and correlation terms
- account for the joint spatial and temporal dependence between observations

The main disadvantage of the proposed approach is that it is computationally intensive, even though simple to apply. Nevertheless, as it allows accounting for the detection of multiple breaks simultaneously, it might be less time consuming than other testing techniques that are used iteratively by systematically dividing the tested series into smaller segments when a break is detected, and then perform the test on those segments.

We would like to suggest for future research the application of direct sequential cosimulation (Soares 2001) so that the inhomogeneities detection procedure incorporates covariates, such as altitude and distance from the coastline, which are known to influence the precipitation distribution.

List of Abbreviations

DSS - direct sequential simulation
ECA - European Climate Assessment
pdf - probability density function
pdf - probability density functions
SNHT - Standard normal homogeneity test
SNIRH - National System of Water Resources Information (Sistema Nacional de Informação de Recursos Hídricos)

References

Aguilar E, Auer I, Brunet M, Peterson TC, Wieringa J (2003) Guidelines on climate metadata and homogenization. WMO-TD No 1186, WCDMP No 53, World Meteorological Organization, Geneva

Alexandersson H (1986) A homogeneity test applied to precipitation data. J Climatol 6:661–675

Alexandersson H, Moberg A (1997) Homogenization of Swedish temperature data, Part I: Homogeneity test for linear trends. Int J Climatol 17:25–34

Auer I, Böhm R, Jurkovic A, Orlik A, Potzmann R, Schöner W, Ungersböck M, Brunetti M, Nanni T, Maugeri M, Briffa K, Jones P, Efthymiadis D, Mestre O, Moisselin J-M, Begert M, Brazdil R, Bochnicek O, Cegnar T, Gajic-Capka M, Zaninovic K, Majstorovic Z, Szalai S, Szentimrey T, Mercalli L (2005) A new instrumental precipitation dataset for the greater alpine region for the period 1800–2002. Int J Climatol 25:139–166

Buishand TA (1982) Some methods for testing the homogeneity of rainfall records. J Hydrol 58:11–27

Costa AC, Soares A (2006) Identification of inhomogeneities in precipitation time series using SUR models and the Ellipse test. In: Caetano M, Painho M (eds.) Proceedings of Accuracy 2006 - 7th International Symposium on Spatial Accuracy Assessment in Natural Resources and Environmental Sciences. Instituto Geográfico Português, pp 419–428

Kendall MG (1975) Rank correlation methods. Charles Griffin, London

Mann HB (1945) Mann HB (1945) Non-parametric test against trend. Econometrika 13:245–259

Peterson TC, Easterling DR, Karl TR, Groisman P, Nicholls N, Plummer N, Torok S, Auer I, Boehm R, Gullett D, Vincent L, Heino R, Tuomenvirta H, Mestre O, Szentimrey T, Salinger J, Forland EJ, Hanssen-Bauer I, Alexandersson H, Jones P, Parker D (1998) Homogeneity adjustments of in situ atmospheric climate data: A review. Int J Climatol 18:1493–1517

Pettit AN (1979) A non-parametric approach to the change-point detection. Appl Stat 28:126–135

Soares A (2001) Direct Sequential Simulation and Cosimulation. Math Geol 33:911–926

Von Neumann J (1941) Distribution of the ratio of the mean square successive difference to the variance. Ann Math Stat 13:367–395

Wald A, Wolfowitz J (1943) An exact test for randomness in the non-parametric case based on serial correlation. Ann Math Stat 14:378–388

Wijngaard JB (2003) Homogeneity of daily 'European Climate Assessment and Dataset' series. In: World Meteorological Organization (ed) Proceedings of the Second Seminar for Homogenization of Surface Climatological Data. WMO-TD No 962, WCDMP No 41, WMO, Geneva, pp 143–149

Wijngaard JB, Klein Tank AMG, Können GP (2003) Homogeneity of 20th century European daily temperature and precipitation series. Int J Climatol 23:679–692

Bayesian Classification of a Meteorological Risk Index for Forest Fires: DSR

R.M. Durão, A. Soares, J.M.C. Pereira, J.A. Corte-Real and M.F.E.S. Coelho

Abstract The Daily Severity Rating (DSR), a meteorological rating for assessing the risk of fires by using the forest fire index, is calculated on a daily basis in two sequential steps: first it is forecast for the day of interest, within a day of advance, and for a limited set of control points, using the Canadian Forest Fire Weather Index System (CFFWIS). Afterwards the forecast values are interpolated (ordinary kriging) into a regular grid of points covering Portugal.

In this chapter we propose a model for fire risk assessment in Portugal that is based on the conditional probability of fire, $I(x)$, as given by the class of DSR predicted for that specific period of time – $p(I(x)|R(x))$. The evaluation of this a posteriori $p(I(x)|R(x))$ is based on the update for marginal local probability of fire in each chosen county. Mapping of the risk of fire is obtained for the entire country by kriging.

As the definition of DSR classes should be dependent on specific conditions for each county, in the last part of this study the thresholds of DSR classes that can lead to a high risk of fires are calculated for each county. This regional DSR threshold is an indirect measure of other factors that are the cause of fires.

1 Introduction

Fire activity in Portugal presents high spatial and temporal variability for burnt area and fire occurrences. Fire danger rating systems like the Canadian Forest Fire Weather Index System (CFFWIS) transform daily weather observations into relatively simple indices that can be used to predict fire occurrence, behaviour and impact (Stocks et al., 1989). They are used for many purposes including planning for the daily deployment of fire suppression resources and the evaluation of fire management strategies. They can also be incorporated in different types of models to assess the long-term implications of specified fire management policies and fire regimes.

R.M. Durão
Centro de Geofísica de Évora, Universidade de Évora, Portugal
e-mail: rddurao@fc.ul.pt

A. Soares et al. (eds.), *geoENV VI – Geostatistics for Environmental Applications*, 283–294

The daily severity rating (DSR) is a numerical rating of the difficulty in controlling fires, and it is based on the fire weather index (FWI), although it more accurately reflects the expected efforts required for fire suppression. However, the DSR itself is an incomplete measure of seasonal fire activity because it is also dependent on the ignition pattern and the available control resources. DSR basically measures the meteorological factors that can cause the fires.

The idea of this study is to use the DSR values to calculate the risk of fire, taking into account historical data and recent records. In other words, starting with the marginal probability of fire in a given county, one can use DSR to update it and calculate the a posteriori probability of fire given the class DSR forecast for the same county at a given period.

Bayes' rule is applied to obtain the risk of fire as given by the conditional probabilities at the individual county locations. In the second step these conditional probabilities are interpolated (ordinary kriging) for the entire country.

The definition of DSR classes is dependent on the specific conditions of each county. For example a high DSR does not mean the same in all counties in terms of the risk of fires. Hence in the last part of this chapter, thresholds of DSR classes that can lead to a high risk of fires are calculated for each county. This regional DSR threshold can be interpreted as an indirect measure of other factors that can cause the fires apart from meteorology. The ultimate purpose of this model is to provide a tool for a daily updating evaluation of the risk of fires given the classes of DSR predicted for that period.

2 Data and Methods

2.1 Canadian Forest Fire Weather Index System

The CFFWIS consists of seven components that account for the effects of fuel moisture and wind on fire behaviour. This system uses daily weather observations or forecasts to calculate the moisture content of several fuel types and size classes, and combines them into indices of fire danger related to the potential rate of spread of fire, heat release and fire line intensity.

The FWI system (Fig. 1) depends on daily measurements of air temperature (^{o}C), relative humidity (%), 10-m open wind speed (km/h) and 24-h accumulated precipitation (mm). The first three components, the fuel moisture codes, are numeric ratings of the moisture content of litter and other fine fuels, and the average moisture content of deep and compact organic layers. The remaining four components are fire behaviour indices, which represent the rate of fire spread, the fuel available for combustion and the frontal fire intensity; their values increase as the fire danger increases. Therefore the final component, the DSR index, is an overall measure of fire danger.

The DSR, proposed by Williams (1959), is a more linear indicator of fire control effort than the FWI and thus is preferred for averaging through time and across sites,

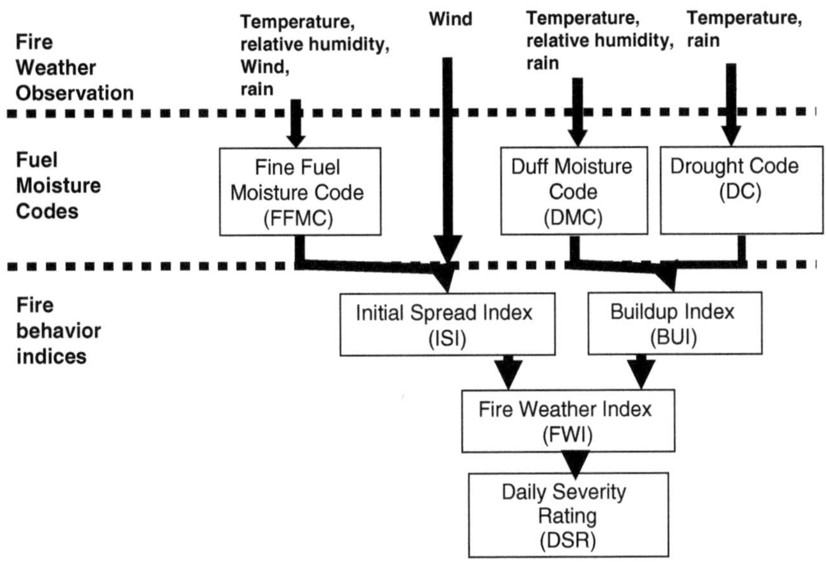

Fig. 1 Structure of the fire weather index FWI System (adapted from Pyne et al., 1996)

being used in the present study. This index is a numeric rating of the difficulty in controlling a fire and it is based on the FWI but reflects more accurately the expected efforts required for fire suppression.

DSR values result from a deterministic model developed first by Williams (1959) and modified later by Van Wagner (1970), with the following expression:

$$\mathrm{DSR} = 0.0272(\mathrm{FWI})^{1.77} \tag{1}$$

Equation (1) is valid for a regional scale. Consequently, the DSR averaged for a whole season is termed the seasonal severity rating (SSR) and is used as an objective measure of fire weather from season to season, or fire climate from region to region (Van Wagner, 1987). This severity rating technique allows the integration of fire severity over periods of various lengths from daily (DSR) to seasonal (SSR) values.

2.2 Meteorological Data

TThe present analysis was applied to the so-called fire season in Portugal (starting on 1 May and ending on 30 September), for the 2-year period 2003–2004. Meteorological data to calculate DSR values were obtained from 15 widely separated meteorological stations representing the counties where the fire occurrence is being studied.

2.3 Forest Fires Database

The fire database for this study was provided by the Portuguese Governmental Forest Service (Direcção Geral de Recursos Florestais, DGRF). Detailed statistics for forest fires in Continental Portugal are available since 1980. The fire database contains all the information about each fire in Continental Portugal between the years1980 and 2004.

The data set includes information on fire location organized by district, county and parish, date and time of ignition and extinction and burnt area type (forests, shrublands and agricultural crops). The number of fire days for each county is read from the database records, which contains the available information. The register *Date of Ignition* is the data field used for counting the number of fire days, per parish of each county. The number of fire days per county is then obtained by summing up the output values of the function $I(x)$, defined as follows:

$$I(x) = 1, \text{if a fire occur at location x in a given day;}$$
$$= 0, \text{otherwise.}$$

But, to be precise, this summation should be done after confirming that a *Date of Ignition* really is paired with a measurable value of burnt area. Hence $I(x) = 1$ only if it is associated with a burnt area value greater than 1 ha. The spatial and time homogeneity of the database must also be subjected to validation techniques, in order to refine the results.

The number of fire days and DSR values, for each county, present a very strong local behaviour that is characteristic of each region. Therefore we must expect that, even having a region classified as a class at high meteorological risk of fire, different DSR threshold values can be found in each county due to the specific local area under study. From the above, it is clear that this index shows a great sensitivity to regional fire climates, presenting a regional behaviour pattern too.

We define fire climate in each region as the data resulting from the integration of weather variables over a long period of time, affecting accordingly the fire behaviour. DSR classes considered in this study were estimated from the FWI threshold values for each Portuguese county (Viegas et al., 2004) according to equation (1). The generic defined classes of meteorological risk of fire for Portugal (numbering five) are low, moderate, high, very high and extreme risk (Table 1).

2.4 Proposed Model for the Risk of Fire

2.4.1 Risk Given by Conditional Probability of Fires

The probability (a priori) of fire $p(I(x))$ at a given county location is estimated using historical data. In this illustrative example we have used the records of the 2003–2004 fire seasons. The objective is to calculate the risk of fire, given the predicted meteorological risk and the DSR for a given time period. This risk can be

Table 1 Classes of Risk of the DSR Index, and corresponding thresholds for the 15 Portuguese counties in study, adapted from (Viegas *et al.*, 2004)

Districts	Low	Moderate	High	Very High	Extreme
V.Castelo	<2	3	11	23	>23
Porto	<1	3	8	19	>19
V.Real	<3	5	11	28	>28
Bragança	<7	11	23	33	>33
Viseu	<3	8	23	50	>50
Guarda	<1	3	8	28	>28
Coimbra	<3	6	11	23	>23
Leiria	<3	8	11	28	>28
C.Branco	<5	15	23	38	>38
Santarém	<8	13	28	38	>38
Setúbal	<11	19	33	50	>50
Portalegre	<15	28	44	57	>57
Évora	<19	28	44	57	>57
Beja	<19	28	44	57	>57
Faro	<11	19	38	57	>57

assessed using the conditional distribution $p(I(x)|R(x))$, where $R(x)$ is the DSR's class predicted for county x.

The idea of the proposed model is to use the more recent data, of fires and DSR, to update the a priori probability $p(I(x))$ through Bayes' formalism:

$p(I(x)|R(x)) = p(I(x).R(x))/R(x)$ where the joint probability is given by:

$p(I(x).R(x)) = p(R(x)|I(x))\,p(I(x))$, which leads to classical rule that up-dates the a priori probability $p(I(x)$ into a posteriori

$p(I(x)|R(x))$ through what is called a (normalized) likelihood function $p(R(x)|I(x))/R(x)$:

$$p(I(x)|R(x)) = [p(R(x)|I(x))/R(x)].p(I(x)) \quad (2)$$

In this case study we have categorized the DSR into just two classes: high risk (HR) and low risk (LR). The HR class is equivalent to the DSR high risk class in each county and includes the high, very high and extreme risk of the DSR classes (Table 1). The LR class includes the complementary DSR classes: moderate and small risk (Table 1). In order to make further result analysis easier, the threshold values of these two classes were named as official threshold values (Table 1).

Conditional probabilities $p(I(x)|R(x))$ for the classes LR and HR were estimated for each fire season according to equation (2). These a posteriori probabilities have been interpolated by ordinary kriging for the entire country, enabling corresponding mapping.

2.4.2 Regional Threshold Values of DSR

Note that the likelihood function $p(R(x)|I(x))$ of equation (2), calculated with very recent data, is also dependent on the specific conditions of each county. In other words, the same risk class $R(x)$ of DSR does not have the same meaning, in terms of the risk of fire, for all the chosen counties. The definition of a high and a low meteorological risk, HR and LR, must account for the specific conditions of the county located in x. Hence we propose to calculate, for each county, the threshold of DSR that splits HR and LR in such a way that it leads to high/moderate a posteriori risk of fire:

$$p(I(x)|R(x)) >= .65.$$

The regional DSR thresholds are an indirect measure, apart from meteorology, of factors (anthropogenic, fuel load type, topography) that can cause the fires. In other words, the higher the threshold, the greater are the other factors that influence the risk of fire. These DSR thresholds are then interpolated for the entire country.

3 Results

The model described so far, in fact, can be an analytic tool for fire studies. For testing the model's accuracy, a forecast map for 26 June 2006 was generated. Note that according to Bayes' formalism, daily records are fundamental data for the updating process. We must input actual daily values, in order to increase the accuracy of the next map on an incremental basis.

The starting data for the series of maps came from the historic fire occurrence events in DGRF's database. All the model's outputs were interpolated by ordinary kriging for a regular grid of points covering the entire country. We present three risk maps for the 2004 fire season and two risk maps for the forecast day, 10 July 2006, of the:

a. seasonal meteorological risk of fire (SSR);
b. conditional probability values $p(I|\text{HR})$ – the probability of having a fire occurrence given an a priori high risk class;
c. high risk regional threshold values, which have the conditional probability $\geq .65$;
d. forecast map of $p(I|\text{HR})$ values for 10 July 2006;
e. associated regional threshold values map for $p > .65$ (10 July 2006).

The first analysis of the meteorological risk of fire for the 2004 fire season was obtained with the SSR index. The main idea is to know the spatial pattern of the meteorological seasonal risk of fire in Portugal, before applying our proposal method. Then the spatial distribution of the SSR values in Fig. 2 shows the following pattern: the meteorological risk of fire increases from the coast to the interior of the country, being very high and extreme in the southeast of Continental Portugal. But the real risk of fire in Portugal can't be explained by using this map. The real spatial pattern is symmetric to this one, because we have most of the fires and burnt areas in northern Portugal, where the majority of the forests and shrublands are located.

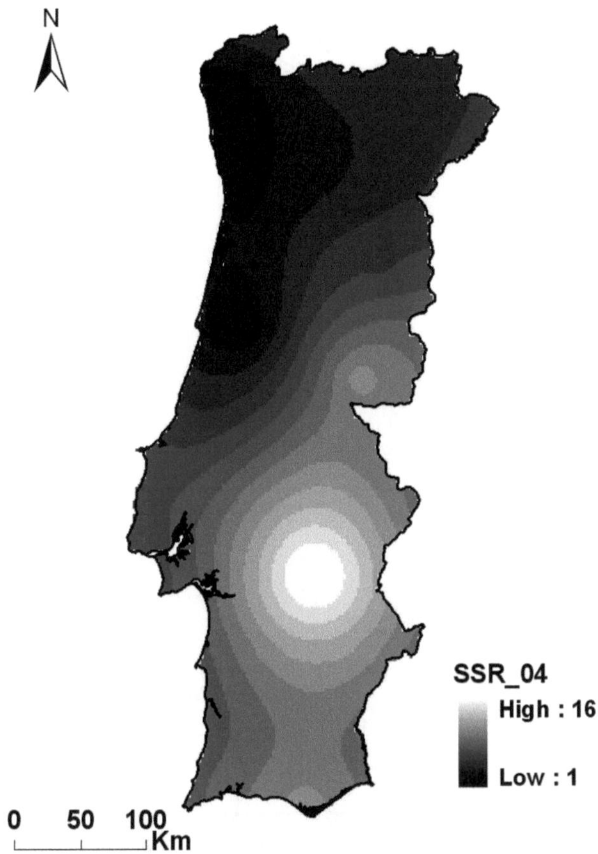

Fig. 2 Spatial distribution of the seasonal severity rating index (SSR) for the 2004 fire season

The map in Fig. 3 shows the probability of having fire occurrence given the predicted HR DSR class for all counties, with $p(I|\text{HR})>= .65$. Comparing the present pattern with the previous one the regions with a higher probability of fire are now located in the northeast, the northwest, the centre and the south coast of the country. We already have the main probability of fire occurrence in the northern part of Portugal. This behaviour corresponds with what has been published in 2004. The great majority of fires and burnt areas took place north of the Tejo River and in Algarve (southern coast) in 2004. Therefore besides having data for only 15 counties, this model result fits quite well to what happened and was officially published for 2004.

Figure 4 shows the range of regional cut-off values for which the probability $p(I|\text{HR})$ is greater than or equal to .65. As the HR definition is dependent on the threshold DSR value, low values meant a very strong sensitivity of the risk of fire to the meteorological conditions. On the other hand, a high value meant that there are other prevailing factors (anthropogenic, fuel load types, topography, controlled fire works) that could explain the risk of fire.

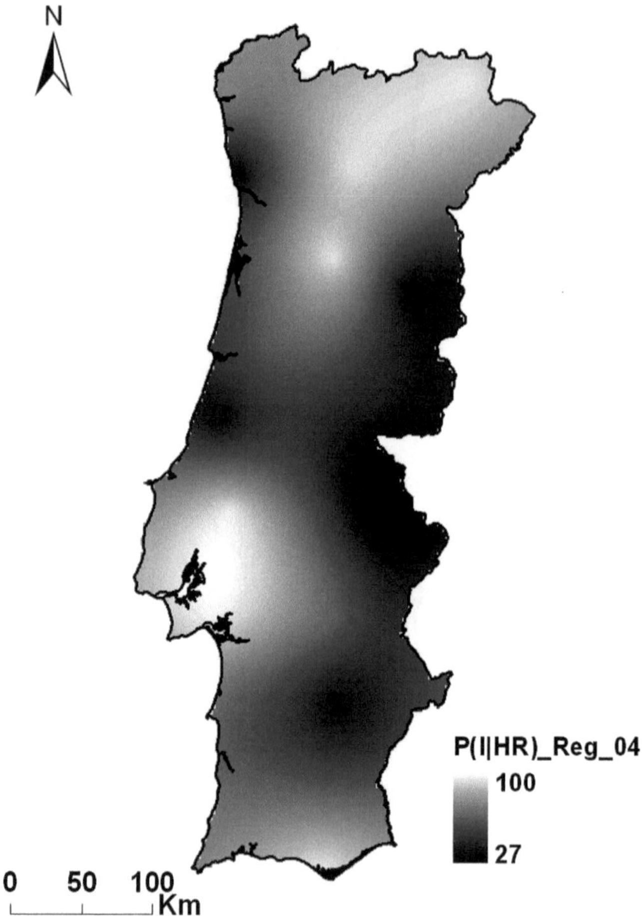

Fig. 3 Spatial distribution of *P (I|HR)* for the 2004 fire season, assuming high risk classes were predicted for all counties

Figure 4 represents the spatial pattern of the regional cut-off values, indicating that in the north of Portugal the risk of fire can be explained by the meteorological conditions due to the observed low cut-off values of the DSR index. For the chosen day, 10 July 2006, forecast with high meteorological fire risk, the method gave an estimation of fire occurrence, due to the value of $p(I|\mathrm{HR})$.

The forecast probability presented in Fig. 5 has greater values in the northern area of the country (around 100%), meaning that a potential fire may occur according to what was predicted for that summer day. In the southern area of the country, where probabilities were lower, near 10%, we cannot give a no-fire forecast.

Finally Fig. 6 illustrates the associated regional cut-off values. The spatial pattern of the associated regional cut-off values (Fig. 6) shows that the western half of the

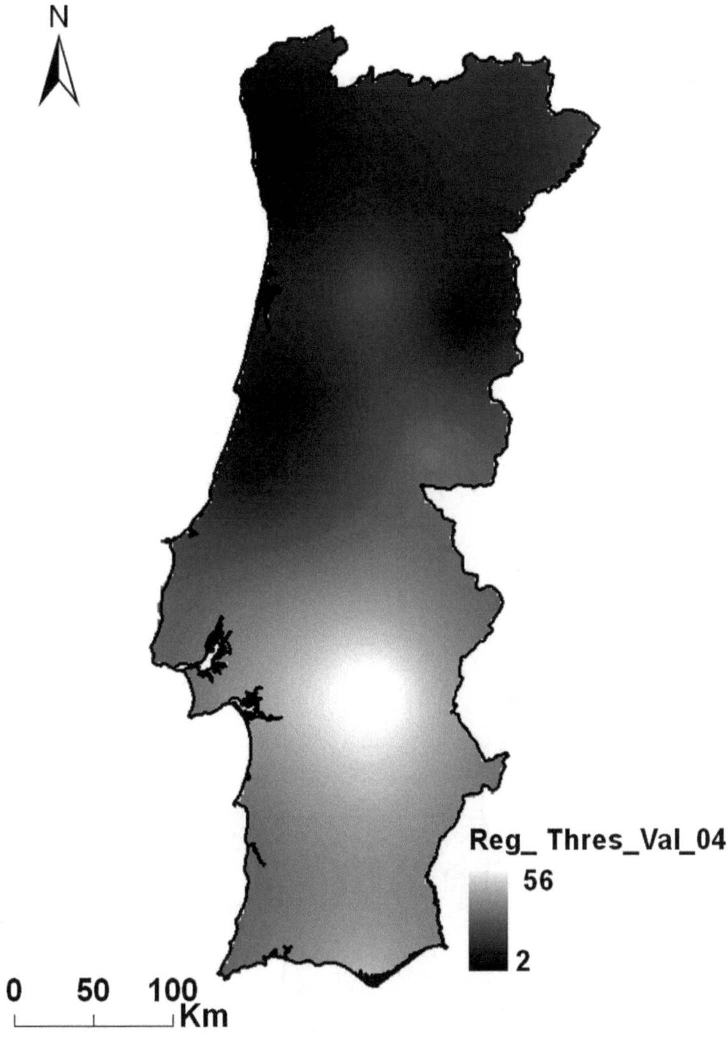

Fig. 4 Spatial distribution of the regional cut off values for the 2004 fire season

country is highly sensitive to the meteorological factors and in the eastern part the risk of fire can be explained by the prevailing factors.

4 Discussion and Conclusions

In this chapter we propose a simple model for fire risk assessment in Portugal based on a constant update of a priori local probability of fire, by using Bayes' rule. The proposed model shows an improvement regarding the simple use of local marginal probabilities and the meteorological risk of fire, DSR.

Fig. 5 Spatial distribution *p(I|HR)* forecasted for the 10 July 2006

The strong correlation between burnt area and meteorological factors reveals a weakness in the National Defence Fire System, namely in prevention, detection and suppression (Pereira et al., 2006). Inter-annual variability of meteorological conditions fully explains about two-thirds of the burnt area variability.

The great majority of burnt areas occurs under severe and extreme conditions; 10% of summer days are responsible for 80% of the burnt area. This means about 12 days a year (Pereira et al., 2005).

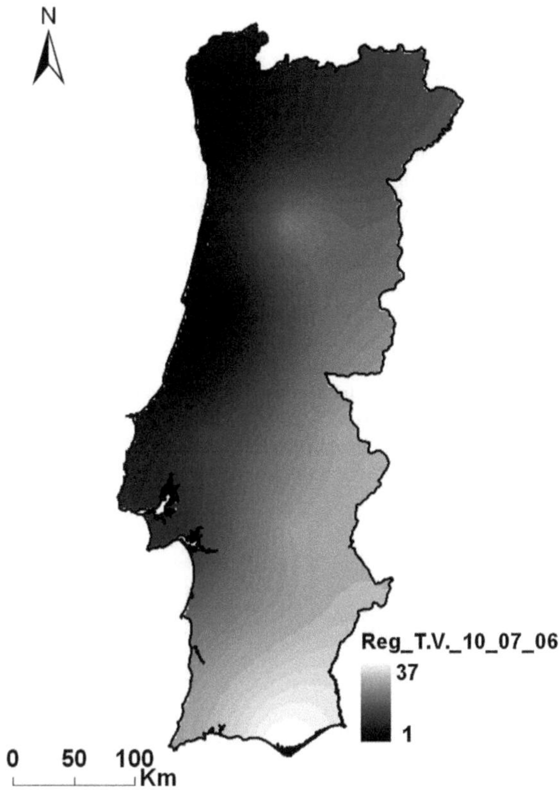

Fig. 6 Spatial distribution of the thresholds values for 10 of July 2006

The relationship between "number of fires" and "burnt area" has a broad range of variation; therefore the proposed method must "include" the measurement of burnt area in each Portuguese county.

5 Further Work

We intend to do this re-classification for all the counties of the 18 districts of Continental Portugal, thereby producing more realistic and helpful risks maps that can be updated on a daily basis.

Acknowledgments The fire database was provided from the Portuguese Governmental Forest Service (Direcção Geral dos Recursos Florestais, DGRF) and was kindly provided by Eng^{o} Miguel Cruz to whom we convey our thanks.

References

Pereira MG, Trigo RM, daCamara CC, Pereira JMC, Leite SM (2005) Synoptic patterns associated with large summer forest fires in Portugal. Agric Forest Meteorol **129:**11–25

Pereira JMC, Carreiras JMB, Silva JMN (2006) Alguns conceitos básicos sobre os fogos rurais em Portugal. In: Pereira JS, Pereira JMC, Rego F, Silva e JMN, Silva TP (ed) Incêndios florestais em Portugal: caracterização, impactes e prevenção ISAPress, Lisbon, pp 133–161

Pyne SJ, Andrews PL, Laven RD (1996) Introduction to Wildland Fire, John Wiley & Sons, Inc, p 760

Stocks BJ, Lawson BD, Alexander ME, Van Wagner CE, McAlpine RS, Lynham TJ, Dube DE (1989) Canadian forest fire danger rating system: an overview. The Forest Chronicle

Van Wagner CE (1970) Conversion of Williams' severity rating for use with the Fire Weather Index. Can For Serv Petawawa Forrest Exp Sta Inf Rep PS-X-21, p 5

Van Wagner CE (1987) Development and structure of the Canadian forest fire weather index system. Canadian forestry service, Forestry technical report 35, Ottawa, p 37

Viegas DX, Reis RM, Cruz MG, Viegas MT (2004) Calibração do Sistema Canadiano de Perigo de Incêndio para Aplicação em Portugal. Silva Lusitana **12(1):**77–93

Williams DE (1959) Fire season severity rating. Can Dep Northern Aff Nat Res, Forest Res Div tech Note 73, p 13

Part IV
Remote Sensing

Super Resolution Mapping with Multiple Point Geostatistics

A. Boucher

Abstract Super-resolution is the process of providing fine scale land cover maps from coarse-scale satellite sensor information. Such a procedure calls for a prior model depicting the spatial structures of the land cover types. When available, an analog of the underlying scene (a training image) may be used for such a model. The snesim (single normal equation simulation) algorithm allows extracting the relevant pattern information from the training image and uses that information to downscale the coarse fraction data into a simulated fine scale land cover scene.

First, the coarse fraction observed image is downscaled using block indicator Kriging (BIK), with the fine scale indicator variograms computed from the training image. The resulting downscale fractions at any given location are interpreted as a prior probability of having a specific land cover at that location. That prior probability is then merged with a probability lifted from the training image. That latter probability is made conditional to any available fine scale data and previously simulated class data along the simulation path. Land cover types are drawn from the resulting posterior distribution. A servo-system keeps track of the number of simulated classes inside each coarse fraction and assures exact reproduction of the coarse fraction data.

By repeating the process with a new path visiting the simulation grid, one can generate several super resolution maps and explore the space of uncertainty for the fine scale land cover. The proposed snesim super resolution mapping algorithm allows to i) exactly reproduce the coarse fraction, ii) inject the structural model carried by the training image, and iii) condition to any available fine scale ground observations. A case study is provided to illustrate the proposed methodology using Landsat TM data from SE China.

A. Boucher
Department of Geological and Environmental Sciences, Stanford University, CA, USA
e-mail: aboucher@stanford.edu

A. Soares et al. (eds.), *geoENV VI – Geostatistics for Environmental Applications*, 297–305

Introduction

Satellite sensor images often provide spatial resolutions much coarser than the extent of land cover patterns, leading to mixed pixels containing multiple land cover classes. Spectral unmixing procedures (Tso and Mather 2001) only determine the fractions of such classes within a coarse pixel without locating them in space. Super-resolution (or sub-pixel) mapping aims at providing a fine spatial resolution map of class labels, one that displays realistic spatial structures and reproduces the coarse spatial resolution fractions.

Traditional methods of super resolution imaging aims at generating a single map by maximizing the spatial continuity of the fine-scale land cover types (Atkinson et al. 1997; Verhoeye et al. 2001; Atkinson 2001; Tatem et al. 2001; Mertens et al. 2003). As an alternative of assuming maximum spatial continuity, Tatem et al. (2002) and Atkinson (2004) use some structural prior information, such as an indicator variogram, to generate a single super-resolution map. Boucher and Kyriakidis (2006) frame super-resolution mapping in a inverse theory perspective and propose to generate multiple maps reproducing an *a priori* structural model. These multiple alternatives generated with the sequential simulation formalism provide an assessment of uncertainty.

This paper uses the probabilistic inverse theory approach to super-resolution modeling as described in Boucher and Kyriakidis (2006). It is understood that there are many solutions that honor both the satellite coarse fractions and the structural model of the fine scale fraction.

The traditional criterion of maximizing the spatial correlation of land cover types is, thus, replaced by reproducing spatial patterns as given by a chosen structural model. Boucher and Kyriakidis (2006) use the indicator variogram as a structural model and simulate the land covers with a sequential indicator simulation algorithm (Journel and Alabert 1989). Instead of variograms, the present work retrieves the relevant patterns from a training image (Strebelle 2002) and performs multiple-point simulation.

A training image (*Ti*) is a rasterized depiction of the properties of interest; it is not conditional to any local data but must depict the relevant patterns that are expected to pertain to the actual true and unknown image. An unconditional simulation done with a variogram-based algorithm could be used as the training image. However, a training image is best used when the variogram alone fails to capture the important patterns of a given spatial property.

Super-Resolution Imaging

At the fine scale, denote $c(\mathbf{u})$ the class at location $\mathbf{u}$; $c(\mathbf{u})$ can take one of the K mutually exclusive labels, $c(\mathbf{u}) = k$, with $k = 1,\ldots,K$. Furthermore, consider the binary class indicator: $I_k(\mathbf{u}) = 1$ if $c(\mathbf{u}) = k$, and 0 otherwise. At the coarse scale, denote $a_k(\mathbf{V})$ the fraction of class k within the coarse pixel $\mathbf{V}$. The fraction $a_k(\mathbf{V})$ is the average of the indicator values contained within the coarse pixel:

$$a_k(\mathbf{V}) = \frac{1}{n} \sum_{\mathbf{u} \subset \mathbf{V}} I_k(\mathbf{u}) \tag{1}$$

The aim of super-resolution imaging is to find a set of $c(\mathbf{u}_i)$ for $i = 1,\ldots,N$ covering the entire image, such that both the coarse fractions $a_k(\mathbf{V})$, $k = 1,\ldots,K$ and the *a priori* structural model are honored. The coarse fractions are obtained by spectral unmixing of the raw measurements of the satellite; see Tso and Mather (2001) for more details on spectral unmixing. In this work, it is assumed that the unmixing had been previously done to satisfaction.

Method

Sequential Simulation

Let $C(\mathbf{u})$ be a categorical random function that can take K values $k = 1,\ldots, K$. The multivariate conditional probability that the random variables $C(\mathbf{u}_i) = k$ for $i = 1,\ldots,N$ is given by a recursive Bayes relation:

$$\Pr\left\{c(\mathbf{u}_1) = k, \cdots, c(\mathbf{u}_N) = k\right\} = \Pr\left\{c(\mathbf{u}_1) = k\right\} \cdot \prod_{i=2}^{N-1} \Pr\left\{c(\mathbf{u}_i) = k|(i-1)\right\} \tag{2}$$

where $|(i-1)$ refers to conditioning of the ith nodes to the $i-1$ previously simulated classes. The often intractable multivariate probability mass function is transformed into a product series of univariate mass probability functions conditional to increasingly larger neighborhood. The sequence, $i = 1,\ldots,N$ is called the simulation path.

The snesim algorithm is used to infer $\Pr\left\{c(\mathbf{u}_i) = k \mid (n)\right\}$ from a training image.

Snesim Algorithm

The Single Normal Equation simulation (Strebelle 2002) is a multiple-point sequential simulation algorithm (Deutsch and Journel 1997) for categorical variables. The structural model consists of a search tree structure that records the frequency of patterns encountered in a training image.

Kriging-based geostatistics uses the indicator variogram to model two-point correlations between categories spread in space as, for example, facies, rock type and land cover types. A training image controls in addition the geometric patterns and their frequency of occurrence. As opposed to the analytical expression of a variogram, a *Ti* is an image, a rasterized depiction of the phenomena. Most importantly, a *Ti* can encompass patterns that cannot be modeled with variogram, such as curvilinear features.

Estimating Conditional Probability with Snesim

The conditional probability density function $p_k^{Ti}(\mathbf{u}_i|(n)) = \Pr\{z(\mathbf{u}_i) = k \mid (n)\}$ for $k = 1,\ldots,K$ is calculated by first counting how many times the conditioning data event (n) is found in the training image. The probability of having class k is then the proportion of these replicates that have class k as center value. To avoid having to re-scan the training image at each new node along the simulation path, all the patterns found in the training image are stored in a search tree for quick retrieval during simulation (Strebelle 2002).

When the conditioning event (n) is not found in the search tree, the farthest away conditioning datum is dropped until the reduced-conditioning event is found often enough to allow retrieving the conditional probability.

Downscaling with Indicator kriging

The snesim algorithm does not readily integrate information on different supports. Thus, it cannot handle directly the information contained in the coarse fraction. Instead, the downscaling from coarse fraction $a_k(\mathbf{V})$ to the BIK-derived probability $p_k^{IK}(\mathbf{u}) = \Pr\{c(\mathbf{u}) = k\}$ is done with block indicator kriging (BIK) (Goovaerts 1997; Boucher and Kyriakidis 2006). The fine-scale variograms $\gamma_k(\mathbf{h})$ are obtained from the training image, the fine-to-coarse and the coarse-to-coarse variogram are determined by the process of regularization (Journel and Huijbregts 1978).

Since the K indicator variograms are a subset of the information contained in the training image, the downscaling via block indicator kriging is consistent with the information contained in the training image

Consider that the local estimation neighborhood includes N coarse fraction data $\mathbf{a}_k$ and G fine-scale class data $\mathbf{j}_k$, the latter known from a ground survey. The downscaled probability of having class k at location $\mathbf{u}$ is

$$p_k^{IK}(\mathbf{u}) = \boldsymbol{\eta}^T \mathbf{a}_k + \boldsymbol{\lambda}^T \mathbf{j}_k + \pi_k \left[1 - \boldsymbol{\eta}^T \mathbf{1}_N - \boldsymbol{\lambda}^T \mathbf{1}_g\right] \tag{3}$$

where $\boldsymbol{\eta}$ and $\boldsymbol{\lambda}$ are the vector of kriging weights for the coarse fractions and the fine scale hard data (if any); π_k is the marginal probability for class k and $\mathbf{1}_N$ is a vector of 1 of length N. The weights are found by solving the following BIK system.

$$\begin{bmatrix} \Gamma_k^{u,u} & \Gamma_k^{u,V} \\ \Gamma_k^{u,V} & \Gamma_k^{V,V} \end{bmatrix} \begin{bmatrix} \boldsymbol{\lambda} \\ \boldsymbol{\eta} \end{bmatrix} = \begin{bmatrix} \boldsymbol{\gamma}_k^{u,u} \\ \boldsymbol{\gamma}_k^{u,V} \end{bmatrix} \tag{4}$$

where $\Gamma_k^{u,u}$, $\Gamma_k^{u,V}$, and $\Gamma_k^{V,V}$ are respectively the fine data to-fine data variogram matrix, the fine data-to-coarse data variogram matrix and the coarse data-to-coarse data variogram matrix; $\gamma_k^{u,u}$ $\gamma_k^{u,V}$ are the fine unknown value -to-fine data and fine unknown value-to-coarse data variogram vector. The resulting values $p_k^{IK}(\mathbf{u})$ may need to be corrected to ensure that they are valid probabilities summing to 1.

Honoring Coarse Fractions

The snesim algorithm drawing from the previous pre-posterior probabilities does not ensure exact reproduction of the coarse fractions. A servo-system is added to steer the simulation such that the right proportion of classes are simulated inside each coarse fraction pixel (Boucher and Kyriakidis 2006).

Define $p_k^R(\mathbf{V})$ as the running probability that any fine-scale locations within pixel $\mathbf{V}$ belongs to class k given the classes already simulated within $\mathbf{V}$. At every point along the path, the final posterior probability is obtained by combining with the tau model (Journel 2002) the running probability $p_k^R(\mathbf{V})$, the BIK derived probability $p_k^{IK}(\mathbf{u})$ and the snesim-derived probability $p_k^{Ti}(\mathbf{u}|(n))$:

$$p_k(\mathbf{u}) = \left[1 + \left(\frac{1 - p_k^{IK}(\mathbf{u})}{p_k^{IK}(\mathbf{u})}\right) \cdot \left(\frac{1 - p_k^{Ti}(\mathbf{u})}{p_k^{Ti}(\mathbf{u})}\right) \cdot \left(\frac{1 - p_k^{R}(\mathbf{u})}{p_k^{R}(\mathbf{u})}\right) \left(\frac{1 - \pi_k}{\pi_k}\right)^{-2}\right]^{-1} \tag{5}$$

The reader is referred to Boucher and Kyriakidis (2006) for more detailed discussion on the servo-system and its properties.

Structured Path

The sequential simulation paradigm has no constraint on the order in which the nodes are to be simulated. With the addition of the servo-system, this order becomes consequential because artifacts get created. To reduce any possible discontinuities created by the servo-system, the algorithm is implemented with a stratified and structured path. The coarse pixels V, which are the most informed are visited first, all the fine scale pixel within that coarse pixel V are then simulated before moving to the next coarse pixel. Shannon's entropy $H(\mathbf{V})$ provides the information measure.

$$H(\mathbf{V}) = -\sum_k a_k(\mathbf{V}) \log a_k(\mathbf{V}) \tag{6}$$

The coarse pixels are sorted by this information measure; the most informed (lower entropy) decile of the pixels is visited randomly, followed by the second most-informed decile up to the least informed one.

Within each coarse pixel V, the above path definition is applied for simulation of the fine scale location. The most informed fine scale locations, based on the entropy of the BIK-derived probability, are visited and simulated first followed by the least informed.

Algorithm

The super-resolution algorithm integrating the structured path, the BIK-derived probability, the *Ti*-derived probability and the servo-system is:

1. Downscale the coarse fraction into the BIK-derived probabilities with indicator Kriging.
2. Define a path visiting the most informed coarse pixel first.
3. For each coarse pixel **V** :

   a. Define a path visiting the fine scale locations within **V**.
   b. For each fine scale location **u** :

      i. Compute the running correction probability $p_k^R(\mathbf{V}),\ k = 1, \ldots, K$.
      ii. Retrieve the conditional probabilities from the search tree $p_k^{Ti}(\mathbf{u}|(n)),\ k = 1, \ldots, K$
      iii. Integrate the three probabilities $p_k^R(\mathbf{V})$, $p_k^{Ti}(\mathbf{u}|(n))$, and $p_k^{IK}(\mathbf{u})$ into a posterior probability using the tau-model, see relation (4).
      iv. Draw from that posterior and add the simulated classes to the conditioning data set.

4. Repeat from step 2. for a new super-resolution realization.

A Landsat Application

The proposed multiple-point super-resolution algorithm is applied on a South-East China Landsat dataset. A 495 × 495 pixels reference map consisting of three land types : vegetation, urban and soil (Fig. 1) is derived from a Landsat image. The reference is then upscaled into coarse fractions for testing purposes, the three coarse fractions map, one for each land cover type are shown in Fig. 2. Each fine scale pixel is 30 m wide and a coarse pixel is 450 m wide. The coarse fraction pixel is the average of the $15 \times 15 = 225$ indicator attributes within that coarse fraction (Eq. 1).

For this application the reference (Fig. 1) is used as training image. This recursive procedure serves only to validate the algorithm by providing the exact structural model; for any real application the training image would be have to be obtained from high spatial resolution imagery either from somewhere else or from the past.

The BIK derived probabilities from Eq. 2 are shown in Fig. 3. The required fine-scale variogram was taken from the training image. These probabilities are then used as soft information and to determine the simulation path.

Two super-resolution realizations are shown in Fig. 4. They are to be compared with the reference in Fig. 1. By virtue of the servo-system, the coarse fractions are reproduced exactly up to rounding errors. Interestingly, the variograms and cross-variograms, while not explicitly used in the drawing part of the simulation process, are also reproduced as shown in Fig. 5.

Fig. 1 Reference land cover map. Light gray is vegetation, mid-gray is urban, black is soil. The map has size 495 × 495 pixels and each pixel is of size 30 m

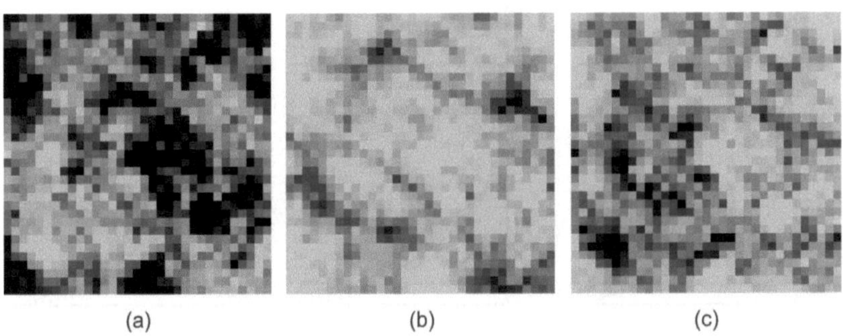

Fig. 2 Fractions associated with the reference land cover map. Lighter tones indicate low fraction, darker tones indicate high fraction. Each coarse pixel correspond to 15 × 15 = 225 fine scale pixels. These coarse maps have size 33 × 33 pixels and each pixel is of size 450 m

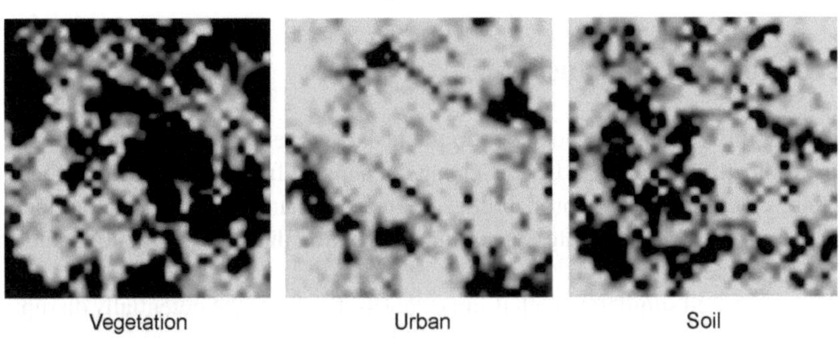

Fig. 3 BIK-derived probability map obtained with block indicator Kriging. Lighter tones indicate low fraction, darker tones indicate high fraction. The map has size 495 × 495 pixels and each pixel is of size 30 m

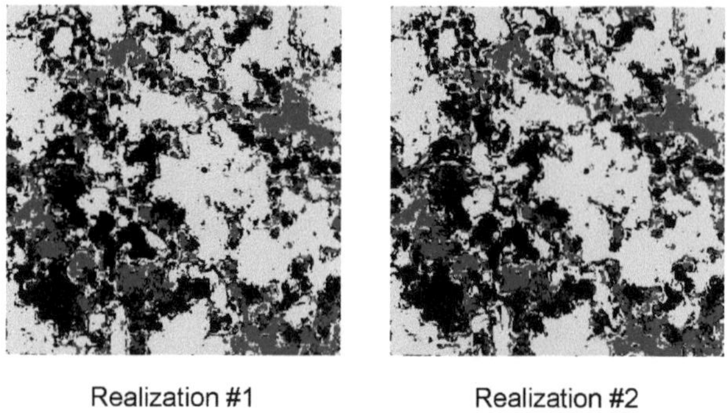

Fig. 4 Super-resolution maps from the proposed algorithm. Light gray is vegetation, mid-gray is urban, black is soil. The maps has size 495 × 495 pixels and each pixel is of size 30 m

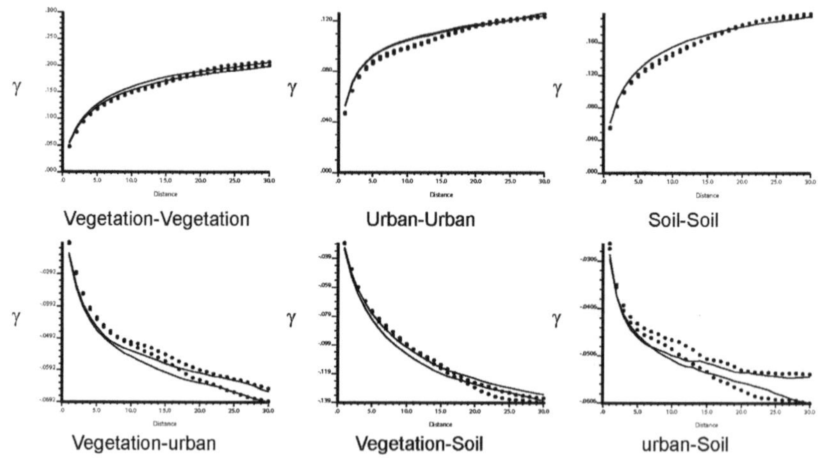

Fig. 5 Comparison of the experimental indicator variograms and cross-indicator variograms of the spatial patterns in both the reference and the simulated images. The solid lines represent the N-S and E-W experimental variograms for the reference image, the black dots are the N-S and E-W experimental variograms of the simulation shown in Fig. 4

Discussion and Conclusion

Super-resolution imaging is equated to the task of generating fine-scale land cover maps. Indicator Kriging handles the change of support from the coarse fraction into the fine scale BIK-derived probabilities, then the snesim algorithm in conjunction with a servo-system generates maps that reproduce the coarse fractions and the structural model as displayed by a *Ti*.

The controlled simulation path is important as it reduces the potentially disruptive behavior of the servo-system. By visiting the least-informed locations last, the servo-system affects mostly the locations that are least informed by the satellite images.

This methodology is an extension of the work in Boucher and Kyriakidis (2006) where indicator variograms provide the structural model. When an analog image for the fine scale spatial distribution is available, the proposed algorithm allows taking that image as structural model instead of retrieving from it only two-point summary statistics that may not be enough to describe complex land cover patterns. Another advantage is that the snesim algorithm implicitly takes into account the cross-correlation between categories and does not generate order relations violations. However, the proposed implementation is slower and does encounter problems if the number of categories is large.

The proposed algorithm is not restricted to satellite sensor imagery, and could be applied to any task of downscaling coarse proportions into fine-scale maps of categorical attributes.

References

Atkinson PM (2001) Super-resolution target mapping from soft-classified remotely sensed imagery. In: Proceedings of the 6th International Conference on Geocomputation. University of Queensland, Brisbane, Australia, September 24–26

Atkinson PM (2004) Super-resolution land cover classification using the two-point histogram. In: Sánchez-Vila X, Carrera J, Gómez-Hernández J (eds) GeoENV IV: Geostatistics for environmental applications, Kluwer, Dordrecht, pp 15–28.

Atkinson PM, Cutler MEJ, Lewis H (1997) Mapping sub-pixel proportional land cover with AVHRR imagery. Int J Remote Sens 18:917–935.

Boucher A, Kyriakidis PC (2006) Super-resolution mapping with indicator geostatistics. Remote Sens Environ 104:264–282.

Deutsch CV, Journel AG (1997) GSLIB: Geostatistical software library and user's guide, Second Edition, Oxford University Press, pp 369

Goovaerts P (1997) Geostatistics for natural resources evaluation. Oxford University Press, New York

Journel AG (2002) Combining knowledge from diverse sources: An alternative to traditional data independence hypotheses. Math Geol 34:573–596.

Journel AG, Alabert F (1989) Non-gaussian data expansion in the earth sciences. Terra Nova 1:123–134.

Journel AG, Huijbregts CJ (1978) Mining Geostatistics. Academic Press, San Diego

Mertens KC, Verbeke LPC, Ducheyne EI, Wulf RRD (2003) Using genetic algorithms in sub-pixel mapping. Int J Remote Sens 24:4241–4247

Strebelle S (2002) Conditional simulation of complex geological structures using multiple-point statistics. Math Geol 34:1–21.

Tatem AJ, Lewis HG, Atkinson PM, Nixon MS (2001) Super-resolution target identification from remotely sensed images using a Hopfield neural network. IEEE Trans Geosci Remote Sen 39:781–796

Tatem AJ, Lewis HG, Atkinson PM, Nixon MS (2002) Super-resolution land cover pattern prediction using a Hopfield neural network. Remote Sens Environ 79:1–14

Tso B, Mather P (2001) Classification methods for remotely sensed data. Taylor and Francis, London

Verhoeye J, De Wulf R (2001) Land cover mapping at sub-pixel scales using linear optimization techniques. Remote Sens Environ 79:96–104

Super-Resolution Mapping Using the Two-Point Histogram and Multi-Source Imagery

P. M. Atkinson

Abstract A new method for super-resolution classification from remotely sensed imagery is presented. The method allows prediction of a super-resolution (sub-pixel) land cover map from a coarse spatial resolution (original pixel) land cover proportions image and an intermediate spatial resolution panchromatic (Pan) image. The method is based on spatial simulated annealing and combines two objectives: (i) to match a prior sub-pixel two-point histogram obtained from some training image and (ii) to match the predictions made of an intermediate spatial resolution panchromatic image from the super-resolution classification via a forward model to an observed panchromatic image. The method is demonstrated on simulated remotely sensed imagery. The main advantage of the Pan image is to fix locally the outcome of the two-point histogram objective function.

1 Introduction

Super-resolution classification, also referred to as sub-pixel classification, is the classification of land cover from remotely sensed imagery at a spatial resolution that is finer than that of the original multi-waveband imagery. Recently, a series of papers has emerged in which the more specific objective is to take the output from a land cover proportions prediction algorithm (e.g., mixture model, neural network, fuzzy *c*-means) (Adams et al. 1985, Atkinson et al. 1997, Bezdek et al. 1984, Brown et al. 1999), and transform these land cover proportions per original pixel into hard land cover class predictions per sub-pixel, where sub-pixels are nested within original pixels.

Many research and commercial projects in remote sensing have as their goal the prediction of land cover proportions within pixels. Proportions prediction has several advantages over hard classification (i.e., allocation of each pixel to one class only), not least that the information content of the resulting proportions prediction image has potential to be greater than for a hard classification (Thomas et al. 1987).

P. M. Atkinson
School of Geography, University of Southampton, Highfield, Southampton, SO17 1BJ, UK
e-mail: pma@soton.ac.uk

A. Soares et al. (eds.), *geoENV VI – Geostatistics for Environmental Applications*, 307–321

However, even with land cover proportions prediction some information is lost. Specifically, no information is provided on where within each pixel the land cover actually exists: all that is predicted is the set of proportions. Since in the majority of cases hard classes can be assumed to exist on the ground, it is reasonable to attempt to map the hard land cover classes within each pixel. This is the goal of super-resolution classification (Atkinson 1997).

Previously, several basic classes of algorithm have been presented for super-resolution classification, including contouring (Foody 1998), vector segmentation (Steinwendner et al. 1998), simulated annealing (SA) and related approaches (Atkinson 2004, 2005, Thornton et al. 2006), genetic algorithms (Mertens et al. 2003), standard feed-forward neural networks to estimate wavelet coefficients (Mertens et al. 2004), and the Hopfield neural network (HNN), (Hopfield and Tank 1985, Tatem et al. 2001a, 2001b, 2002, 2003). The spatial optimization algorithm presented here falls into the SA framework.

Recently, the work of Tatem et al. (2001a, 2001b, 2002, 2003) was extended to allow inclusion in the optimization algorithm of ancillary data defined at a spatial resolution in-between that of the original coarse spatial resolution of the land cover proportions image and the fine spatial resolution of the predicted sub-pixel land cover map (Nguyen et al. 2005, 2006). Examples were demonstrated using LiDAR height data (Nguyen et al. 2005) and a multi-waveband image (Nguyen et al. 2006), both defined at an intermediate spatial resolution. The ancillary data sets were included through definition of a forward model that allowed prediction of the ancillary variables at intermediate spatial resolution from the predicted sub-pixel land cover map. Once these variables were predicted (at iteration t) the predicted intermediate spatial resolution images could be compared to the observed images and the differences used to direct any future alterations to the predicted sub-pixel map (i.e., at iteration $t+1$).

Atkinson (2004) first presented an algorithm for super-resolution classification based on the two-point histogram control statistic. The algorithm allowed matching of the empirical two-point histogram of the predicted sub-pixel map to a prior empirical two-point histogram obtained from training data. The algorithm is, thus, aimed at pattern-matching. This goal is quite distinct from the goal of spatial clustering in which the spatial correlation between neighbouring sub-pixels is *maximised*. Such an algorithm was presented by Atkinson (2005). The algorithm of Atkinson (2004) was updated here to a full SA framework and combined with forward modelling to allow use of secondary or ancillary data in the pattern-matching process.

2 Methods

It was assumed that the overall aim is to classify land cover at a (relatively) fine spatial resolution from a multispectral (MS) remotely sensed image available at a (relatively) coarse spatial resolution and a panchromatic (Pan) image available at an intermediate spatial resolution (i.e., between that of the MS and predicted images).

It is further assumed that the MS image has been, or can be, used to predict the proportions of each land cover class within each pixel (e.g., through application of a proportions prediction algorithm).

In this paper, two goal functions were combined to create a single spatial optimization algorithm for predicting land cover at a fine spatial resolution from a coarse spatial resolution remotely sensed image of land cover proportions and a Pan image at intermediate spatial resolution. The first goal was to match the spatial pattern (defined at a fine spatial resolution) of land cover classes in a training image as represented by the two-point histogram. The second goal was to match the brightness values in a Pan image available at an intermediate spatial resolution. The second goal required the introduction of a forward model that predicted Pan brightness at an intermediate spatial resolution from the predicted land cover map at a fine spatial resolution.

2.1 Two-Point Histogram Goal

The two-point histogram goal for use in a spatial simulated annealing framework was presented in Deutsch and Journel (1998). A similar pixel-swapping algorithm based on the two-point histogram was presented in Atkinson (2004). Given a random variable (RV) Z that can take one of $k = 1, \ldots, K$ outcomes the two-point histogram for a given distance and direction vector (or lag) $\mathbf{h}$ is the set of all bivariate transition probabilities:

$$p_{k,k'}(\mathbf{h}) = \Pr\left\{ \begin{array}{l} Z(\mathbf{u}) \in \text{category } k, \\ Z(\mathbf{u}+\mathbf{h}) \in \text{category } k' \end{array} \right\} \tag{1}$$

independent of $\mathbf{u}$, for all $k, k' = 1, \ldots, K$. The two-point histogram is linked to the indicator variogram, but provides twice the information. The control statistic was estimated for a window of fixed size.

The objective function for the two-point histogram control statistic is defined as:

$$O_{tph} = \sum_{\mathbf{h}} \left(\sum_{k=1}^{K} \sum_{k'=1}^{K} \left[p_{k,k'}^{training}(\mathbf{h}) - p_{k,k'}^{realization}(\mathbf{h}) \right]^2 \right) \tag{2}$$

where $p_{k,k'}^{training}(\mathbf{h})$ are the target bivariate transition probabilities, for example, calculated from a training image and $p_{k,k'}^{realization}(\mathbf{h})$ are the corresponding transition probabilities of the realization image (i.e., the current image being altered iteratively through application of the algorithm).

A potential problem is that $p_{k,k'}(\mathbf{h})$ is not standardized for the proportional areal cover of a given class. Thus, if the proportional coverage of a given class differs greatly between the training image and the current realization image then the algorithm will be unable to provide a good match between the global functions. In this paper, $p_{k,k'}(\mathbf{h})$ was standardized as follows:

$$p'_{k,k'}(\mathbf{h}) = \frac{p_{k,k'}(\mathbf{h})}{\left|\sum_{h_x=-W}^{W}\sum_{h_y=-W}^{W} p_{k,k'}(h_x, h_y)\right|_{\max}} \tag{3}$$

Where h_x is the separation in x and h_y is the separation in y such that (h_x, h_y) sweeps the window of size $2W+1$ over which the two-point histogram is calculated. This effectively removes the attribute information such that the pattern-matching was based only on information related to the spatial scale and pattern of the land cover objects in the two images.

2.2 Panchromatic Constraint

The panchromatic constraint for use in a spatial SA framework is presented in this section. The constraint function depends on definition of a forward model that allows prediction of brightness in the waveband, and at the intermediate spatial resolution, of the Pan image from the sub-pixel land cover classes predicted by the algorithm. Mean brightness values $\bar{R}(k)$ were defined per land cover class k. These could be obtained from training data where the algorithm is applied to real imagery. This allowed prediction of the spectral brightness of each sub-pixel $z_{Pan}(\mathbf{x}_{ij})$ as follows:

$$z_{Pan}(\mathbf{x}_{ij}) = \bar{R}(k(\mathbf{x}_{ij})) \tag{4}$$

where $\overline{R}(k(\mathbf{x}_{ij}))$ represents the brightness of each sub-pixel at location $\mathbf{x}_{ij}$ of class k. Since the Pan image represents a single waveband it was necessary to define only one brightness value per class. The change in spatial resolution (from fine to intermediate) was achieved through a simple square wave response transfer function:

$$z^*_{Pan}(\mathbf{x}_{mn}) = \frac{\sum_{i=1}^{I}\sum_{j=1}^{J} z_{Pan}(\mathbf{x}_{ij})}{IJ} \tag{5}$$

where $z^*_{Pan}(\mathbf{x}_{mn})$ is the brightness z predicted in the Pan waveband for pixel location x_{mn}, and i and j sweep the Pan pixel m, n and I and J are the dimensions (in sub-pixels) of the Pan pixel in the x and y directions, respectively. That is, sub-pixel brightnesses were averaged within each intermediate Pan pixel. When dealing with real images alternative transfer functions could be applied readily to account for the sensor point spread function.

Once predicted, the Pan brightness values may be compared with the actual observed Pan brightness values in the following objective function:

$$O_{Pan} = \sum_{m=1}^{M}\sum_{n=1}^{N}\left[z_{Pan}^{training}(\mathbf{x}_{mn}) - z^*_{Pan}(\mathbf{x}_{mn})\right]^2 \tag{6}$$

where m and n are the number of Pan pixels along the image edge.

2.3 Spatial Simulated Annealing

A standard spatial simulated annealing algorithm was employed (Deutsch and Journel 1998). The objective function was defined as

$$O = w_1.O_{tph} + w_2.O_{Pan} \tag{7}$$

where w_1 and $w_2 = (1 - w_1)$ are weights. Since the magnitudes of O_{tph} and O_{Pan} may vary greatly the weights need to be chosen carefully to achieve efficient convergence. Here, $w_1 = 0.99$ was selected through experimentation and comparison of O_{tph} and O_{Pan}. An alternative would be allow the weights to vary iteratively based on the relative magnitudes or variances of O_{tph} and O_{Pan}.

An annealing schedule was defined based on an exponential decay function:

$$s(t) = \exp(-t/r) \tag{8}$$

where t is the iteration number and r is a non-linear parameter defining the rate of decay. Where the objective function was not decreased through a valid swap then the swap was accepted only if $v < s(t)$ where v is a draw from $V \sim \mathrm{U}(0, 1)$.

In the present algorithm, the scheme for altering the sub-pixel values was based on pixel-swapping rather than a change in the attribute value (as for HNN, ANNEAL.for; Deutsch and Journel 1998). It is important that the solution (the predicted sub-pixel land cover map) is constrained to match the original pixel-level proportions used as input. In the present case, sub-pixel swapping was limited strictly to within the original (coarse spatial resolution) pixels. This means that the number of sub-pixels allocated to each class per-pixel cannot be changed from the initial number as determined by the pixel land cover proportions. For example, in a pixel of 8 by 8 sub-pixels (for which proportions are predicted as woodland (50%), grassland (25%) and built-land (25%)) there will be 32 sub-pixels of woodland, 16 of grassland and 16 of built-land. The initial fine spatial resolution land cover map to be presented to the optimization algorithm was created by distributing spatially the required number of sub-pixels randomly within each pixel.

The sub-pixel pairs considered for swapping at each iteration were selected per original (coarse spatial resolution) pixel. Each coarse pixel was visited in turn and α pairs of sub-pixels were selected randomly from within each pixel. The number α was either (i) fixed (e.g., equal to the number of sub-pixels on a pixel side) or (ii) determined based on the number of iterations:

$$\alpha(t) = n_{sp}n_{sp}ceiling(t/T) \tag{9}$$

where ceiling represents the rounding up of a non-integer value, n_{sp} is the number of sub-pixels along the side of a pixel and T is the number of iterations.

2.4 Efficiency

As for the algorithm in Atkinson (2004), two checks were added to the algorithm to increase its efficiency. First, it was found that many pixels contained only one land cover class. Such pixels were ignored. It should be noted, however, that sub-pixels within such pixels may be compared with sub-pixels within adjacent pixels because the two-point histogram was computed over a window that may straddle across pixel boundaries. Second, sub-pixels were compared only if their classes were different. While not fast, the current implementation in SPlus was sufficient to demonstrate the utility of the optimization algorithm on simulated images. It is anticipated that the algorithm will be written in C or C++ in the future for operational use.

3 Data

A simple Boolean simulation was used to provide an image with which to test the spatial optimization algorithm. First, $n_w = 3$ rectangles of varying height and width $r \sim$ U ($\min_r$, $\max_r$) were drawn at locations $l \sim$ U ($\min_l$, $\max_l$) where $\min_r = 1$, $\max_r = 3$, $\min_l = {}^1\!/_6\,(n_p.n_{sp})$ and $\max_l = {}^2\!/_3\,(n_p.n_{sp})$ sub-pixels, where n_p is the number of pixels along the image edge and n_{sp} is the number of sub-pixels along a pixel edge. These n_w rectangles simulate class B, say woodland (Fig. 1). Second, $n_b = 7$ rectangles of varying height and width r ($\min_r = 1$ and $\max_r = 2$) were drawn at locations l with $\min_l = {}^1\!/_3\,(n_p.n_{sp})$ and $\max_l = {}^5\!/_6\,(n_p.n_{sp})$. These rectangles simulate class C, say built-land. The third class, the background, simulates class A, say grassland.

The spatial resolution of the sub-pixel map of land cover (Fig. 1a) was coarsened by a zoom factor of 8 to create an image of 5 by 5 pixels representing proportional land cover (Fig. 1c-e). Fig. 1c-e simulates the output from a soft classifier or proportions prediction algorithm applied to a remotely sensed image. It is worth noting that a map of proportional land cover is usually the end-point for remotely sensed classification: here it represents the starting-point. Fig. 1b shows the (assumed known) Pan image that corresponds to the unknown target land cover map at the sub-pixel scale (Fig. 1a).

4 Analysis

4.1 Initialization

The land cover proportions shown in Fig. 1c-e were used to allocate hard land cover classes to sub-pixels randomly, that is, under the constraint that the proportional cover per class at the sub-pixel scale must match that at the pixel-scale. The initial random allocation is shown in Fig. 1f. The resultant initial Pan image is shown in Fig. 1g.

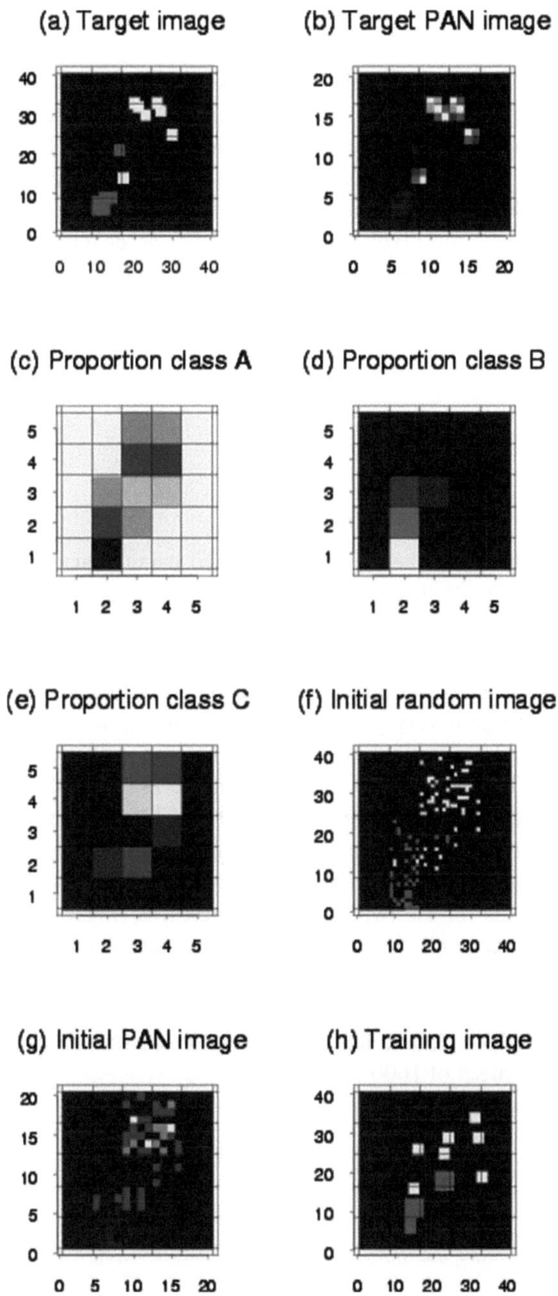

Fig. 1 (a) Target image, (b) target Pan image, (c) class A proportions, (d) class B proportions, (e) class C proportions, (f) initial random image, (g) initial Pan image, (h) training image

4.2 Training

When dealing with real imagery, the target two-point histogram would be obtained from an image with the desired super-resolution. An example would be super-resolution classification of a large Landsat Thematic Mapper (TM) image (spatial resolution of 30 m by 30 m) via training with a classified IKONOS image (spatial resolution of 4 m by 4 m) of limited spatial extent. To test the present algorithm, a training image (i.e., sub-pixel land cover map) was generated using the same Boolean model as was used to derive the target image. The training image is shown in Fig. 1h. It can be seen that while different to the target (Fig. 1a) on a per-sub-pixel basis, it has a similar spatial character.

4.3 Optimization

The first 100 iterations of the algorithm, applied using Equation 9 to determine $\alpha(t)$, are shown in Fig. 2. The (i) current sub-pixel land cover map, together with (ii) the predicted Pan image using the forward model, (iii) the difference between the predicted and observed Pan images, and (iv) the two-point histogram for class B are shown at every twenty iterations in Fig. 2. As can be seen from the Figure, the match to the Pan image is very good after only 100 iterations.

The final predicted sub-pixel land cover map produced after 200 iterations is shown in Fig. 3. The prediction of the Pan image obtained through application of the forward model to this image is a perfect match to the observed Pan image. While a reasonable match to the target sub-pixel land cover class image, there are some discrepancies. Fig. 4 shows the difference image between the predicted and target sub-pixel maps for the two classes of interest B (Fig. 4a) and C (Fig. 4b). White represents over-prediction (i.e., an error of commission), while black represents under-prediction (i.e., an error of omission). The errors for each class were as follows: A = 1.5%, B = 0.75% and C = 0.75%, where, for example, 0.75% includes six sub-pixels falsely included and a matching set of six sub-pixels erroneously excluded from a total set of 1600 sub-pixels.

4.4 Comparison

Figure 5 shows the result of applying the algorithm with $w_1 = 1$ (i.e., ignoring the Pan constraint). The result is similar in terms of spatial character. However, the per-sub-pixel accuracy of the predicted land cover map is much less (Fig. 6). The errors for each class were as follows: A = 11.13%, B = 5.38% and C = 5.63%, iluustrating well the localising effect of the Pan constraint.

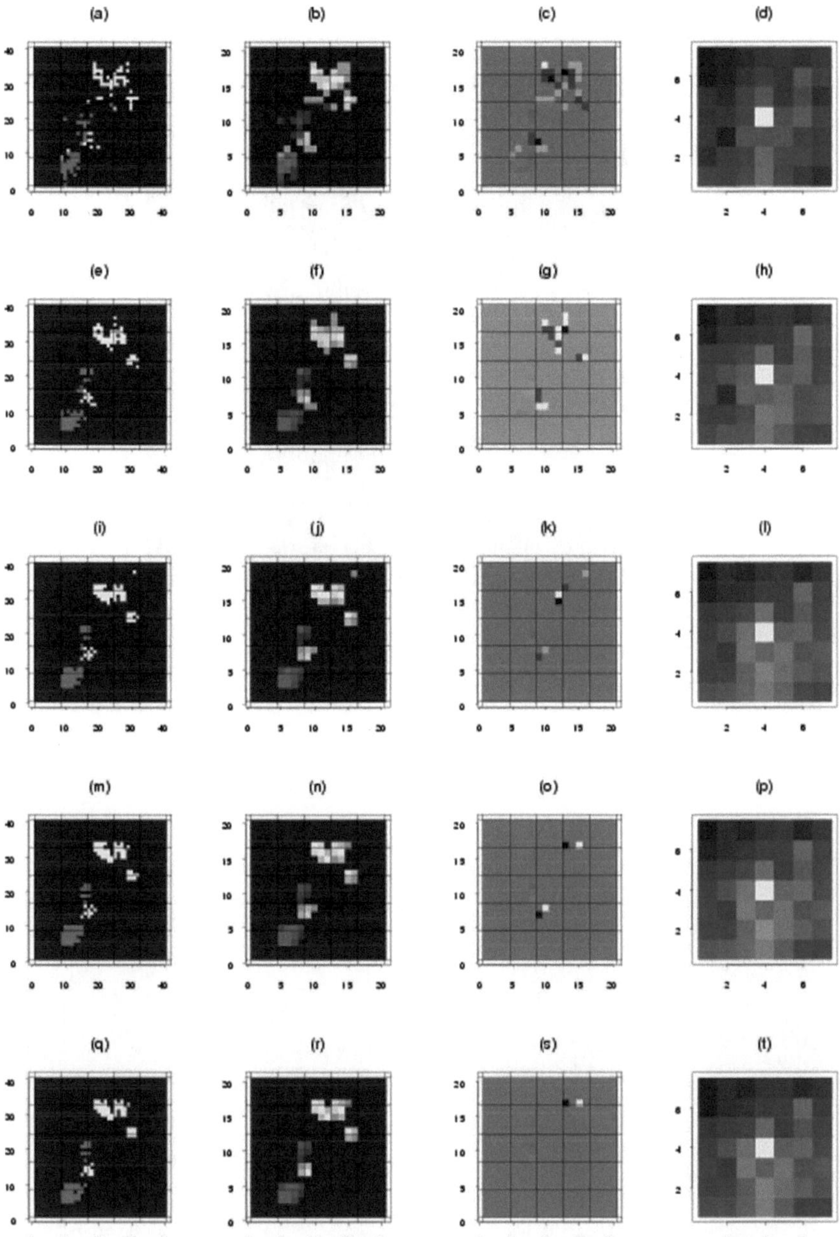

Fig. 2 Spatial simulated annealing model results using two-point histogram goal and Pan constraint: (a, e, i, m, q) current sub-pixel land cover class prediction, (b, f, j, n, r) current forward predicted Pan image, (c, g, k, o, s) relative Pan error and (d, h, l, p, t) two-point histogram for class B. The rows represent 20, 40, 60, 80 and 100 iterations from top to bottom

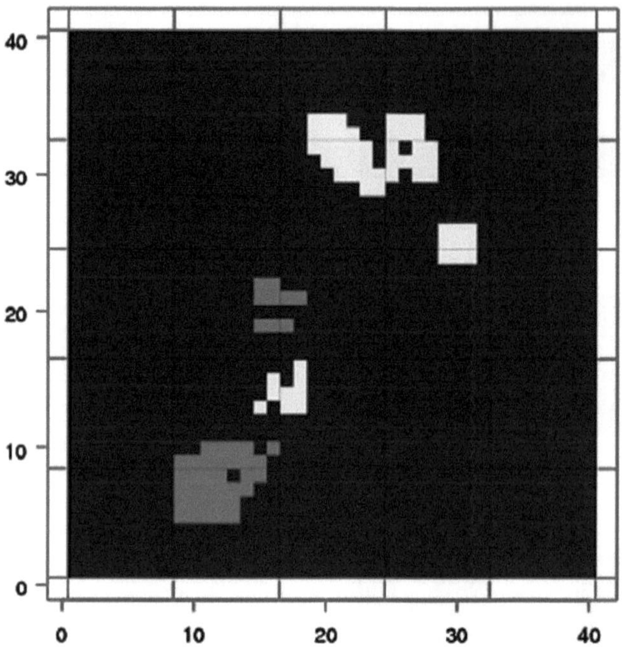

Fig. 3 Final predicted sub-pixel land cover classification after 200 iterations

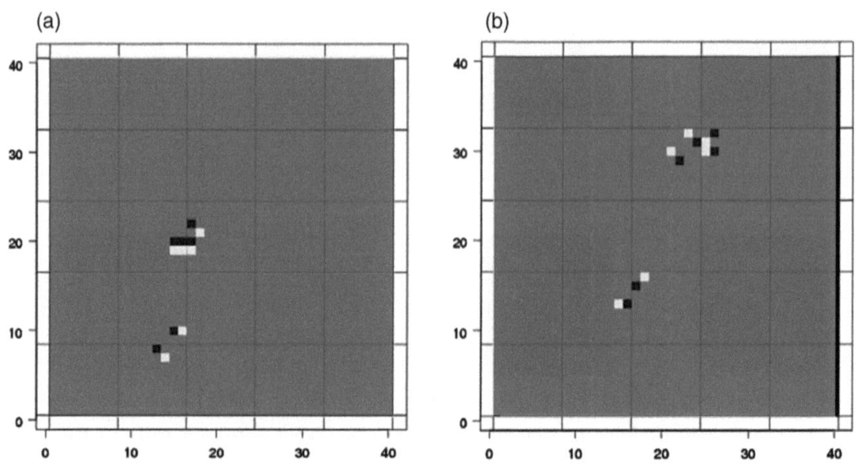

Fig. 4 Error image corresponding to Fig. 3: (a) class B and (b) class C

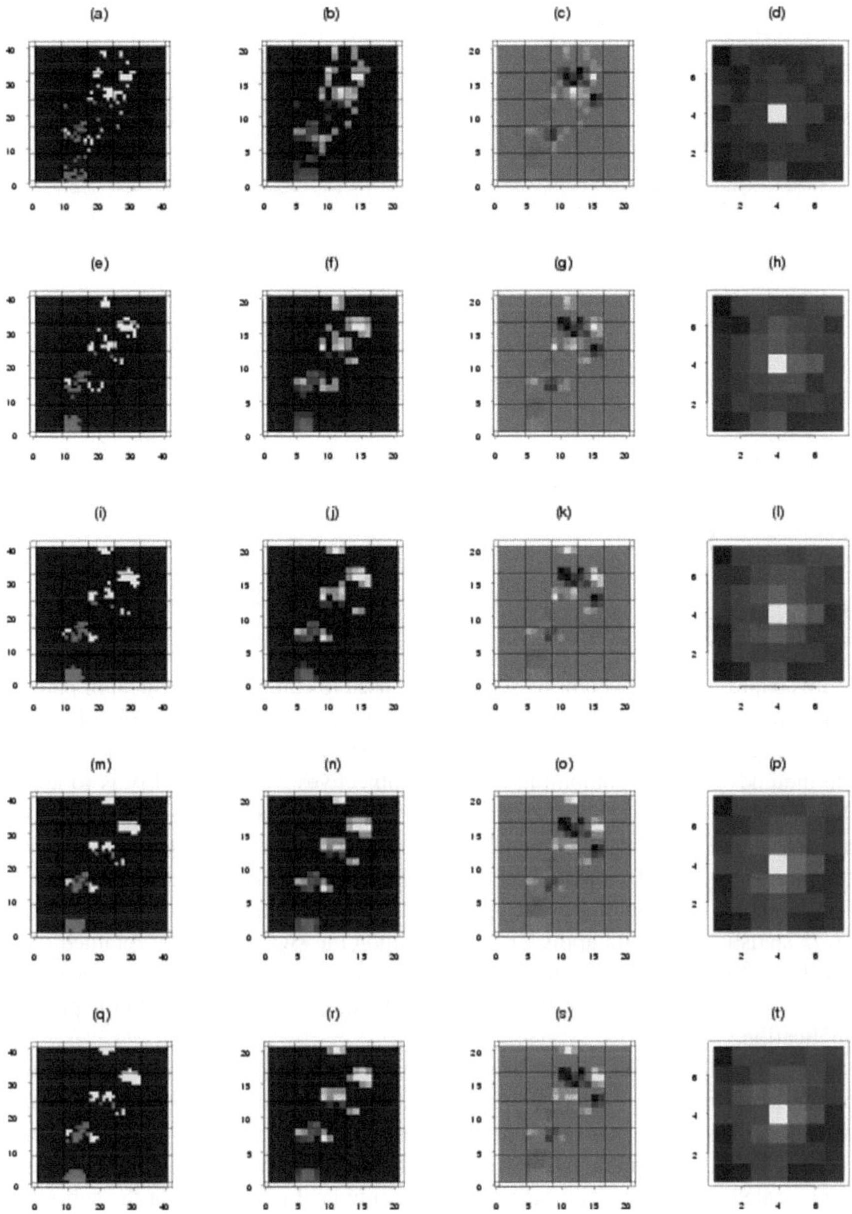

Fig. 5 Spatial simulated annealing model results using only the two-point histogram goal: (a, e, i, m, q) current sub-pixel land cover class prediction, (b, f, j, n, r) current forward predicted Pan image, (c, g, k, o, s) relative Pan error and (d, h, l, p, t) two-point histogram for class B. The rows represent 20, 40, 60, 80 and 100 iterations from top to bottom

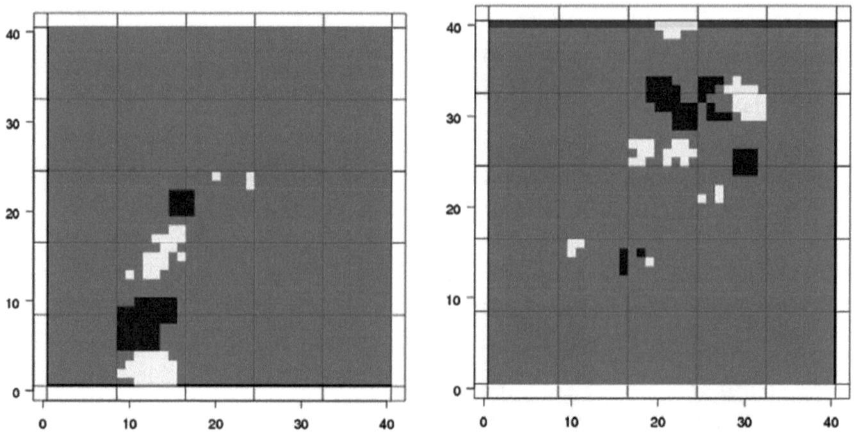

Fig. 6 Error image corresponding to spatial simulated annealing model results using only the two-point histogram goal: (a) class B and (b) class C

5 Discussion

In this paper, a simple weighting scheme was used to combine (i) the goal of pattern matching via the two-point histogram with (ii) the constraint of a forward model predicting the brightness of an observed panchromatic image. Many other possible methods exist for combining these two objectives. One possibility is to apply the two objective functions O_{tph} and O_{Pan} (Equations 2 and 6, respectively) independently and interleaved with each other. Another possibility is to apply the two functions in a nested sequence such that objective 2 is dependent on objective 1. For example, one could apply O_{Pan} and if the swap leads to a reduction in the O_{Pan} statistic one could apply O_{tph}, whereupon the swap would be retained only if it leads to a reduction in the O_{tph}statistic (ignoring the possibility of accepting "bad" swaps for ease of exposition). These schemes have some merit in the present application.

The overall aim of this paper was to combine two objectives for super-resolution or sub-pixel mapping of land cover from remotely sensed imagery. However, the two objectives chosen are quite different in their nature: pattern-matching is not point-specific, while point prediction is point-specific. Further, these differences are complicated by the fact that the pattern matching goal is executed at the sub-pixel level whereas the panchromatic constraint is applied at the intermediate Pan pixel level. It is interesting to consider the circumstances under which such a method be most applicable. There seem to be two distinct scenarios that have to do with the H-resolution and L-resolution cases (Woodcock and Strahler 1987, Atkinson and Tate 2000). It is well-known that for pattern matching to be appropriate, the size of the objects (or more generally the frequency of spatial variation) should be small relative to the scale of sampling afforded by the sensor's spatial resolution. If the objects are larger then the pixels then it is likely that spatial clustering methods that seek to maximise the spatial correlation between neighbouring sub-pixels will be satisfactory. It is interesting to consider this dichotomy of approaches (pattern matching, generally L-resolution *v.* spatial clustering, generally H-resolution) in the context of the zoom factor between the sub-pixel and Pan image spatial resolutions. In the present example, the resulting sub-pixel predictions were at a zoom factor of 8X, and the Pan image was at a zoom factor of 4X, the spatial resolution of the input land cover proportions image. This means that the zoom factor between the sub-pixel prediction image and the Pan image was only 2X such that objects of around 2-to-3 sub-pixels on a side that effectively lead to an L-resolution scenario at the original spatial resolution are likely to lead to an H-resolution scenario at the intermediate Pan spatial resolution. Thus, the appropriateness of pattern-matching as a goal may be questionable in the example given. For smaller objects (i.e., that are smaller than the Pan pixels) pattern-matching may be more appropriate as a goal in combination with the Pan constraint.

One of the primary contributions of this paper was to demonstrate the utility of forward modelling as a suitable approach of wide utility for incorporating ancillary data into super-resolution classification algorithms. In a spatial optimization framework involving iteration to approach a solution, forward modelling allows use of the sub-pixel land cover map at iteration t to predict related variables at different spatial resolutions, some of which may have been observed empirically. Forward modelling, thus, facilitates comparison of the observed variable(s) with predictions of the same variable(s) made based on the sub-pixel classification at iteration t. Differences between the observed and predicted maps at iteration t can be used together with an algorithm such as SA to alter the sub-pixel spatial distribution of land cover classes at iteration $t+1$.

Atkinson (2004) made four suggestions for improvement to the sub-pixel swapping algorithm, three of which remain untackled. The first and simplest suggestion was that the simple sub-pixel swapping algorithm could be extended to include full spatial simulated annealing. That has been done here. Second, it was suggested that local variation in the spatial character of the land cover distribution could be accounted for through local variation in the target two-point histogram. The problem is that the only information on spatial variation in the target land cover distribution exists in the pixel-level land cover proportions whereas information is required at the sub-pixel scale. A solution to this problem may be possible via regularization of the *modelled* sub-pixel two-point histogram, thereby providing a link between the two scales of measurement (Journel and Huijbregts 1978, Jupp et al. 1988). A third suggestion was to modify the algorithm to deal with error in the predicted pixel-level land cover proportions (no remote sensing classification is ever expected to be perfect). A fourth suggestion was to modify the algorithm to deal with the PSF of the sensor. The PSF is commonly shaped like a two-dimensional step function (termed a square wave response) convolved slightly with a smoothing function. All three latter suggestions remain open for future research.

6 Conclusion

A new method for super-resolution land cover classification has been presented that allows the goal of pattern matching at the sub-pixel scale to be combined with the constraint of matching a panchromatic image at the intermediate pixel scale. The algorithm represents a novel framework for super-resolution classification. Further research is required to explore the properties of the algorithm and report on its accuracy relative to other algorithms.

Acknowledgments The author is grateful to various colleagues and PhD students who have collaborated on super-resolution classification, most notably Dr. Andrew Tatem, Professor Mark Nixon, Dr. Hugh Lewis, Dr. Minh Nguyen, Dr. David Holland, Mr. Matthew Thornton and Mr. John Bevington.

References

Adams JB, Smith MO, Johnson PE (1985) Spectral mixture modelling: a new analysis of rock and soil types at the Viking Lander 1 site. J Geophys Res 91:8098–8112

Atkinson PM (1997) Mapping sub-pixel boundaries from remotely sensed images in Innovations in GIS IV. In: Kemp Z (ed). Taylor anld Francis, London, p 166–180

Atkinson PM (2004) Super-resolution land cover classification using the two-point histogram GeoENV IV: Geostatistics for Environmental Applications. Sánchez-Vila X, Carrera J, Gómez-Hernández J (eds) Kluwer, Dordrecht, pp 15–28

Atkinson PM (2005) Super-resolution target mapping from soft classified remotely sensed imagery. Photogram Eng Remote Sens 71: 839–846

Atkinson PM, Cutler MEJ and Lewis H (1997) Mapping sub-pixel proportional land cover with AVHRR imagery. Int J Remote Sens 18: 917–935

Atkinson PM and Tate NJ (2000) Spatial scale problems and geostatistical solutions: a review. Professional Geographer 52:607–623

Bezdek JC, Ehrlich R, Full W (1984) FCM: The fuzzy c-means clustering algorithm. Comput & Geosci 10: 191–203

Support vector machines for optimal classification and spectral unmixing. Ecol Model 120: 167–179

Deutsch CV, Journel AG (1998) GSLIB: Geostatistical Software and User's Guide Second Edition. Oxford University Press, Oxford

Foody GM (1998) Sharpening fuzzy classification output to refine the representation of sub-pixel land cover distribution. Int J Remote Sens 19: 2593–2599

Hopfield J and Tank DW (1985) Neural computation of decisions in optimization problems. Biol Cybern 52: 141–152

Journel AG and Huijbregts C J (1978) *Mining Geostatistics*. Academic Press, London

Jupp DLB, Strahler AH and Woodcock CE (1988) Autocorrelation and regularization in digital images I Basic theory. *IEEE Trans Geosci Remote Sens* 26: 463–473

Mertens KC, Verbeke LPC, Ducheyne EI and De Wulf RR (2003) Using genetic algorithms in sub-pixel mapping. *International Journal of Remote Sensing* 24: 4241–4247

Mertens KC, Verbeke, LPC Westra T and De Wulf RR (2004) Sub-pixel mapping and sub-pixel sharpening using neural network predicted wavelet coefficients. *Remote Sens Environ* 91: 225–236

Nguyen MQ, PM Atkinson and HG Lewis (2005) Super-resolution mapping using a Hopfield neural network with digital elevation data. *IEEE Trans Geosci Remote Sens* 2: 366–370

Nguyen MQ, PM Atkinson and HG Lewis (2006) Super-resolution mapping using a Hopfield neural network with fused images. *IEEE Trans Geosci Remote Sens* 44: 736–749

Steinwendner J, Schneider W and Suppan F (1998) Vector segmentation using spatial subpixel analysis for object extraction. *Int Arch Photogram Remote Sens* 32: 265–271

Tatem AJ, Lewis HG, Atkinson PM, Nixon MS (2001a) Super-resolution target identification from remotely sensed images using a Hopfield neural network *IEEE Trans Geosci Remote Sens* 39: 781–796

Tatem AJ, Lewis HG, Atkinson PM, Nixon MS (2001b) Multiple class land cover mapping at the sub-pixel scale using a Hopfield neural network *Int J Appl Earth Observation & Geoinf* 3: 184–190

Tatem AJ, Lewis HG, Atkinson PM, Nixon MS (2002) Land cover simulation and estimation at the sub-pixel scale using a Hopfield neural network. Remote Sens Environ 79:1–14

Tatem AJ, Lewis HG, Atkinson PM, Nixon MS (2003) Increasing the spatial resolution of landsat TM imagery for land cover mapping in agricultural areas. *Int J Geograph Inf Sci* 17: 647–672

Thomas IL, Benning VM, Ching NP (1987) *Classification of Remotely Sensed Images*. Adam Hilger, Bristol

Thornton MW, Atkinson PM, Holland DA (2006) Super-resolution mapping of rural land cover features from fine spatial resolution satellite sensor imagery. *Int J Remote Sens* 27: 473–491

Woodcock CE, Strahler AH (1987) The factor of scale in remote sensing. *Remote Sens Environ* 21: 311–322

Dating Fire Events on End of Season Maps of Burnt Scars

T. J. Calado and C. C. DaCamara

Abstract Forested areas cover *circa* 38% of Continental Portugal and the observed increasing trend in the extent and severity of wildfires points to the need for accurate and timely knowledge of the total burnt area. The official fire database is the one provided by the Portuguese Forest Service (DGRF) and is based on information supplied by fire and forest services. Since 1990, maps of burnt areas have been yearly produced based on information from Landsat-TM. A recent study for the period 1984–1989 has pointed out severe discrepancies between ground- and satellite-based data, raising the need to devise procedures aiming to correct such discrepancies. The present work represents a first attempt to assess the potential of using NOAA/AVHRR imagery to assigning dates to burnt scars on end of season maps as the ones derived from Landsat-TM. We begin by degrading to the AVHRR scale a MODIS-based end of season map of burnt scars. Degradation was simply performed by computing the fraction of burnt MODIS pixels inside each AVHRR pixel. Then, we built up a neuro-fuzzy model that assigns to each AVHRR pixel the "possibility" of representing a burnt area. The model uses as input pixel values of AVHRR channel 2 and was trained using a composite of minimum values of that channel and the corresponding fractions of burnt MODIS pixels of the degraded end of season map. The model was then applied to individual AVHRR images and further refined in order to eliminate errors associated to contamination by clouds and water bodies, geo-rectification problems and dark backgrounds. It is shown that the refined model underestimated by 11% the total amount of burnt area as obtained from DGRF data and that the differences reduced to 1% when DGRF data were restricted to records greater than 100 ha. The model was then used to assign dates to burnt scars in the MODIS-based end of season map. Obtained results are quite encouraging since deviations (NOAA-MODIS) between –2 (–1) and +1 (0) days represent 85% (70%) of the total and may be attributed to differences in orbital times of passage of NOAA and TERRA/AQUA.

T. J. Calado
Centro Geofísico da Universidade de Lisboa, 1749–016 Lisboa, Portugal
e-mail: mtcalado@fc.ul.pt

A. Soares et al. (eds.), *geoENV VI – Geostatistics for Environmental Applications*, 323–333

Introduction

Mediterranean regions are some of the most affected by wildfires, which have become a major source of concern for environmental security. In Continental Portugal, the observed increasing trend in both the extent and severity of wildfires (Fig. 1) has been linked to demographic and socio-economic changes that took place in the rural areas of the country (e.g. Almeida and Moura, 1992) but is also related to climatic characteristics (warm and dry summers throughout most of the country) associated to the onset of favourable meteorological conditions (e.g. Pereira et al., 2005, Trigo et al., 2006), as well as to climate change (e.g. Pereira et al., 2002). Since wildfire activity is a serious problem that is likely to become even more serious in the next decades, there is a strong need for accurate databases of the number and time of occurrence of wildfire events and the extent of associated burnt areas (e.g. Rego, 1992).

The main source of fire statistics information for Portugal is the official database that is provided by the Portuguese Forest Service (DGRF) based on information supplied by fire and forest services. Each fire record contains information on fire initial location from the district down to the parish scales, date and time of ignition and extinction, total burnt area and type of burnt terrain. According to DGRF (Fig. 1), the total burnt area in Continental Portugal during the period 1980–2005 has reached 3.074 million ha (circa 33% of the surface of the country), and the years of 2003 and 2005 show up as being the most extensively burnt. Remote sensing is another useful source of information (e.g. Chilar, 2000) and instruments of different spatial, temporal and spectral characteristics, e.g. Advanced Very High Resolution Radiometer (AVHRR) on-board National Oceanic and Atmospheric Administration (NOAA) platforms, Thematic Mapper (TM) on-board Landsat and more recently

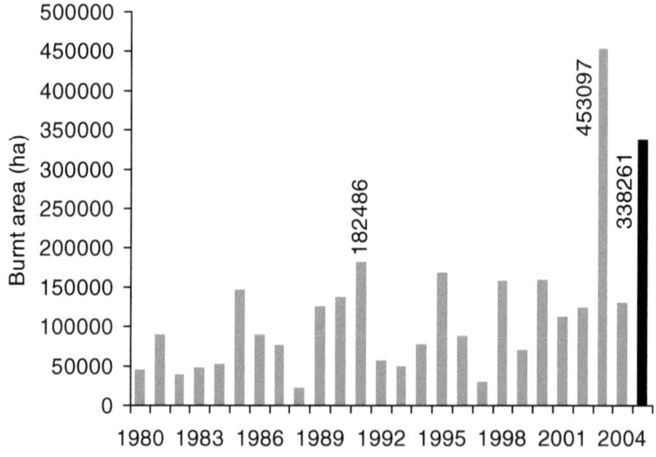

Fig. 1 Yearly amounts of burnt area in Continental Portugal for the period 1980–2005. The year of 2005 is identified by the black bar (*source*: DGRF)

VEGETATION on-board the Système Probatoire d'Observation de la Terre (SPOT) and Moderate Resolution Imaging Spectroradiometer (MODIS) on-board TERRA and AQUA, have been used widely for wildfire monitoring and management. A comprehensive review may be found in Ahern, et al. (2000). Since 1990, the Department of Forest Engineering (DEF) of the Portuguese Institute of Agronomy (ISA) has been producing yearly maps of burnt areas, based on information from Landsat-TM, which have contributed to a better knowledge of the spatial distribution of wildfires in Continental Portugal (Pereira et al., 2000). However no temporal information is provided on the identified fire scars.

Recently, DEF/ISA has started using Landsat-TM imagery to produce end of season maps of burnt areas for the period prior to 1990. A comparison of yearly amounts of burnt area based on DGRF and Landsat information is shown in Table 1 (J.M.C. Pereira, personal communication) for the 6-year period 1984–1989. Serious discrepancies are well apparent between the two datasets. Differences range from a factor of 1.4 to 2.2 and raise the problem of identifying the sources of errors as well as of devising procedures aiming to correcting the errors at different temporal scales, from the monthly up to the weekly or daily levels.

The present work represents a first attempt to assess the potential of using NOAA/AVHRR imagery to assigning dates to burnt scars on a given end of season map as derived from a finer spatial resolution remote-sensing instrument, such as TM. Since Landsat-TM has a revisit interval of 16 days that prevents checking the accuracy of the assigned dates based on NOAA/AVHRR information, we have relied on a sequence of burnt scar maps as obtained from MODIS on-board TERRA and AQUA platforms. The last image of the MODIS series was used to produce an end of season map and the remaining ones were kept for verification purposes only.

We begin by building up a neuro-fuzzy model that assigns to each AVHRR pixel the "possibility" of representing a burnt area. The model uses as input normalized values of AVHRR channel 2 and is applied to individual AVHRR images. The model is further refined in order to eliminate commission errors associated to pixels contaminated by clouds and water bodies, to problems in geo-rectification and to dark backgrounds. The refined model is then used to assign dates to burnt scars in the MODIS-based end of season map. Finally, an assessment is made of the quality of obtained results.

Table 1 Total burnt areas as derived from DGRF and Landsat-TM for the period 1984–1989

Year	DGRF (ha)	Landsat-TM (ha)	Ratio
1984	52,710	116,872	2.22
1985	146,254	291,944	2.00
1986	89,522	113,161	1.26
1987	76,269	137,785	1.81
1988	22,434	31,499	1.40
1989	126,237	204,060	1.62
TOTAL	**513,426**	**895,321**	**1.74**

Data and Procedure

Dataset Characteristics

We used the following two sets of remote-sensed data covering the month of August 2005 (Table 2):

1. MODIS data from TERRA and/or AQUA, consisting of daily vector maps (based on morning and/or afternoon orbits) of burnt scars in Continental Portugal as obtained from visual interpretation and on-screening digitising of RGB colour composites of channels 7 (2.105–2.155 μm), 2 (0.841–0.876 μm) and 1 (0.620–0.670 μm). The spatial resolution is 250 m and the data were kindly provided by DEF/ISA. The image of August 25 was used to simulate the end of season map and the remaining ones were kept for validation purposes.
2. NOAA/AVHRR data, consisting of raster images of the morning and/or early afternoon orbits (Table 2) for channels 1 (0.58–0.68 μm), 2 (0.72–1.10 μm), 3 (3.55–3.93 μm), 4 (10.3–11.3 μm) and 5 (11.5–12.5 μm). The images were kindly supplied by the Portuguese Meteorological Institute and pre-processing was performed using an orbital model, relying on satellite ephemeris data supplied by the NAVY Space Surveillance Centre. In order to obtain precise geo-registration positional accuracy of 1 km RMSE, image navigation was based on data from Digital Chart of the World Database (DCW) to extract identifiable features such as coastlines, water bodies, and rivers and to correlate them with the matching raw image. Adequately navigated and attitude corrected images were then resampled to the Universal Transverse Mercator (UTM) WGS 84 North, zone 29 projection.
3. Fire information for August from DGRF database. These data were used to validate the burnt area estimates as obtained from NOAA and TERRA/AQUA imagery.

Table 2 Data availability for August 2005. Days with both AVHRR and MODIS are in bold

Day of Month	AVHRR (UTC)	MODIS	Day of Month	AVHRR (UTC)	MODIS
4	–	X	**15**	**11:40**	**X**
5	–	–	**16**	**11:17**	**X**
6	–	–	**17**	**10:55**	**X**
7	**11:23**	**X**	18	–	–
8	–	X	19	–	X
9	10:38	–	20	–	–
10	–	–	21	11:03	–
11	–	X	**22**	**10:40**	**X**
12	–	–	**23**	**13:52**	**X**
13	10:46	–	**24**	**13:42**	**X**
14	–	–	**25**	**13:32**	**X**

Degrading the End of Season Map

The different spatial resolutions of MODIS (250 m) and AVHRR (1.1 km) make the combination of information a complex task and suggest using fuzzy techniques to circumvent some of the problems associated to downscaling. Use of fuzzy logics is also quite appealing in identifying burnt area pixels in remote-sensed imagery since fuzzy sets are especially adequate to model verbal expressions such as "burnt areas are generally darker than the background". Accordingly the degradation of the MODIS-based end of season image to the AVHRR scale was performed by counting the number of MODIS burnt pixels inside a given AVHRR pixel and then computing the fraction of burnt pixels. As a result of the degradation process, a number between 0 and 1 was assigned to each AVHRR pixel (Fig. 2), a value of 0 (1) indicating an AVHRR 1.1 km pixel containing no (totally filled by) burnt MODIS pixels. The rationale consists in looking at those fractions (between 0 and 1) as indicating the "possibility" of a given pixel to be identified as a burnt pixel by the AVHRR instrument.

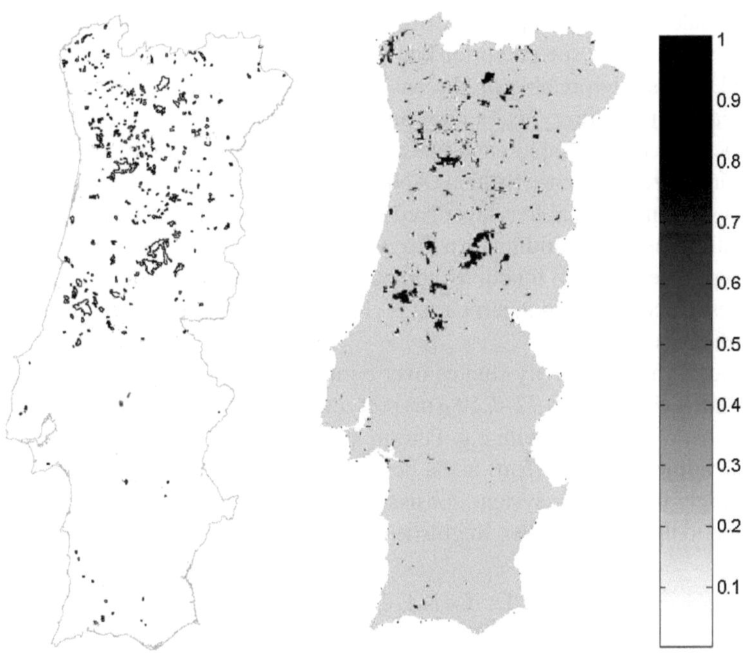

Fig. 2 MODIS-based end of season map (August 25). Left panel: vector image of burnt scars; right panel: grey scale image degraded to NOAA scale,. The scaling bar on the right indicates the fraction of MODIS burnt pixels

Pre-processing of AVHRR Imagery

Before attempting to derive any procedure aiming to identify burnt areas there is the need to remove, from each AVHRR image, any pixel that may not cover land surface alone, i.e. one must identify and then mask all pixels that are likely to be partially or totally contaminated e.g. by water bodies, clouds, cloud shadows or smoke plumes. The water mask was obtained by applying a subtractive clustering method (Chiu, 1994) to the five AVHRR channels of a clear sky image. Cloud masks were obtained by means of strict thresholds that were applied to reflectance (channels 1 and 2) and thermal (channel 5) channels. Dark pixels in the vicinity of cloud pixels that did not remain dark in the following image were identified as cloud shadows and masked.

Finally, for each NOAA image, we normalized the values of five NOAA channels and of NDVI of unmasked pixels over Continental Portugal by subtracting the respective means and then dividing by the respective standard deviations. This normalization procedure was performed with the aim of mitigating the effects of day-to-day variability e.g. due to changes in illumination conditions, viewing angles and surface heating.

Identification of Burnt Areas

The rationale was to build up an algorithm that on the one hand is capable of learning and can use problem specific prior knowledge and on the other will translate into a model that is interpretable on the basis of linguistic terms. This suggests using the so-called neuro-fuzzy techniques and therefore we have adapted a previously developed procedure (Calado and DaCamara, 2002) based on the use of an Adaptive Neuro-Fuzzy Inference System (ANFIS) model (Jang, 1993), i.e. a Sugeno-type fuzzy system in a special five-layer feed forward network architecture. The method essentially consists in building up a Fuzzy Inference System (FIS), i.e. given appropriate input membership functions, the output will be a fuzzy inference curve that translates the concept that a burnt area is characterized by low values of reflectivity (i.e. dark areas).

We performed sensitivity studies over burnt and non-burnt areas and concluded that AVHRR channel 2 (0.72–1.10 μm) presented the greatest discriminating power, a result that is consistent with e.g. Pereira and Setzer (1993) who have shown that the near-infrared (0.7–1.3 μm) is the best spectral region for identifying fire scars. Accordingly the Sugeno system consisted of a single input x (channel 2) and we restricted to the following two linguistic control rules:

$$\text{Rule 1 : if x is} A_1 \text{then} y_1 = m_1 x + b_1$$
$$\text{Rule 2 : if x is} A_2 \text{then} y_2 = m_2 x + b_2$$

where A_1 and A_2 are linguistic terms and the outputs y_1 and y_2 are linear functions of the input. We considered the linguistic terms "is low" (A_1) and "is high" (A_2) and accordingly the expressions "x is A_1" (i.e. "channel 2 is low") and "x is A_2" (i.e.

"channel 2 is high") were quantified by the respective degrees of membership w_1 and w_2 of input x to fuzzy sets A_1 and A_2, which were characterized by Gaussian membership functions:

$$w_i = \exp\left[-(x-c_i)^2 \big/ 2\,(\sigma_i)^2\right] \qquad i = 1, 2$$

where c_i and σ_i are the position and shape parameters of A_i.

The output y of the ANFIS model is finally given by:

$$y = \frac{w_1\, y_1 + w_2\, y_2}{y_1 + y_2}$$

The ANFIS model was trained using as input and output sets respectively the AVHRR composite of minimum values of normalized channel 2 and the corresponding values of fraction of burnt pixels in the degraded MODIS-based end of season map. Scars in the map were identified as contiguous pixels with positive values of fraction of burnt pixels and were labelled from top to bottom and from left to right. The learning dataset S was then subdivided into two subsets S_O and S_E corresponding to odd and even scar labels. Obtained values of root-mean square errors of the ANFIS model were respectively 0.2862, 0.2824 and 0.2846 for sets S_O, S_E and S. Figure 3 presents the corresponding three output curves of the ANFIS models. The similarity among the three fuzzy inference curves is well apparent, an indication of the robustness of the ANFIS model.

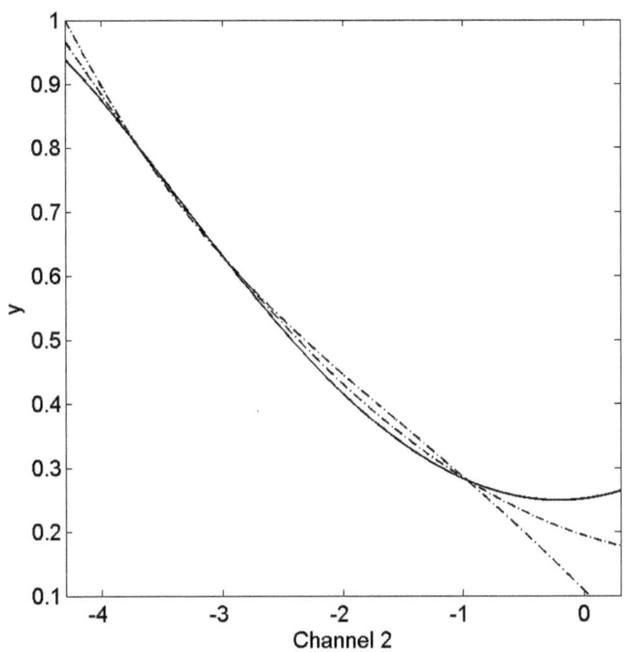

Fig. 3 ANFIS output curves for datasets S_O, S_E (thin lines) and S (thick line)

Results and Discussion

The ANFIS model was applied to every single AVHRR image and defuzzification was performed by stating that a given pixel in a given image is classified as a burnt one if the ANFIS output is greater than 0.55 in that image and in the following one. The threshold of 0.55 was chosen as a compromise between minimizing omission and commission errors. Once classified as burnt, a given pixel remains classified as burnt in all subsequent images. Accordingly, the last AVHRR image contains all pixels that were identified as burnt ones.

A comparison of the last AVHRR image of burnt areas, as obtained from the ANFIS model, with the MODIS-based end of season map pointed to the existence of a rather large number of commission errors. However most of these errors were located at the borders of burnt scars, clouds and water bodies and are attributable to mixed pixels, errors of geo-rectification and non-identified cloud shadows. These errors were readily eliminated by stating that a pixel in a given AVHRR image would remain classified as burnt only if not located in the border of water and/or clouds; otherwise it would be considered as unclassified. However, false alarms in the immediate vicinity of a burnt area were considered as correctly classified as burnt. A second kind of commission error was due to the presence of sparsely covered dark soils that presented radiative signatures similar to burnt areas. In order to mitigate the occurrence of such errors we have introduced the condition that a given pixel was only classified as burnt if normalised values of NDVI in the previous image were positive for that pixel or in any of the surrounding ones. Although this additional condition has excluded some truly burnt pixels, a substantially larger number of false alarms was eliminated.

Validation of results was performed by means of confusion matrices where MODIS burnt scars were considered as representing the "ground data". Confusion matrices were computed for those days where data were available for both AVHRR and MODIS (dates in bold in Table 2). Performance of the classification was then assessed by deriving the Producer's Accuracy (PA) and User's Accuracy (UA), which are measures of omission and commission errors, respectively. Overall values of 70% and 87% were obtained for PA and UA, for the set of days where both MODIS and AVHRR images are available.

Figure 4 presents obtained values of PA and UA for individual pairs of images (left panel) as well as the time evolution of cumulative burnt areas as estimated from DGRF, MODIS and NOAA datasets (right panel). Values of UA in the analyzed pairs of images are quite stable and remain above 80%, an indication of a relatively small number of commission errors. Values of PA are quite low at the beginning of August but show a steep increase, an indication that omission errors rapidly decreased in time. The obtained values for UA and PA are quite encouraging taking into account the differences in the spatial resolution of the AVHRR and the MODIS sensors, as well as the different techniques that were used to identify burnt scars. In relation to the cumulative burnt areas, there seems to be a systematic underestimation of burnt areas by MODIS when compared to DGRF. In the case of NOAA, it is worth stressing that AVHRR underestimates by 11% the amount obtained from

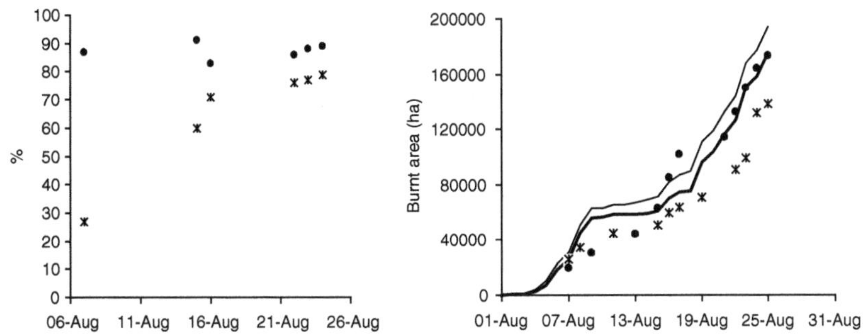

Fig. 4 Evolution of PA and UA (left panel) as obtained from contingency tables for AVHRR and MODIS (taken as the "ground data"). Crosses and dots represent PA and UA, respectively. Evolution of burnt areas (right panel) as derived from three different datasets. Thin and thick curves represent estimates as derived from the DGRF database, the thick curve indicating the results obtained when considering burnt areas larger than 100 ha. Dots and crosses represent estimates as derived from AVHRR and MODIS, respectively

DGRF data and that the difference reduces to 1% when we restrict to DGRF records greater than 100 ha.

Despite the observed differences between AVHRR and MODIS, the obtained results for UA and PA together with the good agreement in trend between the evolution of cumulative burnt areas support the idea that AVHRR is able to identify the radiative characteristics of burnt scars and may be used to assign dates of occurrence to a MODIS-based end-of-season map of burnt scars, i.e. to the last MODIS image of our dataset (the one for August 25).

Assignment of dates to pixels belonging to burnt scars in the MODIS-based map was performed in a rather simple way. First we computed a NOAA composite image containing the times when pixels were classified as burnt for the first time. This map was then upgraded to the MODIS scale using the nearest neighbour technique and vectors of scars were superimposed (Figure 5, leftmost panel).

A MODIS composite with first time of occurrence of burnt pixels inside identified burnt scars was also computed from the MODIS database (Fig. 5, centre left

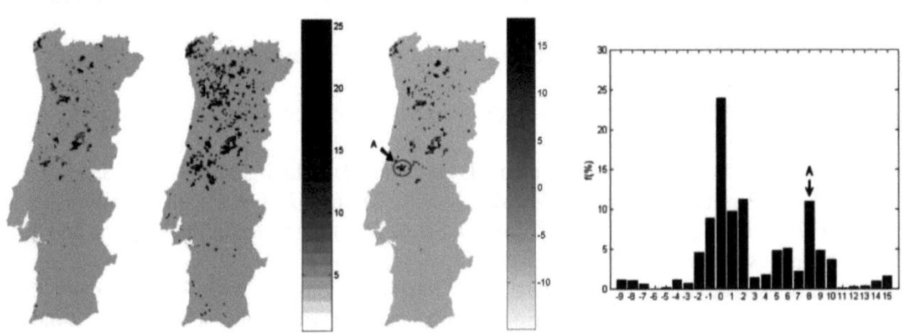

Fig. 5 Initial dates of burnt area pixels as obtained from NOAA (leftmost panel) and MODIS (centre left panel). Respective differences (centre right panel) and histogram of differences (rightmost panel)

panel). An assessment of differences between assigned dates to burnt pixels based on AVHRR and MODIS databases was made by computing deviations of NOAA estimated dates from the MODIS ones (Fig. 5, centre right panel). A histogram of obtained deviations is also shown in Fig. 5 (rightmost panel). The deviation mean (bias) and the standard deviation are 2.7 and 4.5 days respectively. The first quartile, the median and the third quartile are respectively 0, 1 and 6 days. Values of bias and of median reflect the positive skewness of the distribution of deviations, an indication that there is a tendency for NOAA dates to occur later than the MODIS ones. Results are however quite encouraging, especially if one notes that deviations between –2 (–1) and +1 (0) days represent 85% (70%) of the total and may be attributed to differences in orbital times of passage of NOAA and TERRA/AQUA.

Finally we have paid attention to the second maximum in the histogram (8 day differences), which corresponds to a single large burnt area in the AVHRR-based map. Both the second maximum and the corresponding area are marked by an "A" in Fig. 5 and it was found that correct estimates of the date could be obtained by using a lower defuzzification threshold of 0.4. Such lower threshold is attributable to the fact that the date of occurrence of the fire was previous to the initial date of the study period. However, there is the possibility of refining the procedure by assigning optimal thresholds to each scar based on the behaviour of the time series of ANFIS values for burnt pixels inside a given MODIS scar. Such a procedure is currently being investigated.

References

Ahern F, Grégoire J-M, Justice C (eds) (2000) Forest fire monitoring and mapping: a component of global observation of forest cover. European Commission, Joint Research Centre, EUR 19588 EN

Almeida AMSF, Moura PVSV, (1992) The relationship of forest fires to agro-forestry and socio-economic parameters in Portugal. Int J Wildland Fire **2**:37–40

Calado TJ, DaCamara CC, (2002) Burnt area mapping in Portugal with a neuro-fuzzy approach. Proceedings from the EUMETSAT Meteorological Satellite Data User's Conference, Dublin, Ireland, 577–584

Chilar J, (2000) Land cover mapping of large areas from satellites: status and research priorities, Int J remote Sens **21**(6&7):1093–1114

Chiu S, (1994) Fuzzy model identification based on cluster estimation. J Intell Fuzzy Syst **2**:267–278

Jang J-SR, (1993) "ANFIS: Adaptive Network-based Inference System". *IEEE* Trans Syst Man Cybern **23**(3):665–668

Pereira JS, Correia AV, Correia AP, Branco M, Bugalho M, Caldeira MC, Cruz CS, Freitas H, Oliveira AC, Pereira JMC, Reis RM, Vasconcelos MJ (2002) Forests and biodiversity. In: Santos FD, Forbes K, Moita R (eds) Climate Change in Portugal: scenarios, impacts and adaptation measures. Gradiva, Lisbon, pp 363–413

Pereira JMC, Flasse S, Hoffman A, Pereira JAR, González-Alonso F, Trigg S (2000) Operational use of remote sensing for fire monitoring and management: regional case studies. In: Ahern et al. (2000), pp 98–110

Pereira MC, Setzer AW, (1993) Spectral characteristics of deforestation fires in NOAA/AVHRR images. Int J Remote Sens 17:1925–1937

Pereira MG, Trigo RM, DaCamara CC, Pereira JMC, Leite SM (2005) Synoptic patterns associated with large summer forest fires in Portugal. Agric Forest Meteorol **129**(1–2):11–25

Rego FC, (1992) Land use changes and wildfires. In: Teller A, Mathy P, Jeffers JNR (eds) Responses of forest ecosystems to environmental changes. Elsevier Appl Sci London

Trigo RM, Pereira JMC, Pereira MG, Mota B, Calado TJ, DaCamara CC, Santo FE (2006) Atmospheric conditions associated with the exceptional fire season of 2003 in Portugal. Int J Climatol **26**(13):1741–1757

Influence of Climate Variability on Wheat Production in Portugal

C. Gouveia and R. M. Trigo

Abstract In this work we describe the temporal evolution of wheat production and yield in Portugal and the spatial context of this production in Alentejo region (Southern Portugal). Then we have identified the geographical extent of this area and related with wheat yield, using remote sensing data. For this purpose, we have used the normalized difference vegetation index (NDVI), retrieved between 1982 and 1999 from the AVHRR instrument. The year-to-year variations in Portuguese vegetation greenness were estimated and related to national wheat yield. A significant correlation was found over Alentejo region and a validation using Corine2000 land cover map has confirmed the correspondence with arable land code pixels. Finally, we evaluate the relevance of the North Atlantic Oscillation (NAO) pattern in terms of wheat yield. A significant influence of the NAO, associated to spatial patterns of variation of different climatic fields, namely precipitation and radiation, on wheat yield in Alentejo region was found. The most significant monthly correlations were obtained for the two important stages in vegetative cycle, namely: February/March and April to June (NAO/Yield, Precipitation/Yield and Radiation/Yield).

1 Introduction

Wheat production and quality are associated with several factors, e.g. seed variety, soil type and fertilization techniques, which could be considered invariable due the strict regulations imposed by European Community. However, climate is one of the major factors which influence the spatio-temporal distribution of most agricultural systems, which are vulnerable to inter-annual climate variability and, in particular, to extreme events and changes in traditional patterns of regional climate. Regional distribution of temperature and precipitation in Europe are affected by changes in the

C. Gouveia
CGUL, Departamento de Física, Faculdade de Ciências, Universidade de Lisboa, Portugal
e-mail: cgouveia@est.ips.pt

A. Soares et al. (eds.), *geoENV VI – Geostatistics for Environmental Applications*, 335–345

winter atmospheric circulation and relations between them and wheat yield for the European countries were found. However sensibility to winter variability seems to be local and a regional analysis with the aim to clarify the relationship between winter variability and crops are required (Cantelaube et al., 2004). Previous studies have shown that a high North Atlantic Oscillation (NAO) index in winter is associated with better quality of the UK wheat crop (Atkinson et al., 2005) and with better wheat, rye, oats and citrus yields in Iberian Peninsula (Gimeno et al., 2002).

During the last decades, the study of weather conditions and their connection to the plant growth and the crops yield has been very important in agricultural research. At the same time, remote sensing technology has been developing steadily and its products can provide many applications in agriculture, namely crop identification, crop growth monitoring and yield prediction. The most important meteorological variables associated with agriculture production are air temperature (daily maximum and minimum values), solar radiation and precipitation. In particular, solar radiation provides the energy for the processes that drive photosynthesis (Hoogenboom, 2000). Previous studies have shown that cotton and wheat yields for the Canadian Prairies, can be satisfactory predicted combining meteorological and spectral data, (Boken and Shaykewich, 2002).

The majority of wheat in Portugal is sown in October and November and harvested in June and July in the following years, leading to a small production, due to the existence of a very short vegetative cycle. A comprehensive assessment on the Portuguese wheat vegetative cycle and the corresponding relationship with climate variables is given in detail by Sampaio (1990) and Feio (1991). Continental Portugal presents a typical Mediterranean climate with mild and relatively wet winters and dry summers. This situation can be further damaged by bad drainage of soils. For this reason the wheat yield in Portugal is considerably smaller than the wheat yield obtained in the North-western European countries, with cold (but not too wet) winters and relatively wet summers. During the grain filling until the complete grain ripening phase, the Mediterranean conditions can be even worst, due the short period of time between frost episodes and relatively high temperatures at the end of spring (May/June). In this phase it is very important the role of potential evapotranspiration, where large values may lead to weak photosynthetic activity, due the plant spend the majority of this activity to transpire, in way to fit against the warm season and not to produce dry matter. This situation causes the decrease of wheat quality. A "perfect" year for wheat production in Portugal, is one with precipitation in autumn, in way to prepare the soil to catch-crop, followed by low precipitation in early winter, moderate precipitation in late early spring, precipitation in April and some precipitation in May. June must be dry, but not very dry allowing a slow and complete maturation; that origin filled and well formed grains.

The present paper has three main objectives. First, to describe de evolution of wheat production and yield in Portugal and related it with the most important cultivated area with cereals: Alentejo. Secondly, to identify this area, using remote sensing data, and relate this spectral data with wheat yield. Finally we aim to evaluate the relevance of the NAO atmospheric circulation pattern in terms of wheat yield.

2 Data

2.1 Wheat Yield Data

Wheat yield data for Portugal were extracted from the Food and Agriculture Organization (FAO) database for the period 1961–2005. Annual averages of wheat yield in Portugal for the considered period can be observed in Fig. 1. The yield time series has two components: a trend, due essentially to improvements in farm management practices and a weather related component, which explains the yield inter-annual variability on the top of that trend (Cantelaube et al., 2004). Thus, to study the effect of climate related variability on crop yield, the technological driven trend was removed from the raw time series, in order to work on yield anomalies only.

Wheat yield data for different Portuguese regions provided by National Statistic Institute (INE) for the "short" period from 1996 to 2003, this is available for both hard and soft yield wheat. The principal wheat growing area is in the southern sector of Portugal, Alentejo (Fig. 2, left), with more than 80% of the total wheat production, but concentrating more than 95% of the total of hard wheat production (Fig. 2, right).

2.2 Spectral Indices

We have used the monthly NDVI dataset, at 8-km resolution, from the Advanced Very High Resolution Radiometers (AVHRR), provide by the Global Inventory Monitoring and Modeling System (GIMMS) group (Kaufmann et al., 2002). The data for the Iberian Peninsula covers the area between 10 W to 0 E and 35 N to 45 N and respect to the 18-year long period from 1982 to 1999. Details on the quality of GIMMS dataset can be found in Tucker et al. (2005).

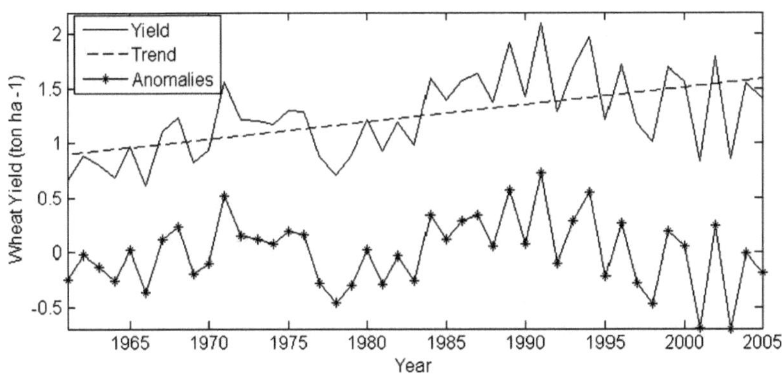

Fig. 1 Time series of wheat yield in Portugal for the period from 1961 to 2005: yield (solid line), general trend (dashed line) and anomalies for detrended time series (line with asterisks)

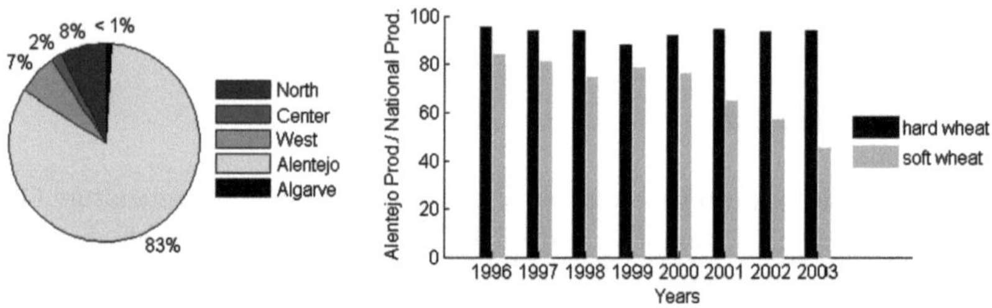

Fig. 2 Wheat yield percentages in Portugal for the period from 1996 to 2003. Left panel: wheat growing areas for different Portuguese areas; Right panel: Percentage of Alentejo's wheat yield for hard and soft wheat

2.3 Large Scale Climatic Data

Meteorological data used in this study are large-scale gridded data retrieved from the Climate Research Unit (CRU) datasets, for the period 1982–1999. A complete description of the method used to derive this monthly high-resolution (10' resolution) climatic database can be found in Mitchell and Jones (2005). Monthly values of cloud cover, temperature and precipitation, were extracted from the CRU database for the window: 35°N–45°N, 10°W–0°E. The cloud cover time series were then converted to net solar radiation fields using the Soil and Water Assessment Tool (SWAT), a method developed by Neitsch et al. (2002).

2.4 North Atlantic Oscillation (NAO)

The North Atlantic Oscillation (NAO) has been recognized for more than 80 years as one of the major patterns of atmospheric variability in the Northern Hemisphere (Walker, 1924). However, only in recent years has this important atmospheric circulation mode become the subject of a wider interest (e.g. Rogers, 1984; Barnston and Livezey, 1987). More recently, the study by Hurrell (1995) had significant impact on the climatological community and has been followed by an increasing number of studies. It is within this context, that several studies have established links between the NAO index and winter season precipitation in Western Europe and, in particular, over the Mediterranean basin (Hurrell 1995; Trigo et al., 2004). This control exerted by NAO on the precipitation field over Europe is likely related to corresponding changes in the associated activity of North Atlantic storm tracks (Serreze et al., 1997; Osborn et al., 1999). The NAO index used in this study was developed by the Climatic Research Unit (University of East Anglia, UK) and is defined, on a monthly basis, as the difference between the normalized surface pressure at Gibraltar (southern tip of Iberian Peninsula) and Stykkisholmur in Iceland (Jones et al., 1997). It should be noticed that the NAO index for winter months presents a positive trend over the last 3 decades of the XXth century; as a consequence its distribution is dominated by positive values, with monthly averages above zero (Jones et al., 1997). Therefore we decided to normalize the entire NAO index on a seasonal basis, having zero mean and standard deviation one. This normalization procedure was based on the computation of seasonal averages and standard deviation between 1982 and 1999.

2.5 Corine Land Cover Map (CORINE2000)

The reference map used here is based on the Corine Land Cover Map (CLC2000), available on a 250m by 250m grid, which has been aggregated from the original vector data at 1:100,000. The Corine Land Cover is a key database for integrated environmental assessment and provides a pan-European inventory of biophysical land cover, using a 44 class-nomenclature. The version used here is available at the Portuguese Environmental Institute web site (http://www.iambiente.pt).

3 Methodology

Monthly and seasonal composites for spring (March, April and May – MAM) were computed for the spectral index (NDVI). As the vegetative cycle usually ends at late June (or early July), in the case of atmospheric variables we have only performed a monthly composite analyses for the first 6 months of the year. We computed a grid point correlation between detrended seasonal and monthly composites of the different variables and detrended wheat yield and seed, for the 18 year study period 1982 to 1999. Furthermore, we computed a simple correlation between monthly composites of NAO index and wheat yield, for the same period. The years characterized by positive or negative NAO index values were also analyzed separately.

The reference map used is the Corine Land Cover Map (CLC2000), originally on a 250m by 250m grid, was re-projected from the CLC2000 to Geographic Coordinates, based on the nearest neighbor scheme. Furthermore, the pixels were geocoded to 1000m grid resolution, based on the most frequent class and considering a pixel to belong to a class if there are at least 9 pixels from this class inside a box of 16 pixels. After this stage all pixels were geocoded to the lower 8000m grid resolution, based on the same scheme (Fig. 4, left). We have found this criterion more appropriated, since the selection of a threshold provides higher confidence in the class label, while maintaining a sufficiently large sample in each class. However, as the second degradation is very strong, we decided to assign a pixel to a specific class if there are inside the considered box, more than 50% pixels from this class.

4 Results

4.1 NDVI and Wheat

Monthly and seasonal composites for spring (MAM) were computed for the spectral index NDVI. Figure 3 shows the grid point correlation between NDVI and wheat yield considering the 18 year period between 1982 and 1999. Pixels over the

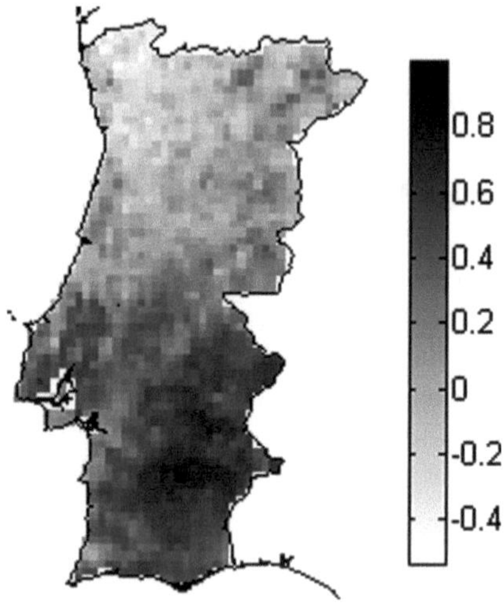

Fig. 3 Patterns of simple correlation between NDVI three months composite and wheat yield in Portugal, for the period of (1982–1999)

Ocean and Spain were intentionally masked and therefore it was only represented the correlation pattern over Portugal. The highest positive correlations were found over the southern region of Alentejo (ranging between 0.6 and 0.8).

4.2 Wheat and Corine2000

In order to identify the predominant class present in this area, we have used the Corine Land Cover Map. It should be stressed that this land cover classification is widely regarded to be a key database for integrated environmental assessment studies, namely those related with agricultural issues. The Corine2000 for Portugal geocoded for Geographic Coordinates and with an 8 km resolution is presented in Fig. 4 (left panel).

The pixels coded as class 12 corresponding to "arable land not irrigated" were selected (Fig. 4: middle panel) and characterized in terms of correlation coefficient values between wheat and vegetation activity (Fig. 4: right panel). The histogram analysis shows that the most frequent correlation class ($0.6<R<0.8$) includes almost 50% of all arable land pixels.

4.3 Wheat and NAO

We have computed the correlation coefficient between monthly NAO index composite and wheat yield (Table 1). The considerable disparities found between monthly correlations can be explained if we take into account the very different requirements of the wheat vegetative cycle, at different stages. The annual evolution, between

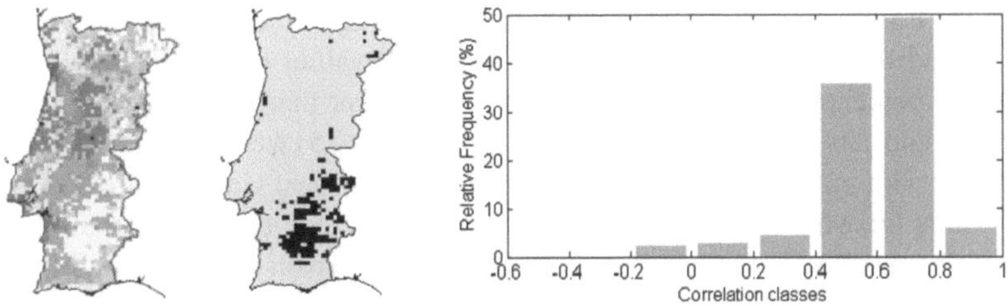

Fig. 4 (left panel) Corine2000 Land Cover classification (44 classes) geocoded for 8 km resolution for Portugal. (middle panel) Pixels coded as Arable Land (not irrigated) class from Corine2000 (code 12) for Portugal. (right panel) Relative frequency of correlation coefficient values between three month composite of NDVI and wheat yield in Portugal, for the pixels coded as Arable Land

Table 1 Correlation coefficient values between monthly NAO index (January to June) and wheat yield. Also shown the correlation coefficient between seasonal NAO index (February and March, FM and April to June, AMJ) and wheat yield

	Jan	Feb	Mar	Apr	May	Jun	FM	AMJ
Yield	0.13	−0.33	−0.19	0.46	0.7	0.50	−0.35	0.56

1982 and 1999, of the Portuguese wheat yield and the two seasonal NAO indices (described in Table 1) can be seen in Fig. 5. Left panel shows the late winter (FM) NAO composite and wheat yield, which present a negative correlation (–0.35), while the right panel presents the late spring NAO composite (AMJ) and wheat yield that reveal a positive correlation value (0.56).

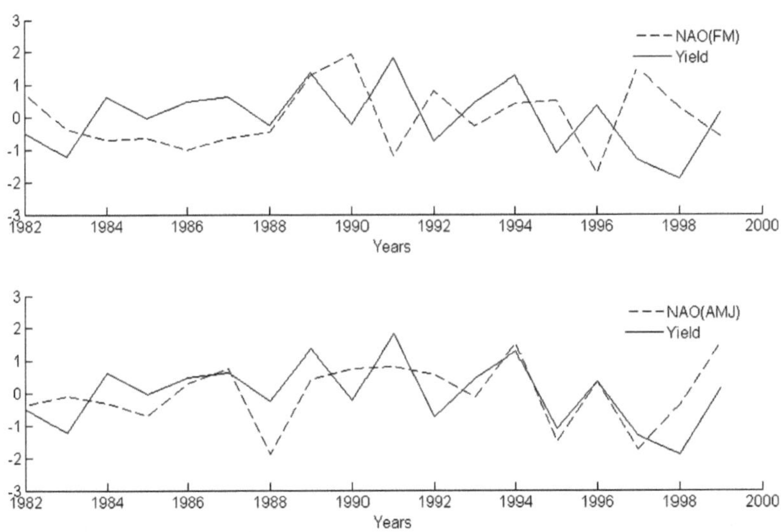

Fig. 5 Seasonal NAO Index (dashed line) and wheat yield over the period 1982-1999 for: left panel: late winter (FM); right panel: early summer (AMJ)

In the following step, all years were divided into three classes according to their winter and spring NAO indices. Higher than usual values of seasonal NAO index (NAO > percentile 75), lower than usual (NAO lower than percentile 25) and the intermediate class (percentile 25 < NAO < percentile 75) were analyzed to identify the years with yield higher or lower than the average wheat yield. Figure 6 presents the results for wheat yield for late winter (left panel) and late spring (right panel). Generally speaking, years characterized by negative (positive) anomalies of wheat yield present positive (negative) NAO index values for late winter. On the other hand, years presenting positive (negative) anomalies for wheat yield are usually characterized by positive (negative) NAO index for late spring. Therefore, it is possible to confirm that late winter and late spring are important moments for the wheat vegetative cycle in Portugal and that the NAO index controls, at least partially, what is happening in both occasions. In summary, a good wheat yield usually corresponds to years characterized by negative values of NAO for late winter and/or positive NAO values for late spring. It is now necessary to show what are the meteorological fields forced by the NAO mode. In the following section we will analyze the relation between wheat and meteorological variables, such as precipitation and radiation for these specific periods.

4.4 Wheat and Meteorological Variables

With the aim of verifying if these specific monthly correlations are due to the different importance of specific meteorological variable, we have computed a spatial

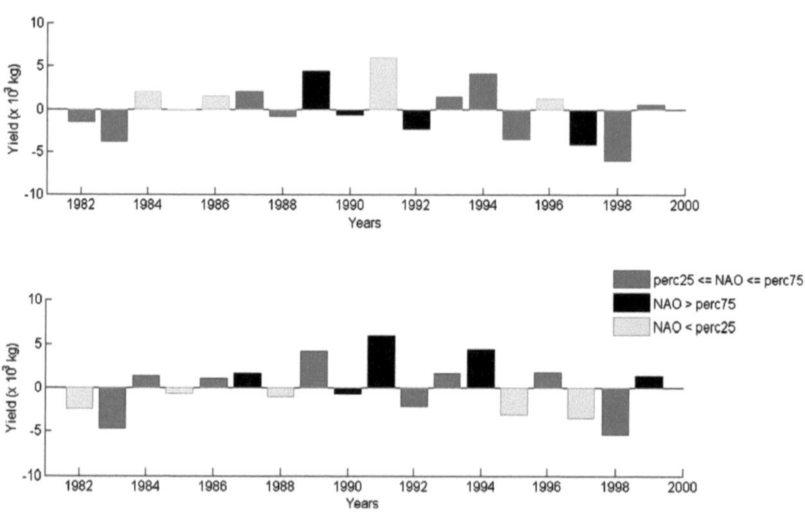

Fig. 6 (left panel) Frequency histograms of annual wheat yield for years with positive late winter (FM) NAO index (black columns), for years with negative seasonal NAO index (light gray columns) and for intermediate years (dark gray columns). (right panel) the same but with late spring (AMJ) NAO index

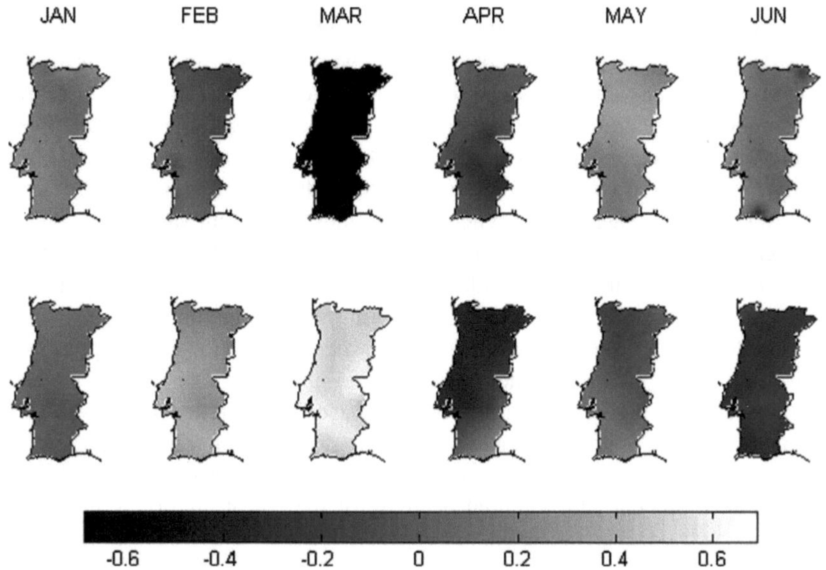

Fig. 7 Patterns of simple correlation between wheat yield in Portugal and two relevant meteorological fields for the period of 1982-1999: top panel: net long wave radiation; bottom panel: precipitation

correlation between wheat yield and monthly averages of two relevant meteorological variables; long wave radiation and precipitation (Fig. 7). Correlation values between monthly net long wave radiation and wheat yield are statistically significant and with a negative signal in March (around –0.6) and positive in June (around 0.3). Interestingly, the spatial pattern of correlation values between monthly precipitation and yield provides results with similar magnitude with higher positive values in March (around 0.6) and negative in June (around –0.3). These results are in good agreement with the most important variables (and timings) that rule the different stages of the wheat vegetative cycle and that have been empirically identified by previous authors. Therefore, a standard good year of wheat production in Portugal is characterized by moderate precipitation in late winter (FM) and early spring, some precipitation May and a dry June. We should stress that the positive impact of NAO on long wave radiation in May and June is in accordance with the wheat necessities of temperature to maturation phase.

5 Conclusions

In this work we have found a strong negative correlation (range from 0.6 to 0.8) between NDVI and wheat yield for the 18 year-long period 1982 to 1999, over the southern part of Portugal (Alentejo). This region corresponds to the area with more than 80% of wheat production, but concentrating more than 95% of the total of hard wheat production. This area was analyzed using the Corine2000 and the pixels designed as "arable land not irrigated" were selected and characterized in terms of correlation values between wheat and vegetation activity. The histogram analysis shows that almost 90% of correlation values belong to the classes between 0.4 and 0.8. In a second step we have shown that a good year for wheat yield is usually characterized by negative values of NAO during late winter and/or positive NAO values for late spring. When the two most important variables that drive the wheat cycle are analyzed (precipitation and radiation), it is possible to verify that if there is a negative NAO index in spring, precipitation in western Iberian Peninsula is higher than usual while the net long wave radiation is lower than average, both of which promote a good annual wheat yield. However to preserve a good wheat yield, the NAO index signal should change in late spring and early summer, inducing higher values of net long wave radiation and lower values of precipitation in the Iberian Peninsula. The low precipitation wheat requirements in this period allow a slow maturation to origin well formed grains and avoid the pests development and potential decrease of the wheat quality as mentioned before.

Acknowledgments This work was supported by the Portuguese Science Foundation (FCT) through project CARBERIAN (Terrestrial Vegetation Carbon Trends in the Iberian Peninsula) PDCTE/CTA/49985/2003. The meteorological large-scale gridded data set was kindly supplied by Climate Research Unit (CRU).

References

Atkinson MD, Kettlewell PS, Hollins PD, Stephenson DB, Hardwick NV (2005) Summer climate mediates UK wheat quality response to winter North Atlantic Oscillation. Agric Forest Meteorol 130(2005):27–37

Barnston AG, Livezey RE (1987) Classification, seasonality and persistence of low-frequency atmospheric circulation patterns. Mon Weather Rev 115:1083–1127

Boken VK, Shaykewich CF (2002) Improving an operational wheat yield model using phonological phase-based Normalized Difference Vegetation Index, Int J Remote Sens 23(20): 4155–4168

Cantelaube P, Terres J-M, Doblas-Reyes FJ (2004) Influence of climate variability on European agriculture —analysis of winter wheat production, Clim Res 27:135–144

Feio M (1991) Clima e Agricultura. *Ministério da Agricultura,* Pescas e Alimentação

Gimeno L, Ribera P, Iglesias R,Torre L,Garcia R, Hernández E (2002) Identification of empirical relationships between indices of ENSO and NAO and agricultural yields in Spain, Clim Res 21:165–172

Hoogenboom G (2000) Contribution of agrometeorology to the simulation of crop production and its applications. Agric Meteorol 103:137–157

Hurrell JW (1995) Decadal trends in the north Atlantic oscillation: regional temperatures and precipitation. Science 269:676–679

Jones PD, Johnson T, Wheeler D (1997) Extension to the North Atlantic Oscillation using instrumental pressure observations from Gibraltar and south-west Iceland. Int J Climatol 17:1433–1450

Kaufmann RK, Zhou L, Tucker CJ, Slayback D, Shabanov NV, Myneni RB, (2002) Reply to Comment on 'Variations in northern vegetation activity inferred from satellite data of vegetation index during 1981–1999' by Ahlbeck JR, J Geophys Res vol 107. no. D11, 10.1029/200IJD001516

Mitchell,TD, Jones PD (2005) An improved method of constructing a database of monthly climate observations and associated high-resolution grids. Int J Climatol 25:693–712

Neitsch SL, Arnold JG, Kiniry JR, Williams JR, Kiniry KW (2002) Soil and Water Assessment Tool theoretical documentation, TWRI report TR-191, Texas Water Resources Institute College Station

Osborn TJ, Briffa KR, Tett SFB, Jones PD, Trigo RM (1999) Evaluation of the North Atlantic Oscillation as simulated by a climate model. Clim Dyn 15:685–702

Rogers JC, (1984) The association between the North AtlanticOscillation and the Southern Oscillation in the Northern Hemisphere, Mon Weather Rev 112:1999–2015

Sampaio AJ (1990) A cultura do Trigo, Ministério da Agricultura, Pescas e Alimentação

Serreze MC, Carse F, Barry RG, Rogers JC (1997) Icelandic Low cyclone activity: climatological features, linkages with the NAO, and relationships with recent changes in the Northern Hemisphere circulation. J Climatol 10:453–464

Trigo RM, Pozo-Vazquez D, Osborn TJ, Castro-Diez Y, Gámis-Fortis S, Esteban-Parra MJ (2004) North Atlantic Oscillation influence on precipitation, river flow and water resources in the Iberian Peninsula, Int J Climatol 24:925–944

Tucker CJ, Pinzon JE, Brown ME, Slayback DA, Pak EW, Mahoney R, Vermote EF, El Saleous N (2005) An extended AVHRR 8-km NDVI dataset compatible with MODIS and SPOT vegetation NDVI data. Int J Remote Sens 26(20):4485–4498

Walker GT (1924) Correlations in seasonal variations of weather. IX Mem Ind Meteorol Dept 24:275–332

Part V
Soil

soil science and environmental engineering (Webster and Oliver 1989; Pachepsky and Acock 1998; Bross et al. 1999) and stochastic simulation is a key tool used in modelling this uncertainty (e.g. Goovaerts 2001). However, these issues have received limited attention to date in relation to the rehabilitation of mined land and related concerns for plant establishment, survival and long term sustainability. Environmental impact mitigation and concurrent reclamation are now commonly regarded as integral components of the mining process (Morrey 1999). Thus, there is an increased need to develop suitable modelling frameworks for the prediction of environmental changes, and hence the assessment of impacts.

A key issue in modelling soil properties, including mine spoil heaps, is that the modelling of several commonly correlated properties of soils are needed. Properties including alkalinity, electrical conductivity, salinity, sodium content, nitrate concentrations and phosphates that may affect plant success (Grigg et al. 2000) show spatial cross-correlations. Techniques to jointly simulate spatial distributions of soil attributes are available (e.g. Gutjahr et al. 1997) and improve the plausibility of resultant models. However, they are computationally intensive. Contributors to complexity include the tedious inference and modelling of cross-correlations and computational inefficiencies, both substantially increasing with the number of variables being jointly-simulated. A practical alternative to the 'direct' joint-simulation of variables is the decorrelation of variables introduced using principal component analysis or PCA (David 1988; Wackernagel 1995). The effectiveness of this approach, in the presence of spatial cross-correlations, is limited because PCA does not eliminate cross-correlations at distances other than zero. To overcome the above limitations, minimum/maximum autocorrelation factors, MAF, (e.g. Desbarats and Dimitrakopoulos 2000) are used to de-correlate pertinent variables into spatially non-correlated factors that are independently simulated and back transformed to correlated attributes. The simulations of MAF are generated with the new fast generalised sequential Gaussian simulation, GSGS, (Dimitrakopoulos and Luo 2004) to provide a substantially more efficient simulation framework.

The following sections will firstly introduce the method of joint simulation of multiple correlated variables based on MAF. A description of the data available follows together with the results of the joint simulation. An application of risk analysis is then presented towards assisting with rehabilitation strategies followed by conclusions.

2 Joint Simulation of Correlated Variables with Minimum/Maximum Autocorrelation Factors

In geostatistical terminology, the attributes of elements in soils are represented by a multivariate stationary and ergodic random function. Consider a multivariate, ℓ dimensional, Gaussian, stationary and ergodic spatial random function $Z(x) = [Z_1(x), ..., Z_\ell(x)]^T$. Minimum/Maximum Autocorrelations Factors are defined as the ℓ orthogonal linear combinations $Y_i(x) = a_i^T Z(x), i = 1, ..., \ell$ of the original multivariate vector $\mathbf{Z}(x)$. MAF are derived assuming that $Z(x)$ is represented by a two-structure linear model of coregionalisation (Wackernagel 1995). The MAF transformation can be rewritten as

$$\mathbf{Y}(x) = \mathbf{A}_{\mathrm{MAF}}\,\mathbf{Z}(x) \tag{1}$$

and the MAF factors are derived from

$$\mathbf{A}_{\mathrm{MAF}} = \mathbf{Q}_2\Lambda_1^{-1}\mathbf{Q}_1 \tag{2}$$

where the eigenvectors $\mathbf{Q}_1$ and eigenvalues $\mathbf{\Lambda}_1$ are obtained from the spectral decomposition of the multivariate covariance matrix $\mathbf{B}$ of $Z(x)$ at zero lag distance. More specifically,

$$\mathbf{Q}_1\mathbf{B}\mathbf{Q}_1^T = \mathbf{\Lambda}_1 \tag{3}$$

and $\mathbf{Q}_2$ is the matrix of eigenvectors from the spectral decomposition

$$\mathbf{Q}_2\mathbf{M}(\Delta)\mathbf{Q}_2^T = \mathbf{Q}_2\left(\frac{1}{2}\left[[\mathbf{\Gamma}_Y(\Delta)]^T + [\mathbf{\Gamma}_Y(\Delta)]\right]\right)\mathbf{Q}_2^T \tag{4}$$

where the matrix $\mathbf{\Gamma}_Y(\Delta)$ is an asymmetric matrix variogram at lag distance Δ for the regular PCA factors $\mathbf{Y}(x) = \mathbf{Z}(x)\mathbf{A}$, where $\mathbf{A} = \mathbf{Q}\Lambda^{-1/2}$. In practice, several Δ lag distances may be used for values lower than the range and the resulting eigenvectors averaged.

Given the MAF transformation above, the joint simulation of multiple correlated variables using the MAF approach proceeds as follows:

i. Normalize the variables to be simulated.
ii. Use MAF to generate the MAF non-correlated factors.
iii. Produce variograms for each MAF.
iv. Conditionally simulate each MAF using a Gaussian simulation method.
v. Validate the simulation of factors.
vi. Back-transform simulated MAF to variables and de-normalize.
vii. Validate the final results.
viii. Generate additional simulations, as needed.

The conditional simulation of a Gaussian random function $\mathrm{Y}(x)$ above is based herein on the decomposition of the multivariate probability density function (PDF) of a stationary and ergodic random function to a product of local conditional distributions

$$f\left(x_1, \ldots, x_N; y_1, \ldots, y_N\right) = \prod_{i=1}^{N} f\left(x_i; y_i \mid (n+i-1)\right) \tag{5}$$

where $f\,(x_1, ..., x_N; y_1, ..., y_N)$ is the pdf, N the number of points discretising the field to be simulated, n the number of available data, and x_i the location of a point in the space considered. Setting $\Lambda_i = \Lambda_0 + y\,(x_\alpha)$, α=1,...,i, where $y\,(x_\alpha)$ is a realization of Y(x) at location x_α, and considering groups of N_p nodes, Eq. (5) is

$$f\,(x_1, ..., x_N; y_1, ..., y_N|\Lambda_0) = \prod_{i=1}^{N_1} f\,(x_i; y_i|\Lambda_{i-1}) \cdot \prod_{i=N_1+1}^{N_1+N_p=N_2} f\,(x_i; y_i|\Lambda_{i-1}) \cdot .... \cdot \prod_{i=N_{k-1}+1}^{N_{k-1}+N_p=N_k} f\,(x_i; y_i|\Lambda_{i-1}) \tag{6}$$

The simulated group of nodes in vector $\mathbf{y}_p$ is then

$$\mathbf{y}_p = \mathbf{C}_{pI}\mathbf{C}_{II}^{-1}\,\mathbf{y}_I + \mathbf{L}_{pp}\mathbf{w}_p \tag{7}$$

where, the covariance matrix $\mathbf{C}_{II}$ is the covariance between the conditioning nodes, $\mathbf{C}_{pI}$ the covariance between the group of nodes to be simulated and the conditioning nodes, and $\mathbf{C}_{pp}$ the covariance between the nodes of the group; $\mathbf{y}_I$ is the data vector contained in Λ_{i-1} and $\mathbf{w}_p$ a standard normal random vector. The generalized sequential Gaussian simulation algorithm (GSGS) from Eq. (6), used to simulate the N nodes in a domain D is:

i. Define a random path visiting each group of N_p nodes to be simulated.
ii. At each group of nodes, use Eq. (7) to generate simulated values and add the values to the data set.
iii. Go to the next group of nodes and repeat the previous two steps.
iv. Loop until all groups of nodes have been visited and the N nodes simulated.

The details of GSGS are described in Dimitrakopoulos and Luo (2004) who also discuss the computational advantages of the method over the regular node-by-node sequential Gaussian simulation. For further discussion, the reader is referred to Appendix A.

3 Data Available

The data used in this study were sampled from a mined waste (spoil) dump of an open pit coal mine in the coalfields of eastern, Australia. They include measurements of electrical conductivity (EC), a measure of salinity, and pH, a measure of acidity.

The dataset is composed of 96 sampled locations on a regular grid with EC and pH measurements in all locations. pH values are converted to concentration of hydrogen ions in solution, [H+]. These data are characteristic of the conditions found in spoil dumps of many coal mines in eastern Australia, in that they are highly alkaline and highly saline. Salinity may be a major impediment to plant establishment and survival, while high levels of pH limit the availability of phosphorus, which in turn limits plant

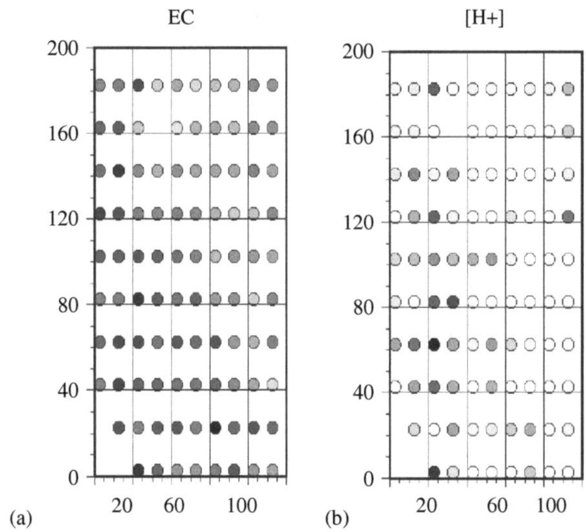

Fig. 1 Measured properties across waste dump at a coal mine. (a) Electrical conductivity, and (b) acidity

growth and sustainability (Grigg et al. 2000). The spatial distribution of both variables is shown in Fig. 1. Dark circles represent high values and light circles represent low values. The heterogeneous nature of the spoil is clearly evident.

4 Joint Simulation of Spoil Parameters: EC and [H+]

4.1 Normal-score Transformation

Following the simulation steps using MAF described earlier, a normal-score transformation is performed on the distribution of EC and [H+] data. Normal score transformations are based on rank ordering of the data and decrease the influence of outliers. This, in turn, assists the inference of the variogram and estimation of covariance matrices in the simulation process that follows.

4.2 MAF Transformation

The transformation matrix A_{MAF} (Eq. (1)) used to generate the min/max autocorrelation factors is shown in Fig. 2 (b). MAF are calculated by multiplying the vector of EC and [H+] by a vector of loadings from the rows of the transformation matrix. It should be noted that the MAF loadings are quite different from the ones derived by PCA, as shown in Desbarats and Dimitrakopoulos (2000). The lag Δ in Eq. (4) used in this example is 65 metres and was derived experimentally by testing several lag

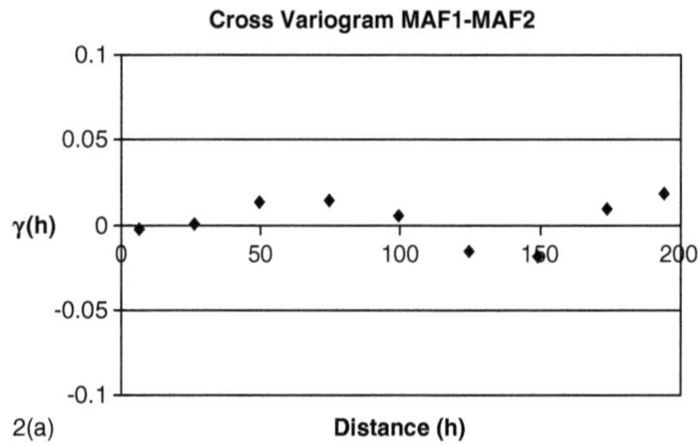

0.756	0.001
0.654	0.999

2(b)

Fig. 2 (a) Cross-variograms of MAF and (b) transformation matrix

distances to assure a suitable decorrelation and stable MAF decomposition. Figure 2 (a) shows the cross-variogram between MAF from the present study and demonstrates variable decorrelation. Experimental variograms and cross-variograms for EC and [H+] are shown and discussed in more detail in a subsequent section.

4.3 Variography of MAF

Variography on each MAF is performed. Figure 3 shows the experimental and model variograms fitted to the both MAF. Note that variogram models for MAF2 are spherical and show clear spatial patterns. MAF1 is modelled as pure nugget. MAF variograms are subsequently used in the simulation of each factor and the validation of the MAF simulation results. It should be noted that MAF variograms are linear combinations of the variograms of the original (normal score) variables.

4.4 Conditional Simulation of MAF

Conditional simulation is performed independently on both MAF using the GSGS algorithm. The simulations are performed on a grid of 5000 nodes within the limits

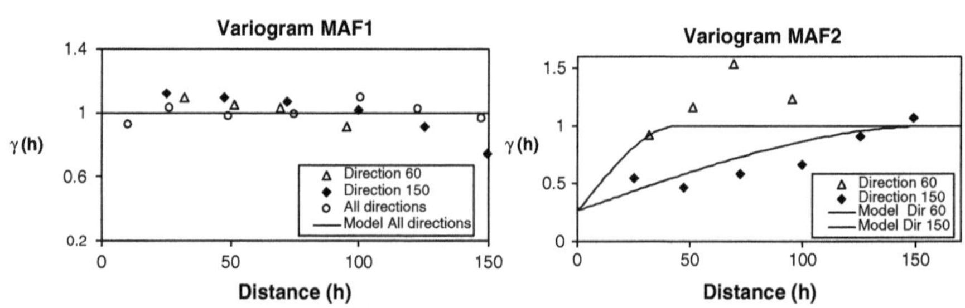

Fig. 3 Experimental and model variograms of MAF

of the waste dump. Thirty simulations are generated in this study and are validated in detail for reproduction of data, histograms and variograms. The validation of the MAF simulations is not presented here as a subsequent section presents the validation of realizations in the data space.

4.5 Back Transformations of MAF

The realizations of MAF were transformed back to simulated normal score variables by multiplying a column vector of simulated MAF in each grid node with the corresponding inverse matrix of the MAF loadings in Fig. 2 (b). Subsequently, the normal score EC and [H+] realizations are back transformed to the data space.

4.6 Validation of the Joint EC and [H+] Simulation Results

Several validation checks are performed to assess the results of the joint simulations of EC and [H+] using the MAF transformations. Validation involves calculation of histograms, experimental variograms and cross-variograms of the simulated realizations to ensure reproduction of original data and their spatial characteristics.

Figure 4 shows plots of variograms and cross-variograms for the original data and conditional simulations. All results suggest that the reproduction of the original data spatial characteristics by the simulated realizations is excellent. Recall that the variograms and cross-variograms of original variables are not directly used in the joint simulation based on MAF, which used the variograms of the independent MAF.

5 Risk Analysis for Decision Making

The main objectives of mined land rehabilitation include a sustainable land use after mining, stability of the land surface, and preservation of water quality. Currently, there are no formal criteria used to assess the success of rehabilitated areas for mines. For open pit coal mines in eastern Australia, such as the case study in the previous section, suitable completion criteria for relinquishment purposes and, more specifically, pasture–based rehabilitation in eastern Australia have been established (Grigg et al. 2001). These criteria suggest the achievement and maintenance of at least 70% vegetation cover because there is considerable evidence, from minesite erosion research and elsewhere, that vegetation cover is the single most important factor affecting soil loss in rehabilitated pastures.

Analysis of data from monitoring studies upon which rehabilitation criteria were based, indicate a strong influence of average salinity (EC) on pasture performance, indicated by the amount of total dry matter. An exponential improvement in DM is evident for decreasing salinity levels. The amount of DM is also related to ground cover and an exponential increase in ground cover is evident with increasing DM.

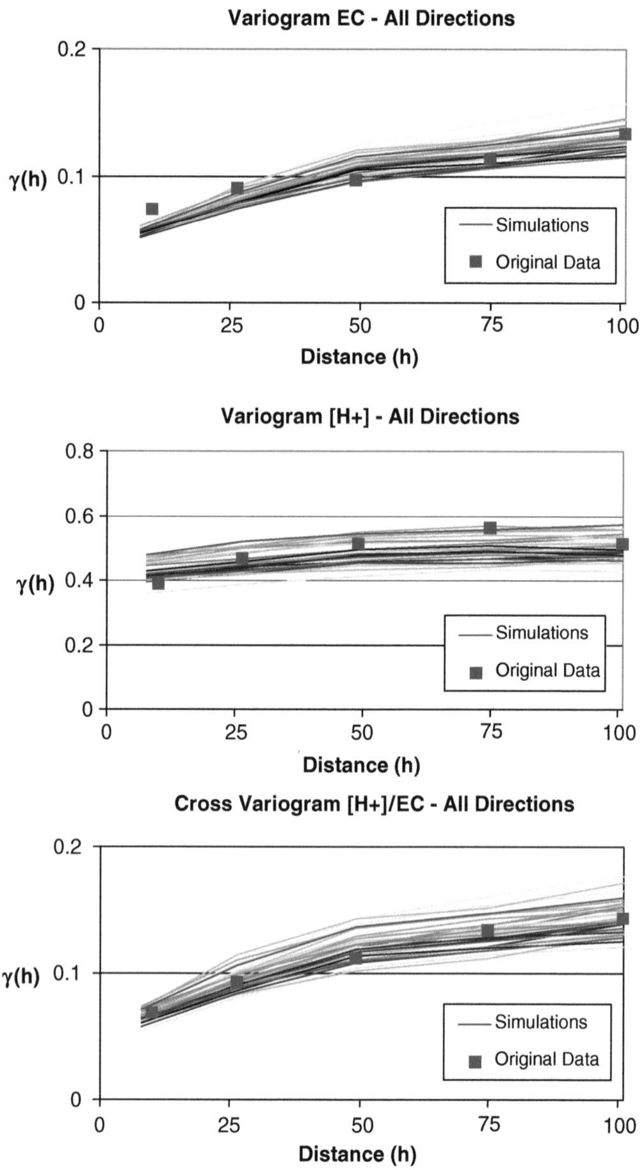

Fig. 4 Variograms and cross-variograms of simulations backtransformed to the data space, compared to experimental variograms of original data

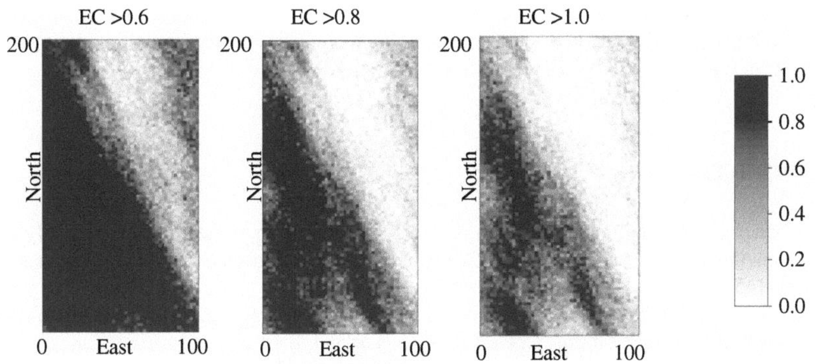

Fig. 5 Probability maps for electrical conductivity

These relationships indicate that an EC of 0.6 would permit development of sufficient dry matter to achieve ground cover of 70% (Grigg et al. 2001).

It has been shown herein that it is possible to simulate spoil parameters throughout the waste dump and that these simulations enable us to quantify the variability of certain parameters. These simulations can then be used to assess the probability, or risk, that EC will exceed 0.6, the cut off value as discussed above to ensure 70% vegetation cover. This is determined from the number of simulations in which the generated value for a given location is above the cut off value. In this example, it was determined that a cover of 70% was required for successful rehabilitation. Other goals may be tested, however, by assessing different cut off levels for EC. Figure 5 shows three cut off values for EC, 0.6, 0.8 and 1.0, which related to goals for ground cover of approximately 70%, 60% and 50% respectively.

The significance of probability maps is their ability to display the risk associated with the rehabilitation goal, and enable decision makers to choose a level of risk that is appropriate and identify areas that may require special attention, as well as in some cases identify areas that will not require any remediation.

6 Conclusions

This study presents a new approach towards the quantification of uncertainty in soil properties and risk assessment for the purpose of land rehabilitation. The approach presented enables the computationally efficient joint simulation of variables and eliminates the necessity for laborious calculations. These methods include the techniques of MAF that decorrelates variables prior to simulation, and GSGS to quantify the variability of soil properties. The above techniques are shown to facilitate risk analysis and assist decision making in mine site rehabilitation in a case study from a coal mine in eastern Australia.

Acknowledgments Funded of this study was provided by the Queensland Department of State Development, Australia. The measured spoil electrical conductivity and pH data from a mined waste dump was provided by the Centre for Mined Land Rehabilitation; Dr A Grigg and Dr D Mulligan provided useful discussions regarding the impact on plant growth of mining waste products in soil.

References

Bross MJ, Aarts L, van Tooren CF, Stein A (1999) Quantification of the effects of spatially varying environmental contaminants into a cost model for soil remediation. J Environ Manage 56: 133–145

Benndorf J (2005) Efficient sequential simulation methods with implications to long-term production scheduling. MPhil thesis, University of Queensland, Brisbane

David M (1988) Handbook of applied advanced geostatistical ore reserve estimation. Elsevier, Amsterdam

Davis MW (1987) Generating large stochastic simulation – The matrix polynomial approximation method. Math Geol 19:99–107

Desbarats AJ, Dimitrakopoulos R (2000) Geostatisical simulation of regionalized pore-size distributions using min/max autocorrelation factors. Math Geol 32:919–942

Dimitrakopoulos R, Luo X (2004) Generalized sequential Gaussian simulation on group size ν and screen-effect approximations for large field simulations. Math Geol 36:567–591

Goovaerts P (2001) Geostatistical modelling of uncertainty in soil science. Geoderma, 103:3–26

Grigg AH, Shelton M, Mullen B (2000) The nature and management of rehabilitation pastures on open-cut coal mines in central Queensland. Tropical Grasslands 34:242–250

Grigg AH, Emmerton BR, McCallum NJ (2001) The development of draft completion criteria for ungrazed rehabilitation pastures after open-cut coal mining in central Queensland. Final Report ACARP Project C8038, Centre for Mined Land Rehabilitation, Brisbane

Gutjahr A, Bullard B, Hatch S (1997) General joint conditional simulations using a fast Fourier transform method. Math Geol 29:361–389

Morrey DR (1999) Principles of economic mine closure, reclamation and cost management – Remediation and management of degraded lands. Lewis Publishers, Boca Raton

Pachepsky Y, Acock B (1998) Stochastic imaging of soil parameters to assess variability and uncertainty of crop yield estimates. Geoderma 85:213–229

Wackernagel HJ (1995) Multivariate geostatistics. Springer, Berlin

Webster R, Oliver MA (1989) Optimal interpolation and isarithmic mapping of soil properties: VI. Disjunctive kriging and mapping the conditional probability. J Soil Sci 40:497–512

APPENDIX A – Some Computational Aspects of the GSGS Simulation Method

Dimitrakopoulos and Luo (2004) suggest a generalisation of the well known sequential Gaussian simulation method (SGS), termed generalised sequential Gaussian simulation (GSGS), summarized in Equations 5, 6 and 7. A key practical reason for developing the approach is its computational efficiency.

GSGS is founded upon the observation that adjacent nodes to be simulated share a common neighbourhood, thus GSGS is constructed to simulated groups of clustered nodes simultaneously instead of one node at a time. The use of groups of nodes amounts to the decomposition of the multivariate probability density function of a stationary random function $Y(x)$ into groups of products of univariate conditional probability density functions (Equation 6). This group decomposition is general and includes as "end member cases," (a) the common SGS, where each group has one node only, and (b) the LU simulation method (Davis 1987), where all nodes to be simulated are in one group.

Theoretical comparisons in terms of calculated computing costs are detailed in Dimitrakopoulos and Luo (2004), who show the link of the optimal group size for GSGS to computational efficiency, and substantial computational efficiencies. For example, when the neighbourhood for SGS is 100 grid nodes, the use of GSGS on a group of 80 nodes can be nearly 50 times faster than the regular SGS, while the same image is generated.

Practical comparisons of the computational efficiency of GSGS are detailed in Benndorf (2005), who demonstrates that GSGS can improve computational efficiency substantially, particularly when the size of a field to be simulated increases. Figure 6 comes from Benndorf's study and shows standardized computing times for different sizes of simulated fields, ranging from 100,000 nodes to 14 million nodes, and for different sizes of groups. The latter groups vary from $1 \times 1 \times 1$ or SGS, to a $4 \times 4 \times 2$ group of nodes. Figure 6 concludes that when simulating small realizations, say less than one million nodes, there is limited benefit of using GSGS with any of the group sizes considered. In this case, the runtime of the algorithm can be reduced, by up to about 30 % compared with SGS, when using small size groups. In addition, Figure 6 suggest that when simulating large realizations containing several millions of nodes, runtime can be reduced substantially; in the example, this reduction is up to 20 times for the case of GSGS with a group size of $4 \times 4 \times 2$ nodes.

For further details on GSGS the reader is referred to the above mentioned publications.

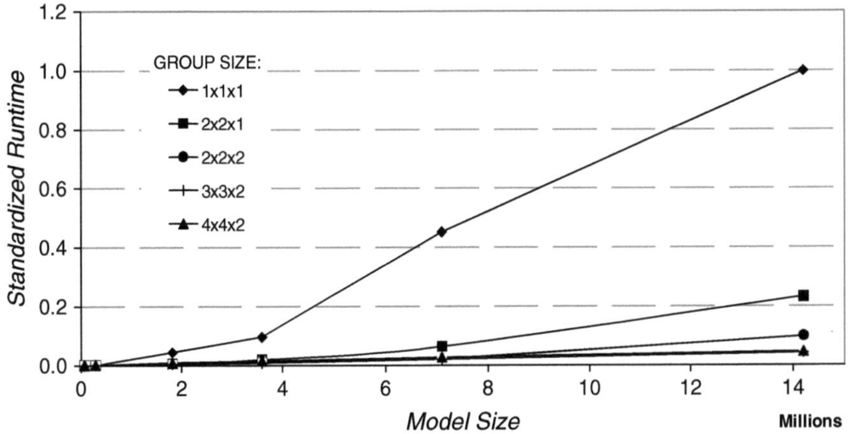

Fig. 6 Experimental runtimes (standardized) of GSGS for different model (realization) sizes and group sizes (Benndorf, 2005)

Temporal Geostatistical Analyses of N_2O Fluxes from Differently Treated Soils

J. M. M. Avalos, A. Furon, C. Wagner-Riddle and A. P. González

Abstract Both CO_2 and N_2O fluxes from soils can contribute in a relatively important way to the greenhouse effect so the study of these variables is highly relevant. The N_2O flux from soil presents high spatial and temporal variability. Application of geostatistical techniques to daily data of N_2O fluxes may reveal underlying temporal structures in this variable. The aim of this study was to analyse patterns of temporal dependence in soil N_2O fluxes. The experimental design consisted of four 100 m x 150 m plots at the Elora Research Station (Ontario, Canada). The data set corresponded to the period from January 2001 to December 2004. An exponential model was fitted to 15 out of 16 series. The models fitted to the experimental variograms consisted of two structures; a nugget effect and an exponential or spherical model depending on the analysed data series. The nugget effect ranged from 0 to 70% of the sill depending on the plot and year analysed. From the models fitted to the experimental variograms a temporal structure was identified, which ranged between 9 and 85 days, depending on the year and plot analysed. Conditional simulation was applied in order to estimate N_2O fluxes for days with missing data, proving to be an appropriate tool for achieving this purpose.

Introduction

Understanding energy and mass fluxes through ecosystems is central to many questions in global change biology and ecology. Ecosystem respiration is a critical component of the carbon cycle and might be important in regulating biosphere response to global climate change (Tate, 2000).

Both CO_2 and N_2O emissions from soils can contribute in a relatively important way to the greenhouse effect so the study of these variables is relevant (Mosier, 1994; Wagner-Riddle et al., 1997; Tate, 2000). These emissions from soil present high spatial and temporal variability, which needs to be quantified in order to optimize sampling and measurements methods. Application of geostatistical

J. M. M. Avalos
Faculty of Sciences, University of Coruña, A Zapateira 15071, A Coruña, Spain
e-mail: jmirasa@udc.es

A. Soares et al. (eds.), *geoENV VI – Geostatistics for Environmental Applications*, 361–371

techniques to soil gas fluxes data may reveal underlying spatial and temporal dependences in these variables facilitating the assessment of gas fluxes over an entire field over time (Chilés and Delfiner, 1999; Christakos, 2000; Goovaerts, 1997).

Predicting the effect of different variables on N_2O emissions is critical to determining the net benefit in greenhouse gases emissions as the global warming potential of N_2O is 310 times greater than that of CO_2(Mosier, 1994; Benckiser et al., 1996; Baggs et al., 2003). Even a modest increase in N_2O emissions can significantly offset net greenhouse gases reductions.

Nitrous oxide production by soil processes is highly variable (Mosier, 1994; Benckiser et al., 1996; Baggs et al., 2003; Ventera et al., 2004). Soil properties controlling the amount of N_2O emitted by soil microbes include soil redox potential, moisture tension, nitrate and oxygen concentrations, as well as the time of the day and the time of the year when the measurements are taken. Generally, the proportion of gaseous nitrogen products composed of nitrous oxide increases with increasing soil acidity, reduced soil temperature, and augmented soil nitrate levels, but the exact ratio of nitrous oxide to dinitrogen for a specific soil system varies with the combination of the soil chemical and physical properties existent therein and is therefore not reliably predicted. Emissions of N_2O have previously been shown to increase after application of inorganic fertilizer (Mosier, 1994). Highest rates of N_2O emissions from fertilized as well as natural ecosystems have often been measured at spring thaw (Wagner-Riddle and Thurtell, 1998).

Geostatistical techniques should facilitate the assessment of the spatial and temporal variability of the soil gas emissions providing a better understanding of these processes (Stacey et al., 2006).

The aim of this study was to analyse the temporal patterns of dependence in soil N_2O fluxes and to apply conditional simulation to estimate N_2O fluxes for days with missing data.

Material and Methods

Data were obtained at the Elora Research Station, 20 Km North of Guelph (Ontario, Canada) using micrometeorological equipment for measuring soil N_2O fluxes. The Elora Research Station is located at 43 deg 39′ N 80 deg 25′ W and at an elevation of 376 m. The soil characteristics were 32% sand, 52% silt, 16% clay, pH(H_2O) = 7, 3.7% total organic C and 0.3% total organic N (Wagner-Riddle and Thurtell, 1998).

Field experiments are part of an ongoing, long term study examining best management practices on nitrogen and carbon cycling. In this study, micrometeorological approaches were used to determine N_2O and CO_2 flux measurements over the entire field throughout the year (Wagner-Riddle et al., 1997; Wagner-Riddle and Thurtell, 1998).

Four plots were studied. Plots 1 and 4 were cultivated using conventional techniques: tillage was done by moldboard ploughing in the fall to a depth of 15 cm and inorganic fertilizer N application was done according to the general N recommendation prior to planting of corn and tillering for winter wheat. Plots 2 and 3 were cultivated using best management techniques: zero tillage, application of fertilizer N for corn according to soil NO_3 test, inclusion of N credits from soybean to winter wheat (applying reduced fertilizer) and use of cover crops over-winter when possible. The same rotation of crops was cultivated on the four plots: corn, soybean and winter wheat.

Nitrous oxide fluxes from each plot were calculated using the flux-gradient method (Wagner-Riddle et al., 1997; Wagner-Riddle and Thurtell, 1998). The four plots were sampled sequentially; this sampling scheme resulted in six hourly concentration differences for each plot during each measurement day. Only concentration gradients measured when the wind direction at the adjacent weather station allowed for a fetch-to-height ratio of at least 50:1 (horizontal distance to height of measurement ratio) were used in the flux calculations. Due to the variable positioning of the sample intakes in the various plots this criteria resulted in a different number of total hourly or daily flux measurements in each plot (Wagner-Riddle and Thurtell, 1998). The N_2O fluxes were expressed in ng N_2O-N $m^{-2}s^{-1}$ and the average for each day was used in the geostatistical analysis. Data were averaged because a different number of measurements were made each day for each plot so we preferred to standardised these measurements.

Initially, data were analysed by descriptive statistics, and the mean, variance, standard deviation, coefficient of variation, maximum value, minimum value, skewness and kurtosis were calculated. This analysis served the purpose of detecting outlier values and the frequency distribution of data.

N_2O flux datasets corresponding to four years (from 2001 to 2004) were analysed geostatistically using GSTAT software (Pebesma, 2000). The temporal autocorrelation was calculated by the semivariogram. Measurements taken at times greater than the range have a random distribution and are therefore independent among themselves.

The dependence relation has been computed according to Cambardella et al. (1994), using the following equation:

$$DR = \frac{C_0 \times 100}{C_0 + C_1}$$

Where C_0 is the nugget effect of the model, C_1 is the sill of the model and DR is the dependence relationship. According to Cambardella et al. (1994) this can be used to classify the spatial and temporal dependence into strong if DR $< 25\%$; moderate for DR between 26% and 75%; and weak if DR $> 75\%$.

Cross-validation techniques were used to assess the goodness of the models. Correlation coefficients between observed and estimated values, mean square error and non-dimensional mean square error were the cross-validation parameters taken into account.

Conditional simulation was used for estimating values of N_2O emissions in those days without measured values. Gaussian conditional simulation was preferred instead of ordinary kriging because the latter smooths the values of the variable whereas Gaussian conditional simulation gives maximum values and so it gives a larger range of prediction. A correlation analysis between measured and mean simulated data was carried out and the mean squared-error of prediction (MSEP) was calculated according to Stacey et al. (2006). Ordinary kriging errors were calculated in order to approach the conditional simulation errors. A mean of 100 realisations was calculated for each dataset.

Results and Discussion

A descriptive statistical summary for all the studied plots and years is shown in Table 1. A high variability is observed in all the data series as the coefficients of variation showed, ranging from 128.8% to 442.6% (Table 1). Plots with different management showed a different number of data.

Data number availability varied from 164 data records for plot 3 in 2003 to 303 data records for plots 1 and 2 in 2001. Lower data numbers corresponded to year 2003. The number of positive values oscillated between 129 for plot 3 in 2003 and 262 data in 2001 for plot 1 (Table 1). Missing values did not appear at the same dates in all the plots and years; these gaps ranged from periods of 1 day to periods of 71 days.

Negative values indicate potential N_2O flux from the atmosphere to the soil. High values of skewness (from 1.67 to 4.73), kurtosis (from 6.87 to 38.76) and variance (from 22.05 to 2736.67) may affect the results of the geostatistical analysis. Figure 1 shows the frequency distributions of the data sets, no transformation of data was performed. Positive value numbers for each plot and year are shown in Table 1. In fact, the highest value for plot 4 in 2003 was 383.33 ng N_2O-N/m^{2}s, while the lowest value for the same data series was –26.36 ng N_2O-N/m^{2}s.

The higher N_2O emission mean values were observed in plot 4 for every year except in 2002 when the highest N_2O emission mean value was observed in plot 1; both plots were managed using conventional techniques.

In Table 2 the parameters of the models which have been fitted to the data series are shown. A model was fitted to all of the analysed datasets but there is no one which fits to all of them. The most common sort of model was the exponential, fitted to 15 out of 16 series; a spherical model was fitted to one of the analysed series.

Cross-validation parameters such as mean square error (MSE) and non-dimensional mean square error (NDMSE) were close to the values indicating good fit of 0 and 1, respectively, for most of the fitted models (Table 2). On the other hand, correlation coefficients for measured and estimated data oscillated between 27 and 91% depending on which plot and which year had been analysed.

Dependence ratio varied from 0% to 70%. According to Cambardella et al. (1994) criteria, a high to moderate dependence relation has been shown in most of the analysed series; however, some series, such as plot 1 during 2003 and plot 4 during 2004, showed a high dependence relation. The magnitude of the nugget effect might be caused by measurement errors, but it also may be a consequence of variability at a smaller scale than the one used in this study suggesting a detailed

Table 1 Descriptive statistics for all the plots and years (data in ng N_2O-N/m^{2}s; C.V. = coefficient of variation)

Plot	Year	N	Positive Values	Minimum	Maximum	Mean	St. Dev.	C.V.	Variance	Skewness	Kurtosis
1	2001	303	262	−9.39	56.78	4.63	5.96	128.8	35.51	3.42	23.26
1	2002	288	246	−12.81	28.02	3.03	4.83	159.4	23.37	1.72	6.90
1	2003	183	167	−13.63	225.46	20.34	38.24	187.9	1461.98	3.46	12.59
1	2004	249	187	−15.01	64.69	3.05	6.95	228.3	48.35	4.73	36.04
2	2001	303	239	−14.66	37.85	2.84	4.79	169.0	22.97	1.69	11.90
2	2002	256	215	−6.61	31.61	2.64	3.57	135.2	12.75	2.61	17.19
2	2003	192	142	−36.51	142.83	4.21	18.64	442.6	347.35	4.23	27.63
2	2004	243	163	−17.12	37.04	2.61	6.73	258.2	45.30	1.86	6.87
3	2001	290	215	−10.27	29.36	2.74	4.70	171.2	22.05	1.67	6.88
3	2002	273	205	−9.33	37.17	2.98	5.73	192.2	32.84	2.32	8.67
3	2003	164	129	−24.63	84.09	4.64	10.76	231.8	115.85	2.87	18.93
3	2004	245	165	−29.11	42.74	2.70	6.90	255.9	47.66	1.88	10.80
4	2001	272	224	−6.44	93.54	4.87	9.13	187.6	83.34	5.10	38.76
4	2002	269	184	−12.84	48.02	2.13	5.51	258.0	30.33	3.09	21.38
4	2003	177	172	−26.36	383.33	25.57	52.31	204.6	2736.67	4.30	20.84
4	2004	250	208	−18.56	48.27	3.85	7.00	181.6	48.95	2.69	11.92

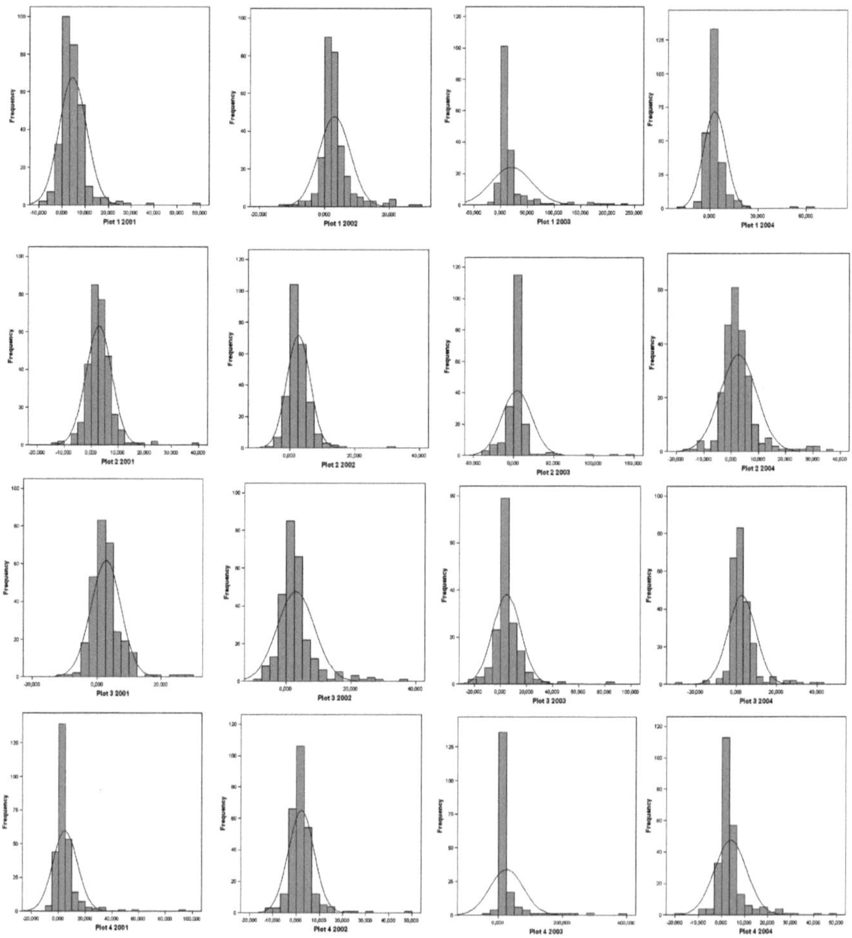

Fig. 1 Histograms for all the studied data sets

study of the N_2O emission variability at a hourly scale is needed. Most of the higher dependence relations were found in those plots managed under best management techniques.

Range values fluctuated between 9 and 85 days, approximately; this indicates a highly variable temporal dependence. The higher range values corresponded to plots 2 and 3 which were cultivated under best management practices. High values of range allowed for a estimation of N_2O fluxes by filling the missing gaps, in some cases large gaps, in the studied datasets.

Figure 2 shows an example of experimental variogram with its best fitted model; in this case, for plot 4 during 2001; this plot was cultivated using conventional practices. The depicted model is the one with the highest range value of all the analysed data series (approximately, 85 days) and is considered to be appropriate for describing the temporal variability of the N_2O emissions during this year and plot as

Table 2 Best fitted model parameters for the 16 N_2O series. Cross-validation parameters are also shown (DR = dependence relationship; NDMSE = non-dimensional mean square error)

Plot	Year	Model	Nugget effect	Sill	Range (days)	DR (%)	R^2	NDMSE
1	2001	Exponential	17.66	19.61	50	47.38	0.59	1.07
1	2002	Spherical	13.02	12.92	34.62	50.19	0.56	1
1	2003	Exponential	0	1957	15.2	0	0.89	1.16
1	2004	Exponential	34.81	24.73	33.15	58.46	0.44	0.99
2	2001	Exponential	17.71	7.59	50	70.00	0.27	1.05
2	2002	Exponential	6.80	5.94	16.64	53.38	0.45	1.08
2	2003	Exponential	109.39	266.20	50	29.12	0.65	1.2
2	2004	Exponential	25.56	25.98	31.65	49.59	0.52	1.04
3	2001	Exponential	14.86	8.88	50	62.59	0.36	1.08
3	2002	Exponential	18.37	19.29	50.41	48.78	0.59	1.01
3	2003	Exponential	76.78	58.86	48.06	56.61	0.29	1.09
3	2004	Exponential	31.17	20.72	19.63	60.07	0.46	1.01
4	2001	Exponential	45.17	59.51	84.63	43.15	0.6	1.02
4	2002	Exponential	14.73	17.98	26.04	45.03	0.54	1.09
4	2003	Exponential	0	3828	16.79	0	0.91	1.13
4	2004	Exponential	2.03	51.44	9.04	3.8	0.75	1.39

shown by the cross-validation parameter values (Table 2). Figure 2 shows, clearly, the moderate magnitude of the nugget effect (43.15 %), according to Cambardella et al. (1994) criteria.

Figure 3 shows the model fitted to N_2O data for the experimental variogram of plot 2 during 2001; this plot was cultivated under best management techniques. The fitted model showed the lowest correlation between observed and estimated data of all the analysed data series, $r^2 = 0.27$ (Table 1). The nugget effect of this model was the highest one (70%) for all the data series. Nevertheless, the range of this model is rather high (50 days) and using this model to simulate the behaviour of plot 2 N_2O emissions during 2001 was decided.

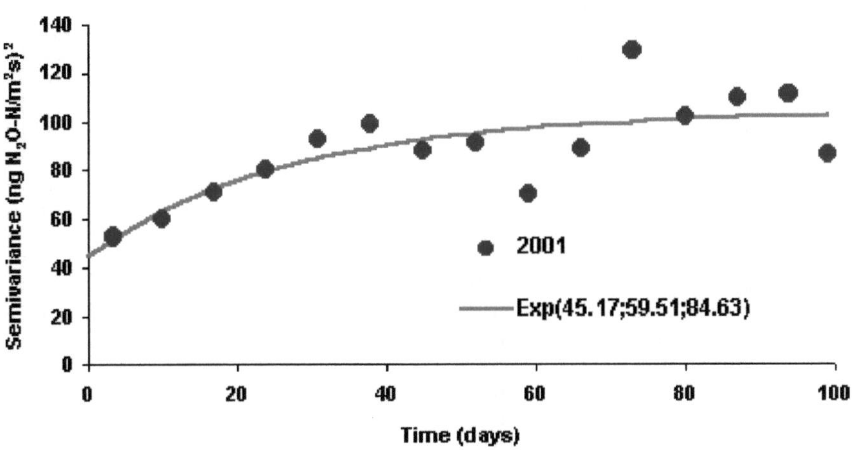

Fig. 2 Experimental variogram and model fitted to N2O data for plot 4 and 2001

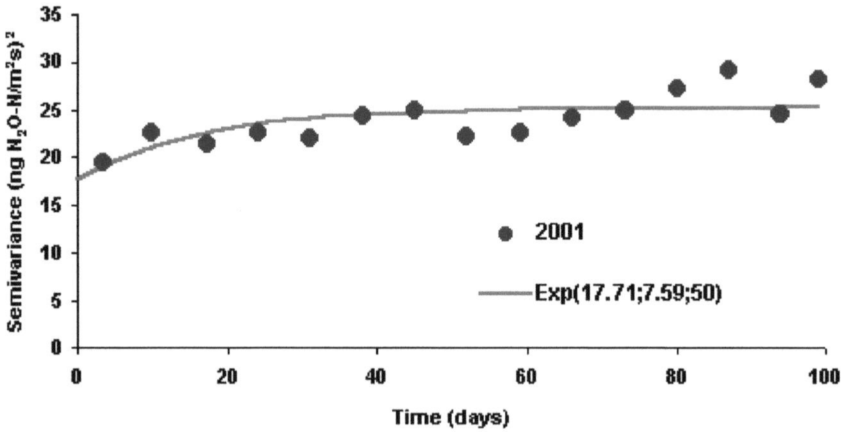

Fig. 3 Experimental variogram and model fitted to N_2O data for plot 2 and 2001

Conditional simulation allowed for estimates of N_2O fluxes for each day and the correlation between the average of the simulated values and measured values was relatively high as shown in Table 3. Mean squared-error of prediction (MSEP) and mean errors are shown, as well. Correlation coefficients showed that most of the fitted models were appropriate for assessing the general behaviour in time of the studied variable; the higher correlation coefficients corresponded to plots managed under conventional techniques.

The extremely high mean error value observed in plot 2 during 2001 was caused by the high nugget effect of the fitted model (70%) which may have affected the Gaussian conditional simulation for that year and plot strongly. Apart from the high

Table 3 Correlation between measured and simulated N_2O fluxes

Plot	Year	Data number	Correlation coefficient	Mean Error	MSEP
1	2001	303	0.712	0.66	17.92
1	2002	288	0.732	0.96	11.15
1	2003	183	0.988	−0.12	37.10
1	2004	249	0.618	9.69	30.60
2	2001	303	0.486	−307.34	17.75
2	2002	256	0.696	−0.14	6.86
2	2003	192	0.805	0.11	131.04
2	2004	243	0.691	0.04	24.58
3	2001	290	0.559	0.58	14.71
3	2002	273	0.713	1.12	16.48
3	2003	164	0.558	1.12	80.66
3	2004	245	0.656	7.92	28.18
4	2001	272	0.715	−0.37	41.08
4	2002	269	0.727	0.01	14.85
4	2003	177	0.990	−0.21	57.80
4	2004	250	0.958	0.58	4.80

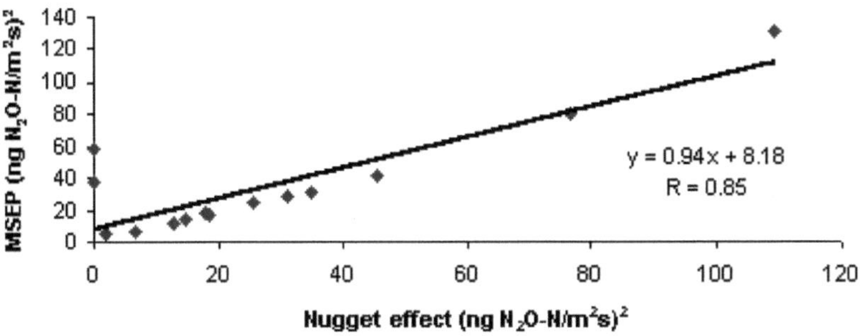

Fig. 4 Relationship between mean squared simulation errors and fitted model nugget effect

nugget effect, this model showed the lowest correlation value ($r^2 = 0.27$) of all the analysed data series influencing the simulation results, as well.

Mean squared-error of prediction were correlated to the magnitude of the fitted model nugget effect ($r = 0.85$), as it is depicted in Fig. 4. This positive correlation showed that low nugget effects gave better simulated values for the analysed variable. Though not so highly correlated ($r = -0.5$), range values had a relatively high importance in the errors of the models, with these errors being higher when range values were lower.

An example of simulated results for a specific data series is shown as an example in Fig. 5.

The general behaviour of N_2O emission data was described properly by this technique. Figure 5 shows that the daily peaks of N_2O emissions were described appropriately when conditional simulation was performed. Nevertheless, the magnitude of these peaks was underestimated in most of the cases. When the correlation coefficients and the range values of the models were higher and their nugget effects were lower, better simulation results were achieved.

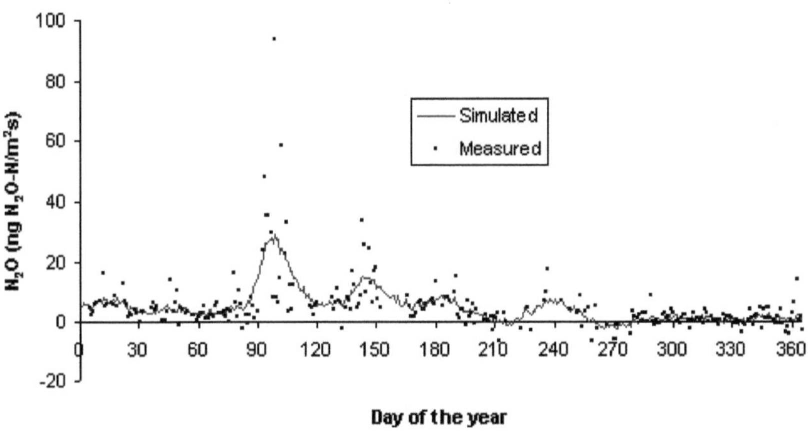

Fig. 5 Measured and simulated values for plot 4 during 2001

Conclusions

A single model for explaining the temporal distribution of the N_2O soil emissions was not identified.

The fitted models explained between 27 and 91% of the temporal dependence of the series depending on the plot and year analysed.

From the models fitted to the experimental variograms, a temporal pattern can be identified, with a range oscillating between 9 and 85 days, depending on the year and plot analysed. Range values suggested no clear differences between both managements for the stabilisation of N_2O emissions.

On the other hand, the dependence relations are, in most of the cases, greater than 25% indicating that a moderate or low relation exists between those data. This might be due to the large variation in N_2O fluxes from soil during the year and might indicate the necessity of studying this variable at a smaller temporal scale.

In general, conditional simulation proved to be a good estimator of the N_2O emission general behaviour in time for those cases when the nugget effect of the model was low or moderate. High range values and low-magnitude nugget effects permitted to obtain good estimations using Gaussian conditional simulation.

A detailed study of the N_2O emission variations during shorter periods of time than a day might improve the knowledge of this variable.

Acknowledgments The stay in Guelph of Dr. Mirás Avalos was supported by Xunta de Galicia (Spain).

References

Baggs EM, Richter M, Hartwig UA, Cadisch G (2003) Nitrous oxide emissions from grass swards during the eighth year of elevated atmospheric pCO_2 (Swiss FACE). Global Change Biol 9:1214–1222

Benckiser G, Eilts R, Linn A, Lorch HJ, Sumer E, Weiske A, Wenzhofer F (1996) N_2O emissions from different cropping systems and from aerated, nitrifying and denitrifying tanks of a municipal waste water treatment plant. Biol Fertil Soils 23:257–265

Cambardella CA, Moorman TB, Novak JM, Parkin TB, Karlen DL, Turco RF, Konopka AE (1994) Field-scale variability of soil properties in central iowa soil. Soil Sci Soc Am J 58:1501–1508

Chilés JP, Delfiner P (1999) Geostatistics. Modeling Spatial Uncertainty. Wiley Series in Probability and Statistics. John Wiley & Sons, Inc New York, p 695

Christakos G (2000) Modern spatiotemporal geostatistics. International association for mathematical geology. Studies in Mathematical Geology No. 6. Oxford University Press, p 288

Goovaerts P (1997) Geostatistics for natural resources evaluation. Applied geostatistics series. Oxford University Press, p 483

Mosier AR (1994) Nitrous oxide emissions from agricultural soils. Nutr Cycling Agroecosystems 37(3):193–200

Pebesma EJ (2000) Gstat User's Manual. Department Physical Geography. Utrecht University, p 100

Stacey KF, Lark, RM, Whitmore, AP, Milne AE (2006) Using a process model and regression kriging to improve predictions of nitrous oxide emissions from soil. Geoderma, 135:107–117

Tate RL (2000) Soil Microbiology. Second Edition. John Wiley & Sons, Inc New York p 508

Ventera RT, Groffman PM, Verchot LV, Magill AH, Aber JD (2004) Gross nitrogen process rates in temperate forest soils exhibiting symptoms of nitrogen saturation. Forest Ecol Manage 196(1):129–142

Wagner-Riddle C, Thurtell GW, Kidd GK, Beauchamp EG, Sweetman R (1997) Estimates of nitrous oxide emissions from agricultural fields over 28 months. Can J Soil Sci 77(2):135–144

Wagner-Riddle C, Thurtell GW (1998) Nitrous oxide emissions from agricultural fields during winter and spring thaw as affected by management practices. Nutr Cycling Agroecosystems 52:151–163

Zinc Baseline Level and its Relationship with Soil Texture in Flanders, Belgium

T. Meklit, M. V. Meirvenne, F. Tack, S. Verstraete, E. Gommeren and E. Sevens

1 Introduction

The total concentration of Zn in soil is composed of a baseline concentration and the superimposed contamination, referred as pollution. Lark (2002) defined the geochemical baseline concentration as a continuous background level that is composed of a native metal content in the soil parent material and diffused sources of pollution. The baseline level of an element is bound to time and area and it does not necessarily reflect the "natural" background level (Kabata-Pendias and Pendias 1984). Tack et al. (2005) described the baseline level as the concentration that is commonly found in the majority of the soils in a study area. Due to the dynamics of the soil environment and the anthropological influences, it is difficult to derive the true "natural" background level of an element. However, it is possible to determine the concentration level that is found most commonly in the study area and use it as a reference value for pollution studies.

In the frame of soil pollution studies, the Public Waste Agency of Flanders, Belgium (OVAM) has been collecting georeferenced soil samples. Over five years a wealth of information was obtained, including about 50,000 observations of soil Zn content. These data were used to determine the range of Zn baseline concentration in the top 50 cm of soil in Flanders, and to map its distribution. Finally an attempt was also made to link the baseline map with other soil properties.

2 Study Area and Data Set

This study covers the entire region of Flanders (13,677 km^2), the northern part of Belgium. The soils are dominantly developed in eolian or marine sediments of Holocene and Pleistocene age (Van Meirvenne and Van Cleemput 2005). The

T. Meklit
Department of Soil Management and Soil Care, Ghent University, Coupure 653, 9000 Gent, Belgium
e-mail: meklit.tariku@ugent.be

A. Soares et al. (eds.), *geoENV VI – Geostatistics for Environmental Applications*, 373–383

northern part of the region is dominated by acid, humus rich sandy soils. Finer wind-blown sediments were deposited in the southern, more elevated, parts resulting in loamy and silty soils, originally lime rich, but now decalcified in the top layers. The polder and river alluvial areas contain water-deposited clayey soil.

Between 1996 and 2005, OVAM has developed a large database on soil heavy metal concentrations, including zinc. In addition to the geographical coordinates, every sample was characterized by its sampling depth and date. Since the sampling depth varied and since there were sometimes multiple sampling at the same location, the database had to be screened carefully. For environmental reasons related to soil use, we targeted the top 50 cm of soil. Our selection criterion was: only records with an average sampling depth (average of upper and lower sampling depths) located within the top 50 cm and an upper sampling depth within the top 20 cm were retained. The latter criterion was added because if soil is polluted, most of it will be concentrated in the top layer. When there were multiple samples taken at different sampling dates but with identical coordinates, the most recent observation was retained.

3 Theoretical Background

3.1 Identification of the Baseline Range

Measurements that are considered to belong to the baseline data can be identified based on the statistical distribution of the whole data set when plotted on a normal probability plot (Matschullat et al. 2000). In pollution studies it is custom to transform the data logarithmically first. On such a probability plot, a normal distribution appears as a straight line and a deviation from this linearity is an indication for the presence of multiple populations in the data set (Chambers et al. 1983). A threshold value is identified by removing larger data values until the smallest absolute skewness of the remaining smaller values is obtained (Tack et al. 2005). The point where the data are found to be least skewed is then taken as the threshold value to define the upper limit of the baseline data. The larger and more deviating values can be considered to be a result of human induced additions, like pollution.

3.2 Simulations

Geostatistical simulation focuses on the reproduction of the spatial variability by drawing multiple, equally probable realizations from a random function. A simulated value for every location is conditioned by not only the neighboring available data, but also with previously simulated values. It allows the assessment of the uncertainty of the prediction locally and jointly over multiple locations from a multiple of realizations that are conditioned by the locally available information (Goovaerts 2000). We selected Sequential Gaussian simulation (SGS) to analyze our Zn data.

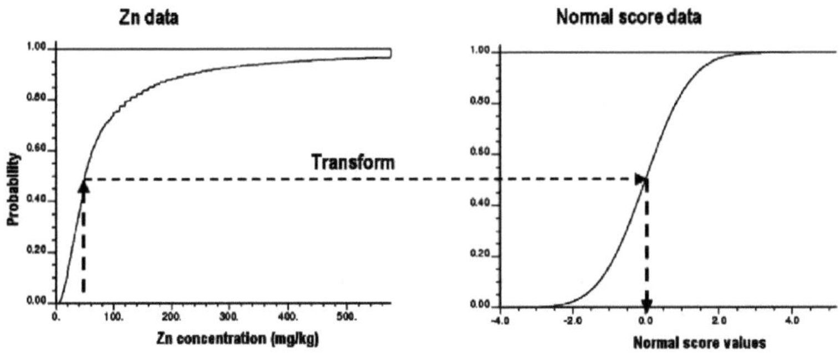

Fig. 1 Transformation of original z-values (Zn) into normal scores

The algorithm of SGS proceeds along a number of consecutive steps:

1. Normal score transformation of the data (Fig. 1) so that the global stationary histogram is Gaussian with mean zero and standard deviation one.
2. Model the variogram of these normal score transformed.

$$\gamma(\mathbf{h}) = \frac{1}{2N(\mathbf{h})} \sum_{\alpha=1}^{N(\mathbf{h})} \left\{ s(\mathbf{x}_\alpha) - s(\mathbf{x}_\alpha + \mathbf{h}) \right\}^2$$

Where:

$\mathbf{h}$ is the lag distance between the normal scores $s(\mathbf{x}_\alpha)$ and $s(\mathbf{x}_\alpha + \mathbf{h})$
$\gamma(\mathbf{h})$ is the variogram for distance lag $\mathbf{h}$
$N(\mathbf{h})$ is number of pairs separated by lag $\mathbf{h}$.

3. Perform L simulations of normal scores as follows, with "l" being the realization number ($l = 1, .., L$):

   - Define a random path visiting each of the m unsampled locations to be simulated only once.
   - At each unsampled location $\mathbf{x}_0$, estimate the parameters (mean and variance) of the Gaussian conditional cumulative distribution function (ccdf) by simple kriging using the model of the normal score variogram and the mean value of the normal scores. The conditioning information consists of n neighboring data of both original normal score data s($\mathbf{x}_i$) and $s^{(l)}(\mathbf{x}_0)$) simulated at previously estimated locations.
   - Randomly draw a simulated value $s^{(l)}(\mathbf{x}_0)$ from this ccdf, and add it to the data set.
   - Proceed to the next location along the random path, and repeat the two previous steps.
   - Loop until all m locations are simulated.
   - Proceed with the next simulation by repeating the previous steps, until all L realizations are available.

4. The results are finally back-transformed to the original variable space by applying the inverse of the normal score transform to the simulated s-values.

3.3 Zn Baseline Versus Soil Texture

The baseline level of an element in the soil is influenced by soil properties like soil pH, organic matter content, soil texture, and CEC (Metwally et al. 1993; White et al. 1997; Almas et al. 2001). Soil pH or organic matter content can vary importantly on a local scale, even within a single parcel while soil texture has often a regional pattern. Since the scale of this investigation is regional, soil texture is expected to have an influence on the Zn baseline level. The relationship between the Zn baseline and soil texture was therefore further investigated. First the correlation coefficient was calculated between Zn and silt and between Zn and sand on a point basis. A fuzzy set map comparison was also conducted for Zn and silt and Zn and sand maps using the Map Comparison Kit (MCK) (Hagen, 2003; Visser and De Nijs, 2006). MCK enables to compare maps in accordance with intuitive criteria.

3.3.1 Map Comparison

The fuzzy kappa algorithm of MCK was used for a fuzzy set map comparison where maps were compared with two fuzzy similarity indices. The first one was the local similarity index or a similarity map and the other one is overall similarity index or a fuzzy kappa value ($\boldsymbol{K}_{Fuzzy}$) (Hagen et al. 2005).

Fuzzy set map comparison requires a cell to have a fuzzy representation. In a classical categorical maps cells are defined crisply; meaning a cell belongs fully to one category, but in the fuzzy representation every cell is defined by a degree of membership to a particular category. In order to go from a crisp to a fuzzy representation of a cell a map need to be interpreted in the way that every cell is partially defined by categories in its neighbourhood. This requires considering the size of the neighbourhood, listing down all the categories in the neighbourhood, the distance of every cell in the neighbourhood from the central cell of interest and defining a distance decay function. By combining all the above factors each cell in the neighbourhood is defined by the degree of membership that ranges between [0,1]. The membership contribution of cells was then combined using the union operation of fuzzy set theory which is simply the maximum of individual membership. Finally we obtain a map where every cell is represented by a fuzzy vector from which the above two fuzzy similarity indices were derived.

Local similarity index was obtained by two way cell-to-cell comparison; fuzzy vector of a cell on the first map to a category vector of the same cell in the second map and the inverse. The local similarity index (s_l) for that particular cell is then the minimum of (intersection rule of fuzzy set) the result of the two way comparison. The range of the similarity index is [0,1] where 0 is for total disagreement and 1 is for identical cells. If every cell is represented by the local similarity index value we obtain a similarity map where visual observation of areas of high and low similarities is possible.

The overall similarity index is represented by a fuzzy kappa ($\boldsymbol{K}_{Fuzzy}$) which is the average similarity overall cells was corrected for the expected over all similarity (Hagen et al. 2005).

$$S = \frac{\sum_{l=1}^{n} S_l}{nc}$$

s is average similarity over all cells, s_l is local similarity index and nc is total number of cells.

$$E = \frac{\sum_{l=1}^{n} E(S_l)}{nc}$$

E is expected over all similarity, $E(s_l)$ is expected local similarity, and nc is total number of cells.

$$\text{FuzzyKappa} = \mathbf{K}_{\text{Fuzzy}} = \frac{S - E}{1 - E}$$

4 Results

4.1 Baseline Zn Data

Initially, 49,607 soil Zn measurements were obtained from OVAM of which 12,646 samples met our criteria of depth of the soil. After declustering, accounting for a preferential sampling in areas with high Zn contents, the regional mean concentration was 170.7 mg kg^{-1} and the median 50.0 mg kg^{-1} (Fig. 2, left). The data ranged from 0 to 63,000 mg kg^{-1} and the standard deviation was 1155.4 mg kg^{-1} which is 6.7 times larger than the mean. Obviously the distribution is strongly positively skewed with 5% of the observations exceeding 600 mg kg^{-1}.

The natural logarithm of these Zn values was plotted on a probability plot (Fig. 2, right). This distribution shows two deviations from a straight line. The first deviation was related to the analytical detection limits (several analytical methods were used

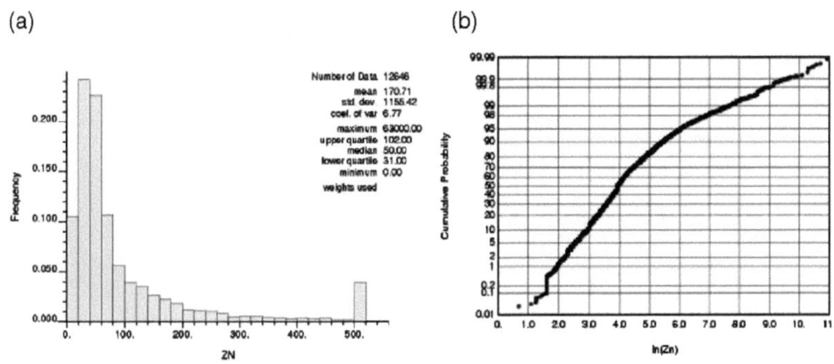

Fig. 2 Declustered histogram of Zn (a) and probability plot of ln(Zn) (b)

Table 1 Skewness for different ranges of data as defined by different upper limit

Upper limit (ln (Zn))	Data range (mg/kg)	Skewness
4.17	5 – 65	0.046
4.14	5 – 63	0.021
4.13	5 – 62	0.007
4.11	**5 – 61**	**0.002**
4.09	5 – 60	–0.013
4.08	5 – 59	–0.018
4.03	5 - 56	–0.047

by the many laboratories providing these data) which is most commonly taken at a Zn level of 5 mg/kg (corresponding to ln(Zn) = 1.6 (Tack et al. 2005). The second deviation was observed around ln(Zn) = 4.

In order to determine the threshold value that delineates the range of baseline measurements, the coefficient of skewness was calculated consecutively for different ranges of data by removing larger values (Table 1).

The minimum absolute skewness was obtained with an upper limit at ln(Zn) = 4.11, or at a Zn concentration of 61 mg/kg. Hence, the range of the data between 0 and 61 mg/kg was considered to define the Zn baseline for the top 50 cm of the soils in Flanders. The result obtained is almost identical to the baseline limit defined by OVAM for a so-called "standard soil" (containing 10% of clay and 2% organic matter), which was 62 mg/kg (Heymann & Smout 2001).

Data with values above the threshold of 61 mg/kg were found to be clustered in major cities and industrialized areas. Based on their location it is logical to assume a relationship between intensive human activity and elevated level of Zn. The variograms (Fig. 3) computed using ln-transformed values of the whole dataset, baseline and the pollution data, showed that the spatial autocorrelation which existed between the baseline measurements was obscured by the random nature of the pollution data.

4.2 Estimation of Baseline Zn

Our data set contained 6964 observations within the range 0–61 mg/kg which we will call "baseline Zn data". Figure 4 shows their locations and Fig. 5 shows the histogram of these data, as well as the variogram of their normal scores. The variogram

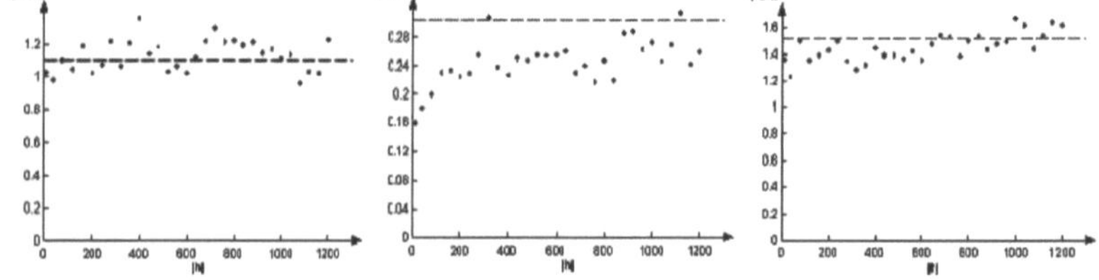

(a) (b) (c)

Fig. 3 Experimental variogram of the whole dataset (a), baseline data (b) and values above baseline data (c) (dashed line represent data variance)

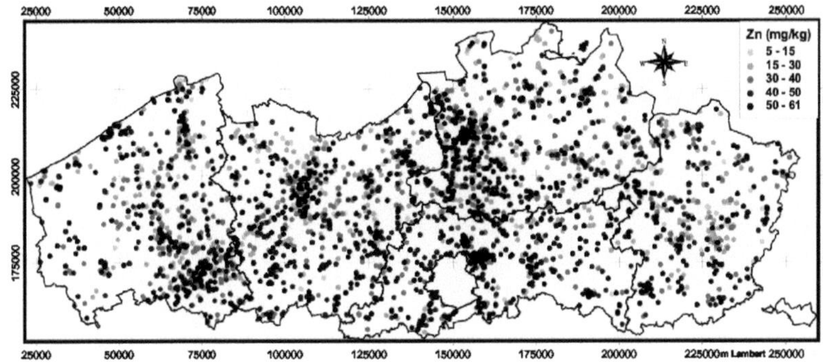

Fig. 4 Locations of the 6 964 baseline Zn data

was modeled by a nested exponential model containing a nugget and a short range (450 m) plus a long range (9500 m). The nugget accounts for 59% of the total variance, the first structure for 25%.

The baseline Zn content was simulated using 500 realizations from which, after back transformation, the E-type estimates were obtained. The map with estimated baseline Zn values (Fig. 6a) shows a regional pattern that is visually very similar to the map of soil textural classes (Fig. 6b). The northern part of Flanders, which is dominated by acid sandy soils, has generally low Zn baseline levels while larger estimates of baseline Zn were obtained in the southern part where more calcareous loamy and silty soils are dominant.

4.3 Map Comparison

The Pearson correlation coefficient between the predicted baseline Zn and the sand or silt fraction of the soil was 0.73 and -0.72 respectively (Fig. 7), indicating a fairly strong linear relationship.

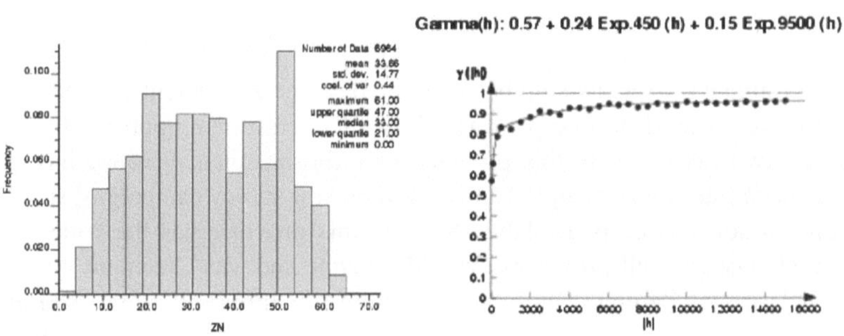

(a) (b)

Fig. 5 Histogram of the baseline Zn data (a) and variogram of the normal score baseline Zn data with fitted model (b)

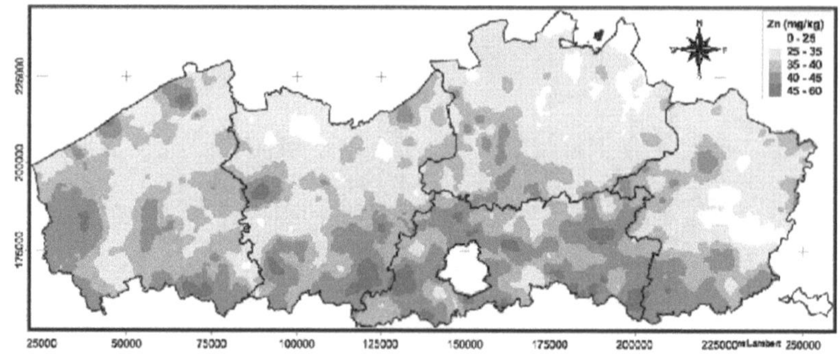

(a)

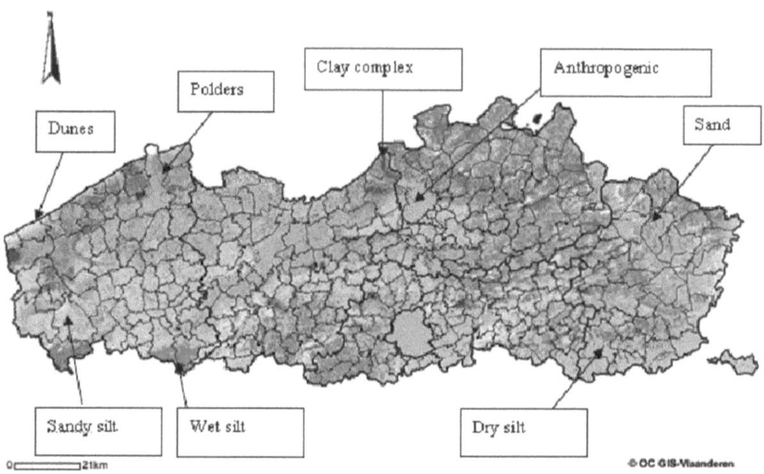

(b)

Fig. 6 Estimated baseline Zn (a) and soil texture (b) maps for the region of Flanders (East Flanders is the second province from the left)

To investigate the relationship between the baseline Zn content and the textural fractions, we focused on one province: East-Flanders (area: approx. 3000 km^2). There are two reasons why this province was chosen. First, because East Flanders has been intensively sampled during several soil survey campaigns; so a large databank of soil texture is available. Second, this province has the widest range of textural classes of all provinces (Van Meirvenne and Van Cleemput 2005). All textural classes of Belgium can be found within this province; which also means that the result obtained for this province can be extended to the other provinces.

Figure 8 shows maps of baseline Zn, silt and sand as obtained from MCK. While producing these maps, there were no reference values to decide on the range and number of classes to define a legend for every map. Therefore we used the scaling

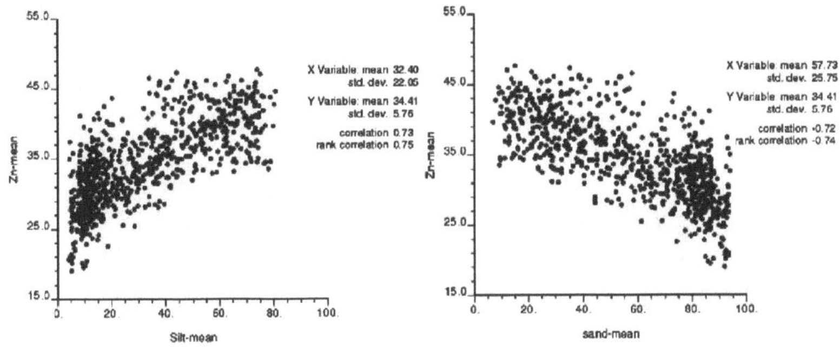

(a) (b)

Fig. 7 Scatter plots of the baseline Zn versus silt (a) and sand (b) content

method of MCK that gives the optimum number of classes as measured by the effectiveness of classification (user manual MCK, 2006).

To proceed with the map comparison, all maps have to be displayed with the same number of classes. Although all the three maps have equal numbers of classes, six, the values of the upper and lower boundaries of each class were different in every map so a numerical map comparison was not possible. For a categorical map comparison, numerical intervals were converted to categorical legends. For instance, for the first category a class name of "class1" was used to represent the interval of 16–25 mg/kg of Zn, 0–4% of silt and 3–10% of sand. The other classes were also transformed in a similar way. Then, a map comparison was conducted using the fuzzy kappa algorithm. The radius of neighborhood was set at 4 km and an exponential distance decay function was used. The resulted similarity maps for Zn and silt and Zn and sand are shown in Fig. 9.

In Fig. 9a high similarity index, ≥ 0.8 was obtained in the southern part. The soil in this area is silty and it has high Zn baseline level that indicated a strong spatial distribution similarity between high Zn baseline concentration and high silt content of the soil.

The legend of the Zn and sand comparison map (Fig. 9b) was obtained as one minus the similarity index, since Zn and sand have an inverse relationship. Except

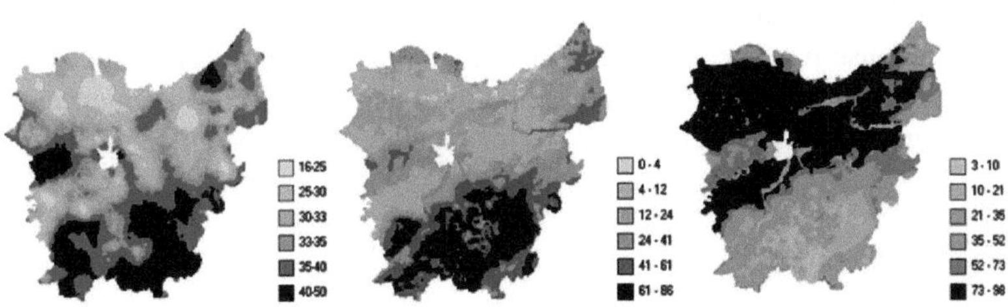

(a) (b) (c)

Fig. 8 Baseline Zn (mg kg^{-1}) (a), silt (%) (b) and sand (%) (c) of the province of East Flanders

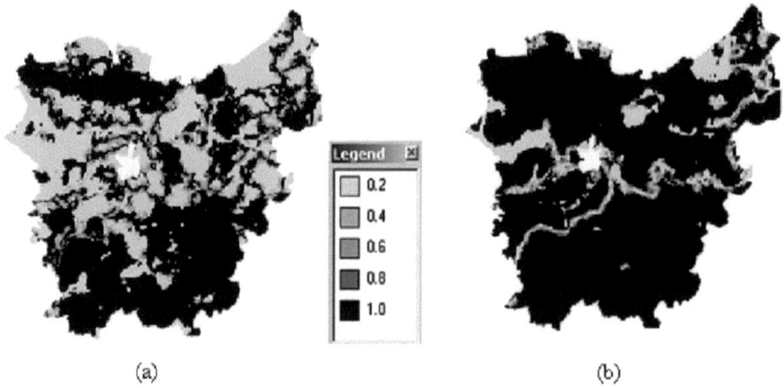

Fig. 9 Fuzzy similarity maps for Zn and silt (a); and Zn and sand (b). For map (b) legend has to be read as 1 minus similarity index

for areas with polder and clay soils, Zn and sand comparison yields high similarity index. The calculated K_{Fuzzy} value for Zn and sand maps was 0.96 while for Zn and silt it was 0.56 confirming a larger over all similarity between baseline Zn and sand distribution. From the result it was concluded that In Flanders sand fraction can be used as indication for the Zn baseline level.

5 Conclusions

The result of the geostatistical analysis of 6,964 point measurements from the region of Flanders indicated that the range of baseline level of Zn is between 0 and 61 mg kg^{-1}. The distribution of the baseline Zn shows a similar regional pattern with the soil textural map of the area. The level of baseline Zn was found to be low in sandy areas and it increase in the south where the soil is silty. As confirmed by large coefficients of correlations, Zn baseline content was strongly positively correlated with silt content of the soil and inverse relation was observed with sand content. Zn and silt as well as Zn and sand maps were further compared with MCK where a high local as well as overall similarity indexes were obtained for baseline Zn and sand content of the soil. The result suggested sand content to be used as an indicator for the baseline Zn concentration.

The relationship between baseline Zn and soil texture suggests differences either in geologically deposited level of Zn or in the Zn mobility as a consequence of soil texture related soil properties, such as pH for which further research is recommended.

References

Almas AR, Singh BR (2001) Heavy metals in the environment; plant uptake of cadmium-109 and Zinc-65 at different temperature and organic levels. J Environ Qual 30:869–877

Chambers, John, William C, Beatkleiner, Paul T (1983) Graphical methods for data analysis, Wadsworth

Goovaerts P (2000). Goovaerts P (2000) Estimation or simulation of soil properties? An optimization problem with conflicting criteria. Geoderma 97(2000):165–186

Hagen A (2003) Fuzzy set approach to assessing similarity of categorical maps. Int J Geogr Inf Sci 17(3):235–249

Hagen A, Straatman B, Uljee I (2005) Further developments of a fuzzy set map comparison approach. Int J Geogr Inf Sci 19:769–785

Heymann J, Smout L (2001) Milieuwetboek, Afval & Water. Kluwer Academic Publishers, pp 476–477

Kabata-Pendias A, Pendias H (1984) Trace elements in soils and plants. Boca Raton, Florida, CRC Press

Lark RM (2002) Modeling complex soil properties as contaminated regionalized variables. Geoderma 10:173–190

Map comparison kit 3-user manual (2006) The research institutes for knowledge systems (RIKS bv), Maastricht, The Netherlands

Matschullat J, Ottenstein R, Reimann C (2000) Geochemical background - can we calculate it? Environ Geol 39:990–1000

Metwally AI, Mashhady AS, Falatah AM, Reda M (1993) Effect of pH on zinc adsorbtion and solubility in suspensions of different clays and soils. J Plant Nutr Soil Sci 156:131–135

Tack FMG, Van Haesebroeck T, Verloo MG, Van Rompaey K, Van Ranst E (2005) Mercury baseline levels in Flemish soils (Belgium). Environ Pollut 134:173–179

Van Meirvenne M, Van Cleemput I (2005) Pedometrical techniques for soil texture mapping at a regional scale. In: Grunwald S. (ed) Environmental soil-landscape modeling.geographical information technologies and pedometrics. CRC Press, Taylor & Francis Group. Boca Raton, FL, USA, pp 323–341

Visser H, De Nijs T (2006) The map comparison kit. Environ Model Software 21:346–358

White JG, Welch RM, Norvell WA (1997) Soil zinc map of the USA using geostatistics and geographic information systems. Soil Sci Soc Am 61:185–194

Assessing the Quality of the Soil by Stochastic Simulation

A. Horta, J. Carvalho and A. Soares

Abstract For the assessment of soil quality, this chapter proposes a methodology relying on the stochastic simulation and co-simulation of relevant variables representing soil quality. Taking into consideration possible quality thresholds for each variable, we applied loss functions to the simulated maps and combined them to identify areas with different soil quality dynamics. This methodology was applied to an area in the southeast of Portugal, to the left margin of the Guadiana River, classified with a high susceptibility index to the desertification phenomenon.

Introduction

The Importance of Soil Quality Assessment

Soil quality has recently been defined as the "capacity of a specific kind of soil to function, within natural or managed ecosystems boundaries, to sustain plant and animal productivity, maintain or enhance water and air quality and support human health and habitation". In the context of the recent developments to produce a European soil policy, there has been an increased demand to establish criteria to determine soil quality and to develop indices that may be used to rank and compare the quality of soils at different locations (Rosa, 2005).

Soil quality is controlled by inherent soil properties governed by the factors affecting soil formation. Its indicators refer to measurable soil attributes that influence the capacity of soil to produce crops or perform its environmental functions (Arshad and Martin, 2002).

Whilst it is fairly straightforward to quantify the fitness of air for breathing and the fitness of water for drinking, it is much more difficult to identify similar criteria to be applied to soil. In part, this arises because of the wide range of uses to which soils are put and in part because of the complexity of the soil and the possibility

A. Horta
Environmental Group of the Centre for Modelling Petroleum Reservoirs – CMRP, Instituto Superior Técnico Technical University of Lisbon, Portugal
e-mail: ahorta@ist.utl.pt

A. Soares et al. (eds.), *geoENV VI – Geostatistics for Environmental Applications*, 385–396

that changes in the soil may be slow and may occur only when some threshold is reached (Nortcliff, 2002).

Values for a selected soil indicator must be maintained between critical limits for normal functioning of a healthy soil ecosystem. Selection of critical limits for soil quality indicators poses several difficult problems (Arshad and Martin, 2002). Soils frequently perform several functions simultaneously and so the critical or threshold values for each property will probably vary depending upon function (Nortcliff, 2002).

However, instead of establishing a fixed critical threshold, one can assess soil quality through an uncertainty evaluation of the spatial variability of a relevant soil property. This can be achieved through the application of geostatistical methods, commonly used for the characterization of soil properties (Webster and Oliver, 2001).

Objectives

The aim of this chapter is to present a new methodology, based on geostatistical models, for the assessment of soil quality. This can be achieved through the development of a tool for the identification of areas susceptible to a decrease in soil quality. The location of susceptible areas is important for planning remedial actions for preventing critical situations of soil degradation.

With this objective, we used stochastic simulation and co-simulation to generate images of the spatial distribution of soil quality indicators in the study area. Each simulated image is evaluated using loss functions to identify areas with different soil quality dynamics. Loss functions are employed to quantify the decrease of quality when there is a variation in the content of the variables chosen as indicators.

Application

The methodology presented in this chapter was developed within a project framework that aims to characterize soil quality susceptibility in an area classified as affected by the desertification phenomenon (Rosário, 2004). This area is located in the southeast of Portugal, to the left margin of the Guadiana River (Alentejo region), and has 290,000 ha (Fig. 1).

The soil sampling strategy was divided into the following steps:

i. An initial campaign to collect 100 soil bulk samples (0.2-m to 0.7-m support) in the entire study area (Fig. 1) to evaluate the horizontal variability of soil quality indicators.
ii. A second campaign to collect samples in each layer of the soil profile in specific locations to evaluate vertical variability of soil quality indicators (these specific locations are the areas susceptible to a decrease in soil quality).

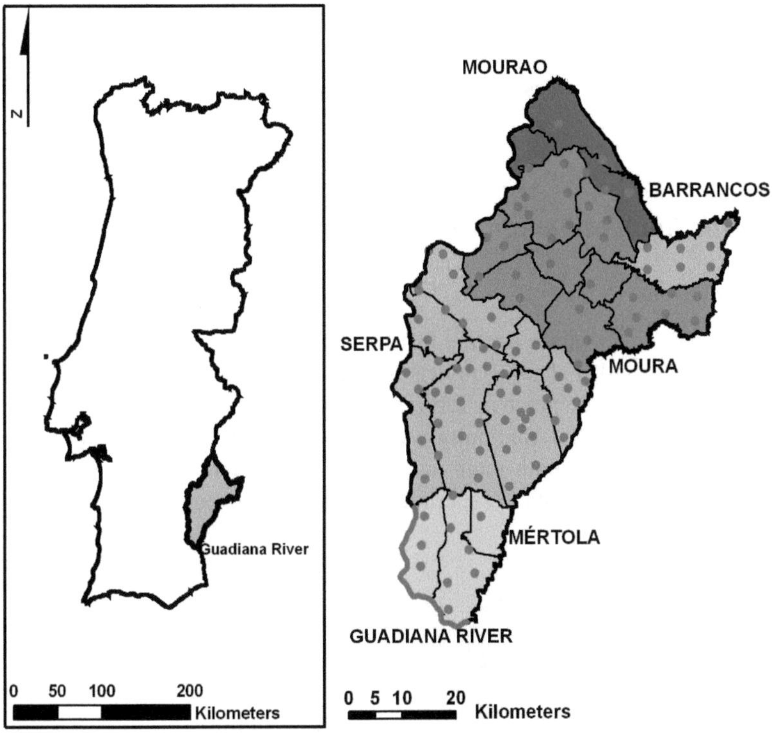

Fig. 1 Study area and first sampling data points distribution

This chapter concerns the geostatistical analysis of the first campaign results to identify critical areas based on two different quality indicators: pH and total organic carbon (TOC). These soil properties have been referred to as chemical indicators to measure soil quality. Soil pH influences most chemical and biological activity thresholds and determines nutrient availability. TOC is correlated with the amount of organic matter in the soil (Rowell, 1994; Costa, 1999).

Methodology

To assess soil quality, the following methodological steps were implemented:

- Stochastic simulation and co-simulation of the variables related to soil quality (pH and TOC) using direct sequential simulation and co-simulation (Soares, 2001).
- Application of impact loss functions to the simulated maps of soil properties so as to obtain the responses related to a change in soil quality.

- Identification of areas with a different soil quality dynamic for a specific indicator by comparing the different responses (feedback scenarios).
- Combination of the areas susceptible to a decrease in soil quality obtained for each indicator.

Stochastic Simulation and Co-simulation

Spatial Analysis

Sample location map values and the histograms obtained for pH and TOC (%) are shown in Fig. 2 (darker shading indicates larger concentrations). Main descriptive statistical parameters are given in Table 1.

Concerning the spatial continuity, variograms calculated for pH and TOC are presented in Fig. 3 (for both main and minor directions). We used the spherical model for the adjustment of the experimental curve and an angular tolerance of 20°.

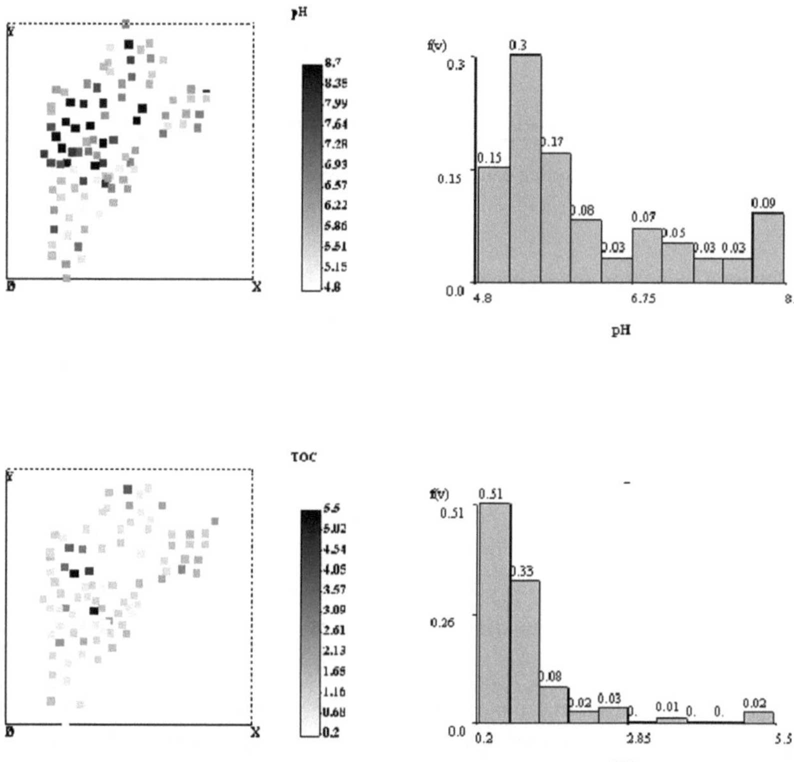

Fig. 2 Samples location map (darker shading indicates larger concentrations) and histogram for pH and TOC

Table 1 Descriptive statistical data

Variables	Mean	Var	Standard Deviation	Coefficient of Variation	Min	Max	Median	S
pH	6.12	1.25	1.12	0.18	4.8	8.7	5.7	1.01
TOC	0.92	0.69	0.83	0.9	0.2	5.5	0.72	3.43

The anisotropy relation is 1.3 for pH and 1 for TOC. Concerning pH, the range obtained is about 12 km whereas for TOC it is 9 km.

Direct Sequential Simulation and Co-simulation

Direct sequential simulation (DSS) and co-simulation (CODSS) are geostatistical stochastic simulation algorithms proposed by Soares (2001). In general terms, DSS enables the simulation of untransformed continuous variables, using simple local kriging estimates of the variable's mean and variance to sample from the global cumulative distribution function. In turn, CODSS enables joint simulation of several variables considering their local correlation.

Although global correlation between variables is about 0.4 there is a strong local correlation between pH and TOC. Therefore, we used local moving windows to calculate local correlation coefficients (Pereira et al., 2000; Carvalho et al., 2006). DSS was used to generate a set of 30 simulated images of pH, reproducing the spatial variability as it was revealed by the variograms and histogram. Afterwards, CODSS was used for the joint simulation of 30 equivalent pairs of images of TOC based on a secondary image of pH. Co-simulation was based on hard data (100 values corresponding to the analytical results obtained for each variable), secondary information given by the

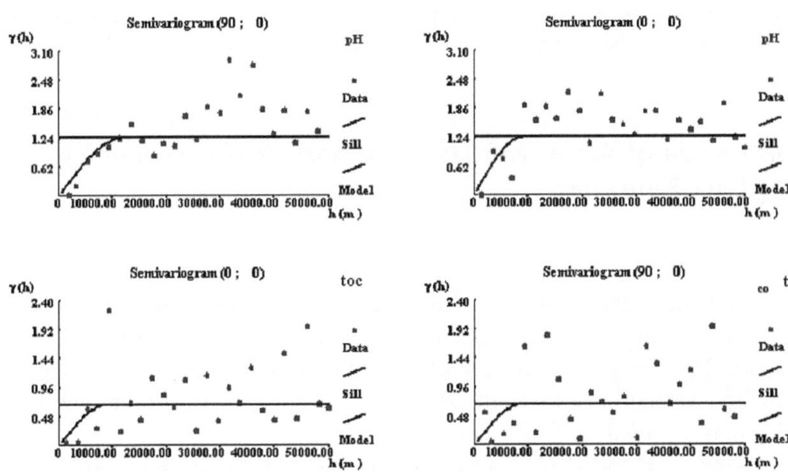

Fig. 3 Experimental variograms for pH and TOC (on the left: main direction, on the right: minor direction)

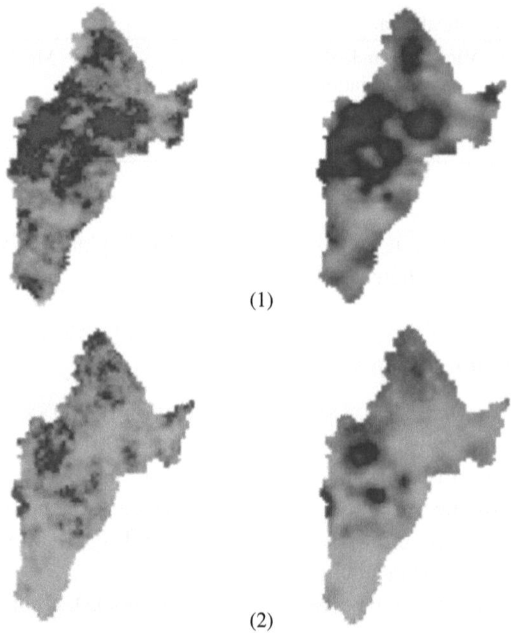

Fig. 4 Simulated image of pH (1) and TOC (4) and the mean of simulated images (right image) (darker shading indicates higher concentrations)

previous simulated images of pH and the local correlation map. For the DSS and CODSS the study area was considered to be a 1000 m by 1000 m grid.

Figure 4 shows an example of the simulated images and the mean simulated maps obtained using DSS (pH) and CODSS (TOC). Figure 5 depicts the local correlation obtained for this pair of variables. For the set of 30 simulations of pH and TOC the histogram and the variogram of the simulated images have matched those of experimental data.

Identification of Areas with Different Soil Quality Dynamics Using Loss Functions

Loss Functions

The methodology defined for identifying areas of poor quality is based on the development of loss functions to be applied to the simulated images of different quality indicators. The impact loss function applied to a given indicator, in the context of soil quality, depends on a given threshold value (t) that represents the limit below which soil is regarded as of inferior quality.

To evaluate the impact of a given soil property z, at the spatial location x, the concept of loss function $L^i(x)$ of $z^i(x)$ (i refers to the simulated image i) was introduced and modelled as follows:

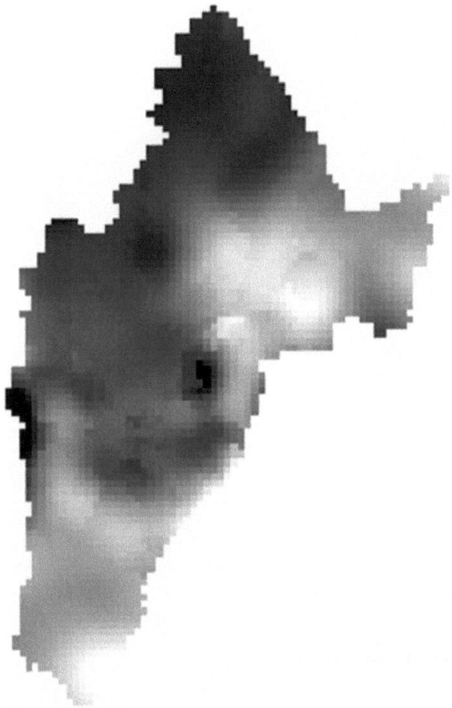

Fig. 5 Local correlation between pH and TOC (darker shading indicates local correlations above 0.7)

$$L^i(x) = \begin{cases} 0, \; if z^i(x) \geq t \\ \dfrac{|z^i(x) - t|}{z^i_{max} - z^i_{min}} \times w, \; if z^i(x) < t \end{cases} \quad (1)$$

where $z^i(x)$ is the simulated image i of soil property z, t is the threshold chosen for the soil property z, z^i_{max} and z^i_{min} are, respectively, the maximum and the minimum values for soil property z in the simulated image i and w is a positive constant.

As shown in Equation 1, the loss is linearly proportional to the negative deviations of z from t and enhanced by w that could be, for example, the cost associated to the loss of soil productivity due to a lack of quality. The application of the proposed loss function to the set of simulated images Z^i is schematically represented in Fig. 6 and the resulting total loss is obtained as follows:

$$L(x) = \sum_{i=1}^{N_s} L^i(x) \quad (2)$$

where N_s is the number of loss maps obtained for the set of simulated images $z^i(x)$.

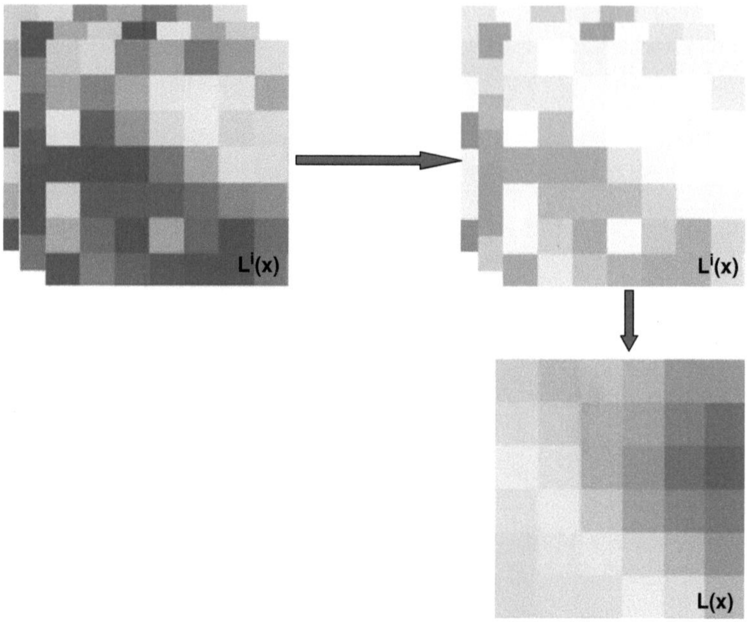

Fig. 6 General application of loss functions

Feedback Scenarios

The application of loss functions will enable us to obtain "feedback scenarios", the designation adopted for the total loss maps $(L(x))$ for a specific t value. To ascertain the pattern of soil quality dynamics, we used two different feedback scenarios: a "best case scenario" $(L_{BCS}(x))$ and a "worst case scenario" $(L_{WCS}(x))$, which resulted from the application of the loss function using two different thresholds $t_1(L_{BCS}(x))$ and $t_2(L_{WCS}(x))$, where $t_2>t_1$. To ascertain the pattern of soil quality dynamics, we used two different feedback scenarios: a "Best Case Scenario" $(L_{BCS}(x))$ and a "Worst Case Scenario" $(L_{WCS}(x))$, that resulted from the application of the loss function using two different thresholds $t_1(L_{BCS}(x))$ and $t_2(L_{WCS}(x))$, where $t_2 > t_1$.

As indicators of soil properties, pH and TOC values are a decreasing function of the intensity of quality degradation; hence a higher content is considered an indicator of higher soil quality (Rowell, 1994; Costa, 1999). Therefore, by using a maximum and a minimum t value we expect to quantify the loss of quality through the comparison between the mentioned feedback scenarios.

Soil Quality Thresholds

It is difficult to identify a unique threshold representing the limit from which there is a decrease in soil quality. In this chapter, we have considered a set of critical values to "measure" a relative trend in the soil quality dynamics.

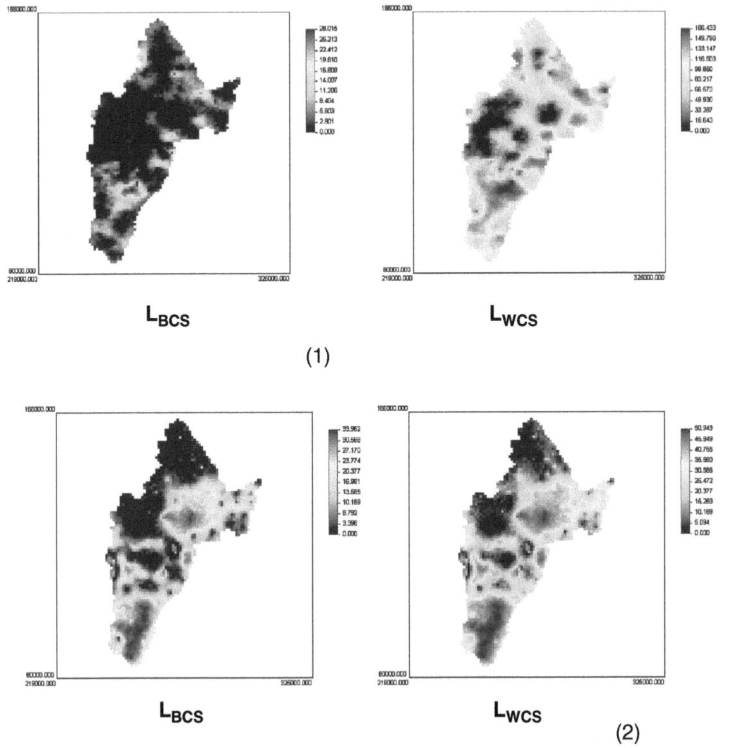

Fig. 7 Feedback scenarios for pH (1) and TOC (3) (darker shading indicates bigger loss values)

For an agro-ecosystem, critical limits will vary from one indicator to another. For some indicators, a 10% increase or decrease may be significant while others may not be affected by a 20% decline. For organic matter, a 15% increase or decrease over the average or baseline value seems reasonable to use as a critical limit (Arshad and Martin, 2002).

Based on this statement, we choose a set of thresholds for each variable considering a 15% variation of the mean value calculated for pH and TOC. Hence, the minimum (t_1) and the maximum (t_2) thresholds used were, respectively, for pH ($t_1 = 5.2$; $t_2 = 7$) and for TOC, ($t_1 = 0.8$; $t_2 = 1.1$). The $L_{BCS}(x)$ and the $L_{WCS}(x)$ obtained for pH and TOC are presented in Fig. 7.

Areas Susceptible to a Decrease in Soil Quality

For the purpose of this chapter, soil quality assessment is achieved through the development of a tool for the identification of areas susceptible to a decrease in soil quality. To obtain these areas we computed the difference between the $L_{BCS}(x)$ and the $L_{WCS}(x)$ as follows:

$$e(x) = L_{WCS}(x) - L_{BCS}(x) \quad (3)$$

This residual $e(x)$ is a measure of the susceptibility to a decrease of soil quality so that the susceptible areas are the ones with $e(x) \neq 0$ and higher values of $e(x)$ are representative of areas with higher susceptibility. The last statement assumes that a higher variability between the loss function values of the two specific feedback scenarios implies a change of quality state. The variation of loss values is directly dependent on the simulated values for the chosen soil quality indicator and on the set of thresholds.

Moreover, if $e(x)$ is equal to zero then at a certain location x, $z^i(x)$ is always higher than the thresholds used to produce the $L_{BCS}(x)$ and the $L_{WCS}(x)$. Thus, soil quality is not affected and this location can be classified as of good quality for the chosen indicator.

But, $e(x)$ is also equal to zero if the calculated values of $L_{BCS}(x)$ and $L_{WCS}(x)$ are equal. In this situation, it is not possible to assess the degree of possible quality degradation.

Results and Discussion

The areas susceptible to a decrease in soil quality obtained using pH and TOC as soil quality indicators are presented in Fig. 8.

We also combined these maps to obtain an average soil quality susceptibility for the study area as shown in Fig. 9.

According to the results obtained, the distribution of the susceptible areas follows almost the same geographic pattern for each indicator. Also, the final set of sensitive areas combining the results of the two soil variables does not show a predominant influence of a particular one, mainly because of the method used for combining the two images in the final quality map. This result can be improved if there is

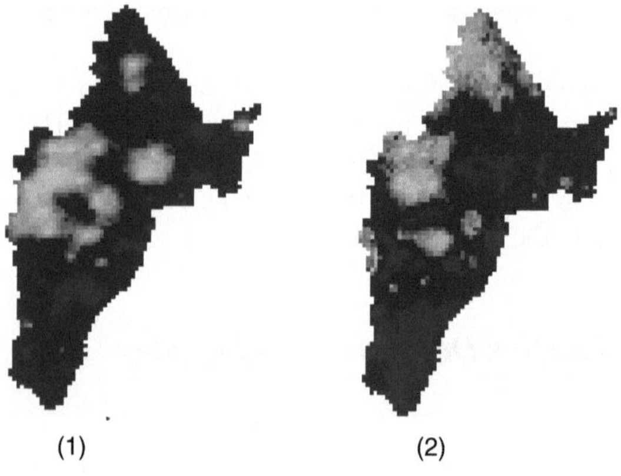

Fig. 8 Areas susceptible to a decrease in soil quality (darker shading) obtained for pH (1) and TOC (2)

Fig. 9 Areas susceptible to a decrease in soil quality (darker shading) using pH and TOC as soil quality indicators

a weighted contribution of each indicator to compute the final map. To do this, it is necessary that more information about the soil indicators be obtained, which is expected in the second stage of experimental work.

In conclusion, we consider that the methodology developed to assess soil quality has shown good results for the identification of susceptible areas due to the combination of geostatistical methods and impact loss functions. Concerning the application presented in this chapter, the results obtained will have to be validated further using additional field information and data concerning other soil quality indicators.

Acknowledgments This chapter was produced in the context of Project "CIDMEG" (POCI/CLI/58865/2004), financed by FEDER through the national Operational Science an Innovation Program 2010 with the support of the Foundation for Science and Technology.

References

Arshad M, Martin S (2002) Identifying critical limits for soil quality indicators in agro-ecosystems, Agriculture, Ecosystems and Environment 88:153–160

Carvalho J, Soares A, Bio A (2006) Improving satellite images classification using remote and ground data integration by means of stochastic simulation. Int J Remote Sens 16:3375–3386

Costa J (1999) Caracterização e constituição do solo, 6ª edição, (Edições Gulbenkian, Lisboa)
de la Rosa D (2005) Soil quality evaluation and monitoring based on land evaluation. Land degradation & Development, 16:551–559
Nortcliff S (2002) Standardisation of soil quality attributes, Agriculture, Ecosystems and Environment 88:161–168
Pereira MJ, Soares A, Rosário L (2000) Characterization of forest resources with satellite SPOT images by using local models of co-regionalization. In: Kleingeld WJ, Krige DG (eds) Geostatistics 2000 Cape Town (Kluwer Academic Publishers, The Netherlands), 1:581–590
Rosario, L., 2004, Indicadores de Desertificação para Portugal, (Direcção-Geral dos Recursos Florestais, Ministério da Agricultura, Desenvolvimento Rural e Pescas).
Rowell D (1994) Soil science methods and applications (Addison Wesley Longman Limited, England)
Soares A (2001) Direct sequential simulation and cosimulation. Math Geol 33(8):911–926
Webster R, Oliver M (2001) Geostatistics for environmental scientists (John Wiley & Sons, Chichester)

Interpolation of Soil Moisture Content Aided by FDR Sensor Observations

K. Vanderlinden, J.A. Jiménez, J.L. Muriel, F. Perea, I. García and G. Martínez

Abstract Automated soil moisture sensors are widely used in agronomy and environmental sciences. However, to sense properly the amount of water retained in the soil matrix, a calibration with gravimetrically observed values is required. This is laborious and not appropriate when the soil physical properties change during the wetting and drying cycles due to swelling and shrinkage, respectively. The objective of this study was to analyze how gravimetric soil moisture, θ, series can be interpolated from a minimum number of field observations, using daily automated sensor output as secondary information. Weekly observed soil moisture data from three depth intervals, and daily data from four sets for FDR sensors in two adjacent plots subject to different tillage practices were used to validate the method. A variogram analysis showed that soil moisture was more persistent in time and with depth in no-tillage and that at least two scales of temporal variation could be distinguished. Kriging with an external drift produced the largest model efficiency when using calibrated sensor measurements as secondary information.

Introduction

There exists an increasing interest in electromagnetic soil moisture monitoring devices for agronomical and environmental applications. These sensors respond to changing soil electromagnetic properties and measure the apparent dielectric permitivity of the soil, which depends to a large extent on the water content (Muñoz-Carpena et al., 2005). In this study a Frequency Domain Reflectometry (FDR) soil moisture sensing system was used, which contained multiple capacitance sensors, installed at different depths in a previously installed access tube. The low frequency at which these instruments operate make them sensible to clay and organic matter content, bulk density, salinity, and temperature (Paltineanu and Starr, 1997; Polyakov et al., 2005). In addition, the sampled volume is generally several orders of magnitude smaller

K. Vanderlinden
IFAPA, Centro "Las Torres-Tomejil", Junta de Andalucía. Ctra. Sevilla-Cazalla km 12.2, 41200 Alcalá del Río (Seville), Spain
e-mail: karl.vanderlinden.ext@juntadeandalucia.es

A. Soares et al. (eds.), *geoENV VI – Geostatistics for Environmental Applications*, 397–407

than the representative elemental volume, which means that small disturbances of the soil around the sensor or a bad soil-sensor contact may have a negative impact on the sensor performance (Evett and Parkin, 2005). For these reasons, numerous studies have shown that it is necessary to realize an independent calibration for each soil horizon against gravimetrically observed data (Paltineanu and Starr, 1997; Fares et al., 2004; Polyakov et al., 2005). This procedure is completely impractical and imposes severe limitations on the use and spread of these sensors for non-research applications. In addition, it seems inadequate under conditions where the soil physical properties change during the successive drying and wetting cycles, due to expansion or contraction of the soil matrix, as occurs in the expansive clay soils in SW Spain, which are of great importance for local dry-land farming.

When sensor performance can not be guaranteed and sensor monitoring has to be accompanied by gravimetric soil moisture determinations, geostatistical interpolation methods can be used to interpolate these data, using the sensor measurements as secondary information. Snepvangers et al. (2003) and Jost et al. (2005) analysed the soil moisture content evolution in a spatio-temporal framework. In both studies, spatio-temporal kriging with an external drift provided accurate estimates of soil moisture, using net-precipitation and water balance modelled soil moisture as secondary variables.

In this study a geostatistical methodology is proposed to estimate the daily gravimetric soil moisture content from weekly observations, during the growing period of a sunflower crop, as an alternative to the traditional use of soil moisture sensors. We compared several procedures for incorporating FDR sensed soil moisture or measured frequency in the estimation as secondary information, with special interest for those that do not require a previous calibration. The method is used to compare the soil water dynamics within the soil profile of a Vertisol subject to no-tillage and conventional tillage management.

Materials and Methods

Moisture Sensing and Measurement Methodology

The soil moisture data were obtained at the Tomejil experimental farm (37° 24' 07" N, 5° 35' 10" W) in Carmona, SW Spain. The soil is classified as a Typic Haploxerert (Soil Survey Staff, 1999), with clay, silt, and sand contents of 67, 23 and 10%, respectively. At the farm, the agronomic and environmental consequences of three different tillage systems are compared in a field experiment that was established more than 20 years ago. Within two adjacent plots of 15 $\times$180 m, subject to no-tillage (NT) and conventional tillage (CT) practices, a system of four multi-sensor soil water monitoring probes (EnviroSCAN, Sentek Sensor Technologies, Stepney, Australia) was installed. These sensors use Frequency Domain Reflectometry (FDR) to measure soil moisture content. Each probe (NT1, NT2 in NT and CT1 and CT2 in CT) comprised five sensors placed at depths of 10, 20, 30, 60 and 90 cm.

A customized data logging system transformed the observed frequency (in soil), F_s, to a scaled frequency:

$$SF = (F_a - F_s)/(F_a - F_w), \qquad (1)$$

where F_a and F_w are the observed frequencies in air and water. The SF is then related with gravimetric soil moisture through the following empirical relationship using non-linear regression (Paltineanu and Starr, 1997; Baumhardt et al., 2000; Fares et al., 2004):

$$\theta_s = \mathrm{a} + \mathrm{b}\ \mathrm{SF}^{\mathrm{c}} \qquad (2)$$

The calibration parameters a, b, and c were estimated individually for each sensor, yielding a total of 20 parameter sets. Hourly measurements were daily averaged and stored on a customized data logger.

Between 12 Dec. 2003 and 15 July 2004, disturbed soil samples were weekly collected and analysed in the laboratory for θ. On each of the 24 sampling dates, three samples were taken in each plot from the 0–20, 20–40, and 40–60 cm horizon, nearby (< 3m) the two probes, yielding a total of 426 θ values (on the first sampling date only two samples were taken from the three horizons).

On 15 April 2004 sunflower was sown on both plots. Harvest took place on 30 Aug. 2004.

Geostatistical Methodology

Temporal experimental variograms of θ were calculated for each horizon (0–20, 20–40 and 40–60 cm) and for the entire profile (0–60 cm), for NT and CT. Since only data from 24 sampling dates were available (which is clearly insufficient to calculate a representative variogram), the three θ values measured at each date were considered as if they were obtained at slightly different moments in time (± 0.1 day). Soil moisture variations at this time scale are very small in comparison to the spatial variability of θ. Since the thermo-gravimetric method is destructive, no consecutive measurements can be obtained at the same point. This means that the nugget effect of the temporal variograms accounts for small-scale spatial variability and procedural errors associated with the termo-gravimetric method.

The experimental variograms were calculated using the traditional equation (Goovaerts, 1997):

$$\gamma^* (h) = \frac{1}{2N(h)} \sum_{i=1}^{N(h)} (\theta(d_i) - \theta(d_i + h))^2, \qquad (3)$$

were $\theta(d_i)$is the gravimetric soil moisture content on day d_i. All variograms were calculated using the VARIOWIN software (Panatier, 1996).

Several methods for incorporating secondary information into the interpolation of a primary variable are well-known in geostatistics (Goovaerts, 1997). In the context of this work this secondary information may take several forms: (i) sensor measured gravimetric soil moisture content, θ_s, (using Eq. (2)), (ii) *SF*, or (iii) F_s. The first option can be readily used as local mean in simple kriging with local varying means (SKlm):

$$\theta^*_{SKlm}(d_0) = \theta_s(d_0) + \sum_{i=1}^{n} \lambda_i^{SKlm} \left(\theta(d_i) - \theta_s(d_i)\right), \tag{4}$$

where $\theta_s(d_0)$ is the sensed gravimetric soil moisture content on a day for which a SKlm estimate is to be made, d_0, $\theta(d_i)$ and $\theta_s(d_i)$ are the laboratory determined and sensed gravimetric soil moisture content for $n/2$days (with laboratory observations) before and after d_0, and λ_i^{SKlm} are the SKlm kriging weights.

However, this procedure does not account for possibly changing relationships in time between θ and the sensor output due to changes in the soil physical properties (shrinking and expansion of the soil matrix) during successive drying and wetting cycles. In addition, a prior calibration according to Eq. (2) is still required, which means that this method will not provide any operational benefits as compared to the traditional sensor use. The second and third option for using secondary information involves kriging with an external drift (KED) and do not require a prior calibration, so that the interpolation can be started with a minimum number of observations. This procedure seems, at least theoretically, more appropriate for this application since it estimates the local means, evaluating at each moment in time the relationship betweenθ and the sensor output (*SF* or F_s):

$$\theta_s\,(d) = a_o\,(d) + a_1\,(d)\,y\,(d)\,, \tag{5}$$

where $y\,(d)$ is the secondary variable (*SF* or F_s) and $a_o\,(d)$ and $a_1\,(d)$ are two regression coefficients that are assumed constant within the search neighbourhood. In what follows, the mean θ_s of the two sensors in each plot is used as secondary information in SKlm and KED_S. In KED_1 and KED_2 the *SF* of sensors NT1 and CT1, and NT2 and CT2 is used, respectively. Block kriging was used in all cases to estimate daily soil moisture content, using the KT3D program from the GSLIB software library (Deutsch and Journel, 1998).

The performance of the different geostatistical procedures for producing soil moisture time series was evaluated using cross-validation, leaving out the data one at a time and re-estimating them by interpolation from the remaining data. Observed and estimated values were then compared using the model efficiency coefficient, E, a parameter that is widely used to evaluate the performance of hydrological models (Nash and Sutcliffe, 1970):

$$E = 1 - \frac{\sum_{i=1}^{n} \left(\theta^*(d_i) - \theta(d_i)\right)^2}{\sum_{i=1}^{n} \left(\theta(d_i) - \bar{\theta}\right)^2} \tag{6}$$

where $\theta(d_i)$,$\theta^*(d_i)$and $\bar{\theta}$are the observed, estimated and mean gravimetric soil moisture values, respectively. If the square of the differences between the estimates and the observations is as large as the variability in the observed data, then $E = 0$, and if it exceeds it, then $E < 0$. The efficiency of the traditional sensor use was calculated using the estimates made with Eq. (2), using the same calibration coefficients for each day, so it may be expected that the efficiency of the sensor estimates will be over-estimated as compared to the cross-validation estimates of the geostatistical procedures.

Results and Discussion

Exploratory Data Analysis

During the entire monitoring period significantly larger θ values were observed in the NT plot. The differences became smaller with depth. After calibrating each sensor according to Eq. (2) a similar pattern could be observed in the θ_s values. Figure 1 shows the evolution of rainfall, P, reference evapotranspiration, ET_0, θ for the 0–20 cm soil horizon, and two-sensor average θ_s and SF for the four sensors at 10 cm depth. Note the differences between the four SF series, indicating that the θ_s are only accurate if each sensor is calibrated individually according to Eq. (2).

The descriptive statistics in Table 1 show that the 71 θ observations in the upper soil horizon of the NT plot ranged from 12 to 38%, with a mean and variance of 28% and 38%^{2}, respectively. The θ values of the CT plot ranged from 12 to 36% with a mean and variance equal to 25% and 43%^{2}, respectively. The larger temporal variability of the soil moisture in the CT plot indicates that this tillage systems induces the soil to respond faster to external forcing (P and ET_0) and that it will be less capable of retaining the stored water. However it is not clear which part of this temporal variability is due to short-range spatial variability or variability due to the sampling and moisture determination method. The results of a one-way analysis of variance (Table 1) show that most of the variability occurs between groups of samples taken on different moments in time (B), even if samples from different depths are considered jointly. These results provide evidence of the higher, with depth decreasing, temporal variability of soil moisture in the CT plot. Only a very small proportion of the total variability is due to differences between observations taken on the same moment in time (W), but at points with slightly different coordinates or at different depths. No clear evolution with depth can be observed. The ratio of both sources of variability (F-value) decreases with depth, due to the decreasing

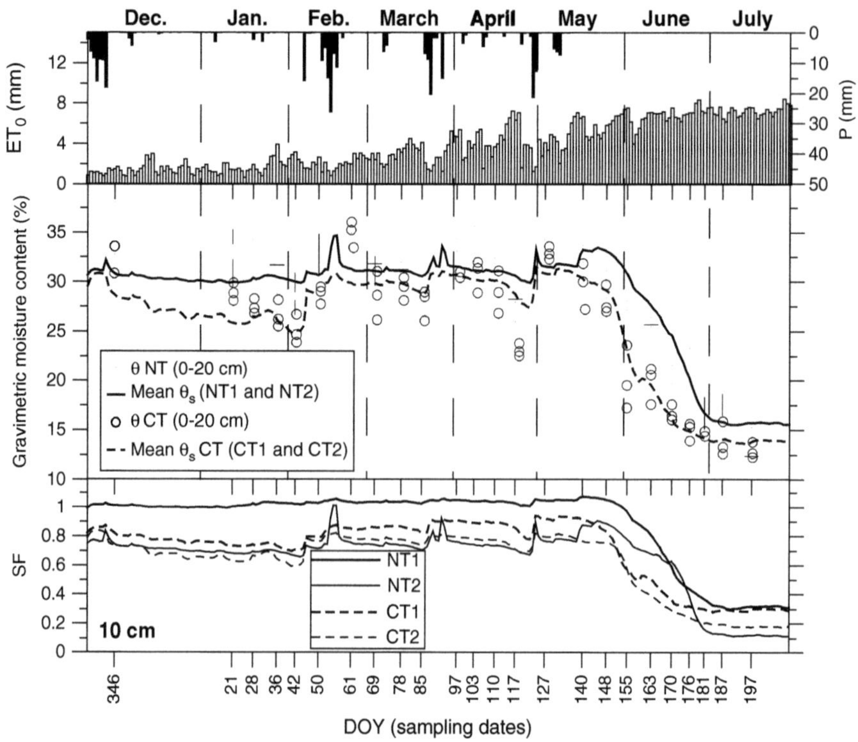

Fig. 1 Evolution of the observed, θ, and sensed, θ_s, gravimetric moisture content in the upper 20 cm of the soil in the no-tillage (NT) and conventional tillage (CT) plots. Scaled Frequency, SF, for sensors installed at 10 cm, rainfall, P, and reference evapotranspiration, ET_0

temporal variability, except for the 40–60 cm horizon in the CT plot, but is clearly higher in CT.

Highly significant ($p < 0.001$) Pearson correlation coefficients were obtained between SF and θ: 0.93 (NT1) and 0.88 (NT2) in NT, and 0.94 (CT1) and 0.95 (CT2) in CT.

Variography

The overall larger variability of the CT observations was also corroborated by the experimental variograms of θ for the different horizons (Fig. 2). In both plots, the upper soil horizon (0–20 cm) showed the largest variability for all the time lags, while the 20–40, the 40–60 and the averaged 0–60 cm horizons showed similar semivariance values, especially at the first lags, where the variability increased gently up to a time-lag of 10 days.

For CT an almost linearly increasing curve was obtained up to a time-lag of approximately 50 days, after which the curve levelled off. The variograms for NT

Table 1 Descriptive statistics for the observed gravimetric soil moisture content (%) in the no-tillage, NT, and conventional tillage, CT, plots for different soil horizons. The last three lines show the results of a one-way ANOVA

	NT					CT				
horizon	0–20	20–40	40–60	0–60	overall	0-20	20–40	40–60	0–60	overall
N	71				213	71				213
m	28.4	28.1	27.7	28.0	28.0	25.2	26.4	26.2	25.9	25.9
median	30.2	29.5	28.7	29.5	29.3	27.3	28.9	28.3	28.2	28.2
s^2	37.8	21.4	17.1	23.6	25.3	42.8	36.8	26.9	33.8	35.4
CV	0.22	0.16	0.15	0.17	0.18	0.26	0.23	0.20	0.22	0.23
min	12.4	14.7	15.5	14.2	12.4	12.4	14.7	15.7	14.4	12.4
max	38.1	34.3	34.7	34.5	38.1	35.8	33.8	32.3	33.0	35.8
range	25.7	19.6	19.2	20.3	25.7	23.4	19.1	16.6	18.6	23.4
B*	110.6	62.0	46.9	69.4	208.3	125.5	107.3	78.721	100.3	300.84
W	2.2	1.6	2.6	1.2	3.0	2.3	2.3	1.5	1.2	3.1
F	50.9	39.5	18.3	58.7	69.2	55.2	46.7	52.5	83.2	96.6

* B: variance between groups of samples taken on different moments in time, W: variance within groups of samples taken on the same moment in time.

showed different scales of variation for lags smaller and larger then 30 days. The first scale is characterized by a steeply increasing variance, while the second scale shows moderate increments. Residual variograms were calculated from the difference between θ and θ_s for use with Eq. (4). Spherical models were fitted to all experimental variograms of the upper soil horizon (Table 2).

Evaluation and Validation of Geostatistical Methods

The efficiency of all the estimation methods was smaller in NT (Fig. 3). According to Eq. (6), the smaller variance observed in the NT plot produces a smaller efficiency if the mean squared error (MSE) does not decreases proportionally. Figure 3 shows that the MSE was larger for the NT system. Kriging with an external drift using

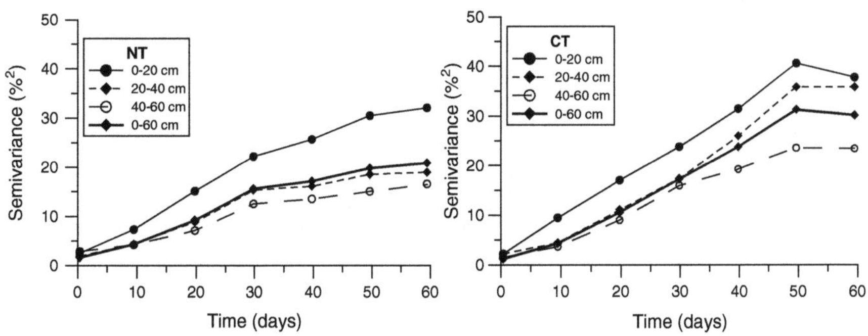

Fig. 2 Experimental variograms of gravimetric soil moisture content, observed in different horizons in the no-tillage (NT) and conventional tillage (CT) plots

Table 2 Parameters of the fitted spherical variogram models for gravimetric soil moisture content of the upper soil horizon (0-20 cm) in the no-tillage (NT) and conventional tillage (CT) plots. The subindex *res* refers to the residual variograms

	$c_0(\%^2)$	c $(\%^2)$	r (days)
NT	1.8	30.7	66.2
CT	1.6	40.6	72.0
NT_{res}	2.0	4.1	17.0
CT_{res}	2.1	3.4	14.0

c_0: nugget, c: contribution, r: range

the mean θ_s as secondary information (KED_S) showed the largest efficiency in both tillage systems. This estimation method also reproduced satisfactorily the mean and the variance of the observations. The sensor and SKlm performed worse in this sense, producing the smallest variance.

When using the SF of a single sensor as secondary information (KED_1 and KED_2) a slightly smaller efficiency than for KED_s was obtained, while the mean and variance were equally well reproduced. This procedure is for practical purposes the most promising since it does not require a previous calibration of the sensor and the estimation of the water content time series can be initiated with a minimum number of gravimetric observations.

Since SKlm uses θ_s as secondary information and the calibrated versions of Eq. (2) explained most of the variance in the observed θ data, its efficiency and MSE were very similar to those obtained by the sensor. Contrary to the cross-validation procedure, sensor performance was evaluated using estimates that used all the available data since they were all used to calibrate Eq. (2) which provided the estimates that were used to calculate the evaluation parameters shown in Fig. 3. It is expected that these results are too optimistic since the geostatistical procedures have been evaluated using cross-validation, leaving out and re-estimating each value at a time. These results indicate that using KED to estimate θ from a minimum number of gravimetric observations is a valid alternative to the traditional use of electromagnetic sensors that require a prior calibration, especially when used in expansive clay soils.

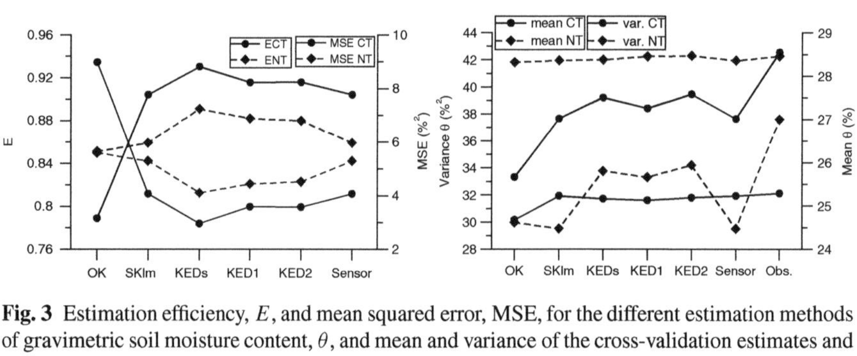

Fig. 3 Estimation efficiency, E, and mean squared error, MSE, for the different estimation methods of gravimetric soil moisture content, θ, and mean and variance of the cross-validation estimates and the observations

Estimation of the Soil Moisture Content

Kriging with an external drift represents the temporal evolution of the soil moisture with more precision and accuracy than the calibrated sensor does. Since kriging is an exact interpolator, it forces the estimation to be equal to the mean of the three observations on each sampling date (Fig. 4).

When using the SF of sensors NT2 and CT2 as secondary information in KED it is not necessary to execute a prior calibration. For each estimation the relationship between θ and SF is evaluated for a short period centred on each estimation date

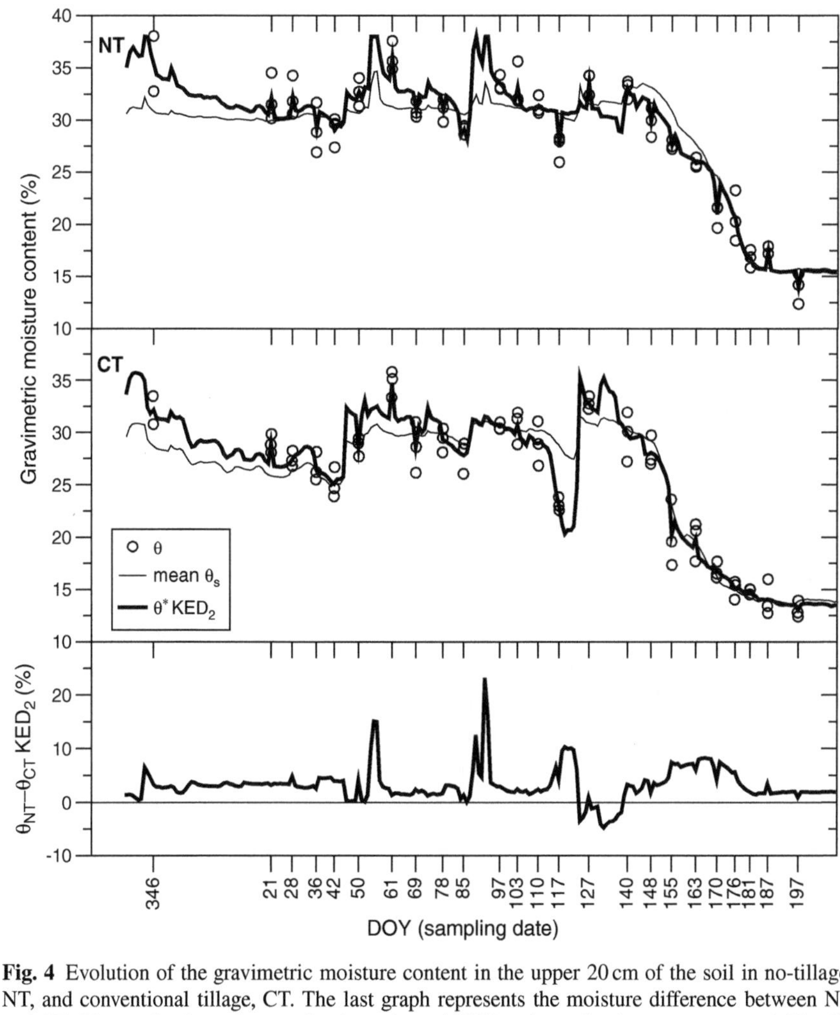

Fig. 4 Evolution of the gravimetric moisture content in the upper 20 cm of the soil in no-tillage, NT, and conventional tillage, CT. The last graph represents the moisture difference between NT and CT. Observed values, θ, sensed values, θ_s, and KED estimated values are compared. The SF of sensors NT2 and CT2 was used as secondary information in KED

and makes it possible to take into account changing soil physical properties, due to processes associated with the wetting and drying cycles, during the monitoring period. In this way a better fit is obtained during periods with important changes in θ, for example after intense rainfall events or at the initial stages of the discharge of the soil profile, especially in NT. However, it is still extremely important that the sensors provide a fast and accurate response to changing soil moisture conditions. Sensors housed in access tubes for which the installation was not optimal (air-gaps between the tube and the surrounding soil or a bad overall tube-soil contact due to soil shrinkage) will provide SF values that, when used as secondary information, will probably not improve the interpolation.

Soil moisture dynamics of both tillage systems are concisely represented in Fig. 4 and interesting features can be clearly appreciated. The intensity of the wet spells in NT is clearly larger and the take-off of the soil moisture discharge form half May until the end of June (from DOY 140 to 180) is retarded in NT. The resulting soil moisture differences between NT and CT can be clearly appreciated on the difference graph in Fig. 4. During the discharge of the soil profile, soil moisture differences of more than 8% are observed in the upper soil horizon at a soil moisture content of 17.5% in CT. These differences are agronomically relevant since the NT soil is able to provide water to the crop during approximately 20 days more.

Conclusions

The necessity to calibrate electromagnetic soil moisture sensors independently for different soil horizons and for each installation imposes severe limitations on its widespread use for practical applications such as irrigation scheduling or nutrient management. The inconveniences related with changing bulk densities were partly resolved by considering soil moisture at a gravimetric instead of a volumetric basis. The use of these sensors require inevitably *in situ* gravimetrical determinations of soil moisture for the entire range of possible values, so it was shown advantageous to take these samples strategically during the monitoring period and to interpolate them using kriging with an external drift, taking into account the scaled frequency or sensed moisture values as secondary information. Also the observed frequency in soil, F_s, could be used so that no prior measurements in air and water have to be made. The use of non-parametric geostatistical methods (e.g. indicator kriging) might be a useful alternative in this case, since these methods are especially appropriate for data with non-Gaussian distributions (e.g. bimodal soil moisture distributions) and since they offer the possibility to estimate the probability of exceeding a threshold value. In numerous agronomical and environmental applications of these sensors the interest is not in knowing the exact value of the soil moisture content, but in detecting the moment in time at which a maximum or minimum soil moisture threshold value is exceeded.

References

Baumhardt RL, Lascano RJ, Evett SR (2000) Soil material, temperature, and salinity effects on calibration of multisensor capacitance probes. Soil Sci Soc Am J 64:1940–1946

Deutsch CV, Journel AG (1998) GSLIB. Geostatistical software library and user's guide, 2nd edition. Oxford University Press, New York

Evett SR, Parkin GW (2005) Advances in soil water content sensing: The continuing maturation of technology and theory. Vadose Zone J 4:986–991

Fares A, Buss P, Dalton M, El-Kadi AI, Parsons LR (2004) Dual field calibration of capacitance and neutron soil water sensors in a shrinking-swelling clay soil. Vadose Zone J 3:1390–1399

Goovaerts P (1997) Geostatistics for Natural Resources Evaluation. Oxford University Press, Nueva York, p 483

Jost G, Heuvelink GBM, Papritz A (2005) Analysing the space-time distribution of soil water storage of a forest ecosystem using spatio-temporal kriging. Geoderma, 128:258–273

Muñoz-Carpena R, Ritter A, Bosch D (2005) Field methods for monitoring soil water status. In: Benedí, Muñoz-Carpena (eds) Soil-Water-Solute process characterization. An integrated approach, CRC Press, Boca Raton, FL, p 778

Nash JE, Sutcliffe JV (1970) River flow forecasting through conceptual models. Part 1. A discussion of principles. J. Hydrol. 10:282–290

Paltineanu IC, Starr JL (1997) Real-time soil water dynamics using multisensor capacitance probes: laboratory calibration. Soil Sci Soc Am J 61:1576–1585

Panatier Y (1996) VARIOWIN: Software for spatial data analysis in 2D. Springer Verlag, New York

Polyakov V, Fares A, Ryder MH (2005) Calibration of a capacitance system for measuring water content of tropical soil. Vadose Zone J 4:1004–1010

Snepvangers JJJC, Heuvelink GBM, Huisman JA (2003) Soil water content interpolation using spatio-temporal kriging with external drift. Geoderma 112:253–271

Soil Survey Staff (1999) Soil Taxonomy. A basic system of soil classification for making and interpreting soil survey. 2ª ed. USDA Agriculture Handbook n 436, Washington

Geostatistics for Contaminated Sites and Soils: Some Pending Questions

D. D'Or, H. Demougeot-Renard and M. Garcia

Abstract This paper addresses three key issues when using geostatistics in soil remediation studies. Firstly, we point out the necessity of using an appropriate model when the contaminant grade distributions are highly skewed due to a large proportion of samples that do not contain the pollutants. In this case, a valuable solution may consist in combining an indicator variable for describing the presence or absence of pollutant and a Gaussian random variable for modelling the transformed contaminant grades at locations where they were present. This model is shown to reduce the uncertainty on the classification of the soils into safe or polluted. Secondly, we address the problem of change of support between the soil samples and the remediation units. A method is proposed to achieve the upscaling. But at the unit scale, average grades above some critical threshold may be explained by the occurrence, at the soil sample scale, either of a large proportion of (possibly moderately) excessive grades or by a few sample with (possibly very) high grades. In one or the other situation, the health or environmental risk is certainly different. The third issue discussed relates to the evaluation of contaminated soil volumes when those soils are affected by numerous contaminants. Multiplying the number of potential contaminants also multiplies the risk for soils to be contaminated. Better integrating the correlation between contaminants then appears essential.

1 Introduction

From their long industrial history, many European countries have inherited numerous abandoned but also contaminated plant sites. Beyond the potential health and environmental risks they involve, these sites constitute valuable lands for new economic and real estate developments. Their rehabilitation calls, however, for remediation before they receive a new occupancy. Due to the complexity of the spatial distribution of soil contaminants as well as to the financial, environmental and health stakes, the management of large contaminated sites cannot be achieved using heuristic methods.

D. D'Or
FSS International r&d 1956, Av. Roger Salengro, 92370 Chaville, France
e-mail: dimitri.dor@fssintl.com

A. Soares et al. (eds.), *geoENV VI – Geostatistics for Environmental Applications*, 409–420

Instead, scientific grounded tools are needed to delineate soil pollution, to quantify contaminated soil volumes along with their uncertainty, and to assess the potential impact of the pollution on neighbouring populations and ecosystems. Such tools should also be able to take into account all available sources of information (laboratory measurements, field observations, soil properties, etc.), risk criteria (depth or occupancy-dependent regulatory thresholds) and support changes (from soil samples to remediation units).

In this context, geostatistics provide helpful and suitable methods for estimating and mapping volumes of contaminated soils, possibly making use of secondary information, and for quantifying local and global uncertainties about contaminant grades and soil volumes. Geostatistical results can then be exploited to classify excavated soils according to remediation channels or to assess remediation costs.

When considering excavation of contaminated soils, a typical geostatistical approach consists of the following tasks.

1. Derive, from uni- and multivariate statistical and exploratory analyses of contaminant grades and available secondary information, the contaminants that could potentially induce health or environmental risks and need to be modelled.
2. Characterise the spatial structures of so identified contaminants by means of conventional variographic tools.
3. Define an excavation grid with blocks corresponding to remediation units.
4. Compute, for each gridblock, the probability that the block grade of at least one contaminant exceeds its critical threshold. This step generally requires prior stochastic simulations of contaminant grades.
5. Classify the blocks as *"safe"* or *"polluted"* and compute the risk of misclassification for decision-making purposes.

If the guidelines for a geostatistical approach are well established, the application of geostatistics to contaminated sites raises non-trivial questions that would call for specific and possibly new geostatistical methods. This paper brings three such key issues into debate: (i) how should we handle skewed distributions of grades, associated with a large proportion (peak) of values below the detection limit, for stochastic simulation purposes, (ii) how should we perform the needed changes of support from punctual data to remediation units, for health and environmental risk assessment, and (iii) which tools should we use for managing soil classification uncertainty when numerous contaminants are to be considered all together.

For each of these questions, the problem is presented and illustrated using a demonstrative example. Possible solutions are then explored and discussed.

2 Skewed Distributions

Distributions of contaminant grades are frequently highly positively skewed. Because detection limits apply to grade measurements, such grade distributions often show a peak at the lowest value (Fig. 1). Grade distributions are far from being normal or

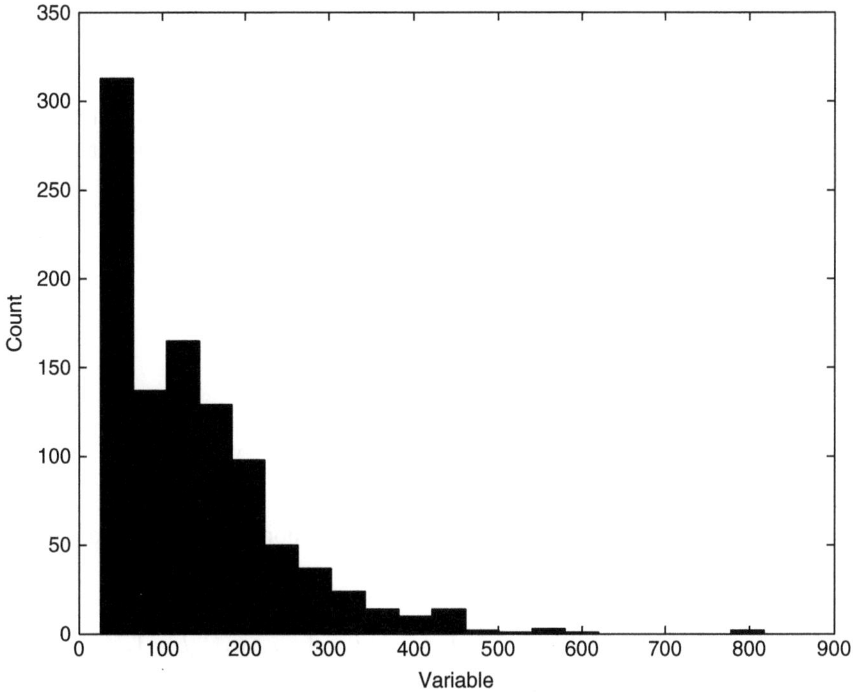

Fig. 1 Example of a typical positively skewed distribution of contaminant grade with a peak of values at the detection limit (here set at 30 ppm) and extreme values

lognormal and may be hard to transform to make them Gaussian as required to carry out the commonly used multivariate Gaussian estimation or simulation methods.

The choice of a suitable method for estimating or simulating contaminant grades is a key and recurrent issue in environmental contamination studies. Saito & Goovaerts (2000) have compared ordinary kriging, lognormal kriging, multigaussian kriging and indicator kriging for the delineation of contaminated areas. Schnabel et al. (2004) used simulation based on log transformed data. None of these techniques are appropriate, however, when the proportion of censored data becomes large.

In this section, we first discuss the use of SGS in the univariate and multivariate cases, and the consequences of correlation between contaminants for soil classification (see section 4). Afterwards, an alternative method based on mixed distributions is proposed and discussed.

2.1 Sequential Gaussian Simulations

The most straightforward simulation method is sequential Gaussian Simulation (SGS) that is easy to implement and use. It generally requires a prior normal score or anamorphosis transformation of data to make them univariate Gaussian. In the multivariate case, the assumption of multigaussianity is seldom checked. A method like the step-wise conditional transformation suggested by Leuangthong & Deutsch (2003) allows to partially overcome this assumption. This method provides a way to transform multivariate distributions showing nonlinear, heteroscedastic or constraint features into a series of univariate uncorrelated Gaussian distributions that may be simulated independently using SGS. After back transformation, the simulated values tend to reproduce the original multivariate distribution. Denote $Z_i,\ i = 1, ..., n$, the n random functions to be simulated and F_i their respective cumulative distribution functions (cdf). The following transformation is considered:

$$
\begin{aligned}
Y_1 &= G^{-1}[F_1(z_1)] \\
Y_2 &= G^{-1}[F_{2|1}(z_2 \mid z_1)] \\
&\vdots \\
Y_n &= G^{-1}[F_{n|1,\dots,n-1}(z_n \mid z_1, \dots, z_{n-1})]
\end{aligned}
$$

where $Y_i,\ i = 1, \dots, n$ are multivariate Gaussian random functions that are independent at lag distance of zero, that is,

$$C_{ij}(0) = C(Y_i(0), Y_j(0)) = 0, \quad i \neq j, \quad i = 1, \dots, n, \quad j = 1, \dots, n$$

Moreover, all multivariate distributions are Gaussian in shape at distance lag $h = 0$. The following three aspects are to be noted for this method:

1. There is no guarantee that $C_{ij}(h) = 0$ for $h > 0$, i.e., the Y_is be spatially independent. This independence must be checked case by case and a valid linear model of coregionalisation (LMC) must be used to fit cross-covariances if necessary.
2. The ordering of the variables has an effect on the covariance models. The most spatially smooth variable is preferably chosen as primary variable.
3. Inference of multivariate distributions requires a large number of data, which are rarely available in environmental studies. Kernel or other smoothing techniques may be used to "fill-in" gaps in the raw-data multivariate distribution.

The SGS method works well in the uni- or multivariate cases at the condition that the distributions are not too asymmetric. Nevertheless, if the proportion of values below the detection limit (DL) becomes important, the transformed distribution remains not Gaussian hence making inappropriate the use of SGS (Rivoirard, 1994).

2.2 Combination of Sequential Indicator and Gaussian Simulations

Let us denote by $Z(x_i)$, the contaminant grade at location x_i. Any location in the site can be considered as possibly containing a contaminant (hereafter referred to as *contaminant-affected*) with a certain probability.[1] This can be modelled using a Bernoulli (or indicator) random variable $I(x_i)$ taking value 0 when $Z(x_i) = 0$ and value 1 when $Z(x_i) > 0$. Then, if the location is effectively contaminated, a continuous distribution may be used to model the distribution of $Z(x_i)$ given the fact that the location is affected or not, $f(z(x_i)|I(x_i))$.

The probability that the contaminant grade is exceeding the regulatory threshold z_t can be computed using the total probability theorem:

$$\begin{aligned} P(Z(x_i) > z_t) = {} & P(Z(x_i) > z_t|Z(x_i) > 0).P(Z(x_i) > 0) \\ & + P(Z(x_i) > z_t|Z(x_i) = 0).P(Z(x_i) = 0) \end{aligned}$$

In this equation, $P(Z(x_i) > z_t|Z(x_i) = 0) = 0$, leading to

$$P(Z(x_i) > z_t) = P(Z(x_i) > z_t|Z(x_i) > 0).P(Z(x_i) > 0) \tag{1}$$

This result is very classical in statistics for modelling mixed distributions or distributions of censored data. It has been used, e.g., by De Oliveira (2004) for modelling spatial rainfall fields. In soil remediation, we couldn't find any reference using this approach.

Practically, the spatial distribution of contaminant grade can be simulated by combining two independent random fields:

- An indicator random field to simulate the presence or absence of a contaminant,
- A continuous random field to simulate grade values > DL where the contaminant is present.

Sequential Indicator Simulations (SIS) and Sequential Gaussian Simulations (SGS) can be used to simulate these two fields. SIS requires a prior indicator transformation of grade data, i.e., $I(z(x_i)) = 1$ if the measured grade at location x_i is greater than DL, otherwise 0. A Gaussian transformation of nonzero (> DL) grade data may also be required to carry out SGS. Combining the two random fields then consists simply of multiplying then. By repeating the exercise with different realizations from the indicator and continuous random fields, local distributions of contaminant grades are obtained with a peak of zero values equal to the local probability of not being contaminant-affected. Possible limitations of the method are (i) poor performances of the Gaussian transformation if the distribution of grades is too asymmetric, and (ii) the application to multivariate cases when several correlated contaminants are to be jointly simulated. Nevertheless, this approach certainly appears as an improvement to single SGS simulations, as shown in the following example.

---

[1] A distinction is made here between *"affected"* and *"contaminated"* (or polluted) soils, given a contaminant. *"Affected soils"* will refer to soils where the contaminant grade is greater than the DL, whereas *"contaminated soils"* will apply to affected soils where the grade exceeds a given regulatory (critical) threshold.

2.3 Comparison of Methods on a Synthetic Example

Both methods, SGS alone and SIS+SGS, have been compared on a univariate synthetic example. Using an exponential variogram model with a nugget effect of 0.1, a sill of 0.725 and a range of 500 m, an unconditional SGS simulation was jointly performed over a 100 by 100 nodes grid (later used as reference) and at 100 randomly sampled locations (later used as the data set see Fig. 2a). The internode distance is equal to 10 m. Then the simulated values were transformed using the normal score back-transform in order to produce a distribution with about 30% of them below the detection limit set at 30 ppm, as shown in Fig. 1. The critical threshold for declaring soils as polluted was set at 200 ppm. The overall probability to exceed this threshold is equal to 15%.

Using the data set, the two SGS and SIS+SGS methods were run each to generate 100 realisations of grade at the nodes of the grid. For SGS+SIS, the two required variogram models were one for the indicator and one for the normal score transform of grade values greater than DL. From this series of realisations, the probability of exceeding the critical threshold was estimated at each grid-node. The probability maps are depicted in Fig. 2c for SGS and Fig. 2d for SIS+SGS. The SIS+SGS probability map appears more contrasted with a large area of very likely non-polluted soils (probability ≈ 0), the rest of the domain being essentially polluted (probability > 0.8 excepted in some narrow zones). On the contrary, the SGS probability map is dominated by intermediate probabilities. According to these probabilities, each node (or associated block) may be classified as polluted, safe or uncertain as follows.

- Polluted if the probability of exceeding the critical threshold is larger than 90%;
- Safe if this probability to exceed the critical threshold is less than 10%;
- Otherwise, uncertain.

Figure 3 shows the soil classification for the reference and the misclassification error maps for both methods. It can be seen that SIS+SGS produces much larger areas with a misclassification error close to zero. Areas with large probability of error are concentrated at the boundaries between polluted and "safe" zones, as expected.

The corresponding volumes of soils, derived from blocks of $10 \times 10 \times 1\,m^3$ centred on the simulated values, are shown in Table 1. They corroborate the observations made above about the maps, i.e., SIS+SGS allows better discriminating contaminated and uncontaminated soils, thus leading to more accurate delineation and estimation of contaminated soil volumes. Without trying to compare the results to a not necessarily significant reference solution (i.e., the non-conditional algorithm-dependent simulation from which data were sampled), it can be noted that SGS+SIS reduces uncertainty by just bringing more statistics into the model.

Besides the classification, volumes of contaminated soils can also be computed directly on each realisation by counting the number of contaminated blocks. Summary statistics are given in Table 2. In this particular example, SGS tends to overestimate the volume of contaminated soils, which also appears much more uncertain.

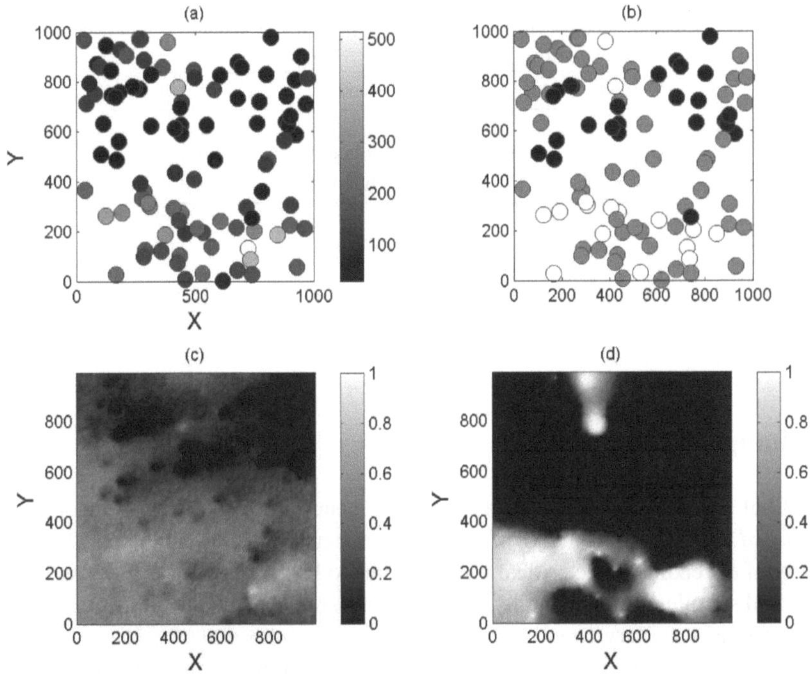

Fig. 2 Top graphs show the data layout with (a) the contaminant grades, and (b) their position against the detection limit (30 ppm) and the critical threshold (200 ppm): black = value below the detection limit, gray = value between the detection limit and the critical threshold, white = value above the critical threshold. Bottom graphs show maps of the probability of exceeding the critical threshold as estimated from (c) SGS and (d) SIS+SGS

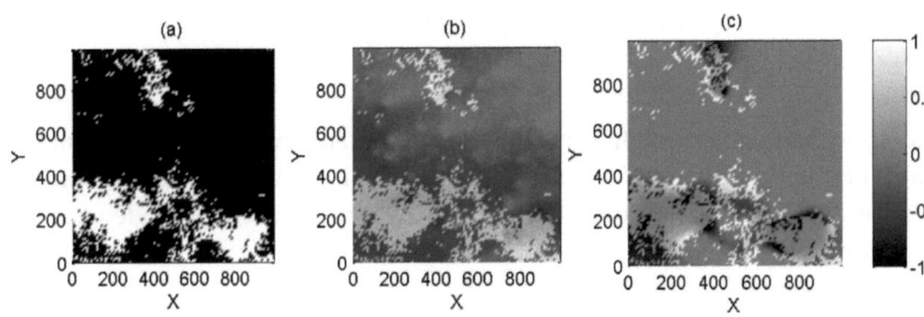

Fig. 3 Soil classification for the reference (a) with the contaminated areas in white. Probabilities of misclassification errors obtained with SGS (b) and SIS+SGS (c). Gray color indicates a probability of misclassification error close to 0. Closer to white, larger is the probability that a location classified as safe is contaminated in reality. Closer to black gives the reverse error

Table 1 Estimated soil volumes (m^3) for each category: contaminated (if the probability of exceeding the critical threshold is larger than 90%), safe (if the probability is less than 10%) and uncertain (intermediate probability)

Method	Polluted	Safe	Uncertain
Reference	201000	799000	–
SGS	0	174800	825200
SIS+SGS	85900	685600	228500

Such a poor performance is certainly to be related to the inability of SGS to simulate the two-population distribution of contaminant grades.

3 Change of Support

The final goal of soil remediation studies is generally to classify into *"safe"* or *"polluted"* remediation units of several cubic meters. To achieve this goal, the probability of exceeding some critical threshold has to be assessed for each remediation unit. Soil samples being generally collected on a small (punctual) support, a change of support is necessary.

It is expected from the change of support an estimation of local conditional cumulative distribution functions (ccdf) of unit (block) averages of contaminant grades. Knowing the local ccdf, it can then be computed the probability that an average unit grade exceeds a critical threshold.

In practice, following Lajaunie & Wackernagel (2000), if $Z_c^k(x), k = 1, \ldots, m$, are independent fine-scale conditional simulations of the unknown spatial distribution of contaminant grade $Z_c(x)$, the average contaminant grade of each remediation unit U is to be approximately calculated on each realisation by numerical integration as:

$$Z_c^k(U) = \frac{1}{|U|} \int_U Z_c^k(x) dx \approx \sum_i w_i Z_c^k(x_i)$$

where w_i is the relative weight (or relative representative volume) of the simulated value $Z_c^k(x_i)$ in the unit and $|U|$ is the volume of the remediation unit. One can then derive for each remediation unit the probability that the average grade exceeds a critical threshold z_t, i.e.,

Table 2 Statistical summary of simulated volumes of contaminated soils (m^3)

Method	SGS	SIS+SGS
Mean	313301	201931
Standard deviation	108040	14150
Variation coefficient	0.35	0.07

$$P(Z(U) > z_t) = \frac{1}{m}\sum_{k=1}^{m} 1_{Z_c^k(U)>z_t}$$

where $1_{Z_c^k(U)>z_t}$ is an indicator variable that equals 1 if $Z_c^k(U) > z_t$, otherwise 0.

This method is simple and thus very attractive. It supposes, however, that average grades are relevant to evaluate the health or environmental risk of contaminated soils at a remediation-unit scale. Actually, fine-scale distributions of contaminant grades, as measured on small soil samples, being often positively skewed, unit averages of contaminant grades can be strongly affected by a few high or extreme values. As an example, consider 64 grade-simulation points in a remediation unit and a critical threshold z_t of 0.2 ppm (risk-based regulatory value for benzene in Walloonia, Belgium). If only one simulated value among the 64 is higher than 20 ppm, then the whole remediation unit is to be classified as contaminated with an average grade exceeding z_t. Should this classification be accepted or should the support be also taken into account? In other words, is a single (fine-scale) high value in a remediation unit enough to deem the whole unit as being contaminated?

In order to take this issue into consideration, an additional classification criterion can be used, based on the proportion of values that exceed z_t in the remediation unit. If this proportion is less than an arbitrary cutoff say of 5 or 10%, the remediation unit is classified as "safe" even if the average contaminant grade is greater than z_t. The probability that the remediation unit U is polluted is then computed as:

$$P(U \text{ is polluted}) = \frac{1}{m}\sum_{k=1}^{m} 1_{((Z_c^k(U)>z_t)\cap(p_c^k(U)>c))}$$

where $p_c^k(U)$ is the proportion of simulated values above z_t and c is the cutoff on this proportion. Interested readers may refer to Kyriakidis (1997) for an example of application of this approach.

4 Multiple Contaminants and Classification Uncertainty

Complex soil pollution involving several contaminants is often encountered in industrial sites. Correlation or principal component analysis may help to identify groups of contaminants to be simulated jointly or partly if one or a few contaminants are enough to recognise all critical soils. More generally, the primary aim of co-simulation is to borrow data from better known contaminants to inform others less sampled. Co-simulating contaminants may then be more or less complex, depending on the simulation method used (see section 2).

Actually, more consequential is the impact of multiple contaminants on the uncertainty about soil classification. It is indeed common practice to declare soils as contaminated where at least one contaminant exceeds its critical grade threshold. The probability that a given remediation unit is contaminated can then be expressed as:

$$P\left(\bigcup_{i=1}^{N} Z_i(x) > z_{t,i}\right) = 1 - P\left(\bigcap_{i=1}^{N} Z_i(x) \leq z_{t,i}\right) \tag{2}$$

where N is the number of contaminants. The second term in the right member of the equation is the probability that the remediation unit is safe, i.e., that ALL (average) contaminant grades are below their critical threshold. Practically, an assumption of independence between contaminants is often made to decompose the joint probability term into a product of probabilities for each contaminant to be distinctively below its critical threshold. Eq. 3 then becomes:

$$P\left(\bigcup_{i=1}^{N} Z_i(x) > z_{t,i}\right) = 1 - P\left(\bigcap_{i=1}^{N} Z_i(x) \leq z_{t,i}\right) = 1 - \prod_{i=1}^{N} P\left(Z_i(x) \leq z_{t,i}\right) \tag{3}$$

By so doing, it can be shown that the soil classification uncertainty is maximum. For example, consider ten independent contaminants each having a small 0.10 probability of exceeding their threshold over a given remediation unit. The overall probability that the unit is contaminated is 0.65, i.e.,

$$P\left(\bigcup_{i=1}^{10} Z_i(x) > z_{t,i}\right) = 1 - \prod_{i=1}^{10} P\left(Z_i(x) \leq z_{t,i}\right) = 1 - (1 - 0.10)^{10} = 0.65$$

Consequences are :

- a remediation unit classified as contaminated, the unit being more likely contaminated than safe (probability of being contaminated > 0.5).
- an important uncertainty about the classification of the unit with a 0.35 probability of error.
- at the site scale, more or less extended zones where the high classification uncertainty makes decision-making difficult.

In these situations, solutions are to be sought to decrease further the classification uncertainty. One would be to augment the number of data to estimate more precisely each spatial distribution of contaminant grade. The higher the number of contaminants to estimate, the greater should be the number of soil grade measurement points. Another solution that is more satisfactory would consist in improving the way soil classification is performed. Taking into account the correlation between contaminants certainly is a first essential improvement on soil classification. This solution requires that correlated contaminants be jointly simulated and that the joint probability in the right term of Eq. 3 be directly computed from so obtained correlated simulations.

To illustrate the impact of considering the correlation between contaminants, different sets of 1000 element long correlated vectors were randomly drawn from standard Gaussian multivariate distributions. Vectors from a same set can be seen as realisations of correlated variables simulated at a particular node location (e.g. normal score transforms of contaminant grades). Each vector set is characterised by a number of correlated variables (from 2 to 10 by step of 2) and a correlation coefficient (from 0 to 1 by step of 0.2). This makes a total of $5 \times 6 = 30$ vector sets. For each vector set, the probability that one of the variables (contaminants) exceeds the 90 percentile was computed, this percentile being equivalent to a critical threshold for contaminant grades. The calculation was repeated by considering the contaminants alternatively independent (using Eq. 3) or correlated. Results are given in Fig. 4.

Without taking into account the correlation in the computation of probabilities (see Fig. 4a), similar probabilities are logically obtained for a same number of variables but different correlation coefficients (columns of relatively uniform colours). With the number of variables increasing, the probabilities tend, however, to increase from 0.2 for two variables up to more than 0.7 for 10.

As expected from theory, considering correlated variables in the computation of probabilities (Fig. 4b) decreases clearly the probability for increasing correlation coefficients, whatever the number of variables. For a correlation coefficient of 0, the probabilities are the same than in Fig. 4a without correlation. For a correlation coefficient of 1, all probabilities equal 0.1, which is the probability for a single variable (contaminant) of exceeding the (90 percentile) threshold.

Figure 4c shows the differences of probabilities obtained with the two different modes of calculation. Increasing differences are observed with an increasing correlation coefficient or an increasing number of variables. This result is very consequential for contaminated soil classification under conditions of uncertainty. It confirms that the correlation between contaminants is to be considered to discriminate better contaminated soils or remediation units from those safe. Ignoring the correlation also tends to overestimate volumes of contaminated soils as derived from classification criteria.

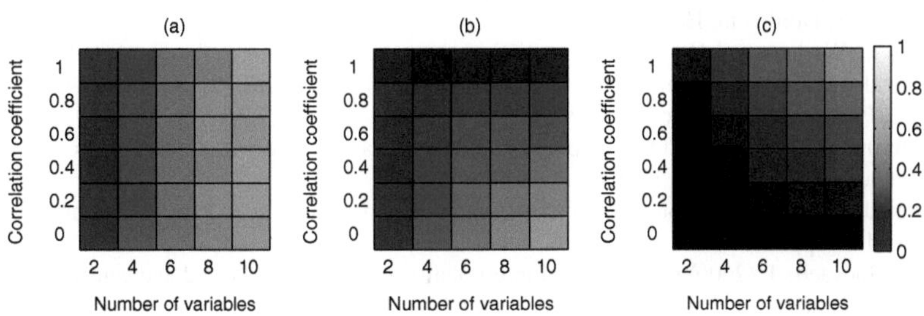

Fig. 4 Probabilities of exceeding the critical thresholds computed (a) by considering independent contaminants, (b) using the actual correlation between contaminants. (c) Differences of probabilities between (a) and (b)

5 Conclusions

Geostatistics is increasingly being recognised as able to provide powerful approaches and tools for risk assessment and decision-making in soil remediation studies. Its application to contaminated soils raises, however, a number of issues or pitfalls that call for specific solutions and possibly new practices. Three such issues have been discussed and illustrated in this article.

One is the stochastic simulation of highly skewed distributions of contaminant grade, possibly associated with a spike of values below a detection limit. Specific simulation approaches are required to reproduce successfully the marginal and spatial (two-population) distributions of grade.

Another issue is the change of support (upscaling) from punctual (soil sample) contaminant grades to grades evaluated at a remediation unit scale. Average remediation-unit grades are just arithmetic means of within-unit fine-scale grades. Average grades above some critical threshold may then be explained either by a large proportion of excessive fine-scale grades or by a few high within-unit grade values. In one or the other situation, the health or environmental risk is certainly different.

The last issue discussed relates to the evaluation of contaminated soils volumes when those soils are affected by numerous contaminants. Multiplying the number of potential contaminants also multiplies the risk for soils to be contaminated. Better integrating the correlation between contaminants and possibly other multivariate statistics then appears essential.

If improperly or poorly addressed, all these issues have the same consequences: a higher uncertainty about soil classification leading to uncertain rehabilitation cost estimates but also hard to define remediation strategies.

References

De Oliveira V (2004) A simple model for spatial rainfall fields. *Stochastic Environmental Research And Risk Assessment,* **18**: 131–140

Kyriakidis P (1997) Selecting panels for remediation in contaminated soils via stochastic imaging. In: Baafi E, Schofield N (eds) Geostatistics Wollongong '96, vol. 2. Kluwer Academic Publishers. Dordrecht, Holland

Lajaunie C, Wackernagel H (2000) Geostatistical approaches to change of support problems theoretical framework. Technical Report N30/01/G, ENSMP - ARMINES, Centre de Géostatistique. Fontainebleau, France

Leuangthong O, Deutsch C (2003) Stepwise conditional transformation for simulation of multiple variables. Math Geol **35**: 155–173

Rivoirard J (1994) Introduction to disjunctive kriging and nonlinear geostatistics Clarendon Press, Oxford 181 pp

Saito H, Goovaerts P (2000) Geostatistical interpolation of positively skewed and censored data in a dioxin-contaminated site. Environ Sci Technol **34**: 4228–4235

Schnabel U, Tietje O, Scholz R (2004) Uncertainty assessment for management of soil contaminants with sparse data. Environ Manage **33**: 911–925

Evaluation of an Automatic Procedure Based on Geostatistical Methods for the Characterization of Contaminated Sediments

G. Raspa, C. Innocenti, F. Marconi, E. Mumelter and A. Salmeri

Abstract This chapter describes a specific procedure that the Italian Central Institute for Marine Research (ICRAM) is researching on, in collaboration with the University of Rome La Sapienza, for the characterization of contaminated sediments. In the first part a description of the procedures developed in these years by ICRAM for a systematic and scientific approach to the characterization of contaminated sediments is provided. In particular it is illustrated how data analysis is performed by means of geostatistics in order to evaluate sediment volumes to be removed. Then, attention is focused on the need to develop an automatic procedure to estimate sediment contamination and the proposed procedure is described in detail. At last, results about applications of the procedure for some case studies are reported and the procedure's future development and progress are discussed.

Introduction

The Italian Ministry of Environment has made the Italian Central Institute for Marine Research (ICRAM) in charge of the environmental characterization of marine and brackish areas located within the contaminated sites of national interest. In compliance with its institutional assignment, ICRAM has defined guidelines and procedural models for a systematic and scientific approach to the characterization strategy, sampling and analytical methodologies, the processing of the characterization data, the evaluation of sediment quality and contaminated sediment management options.

Sediment characterization is realized by applying a site-specific sampling scheme defined on the basis of the information collected about the area and based on the conceptual model of contaminant transfer. As per the area dimension and morphology, sampling stations are disposed on a regular or an irregular grid, uniformly distributed all over the area; for regular sampling, grid size varies from 450 m × 450 m to 150 m × 150 m in more critical areas and is generally reduced to 50 m × 50 m

G. Raspa
University of Rome "La Sapienza", Dept. ICMMPM, via Eudossiana, 18 - 00184 Roma, Italy
e-mail: giuseppe.raspa@uniroma1.it

A. Soares et al. (eds.), *geoENV VI – Geostatistics for Environmental Applications*, 421–441

when specific operations have to take place (such as dredging, building of new docks, construction of confined disposed facilities). For irregular sampling, density continues to be chosen as indicated above, but sampling stations are placed based on an optimization process that permits to minimize the distance between any arbitrary point of the study area and its nearest sampling location (Van Groenigen and Stein 1998; Van Groenigen et al. 2000; Bação et al. 2004).

On the base of the defined sampling grid, surveys longer than or equal to 2 m are realized and core samples are taken approximately of the following sections, starting from the core top: 0–20 cm, 30–50 cm, 100–120 cm, 180–200 cm, another 20-cm section for each core's linear metre beyond 2 m (e.g. 280–300 cm, 380–400 cm) and the section relative to the last core's 20 cm.

The data analysis phase represents a very important step in the process towards the definition of the need for emergency and remediation interventions. Therefore a specialized team of experts has been created. This team, with the support of ISATIS, uses geostatistical methods to estimate the volumes of contaminated sediments.

Thanks to the experience gained in the last few years in the field of sediment characterization, ICRAM has been able to outline some peculiar aspects of a concentration's spatial variability. Sediment contamination generally shows histograms characterized by a strong and systematic asymmetry (Figs. 1 and 2). Due the complexity of the process of contaminant accumulation in the sediment matrix, experimental variograms are generally very irregular (Figs. 3–6) and very difficult to be adapted to a theoretical model; also, it is very hard to see variability structures

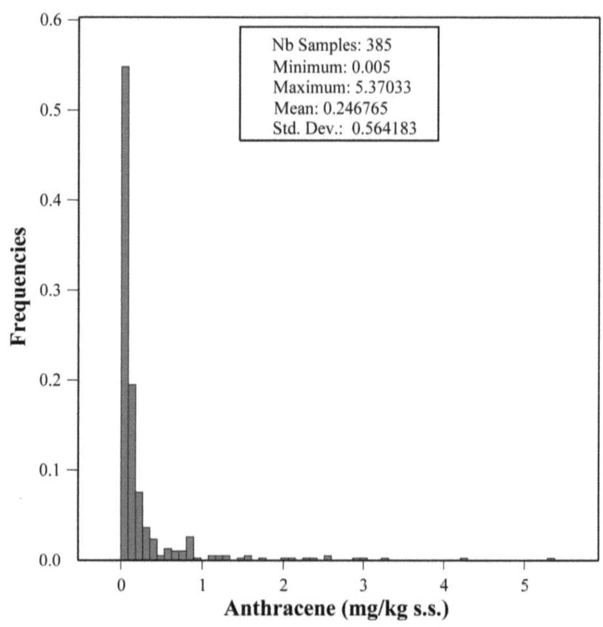

Fig. 1 Anthracene histogram

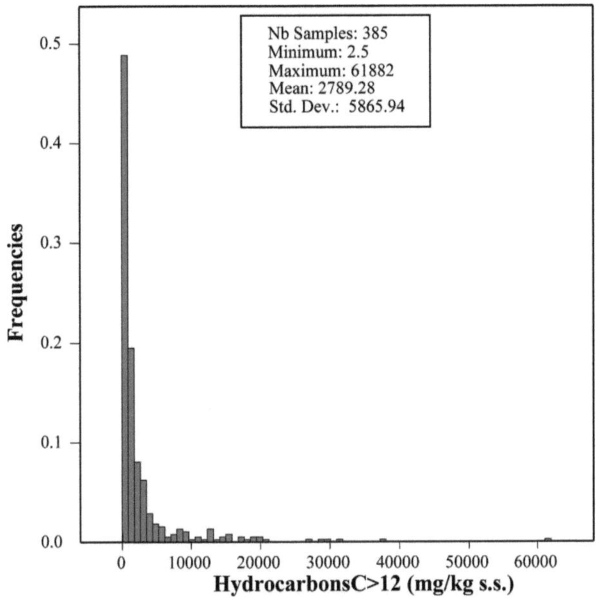

Fig. 2 Hydrocarbons C > 12 histogram

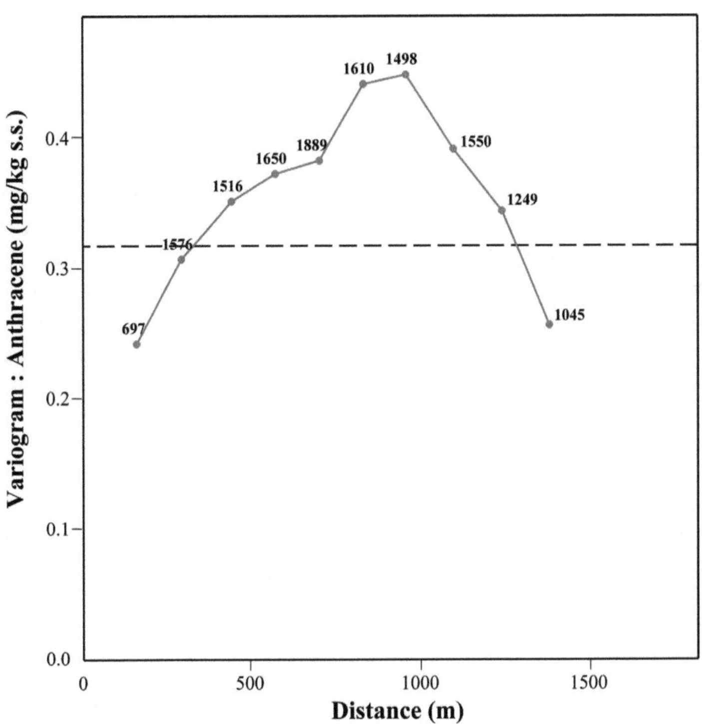

Fig. 3 Anthracene variogram in the horizontal plane

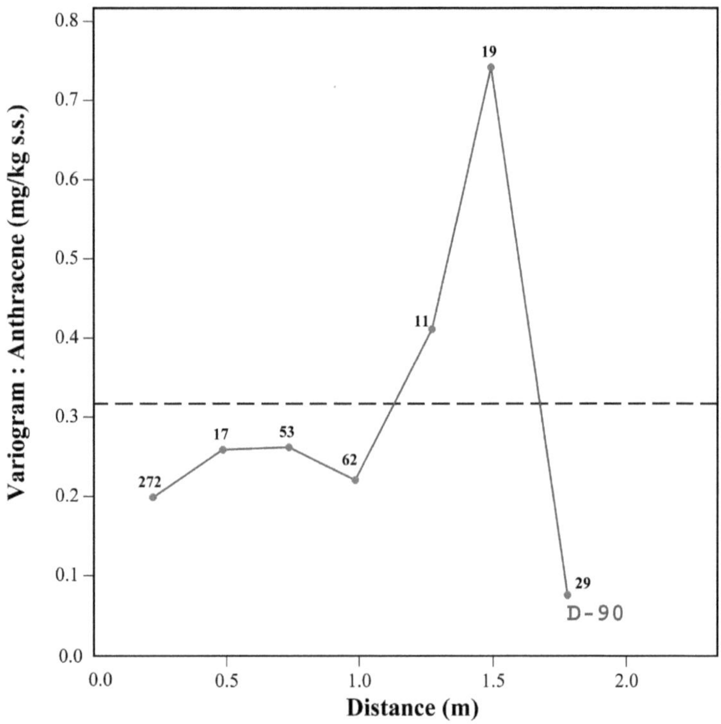

Fig. 4 Anthracene variogram in the vertical direction

by working on data logarithms. Moreover, experimental variograms usually show the existence of considerable low-scale spatial variability. Besides, in reality the contamination's spatial variability depends on space direction, and, usually, the variability in the horizontal plane is different from that in the vertical direction.

Variograms are usually modelled by means of a quasi-stationary nested model, typically composed of a nugget and two spherical variograms, one depending on the vertical component of distance and the other depending on the horizontal one:

$$\gamma(h) = \gamma_0(h) + \gamma_1(h_{xy}) + \gamma_2(h_z)$$

The horizontal component can be isotropic or anisotropic. A model with more than three structures or a too small nugget, due to the presence of many hot spots, may produce a lot of negative estimates, with many resulting problems in the interpretation process.

Data analysis permits the evaluation of sediment quality by producing 3D maps for sediment contamination; by comparing estimated maps with concentration limits, sediment volumes to be remediated are selected and management options are evaluated.

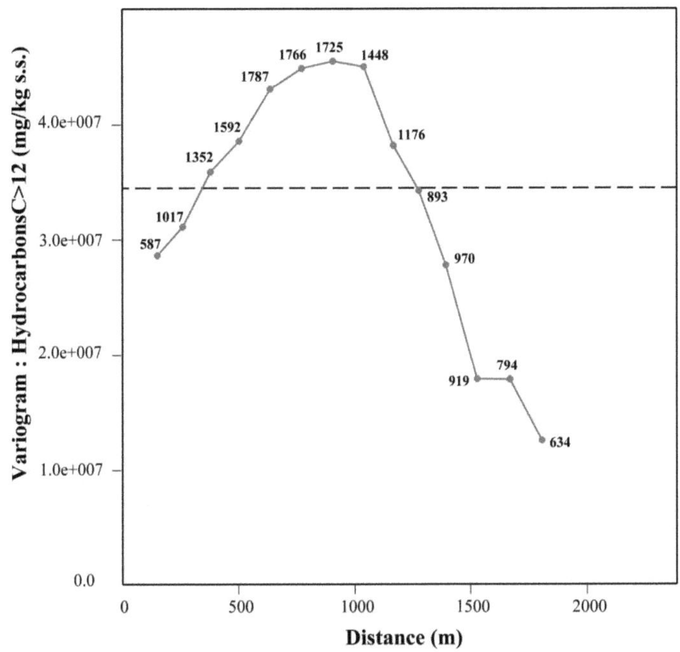

Fig. 5 Hydrocarbons C > 12 variogram in the horizontal plane

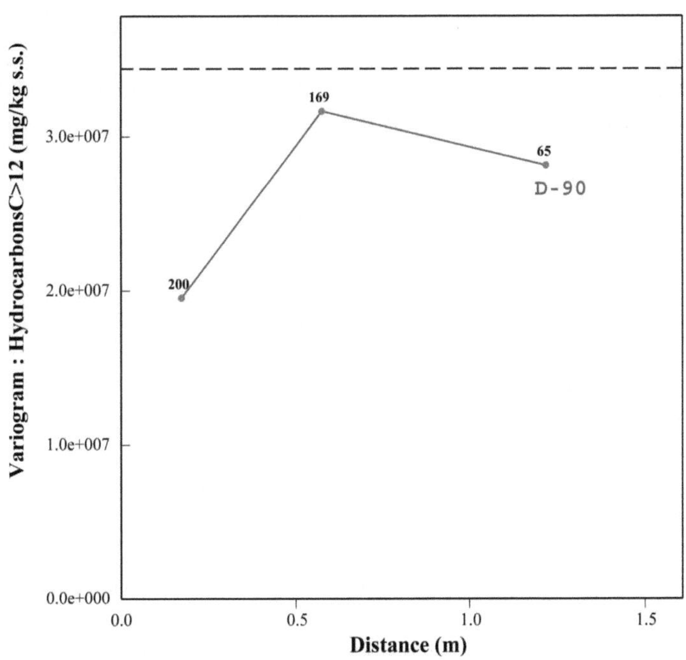

Fig. 6 Hydrocarbons C > 12 variogram in the vertical direction

Automatization of the Estimate Procedure

Generally, the modelling process for spatial variability of contaminated sediments is a very difficult task, in which some amount of subjectivity is unavoidable. However, this should be avoided as much as possible as it involves environmental and sanitary risks. Furthermore, most of the subjects in charge of the definition of the remediation interventions at specific sites (local administrations, port authorities, local agencies, etc.) do not have the geostatistical background or experience in estimating locations and quantities of sediments to be remediated. Therefore ICRAM has decided to develop a specific estimation tool, based on an automatic procedure that eliminates subjectivity but also takes into account the experience gained in these years by ICRAM's geostatistical working team.

This procedure allows the estimation of the quantity of contaminated sediments, starting from the information acquired by means of the above-described sampling strategy; this estimation is realized by dividing the site in 3D blocks and selecting those whose mean concentration exceeds a specific contamination threshold. Blocks mean concentrations are estimated by means of ordinary block kriging (Deutsch 1992).

Parameters

The estimate methodology has been chosen taking into consideration some operative aspects of the problem and, in particular, the necessity of modelling both vertical and horizontal anisotropies. The selected model of spatial variability is composed of four nested structures that allow the modelling of the nugget effect and of both horizontal and vertical anisotropies. In particular, one nugget structure and three 1D linear structures with zonal anisotropy (Chilès and Delfiner 1999) have been used:

$$\begin{aligned}\gamma(h) &= \gamma_0(h) + \gamma_1(h_{xy}) + \gamma_2(h_{xy}\cos\varphi) + \gamma_3(h_z)\\ &= \gamma_0(h) + m_1 h_{xy} + m_2 h_{xy}\cos\varphi + m_3 h_z\end{aligned}$$

where:

h is the module of $\vec{h}$, the vector connecting the estimate point to the sampled data.

$\gamma_0(h)$ is the the nugget structure, with sill C_0 that must be >0 in order to avoid models composed only of the sum of zonal components (Chilès and Delfiner 1999, pp. 96).

$\gamma_1(h_{xy})$ is the linear structure, with slope m_1; it depends on the horizontal component of $\vec{h}$.

$\gamma_2(h_{xy}\cdot\cos\varphi)$ is the linear structure, with slope m_2; it depends on the projection of h_{xy} on the direction forming an angle φ with the east–west direction (counterclockwise from E to W).

$\gamma_3(h_z)$ is the linear structure with slope m_3; it depends on the vertical component of $\vec{h}$.

A linear model is defined by only one parameter (slope), thereby allowing an easier automatic identification with respect to other variogram models (Chilès and Delfiner 1999). As indicated above, parameters defining the variogram model are five: C_0, m_1, m_2, m_3, φ; moreover, it is defined by another parameter that characterizes the estimation neighbourhood, that is the number of sample data (n) used to estimate a point. These six parameters jointly define the estimation model, that is the variogram model and the estimation neighbourhood characteristics that permit the calculation of the estimate. Parameters are obtained from an optimization procedure, based on the minimization of the cross-validation errors. With respect to parameter values obtained from the optimization process, a variogram can be isotropic or anisotropic; in the latter case, it can compete with a model exhibiting geometric anisotropy (Chilès and Delfiner 1999).

A mobile neighbourhood has been used in the estimation procedure; it moves all over the site and its shape and dimension are related to the variogram model and to the number of estimation points. Samples used to estimate a point x_0 are the n ones with the higher weights, with respect to the variogram; this corresponds to the use of a mobile neighbourhood with a shape defined by an iso-variogram surface and dimension so as to include all the n estimation samples.

By expressing $\gamma(h)$ with respect to h and taking Fig. 7 into consideration the following is obtained:

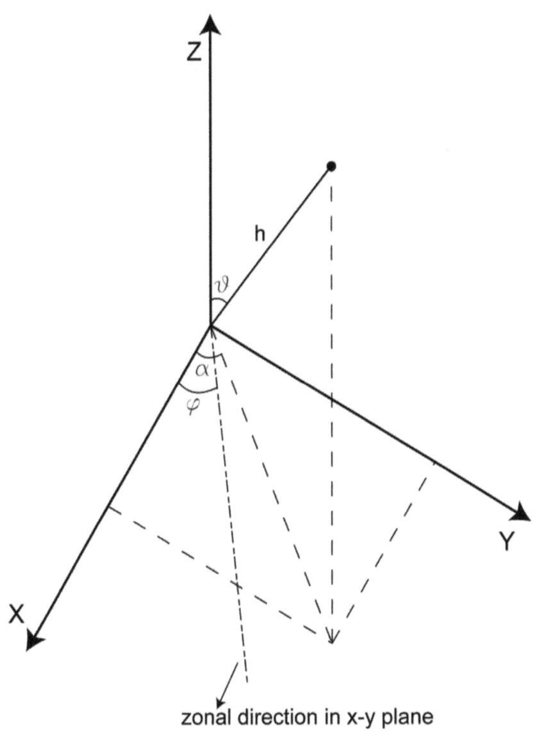

Fig. 7 Variogram components in space

$$\gamma(h) = \gamma_0(h) + m_1 h \sin\vartheta + m_2 h \sin\vartheta \cos(\alpha - \varphi) + m_3 h \cos\theta$$

and by considering $\gamma_0(h) = C_0[1 - \delta(h)]$ with $\delta(h) = \begin{cases} 1 & for h = 0 \\ 0 & for h \neq 0 \end{cases}$

it results:

$$h = \frac{\gamma(h) - C_0}{m_1 \sin\vartheta + m_2 \sin\vartheta \cos(\alpha - \varphi) + m_3 \cos\theta} \qquad \forall h \neq 0$$

where

h is the "structural distance", that is the distance, along $\vec{h}$, between the point to be estimated and the surface on which the variogram has the defined value γ(h).

All measures disposed on an iso-variogram surface have the same weight in the estimation of $z(x_0)$, because their contribution does not depend on the distance from x_o.

In Figs. 8 and 9, the estimation neighbourhood centred in the $x_0$point is represented. In Fig. 8 the neighbourhood section on the horizontal plane with z = 0 is shown, while in Fig. 9 the surface is shown; variogram parameters used to build Figs. 8 and 9 are $\gamma(h) = 30$, γ_0= 3, m_1= 8, m_2= 10, φ = 45Ł, m_3= 4.

The greater m_3 is, that is the more the variability along the z-axis, the more flattened is the iso-variogram surface on the x-y plane.

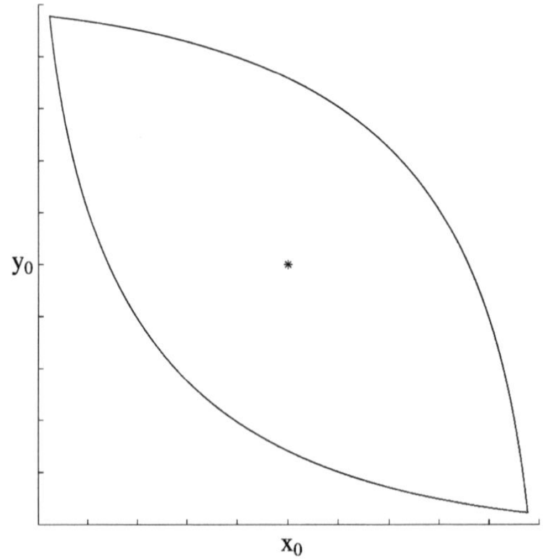

Fig. 8 Neighbourhood section

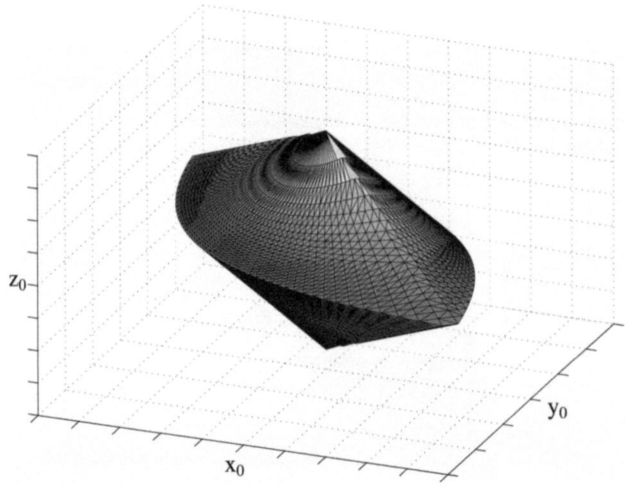

Fig. 9 Neighbourhood axonometric view

Objective Function

The automatic procedure's aim is to select blocks whose mean concentration exceeds the threshold value. The values of the six estimation parameters described above are obtained by an optimization process that minimizes selection errors; since these cannot be calculated for each grid point, where the true value is not known, the selection is evaluated based on the cross-validation data.

Figure 10 shows what is obtained by mapping on a Cartesian plane the couples $\{z_i^*, z_i\}$ relative to all measure points. The cloud of points is more scattered the greater the estimation error. If z_s is the threshold concentration above which the point is evaluated as contaminated, the estimation error produces a selection error. Particularly, as per statistical tests, two types of errors might be present, that is:

- Type I error: It occurs when $z > z_s$ and $z^* < z_s$, that is when a point that should be selected is left on place. This might cause an environmental damage.
- Type II error: It occurs when $z < z_s$ e $z^* > z_s$, that is when a point that should be left in place, is selected. This leads to higher remediation costs.

In this scenario, it is defined as an objective function whose value depends on errors of type I and type II, weighted in a different way, since the possible consequences are different. Moreover, the extracted estimation parameters might produce negative weights in ordinary block kriging. This occurs when data close to the location that is being estimated contain outlying values. When negative weights are applied to high data values, this might lead to negative contaminant estimates (Deutsch 1996). Experience shows that negative values might indeed be produced when estimating

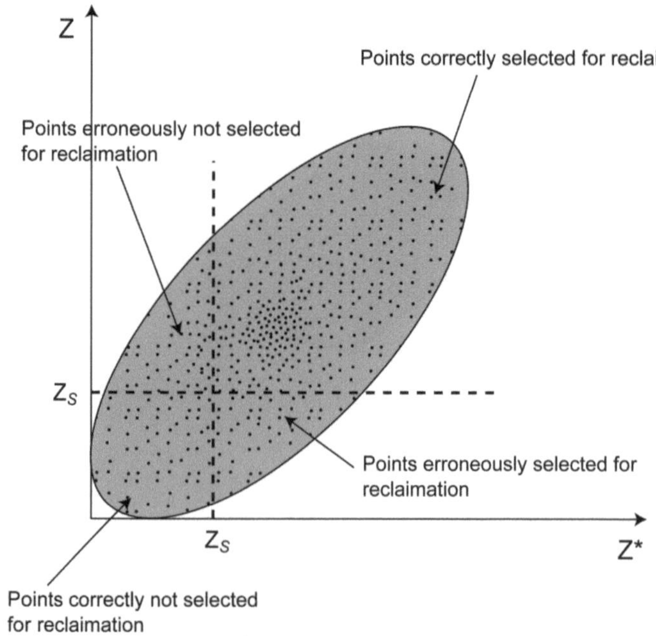

Fig. 10 Scatterplot estimations–measures

contamination in sediments, due to its spatial distribution, which is complex and heterogeneous. In order to avoid negative estimates, the proposed objective function contains a specific term that produces an exponential increase of the function value as the negative estimates increase.

The proposed objective function is:

$$f = \sum_{s=1}^{n_s} \left(\left(r \sum_{i=1}^{n_I} \frac{z_i - z_i^*}{z_i} + \sum_{i=1}^{n_{II}} \frac{z_i^* - z_i}{z_i} \right) e^{\frac{N_{neg}}{N} 100} \right)$$

where

r is the parameter whose value is greater than 1; it assigns higher weight to type I errors than to type II errors; its value must be determined experimentally, by evaluating its influence on the cross-validation results.

n_I number of type I errors.
n_{II} number of type II errors.
n_s number of thresholds $z_s$respect to which the selection is realized.
N_{neg} number of points for which the estimated value is < 0.

The objective function is obtained by summing the selection errors for different thresholds. In order to optimize the objective function with respect to a wide range of limits and to avoid applied thresholds that are lower than the minimum concentration value or higher than the maximum one, the selection errors for three thresholds are calculated, obtained from the 25-, 50- and 75-quantiles of the contaminants' experimental distribution. In order to minimize the selection errors, the objective function has to be optimized by finding its minimum value.

Optimization

The six model parameters (C_0, m_1, m_2, φ, m_3, n) are calculated by minimizing the objective function, using an optimization procedure based on genetic algorithms (GA). Two different genetic optimization procedures have been developed and tested.

Genetic Algorithms

GA were created by J. Holland in the 1960s and were inspired by Darwin's theory of evolution. GA have been used to solve a wide variety of optimization problems, including those for which the objective function is discontinuous, not derivable, stochastic or strongly not linear.

The basic idea is to select the best solutions and to combine them so that they evolve towards an optimal point. The function to be optimized is called the objective function and the variables on which it depends are called genes; the specific sequence of genes forms a chromosome, i.e. an individual that represents a possible solution. Initially many individual solutions are randomly generated to form a starting population. During each successive stop, a proportion of the existing population is selected to breed a new generation. Individual solutions are selected through a *fitness-based* process, where fitter solutions (as measured by a fitness function) are typically more likely to be selected.

The next step is to generate a second-generation population of solutions from those selected through genetic operators: crossover (implemented by combining genes of two chromosomes to form a new individual) and mutation (random change of some genes). In order to obtain a population with a fitness function whose best value doesn't get worse, it is necessary to retain the best individual generated so far from generation to generation; this technique is called elitism. This generational process is repeated until a stop condition is reached.

As already mentioned, two different genetic optimization procedures have been developed and tested; they differ principally in the use of diverse crossover algorithms. In particular:

Procedure (a): the used selection is a rank selection (proposed by James Baker); the individuals' selection probabilities are assigned according to the individuals' rank, which is based on the objective function values. This method avoids excessive preference in the selection of a few individuals with the best fitness, reducing selection pressure when variance on population fitness is high.

Procedure (b): the used selection is a tournament selection (proposed by J. Haataja); it is one of the several existing methods of selection that runs a "tournament" among a few individuals, chosen at random from the population, and selects the winner (the one with the best fitness) for crossover. Selection pressure can be easily adjusted by changing the tournament size. If the tournament size is larger, weak individuals have a smaller chance to be selected.

In order to choose the two algorithms' parameters, a sensitivity analysis on different test functions has been realized; test functions have been obtained from the literature (Winter et al. 1995) and are characterized by a very irregular behaviour. This analysis has led to the choice of the following values: crossover probability equal to 0.8 and mutation rate equal to 0.02. Moreover, elitism is applied, so that the best individual of the population is always carried to the next generation.

Hereafter are reported the results of the application of the two algorithms to the maximization of a test function characterized by some local maximums (Figs. 11–14).

Test function:

$$z = f(x, y) = 10+3(1-x)^2 e^{(-x^2-y^2)} - 10(\frac{x}{5}-x^3-y^5)e^{(-x^2-y^2)} - \frac{1}{3}e^{(-(x+1)^2-y^2)}$$

The realized tests have shown that the two procedures converge and that they can find with good approximation the optimum of very irregular functions, characterized by many local maximums. Therefore they have been used for the optimization of the objective function defined in the automatic procedure.

In the procedure, a solution is represented by a six-component vector, composed of a numerical value for each of the variables that define the variogram model and

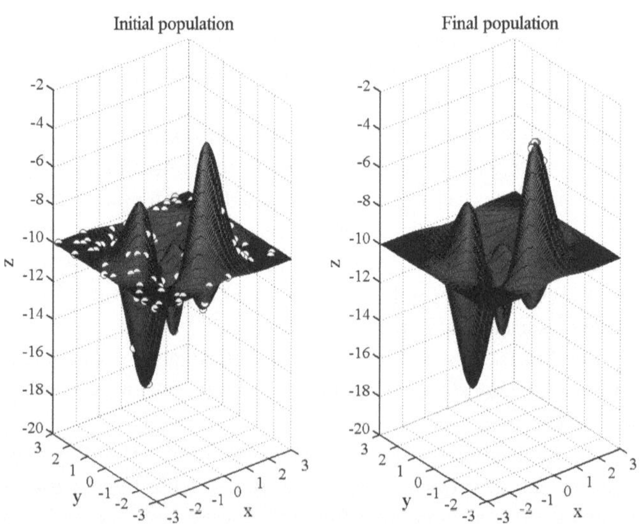

Fig. 11 Function plot and population values (100 individuals); each circle represents an individual (procedure a). Optimum value −3.4494

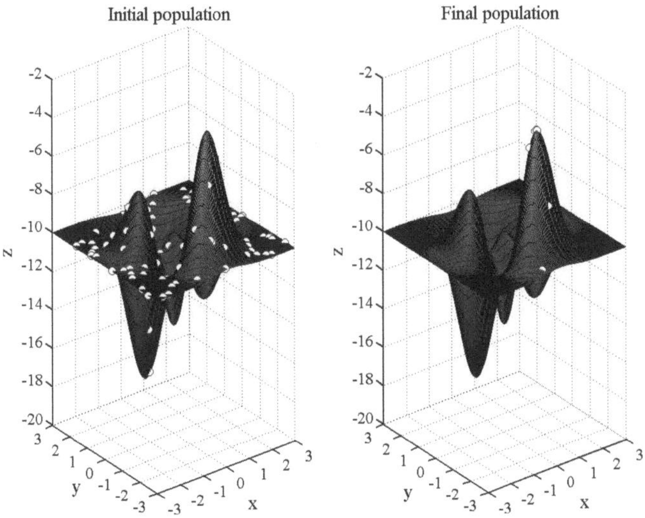

Fig. 12 Function plot and population values (100 individuals); each circle represents an individual (procedure b). Optimum value −3.4489

the neighbourhood. Nugget component varies between a fraction of experimental variance and the variance itself. Parameters m_1, m_2 and m_3 vary between 0 and $p/2$; φ varies between 0 and p, while n varies between 4 and 20 points. The fitness function matches with the above-defined objective function. A population of individuals

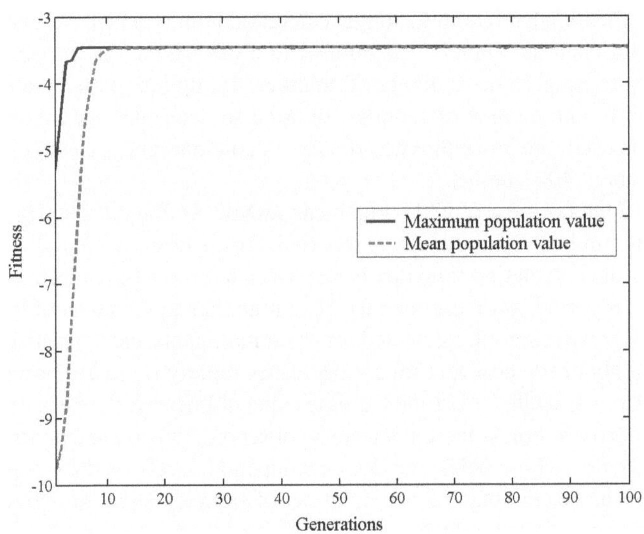

Fig. 13 Fitness versus generations. Procedure a

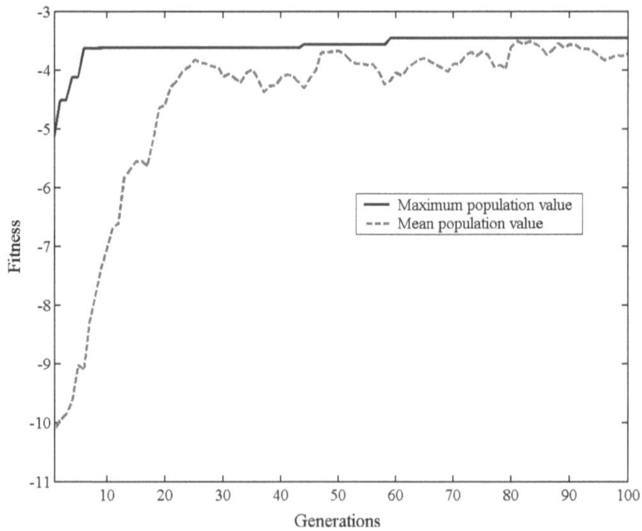

Fig. 14 Fitness versus generations. Procedure b

evolves until a defined end criterion is reached; on the basis of experimental results it has been decided that when the fitness decrement is below 5% for 20 consecutive generations, the procedure be stopped.

Once the end criterion has been defined, the number of individuals comprising the population is determined by taking into consideration two opposite matters: on the one hand, a large population should determine a result nearer to the optimum, but, on the other hand, it requires a longer calculation time (in fact, a cross-validation is realized for each individual). The number of cross-validations executed for each contaminant is equal to the individuals' number, multiplied by the number of generations and by the number of samples. In order to determine the number of individuals to be used, the procedure was run for 28 contaminants, selected from a case study with about 400 samples.

In Fig. 15 the progress of CPU time mean (AMD Athlon XP 2800+) and fitness mean versus number of individuals are reported. For each number of individuals used, the mean of the calculation time that is necessary to obtain the objective function's minimum is reported, as determined for 28 contaminants; the mean of fitness function's value is also reported, calculated for 28 contaminants and normalized to 1.

The data obtained show that fitness decreases rapidly up to 70 individuals, and then more slowly, while CPU time increases almost linearly. Between 100 and 170 individuals, a slight fitness increment can be observed, due to the extreme irregularity of the function whose behaviour does not depend linearly on the GA parameters.

Figure 15 has been obtained for a number of samples equal to 400. In order to define the number of individuals to be used in a more general case, it is necessary to perform some experiments with different number of samples and to express the

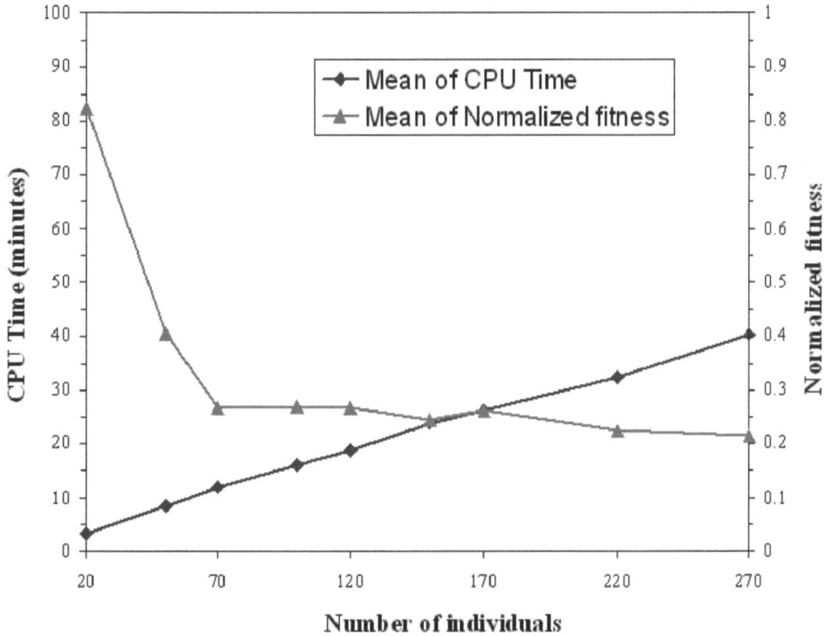

Fig. 15 CPU Time and normalized fitness versus number of individuals

CPU time and the fitness increment with respect to the number of samples and individuals.

In Figs. 16 and 17, the progress of the variance and the mean of cross-validation experimental errors are reported, versus the number of individuals, for 2 of the

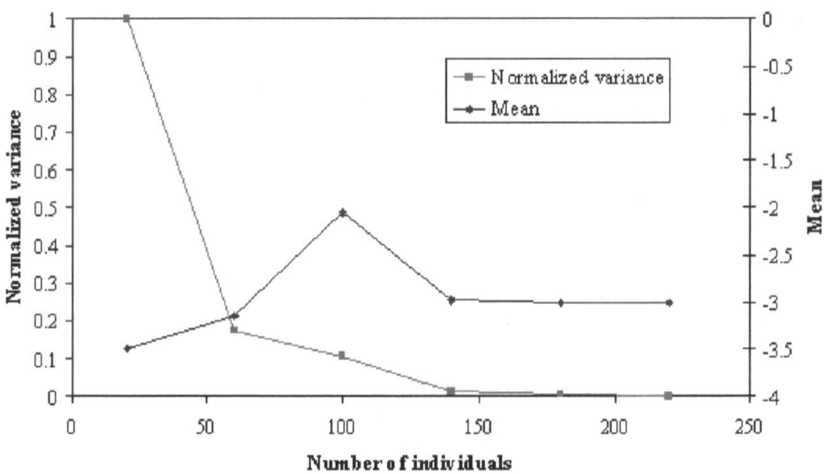

Fig. 16 Normalized variance versus number of individuals (zinc)

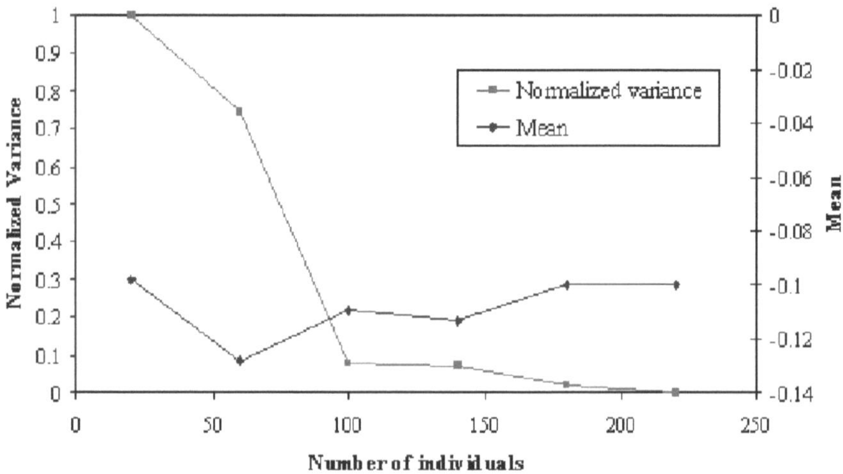

Fig. 17 Normalized variance versus number of individuals (arsenic)

28 contaminants taken into account in the above-described case study; variance has been normalized to 1. In order to obtain the minimum value, for each step the procedure was run 12 times, each time with a different initial configuration. For both contaminants, the experimental variance slowly decreases when the number of individuals increases, even though the objective function is based on the minimization of the selection errors and not on the minimization of the cross-validation experimental variance.

Procedure Steps

Ordinary block kriging method has been implemented to automate the process of reconstruction of 3D contamination, by means of a procedure that consists of the following steps:

1. Procedure starts from a randomly generated population of solutions ($C_0, m_1, m_2, \varphi, m_3, n$).
2. For each individual (solution) the cross-validation is realized and the objective function is calculated; an objective function value is obtained for each individual; the best one is that corresponding to the minimum.
3. A new population of solutions is generated starting from the old one.
4. Point 3 is repeated until convergence criteria is reached.
5. Parameters obtained from the optimization process are used to estimate contaminant concentration on a 3D grid; blocks that need to be remediated are selected on

the basis of the estimated values, by comparing them with the thresholds defined by the user.
6. The procedure exposed is repeated for all contaminants.

Comparison with Interactive Procedure

In this section the results of the cross-validation are reported, obtained for the two contaminants extracted from the above-described case study. A population composed of 100 individuals was used. As indicated above, the procedure's aim is to minimize the fitness function (Figs. 18 and 21).

In the case of zinc, the algorithm stops after 53 generations, with a value of 5962.6. For arsenic, the algorithm stops after 44 generations, with a value of 6202.6. In the above-described cases, the cross-validation results of the automatic procedure are comparable to those obtained with the interactive procedure (Table 1–2 and Figs. 19–20, 22–23).

Conclusions

A specific automatic procedure has been implemented and it has been run for different cases, showing converging results; moreover, cross-validation results are similar to those obtained with the interactive procedure. Before adopting the proposed

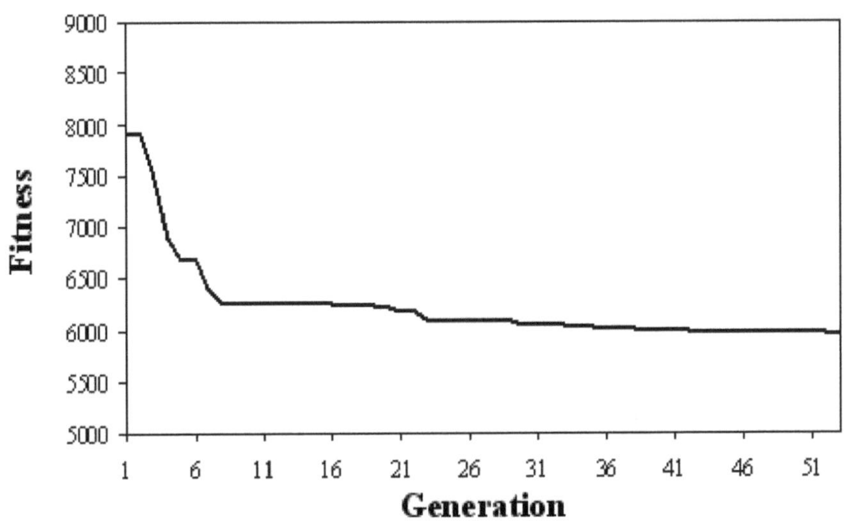

Fig. 18 Fitness value versus number of generations (zinc)

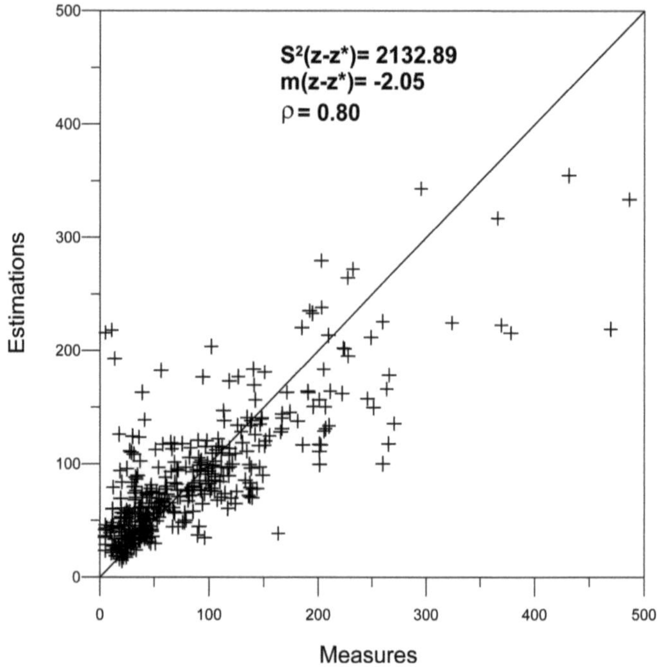

Fig. 19 Scatterplot Zn–Zn* (Interactive procedure)

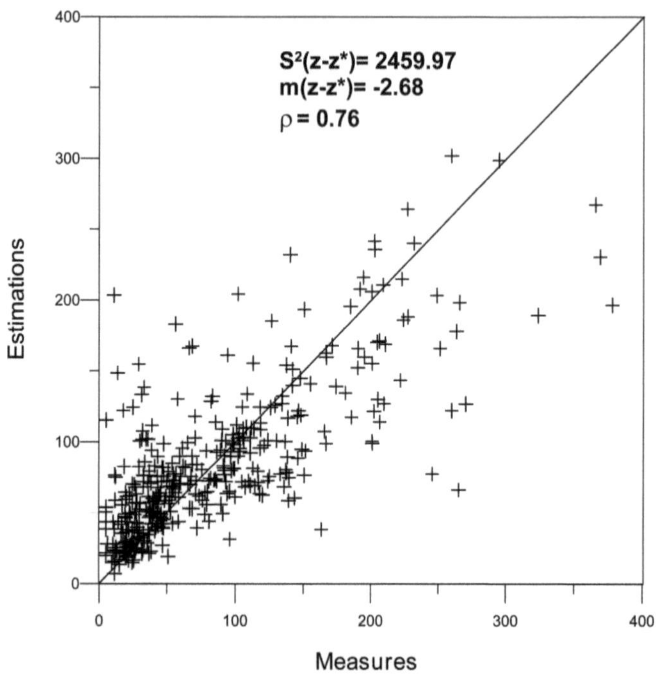

Fig. 20 Scatterplot Zn–Zn* (automatic procedure)

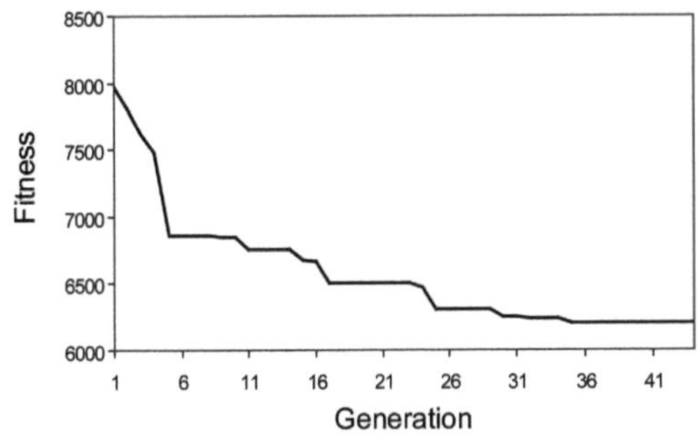

Fig. 21 Fitness value versus number of generations (arsenic)

Table 1 Cross-validation selection errors with respect to three thresholds (Zinc). Interactive procedure (up value) versus automatic procedure (down value)

Limit (mg/kg)	Type I error (%)	Type II error (%)	Correct selection (%)
31.2	2.86	14.81	82.34
	1.04	16.1	82.86
56.2	2.86	19.74	77.40
	2.34	16.62	81.04
114.31	8.83	5.195	85.97
	7.27	5.97	86.75

Table 2 Cross-validation selection errors with respect to three thresholds (arsenic). Interactive procedure (up value) versus automatic procedure (down value)

Limit (mg/kg)	Type I error (%)	Type II error (%)	Correct selection (%)
15.08	4.16	12.47	83.38
	4.94	7.27	87.79
18.64	9.09	15.58	75.33
	8.31	13.77	77.92
23.06	12.99	11.43	75.58
	8.31	9.09	82.6

automatic procedure, some other tests need to be performed with different case studies and different contaminants, taking into account not only the cross-validation results, but also the estimation results and their consequences on the selection of sediments to be remediated, with respect to the results of the interactive procedure. In the next few months, depending on the results from experimentations, possible variants of the procedure will be studied, with respect to both the objective function and the variogram parametrization.

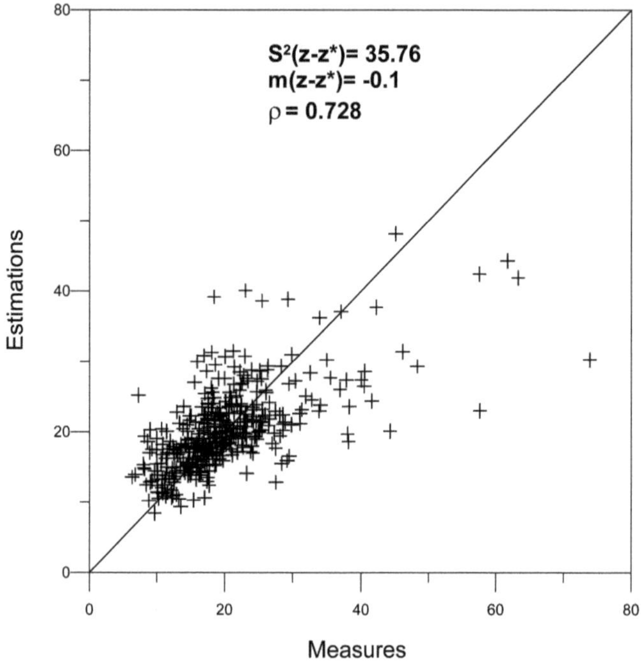

Fig. 22 Scatterplot As–As* (interactive procedure)

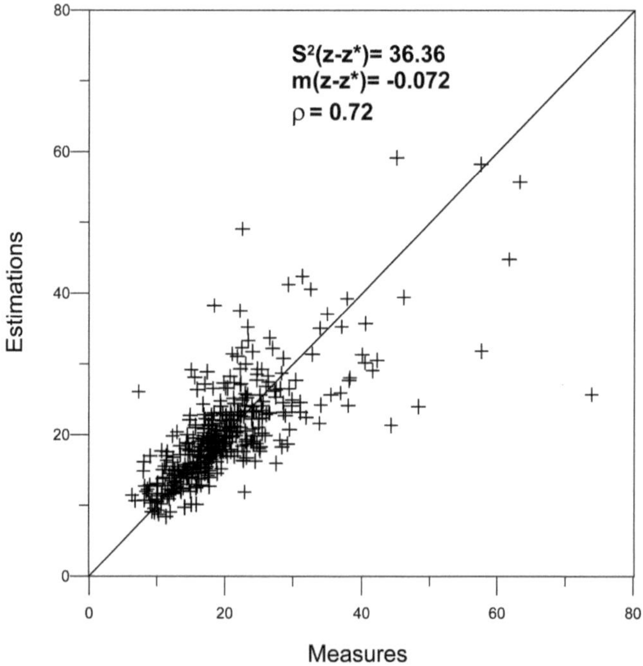

Fig. 23 Scatterplot As–As* (automatic procedure)

References

Bação F, Caeiro S, Painho M, Goovaerts P, Costa MH (2004) Delineation of estuarine management units: evaluation of an automatic procedure. geoENV 2004 Neuchâtel, Pre-Proceedings
Chilès J-P, Delfiner P (1999) Geostatistics. Modeling spatial uncertainty. Wiley series in probability and statistics, pp 95–97
Deutsch CV (1996) Correcting for negative weights in ordinary kriging. Comput Geosci, Elsevier Sci Ltd 22(7):765–773
Deutsch CV, Journel A (1992) GSLIB: Geostatistical software library and user's guide. Oxford University Press, New York
Mitchell M (1996) An Introduction to genetic Algorithms. MIT Press
Van Groenigen JW, Stein A (1998) Constrained optimization of spatial sampling using continuous simulated annealing. J Environ Qual 27:1078–1086
Van Groenigen JW, Pieters G, Stein A (2000) Optimizing spatial sampling for multivariate contamination in urban areas. Environmetrics 11:227–244
Winter G, P'eriaux J, Gal'an M, Cuesta P (1995) Genetic algorithms in engineering and computer science. Wiley series in computational methods in applied sciences

Part VI
Methods

Nonlinear Spatial Prediction with Non-Gaussian Data: A Maximum Entropy Viewpoint

P. Bogaert and D. Fasbender

Abstract We propose to look here at the problem of nonlinear spatial prediction from a maximum entropy viewpoint, where the marginal probability distribution function (pdf) is assumed to belong to the parametric family of exponential polynomials of order p, i.e. the family of maximum entropy solutions under constraints for the p first moments. The general methodology for modeling this marginal pdf is given first, allowing afterwards an estimation of multivariate maximum entropy pdf's that account at the same time for the marginal pdf and a specified covariance function.

As it is notorious that obtaining maximum entropy solutions is computationally heavy, an implementation of the method is proposed using Monte-Carlo integration, with preference sampling as main variance reduction technique. The advantages and drawbacks of using maximum entropy distributions over standard marginal transformation are explained and discussed in the light of a real case study.

Introduction

Let us consider the spatial prediction problem where what is sought for is the conditional pdf $f(z_0|\mathbf{z}_i) = f(z_0, \mathbf{z}_i)/f(\mathbf{z}_i)$ with expectation $\mu_{z_0|\mathbf{z}_i}$ at prediction location $\mathbf{x}_0$ given a set of observed values $\mathbf{z}_i = \{z_1, \ldots, z_n\}$ at surrounding locations. Assuming a second-order stationary random field (RF), it is well-known that $\mu_{z_0|\mathbf{z}_i}$ corresponds to the best linear unbiased predictor (i.e., kriging) as long as the RF is Gaussian (i.e., fully characterized by its mean function $\mu(\mathbf{x})$ and covariance function $C(\mathbf{h})$, where $f(z_0|\mathbf{z}_i)$ is Gaussian). For non-Gaussian RF's however, the best predictor is no longer linear, and one should find other ways of obtaining the conditional pdf. This problem has led to several alternatives ranging from simple ones like, e.g., data transformation to more complex ones like, e.g., disjunctive kriging.

P. Bogaert
Department of Environmental Sciences and Land Use Planning,
Université catholique de Louvain, Belgium
e-mail: bogaert@enge.ucl.ac.be

A. Soares et al. (eds.), *geoENV VI – Geostatistics for Environmental Applications*, 445–455

In this paper, we propose another way of tackling the problem, that relies on a maximum entropy (maxent) viewpoint. The objectives will be threefold. The first one is to briefly present the philosophy of maxent in a spatial prediction context. The second one is to propose a tractable implementation for multivariate pdf's. The last one is to discuss the pros and cons of this approach compared to a very popular and easy to implement one that aims at obtaining the joint distribution through marginal transforms.

Maxent Marginal Density Estimation

Consider a second-order stationary RF, where $\mu(\mathbf{x})$ and $C(\mathbf{h})$ are known or can be inferred with some confidence for the data at hand, so that without loss of generality standardized values can be used instead, with $E[Z_i] = 0$ and $Var[Z_i] = 1\ \forall i = 0, \ldots, n$. What is first sought for is a model for the marginal pdf $f(z)$, that needs to be inferred from data $\mathbf{z}_i$. Classical approaches range from non-parametric ones (e.g., kernel estimates; Silverman, 1983) to parametric ones, that aim at picking the right parametric distribution from which the data are originating. We propose here to adopt the maxent viewpoint, where this pdf is assumed to belong to the exponential polynomial (EP) family of pdf's, i.e.,

$$f(z) = \exp\left(-\sum_{j=0}^{p} \lambda_j z^j\right) = \frac{1}{A}\exp\left(-\sum_{j=1}^{p} \lambda_j z^j\right) \tag{1}$$

where A is a normalization constant so that $\int_{\mathbb{R}} f(z)dz = 1$, and where the λ_j's are uniquely identified by imposing the values of the first p moments $E[Z^j]$ (known experimentally from data $\mathbf{z}_i$). Eq. (1) is the maxent solution, in the sense that it corresponds to the pdf with given moments of order $\leq p$ that maximizes $H(Z)$ (e.g., Zellner and Highfiel, 1988), where

$$H(Z) = -\int f(z)\ln f(z)dz = \sum_{j=0}^{p} \lambda_j E[Z^j], \text{ where } E[Z^j] = \int z^j f(z)dz \tag{2}$$

The Gaussian distribution is a particular case when $p = 2$. As there is no analytic solution when $p > 2$, we have to rely on numerical optimization. We use here the method as advocated by Wu (2003), that consists of fitting iteratively the pdf by progressively increasing the value for p. Technical details of the optimization procedure are discussed at length by Wu (2003) and are omitted here for the sake of brevity. Synthetically, if one defines $\lambda_{j,[m]}$ as the coefficients used at the mth iteration for computing $f(z)$ according to Eq. (1), the general procedure is as follows:

1. set $p = 2$ and $m = 0$;
2. set as initial values $\lambda_{0,[0]} = \ln\sqrt{2\pi}$, $\lambda_{1,[0]} = 0$ and $\lambda_{2,[0]} = 1/2$, so that by definition $Z \sim N(0, 1)$;

3. set $p = p + 2$ and $m = m + 1$;
4. set as initial values $\lambda_{j,[m+1]} = \lambda_{j,[m]}$ when $j \leq p - 2$ and $\lambda_{j,[m+1]} = 0$ when $j = p - 1, p$;
5. solve for the $\lambda_{j,[m+1]}$'s in order to respect the set of constraints $E[Z^j] = \int z^j f(z)dz \; \forall j = 0, \ldots, p$ using a nonlinear optimization procedure, initialized with the values defined in step 4;
6. repeat from step 3 till some optimality criterion is satisfied;

where the optimal p value can either be selected from visual inspection or, more objectively, from statistical testing. I.e., if data are a random sample and $f_p(z)$, $f_{p+2}(z)$ are the EP pdf's of degree p and $p+2$, respectively, the log-likelihood ratio of $f_{p+2}(z)$ vs $f_p(z)$ is asymptotically (i.e., when $n \to \infty$) chi-square distributed with 2 degrees of freedom under the null hypothesis, thus allowing selection of the optimal degree p.

In practice, theoretical moments can be consistently estimated either from raw data with $\widehat{E}[Z^j] = (1/n)\sum_i z_i^j$ or from their histogram counterpart with $\widetilde{E}[Z^j] = (1/n)\sum_{k=1}^m b_k^j n_k$ where n_k is the frequency in the kth bin centered on b_k and $n = \sum_k n_k$, where $\lim_{m\to\infty} \widetilde{E}[Z^j] = \widehat{E}[Z^j]$. If preferential sampling and spatial correlation are issues, various declustering techniques could be applied in order to get an histogram that would be more compatible with the hypothesis of a random sample, as required for likelihood-ratio testing. One may also worry about the quality of the estimates $\widehat{E}[Z^j]$ and $\widetilde{E}[Z^j]$ as j is increased. However, as remarked by Wu and Stengos (2005), the maxent method is equivalent to a maximum likelihood approach where the likelihood function is defined over the exponential distribution and is therefore consistent and efficient.

This maxent approach is flexible enough to account for a wide variety of situations. As we have $\ln f(z) = -\sum_j \lambda_j z^j$ from Eq. (1), as long as the logarithm of the true but unknown pdf can be well approximated over $\mathbb{R}$ by a polynomial of low order, the method will provide very good results. It is worth noting too that we restrict ourselves here to the case of even values for p so that $f(z) \neq 0$ over $\mathbb{R}$ with $\lim_{z\to\pm\infty} f(z) = 0$, but the procedure could be extended as well for odd values if $f(z) \neq 0$ on a semi-infinite or finite interval only. Moreover, as long as the information at hand for estimating this pdf can be written as the expectation of a functional of Z, Eq. (1) is still relevant. This offers the possibility to use, e.g., logarithmic moments $E[(\ln Z)^j]$, or even probability values, as

$$P[a \leq Z \leq b] = \int_a^b f(z)dz = \int_{\mathbb{R}} \delta_{a\leq z\leq b} f(z)dz = E[\delta_{a\leq Z\leq b}]$$

where $\delta_{a\leq z\leq b} = 1$ when $a \leq z \leq b$ and is equal to 0 otherwise.

As is, this solution for estimating $f(z)$ is already useful for data transformation. Indeed, a classical way of handling non-Gaussian data is to (i) transform the original z-values toward marginally $N(0,1)$ y-values using the integral transform theorem where $y_i = G^{-1}(F(z_i))$ with $G(.)$ the $N(0,1)$ cumulative distribution function (cdf), (ii) conduct "optimal" linear prediction using y-values assuming multivariate

Gaussianity (MG) holds, and finally (iii) backtransform the conditional pdf $f(y_0|\mathbf{y}_i)$ towards $f(z_0|\mathbf{z}_i)$ assuming $Y_0|\mathbf{y}_i \sim N(\mu_{y_0|\mathbf{y}_i}, \sigma^2_{y_0|\mathbf{y}_i})$. However, despite its simplicity, this approach suffers from a poorly assessed MG hypothesis. Moreover, as there is no conservation of the maxent property through nonlinear transforms, the corresponding joint pdf $f(\mathbf{z})$ may also contain spurious information that are not supported neither by the marginal pdf $f(z)$ nor by the covariance function $C(\mathbf{h})$. It is thus possible to find other expressions for $f(\mathbf{z})$ that honor $f(z)$ and $C(\mathbf{h})$ too but have higher entropy, hence having higher plausibility.

Maxent Joint Density Estimation

Additionally to the information already brought by $f(z)$, the joint pdf $f(\mathbf{z})$ should also account for cross-moments $E[Z_i Z_{i'}]$ as given by $C(\mathbf{h})$. It is not difficult to prove (details not provided here) that the maxent solution under these extra constraints is a multivariate exponential polynomial given by

$$f(\mathbf{z}) = \exp\Big(-\upsilon_0 - \sum_{i=0}^{n}\sum_{j=1}^{p} \upsilon_{ij} z_i^j - \mathbf{z}'\Upsilon\mathbf{z}\Big) \quad \text{with } \upsilon_0 = \ln A \tag{3}$$

where each set of coefficients $\{\upsilon_{i1}, \ldots, \upsilon_{ip}\}$ has the same meaning (but is of course different) than the set $\{\lambda_1, \ldots, \lambda_p\}$ obtained for $f(z)$, Υ is a symmetric $(n+1) \times (n+1)$ matrix of coefficients with null diagonal elements that accounts for the $E[Z_i Z_{i'}]$'s $(i \neq i')$, and A is a normalization constant.

Solving for the $1 + p(n+1) + p(p-1)/2$ coefficients respecting the set of constraints for the corresponding moments is a difficult task. E.g., with as few as a $n=3$ values in the prediction neighborhood and a degree $p = 4$, this is a 23 parameters optimization problem requiring integration over $\mathbb{R}^4$. It is however possible to obtain solutions through careful numerical implementation. Indeed, Eq. (3) can be reparameterized, with

$$\begin{aligned} f(\mathbf{z}) &\propto \exp\Big(-\sum_i\sum_j \upsilon_{ij} z_i^j - \mathbf{z}'\Upsilon\mathbf{z}\Big) \\ &= \exp\Big(-\sum_i\sum_j \lambda_j z_i^j\Big)\exp\Big(-\frac{1}{2}\mathbf{z}'\mathbf{V}^{-1}\mathbf{z}\Big)\exp\Big(-\sum_i\sum_j \eta_{ij} z_i^j - \mathbf{z}'\mathbf{N}\mathbf{z}\Big) \end{aligned} \tag{4}$$

where $\eta_{ij} = \upsilon_{ij} - \lambda_i$ and $\mathbf{N} = \Upsilon - (1/2)\mathbf{V}^{-1}$ are these new parameters, $\mathbf{V}$ being a given positive-definite matrix. By remarking that

$$\exp\Big(-\sum_i\sum_j \lambda_j z_i^j\Big) = \prod_i \exp\Big(-\sum_j \lambda_j z_i^j\Big) \propto \prod_i f(z_i) \tag{5}$$

where each $f(z_i)$ is the marginal maxent pdf evaluated at z_i, and that

$$\exp\Big(-\frac{1}{2}\mathbf{z}'\mathbf{V}^{-1}\mathbf{z}\Big) \propto g(\mathbf{z}) \tag{6}$$

where $g(\mathbf{z})$ is the joint pdf of the multivariate $N(\mathbf{0,V})$, Eq. (3) becomes

$$f(\mathbf{z}) = \frac{1}{B}g(\mathbf{z})\Big(\prod_i f(z_i)\Big)\exp\Big(-\sum_i\sum_j \eta_{ij}z_i^j - \mathbf{z}'\mathbf{N}\mathbf{z}\Big) \tag{7}$$

Solving for the coefficients is made difficult because of the multivariate integrals. However, expressing moments from Eq. (7) instead of Eq. (3) now gives

$$\begin{aligned} B &= \int g(\mathbf{z})h(\mathbf{z})d\mathbf{z} = E_{g(\mathbf{Z})}\Big[h(\mathbf{Z})\Big] = \lim_{K\to\infty}\frac{1}{K}\sum_{k=1}^{K}h(\mathbf{z}) \\ E[Z_i^j] &= \int z_i^j f(\mathbf{z})d\mathbf{z} = \frac{1}{B}E_{g(\mathbf{Z})}\Big[Z_i^j h(\mathbf{Z})\Big] = \lim_{K\to\infty}\frac{1}{BK}\sum_{k=1}^{K}z_i^j h(\mathbf{z}) \\ E[Z_iZ_{i'}] &= \int z_iz_{i'} f(\mathbf{z})d\mathbf{z} = \frac{1}{B}E_{g(\mathbf{Z})}\Big[Z_iZ_{i'} h(\mathbf{Z})\Big] = \lim_{K\to\infty}\frac{1}{BK}\sum_{k=1}^{K}z_iz_{i'} h(\mathbf{z}) \end{aligned} \tag{8}$$

with $h(\mathbf{z}) = \Big(\prod_i f(z_i)\Big)\exp\Big(-\sum_i\sum_j \eta_{ij}z_i^j - \mathbf{z}'\mathbf{N}\mathbf{z}\Big)$ and where the K sets of $\mathbf{z}$ values are sampled from the multivariate Gaussian $\mathbf{Z} \sim N(\mathbf{0,V})$. This corresponds to Monte-carlo integration with preference sampling and yields an easy way of computing the requested moments. In order to appropriately cover the domain, one can set $\mathbf{V} = c\Sigma$, where $\Sigma = \{C(\mathbf{x}_i - \mathbf{x}_{i'})\}$ with $c \geq 1$ a coverage factor that can be tuned. Using a similar iterative approach as for $f(z)$, the degree p can be progressively increased from 2 up to the degree of $f(z)$. At each iteration, solving for the coefficients can be done using either a direct search (e.g., simplex) algorithm or a Gauss-Newton gradient method for faster convergence.

A Case Study: Cadmium in the Swiss Jura

For the sake of illustration, we will consider here Cadmium concentrations as provided by the Swiss Jura data set (see Atteia et al., 1994), where Fig. 1a shows measurement locations. Prior to subsequent computations, original data (in ppm; see Fig. 1b) have been standardized assuming first-order stationarity (i.e., $\mu(\mathbf{x}) = \mu$ over the whole area), as no obvious spatial gradient effect was observable. The variogram has been modeled using a nugget effect (20% of total variance) and two spherical models with 0.2 km and 1 km range, respectively (see Fig. 1c).

As the histogram exhibits a clear positive skewness, a Gaussian assumption (i.e., $p = 2$) is quite arguable here. The marginal $f(z)$ as estimated from maxent principle is presented in Fig. 2. A log-likelihood ratio test (confidence level $1 - \alpha = 0.99$)

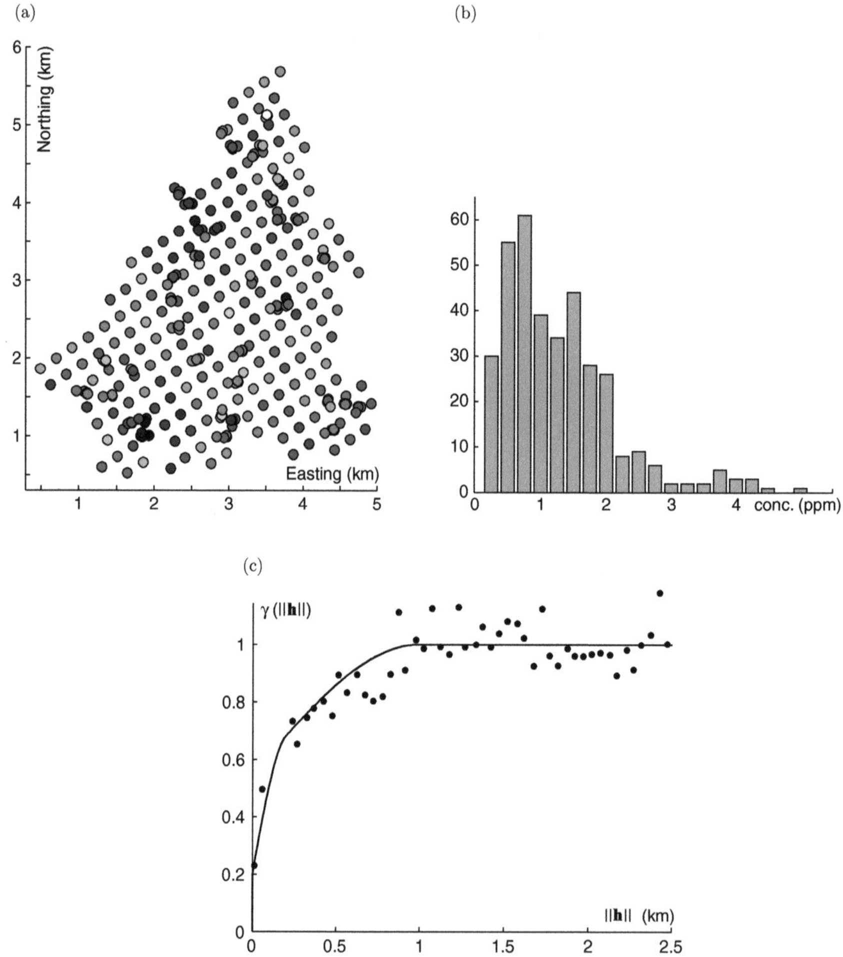

Fig. 1 Locations of the 359 Cadmium measurements (a) with color proportional to soil concentrations (lowest value in black and highest in white), along with histogram in ppm (b) and estimated/modelled normalized semi-variograms (c)

suggests that optimal degree is about $p = 8$, though this is only indicative as the random sample hypothesis does not strictly hold because of the spatial dependence.

Due to the modeled nugget effect, maximum observable correlation is $r = 0.8$ when $||\mathbf{h}|| \to 0$. Let us denote by $\widetilde{f}(\mathbf{z})$ and $\widehat{f}(\mathbf{z})$ the joint pdf's obtained from marginal transformation and entropy maximization, respectively, with $\widetilde{f}(\mathbf{z})$ given from the general formula for monotonic marginal transformation as (see e.g., Papoulis, 1991)

$$\widetilde{f}(\mathbf{z}) = \Big(\prod_i \frac{dy_i}{dz_i}\Big) g(\mathbf{y}) \qquad \text{with } y_i = G^{-1}(F(z_i)) \tag{9}$$

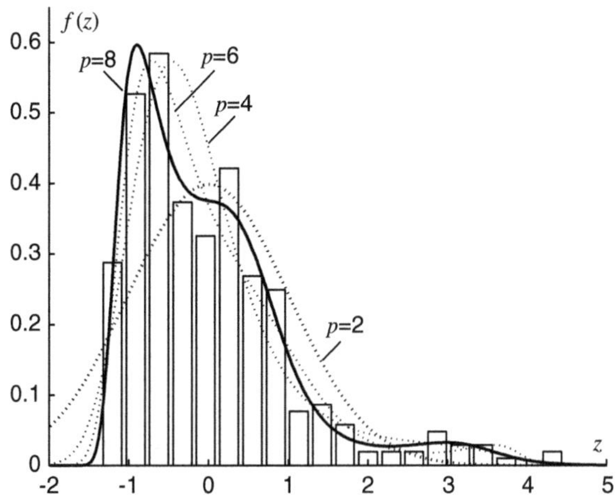

Fig. 2 Scaled histogram and maxent estimated pdf $f(z)$ from standardized values. Solution for $p = 2$ is $Z \sim N(0, 1)$, whereas optimal value is $p = 8$ according to log-likelihood ratio testing

where $G(.)$ is the cdf of the univariate $N(0, 1)$ and $g(.)$ is the pdf of the multivariate $\mathbf{Y} \sim N(\mathbf{0},\mathbf{R})$, with $\mathbf{R} = \{C_{\mathbf{Y}}(||\mathbf{x}_i - \mathbf{x}_{i'}||)\}$ where $C_{\mathbf{Y}}(||\mathbf{0}||) = 1$ (it is worth noting here that $\mathbf{R}$ values on the Gaussian scale are directly computed from the modeled variogram on the original scale using the same transform).

In order to visually assess the differences between $\widetilde{f}(\mathbf{z})$ and $\widehat{f}(\mathbf{z})$, Figs. 3a and 3b show the corresponding pdf's in the bivariate case when $r = 0.8$. As an expected result of the entropy maximization principle, $\widehat{f}(\mathbf{z})$ shows more rounded shape level curves with less peculiarities (e.g., loops) compared to $\widetilde{f}(\mathbf{z})$. As a consequence, the corresponding conditional pdf's $\widetilde{f}(z_0|\mathbf{z}_i)$ may present baseless complex behaviors like, e.g., strong multimodality (see e.g., Figs. 3c and 3d) that are not really supported by information at hand. This in turn may have potentially clear adverse effects for spatial prediction. It is worth remembering here again that both $\widehat{f}(\mathbf{z})$ and $\widetilde{f}(\mathbf{z})$ honor the same marginal $f(z)$ and covariance function $C(\mathbf{h})$, so that any discrepancy between them is only due to the specific way this information (that is the only made available) are processed for estimating the true but unknown $f(\mathbf{z})$. Clearly, differences between $\widetilde{f}(\mathbf{z})$ and $\widehat{f}(\mathbf{z})$ in terms of entropy (i.e., information content) are also expected to increase with correlation, as both methods must obviously yield the same general result $\widetilde{f}(\mathbf{z}) = \widehat{f}(\mathbf{z}) = \prod_i f(z_i)$ when $\mathbf{R}$=$\mathbf{I}$. This is easily seen from Fig. 4a drawn again for bivariate distributions.

Beside the general fact that $\widehat{f}(\mathbf{z})$ conveys less extra information than $\widetilde{f}(\mathbf{z})$, this does not necessarily mean that $\widehat{f}(\mathbf{z})$ will outperform $\widetilde{f}(\mathbf{z})$ for a specific data set. Indeed, the idea of selecting the most entropic joint pdf is to provide a general safeguard against possible blunders. Stated in very simple words, from the concentration theorem (see Jaynes, 1982, 2003), the maxent pdf is the one that maximizes on the average the likelihood of observing a given set of $\mathbf{z}$ values that obey specified

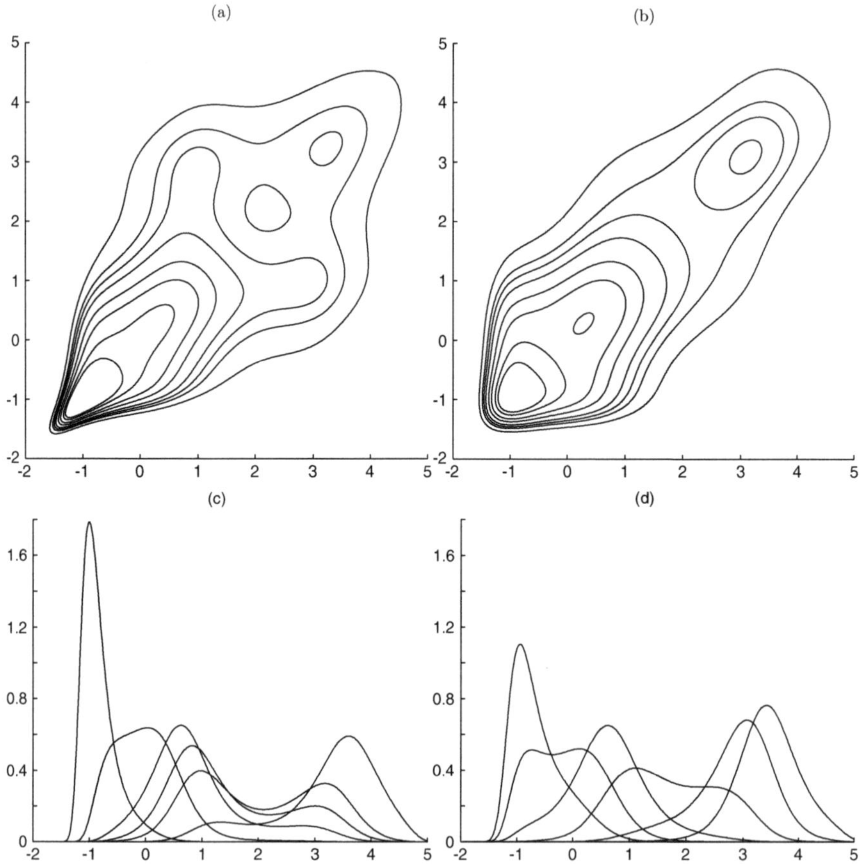

Fig. 3 Bivariate pdf's $f(z_0, z_1)$ as estimated from marginal transform (a) and maxent principle (b) using maximum observable correlation $r = 0.8$. Bottom graphs show a few corresponding conditional pdf's $f(z_0|z_1)$ computed for $z_1 = -1, 0, \ldots, 4$ (c and d)

constraints on the moments. It does not however mean that, for a specific data set, the maxent pdf will systematically yield better results. However, as it is the most generic one, it can at least be used to assess the pertinency of a more specific choice. I.e., in our context, the pertinency of using $\widetilde{f}(\mathbf{z})$ can be assessed based on a comparison with results obtained using the generic $\widehat{f}(\mathbf{z})$, which thus plays the role of a reference here.

In order to illustrate this idea, a cross-validation (leave-one-out) has been conducted using the whole data set. For each location, the respective likelihoods $\widetilde{L}(\mathbf{z})$ and $\widehat{L}(\mathbf{z})$ are computed and compared, the most appropriate pdf for this data set being the one that maximizes it on the average. Results are given in Fig. 4b, where it can be seen that $\widetilde{L}(\mathbf{z})$ gives slightly better results. This can be explained when one look back at Fig. 3, where one can observe that, compared to $\widehat{f}(\mathbf{z})$, the pdf

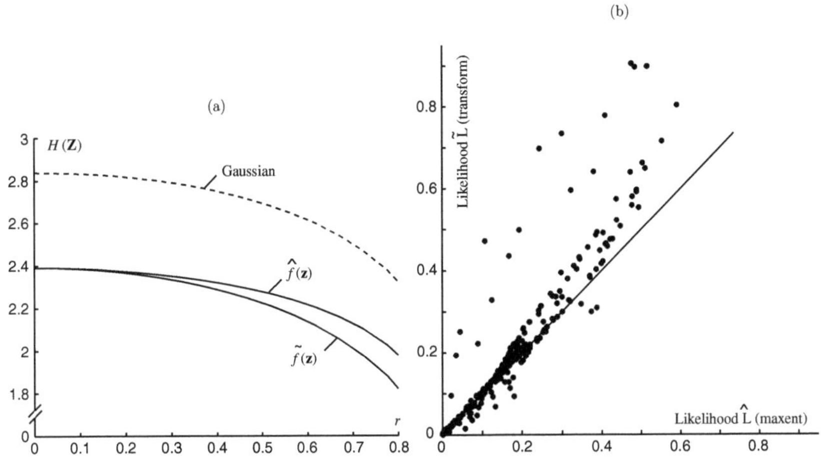

Fig. 4 Part (a) shows entropy $H(\mathbf{Z})$ of bivariate pdf's estimates $\widehat{f}(\mathbf{z})$ and $\widetilde{f}(\mathbf{z})$ over the [0, 0.8] interval of observable correlation values r (entropy for corresponding bivariate Gaussian pdf with same correlation is also provided for comparison purpose). Part (b) shows likelihood values $\widehat{L}(\mathbf{z})$ and $\widetilde{L}(\mathbf{z})$ computed at the 359 cross-validated locations from the corresponding multivariate pdf estimates

$\widetilde{f}(\mathbf{z})$ will tend to give in general narrower (i.e., more assertive) conditional pdf's for low conditioning values, whereas it is the opposite for high conditioning values. As a consequence, if high values tend to be poorly spatially correlated compared to the average whereas it is the opposite for low ones, this tends to favor $\widetilde{f}(\mathbf{z})$. In order to verify this hypothesis, the data set has been split in two subsets according to whether standardized values exceed ($n = 42$) or do not exceed ($n = 317$) a given threshold value, set equal to 1 (see Figs. 5a and 5b; this threshold is equal to 2.15 ppm for raw values as seen on Fig. 1b), as the tail of $f(z)$ seems to suggest a mixture of populations, and the variogram of each data set has been estimated. Though variogram modeling for data above 1 is subject to caution due to the limited sample size, there is little doubt from Figs. 5c and 5d that the essential part of the observed spatial correlation is linked to low values, as high ones seem to lack any dependence pattern. An explanation would be that these data sets are ruled by completely different mechanisms (i.e. high values would be the result of very local and occasional pollution that do not obey any kind of spatial logic and are thus unpredictable from this point of view). As a conclusion, it is here more by chance than as the result of a clear rational choice that $\widetilde{f}(\mathbf{z})$ gives slightly better results than $\widehat{f}(\mathbf{z})$. I.e., for another data set that would exhibit similar histogram and variogram but well spatially correlated high values, $\widetilde{f}(\mathbf{z})$ would have been clearly on the wrong side compared to $\widehat{f}(\mathbf{z})$. Moreover, the benefit of using $\widetilde{f}(\mathbf{z})$ instead of $\widehat{f}(\mathbf{z})$ must be temperate by the globally limited quality of the predictions whatever the method used, according to Table 1 showing that a marginal transform does not bring any clear improvement compared to kriging conducted using the raw data. Splitting predicted values according to the same threshold shows also that very

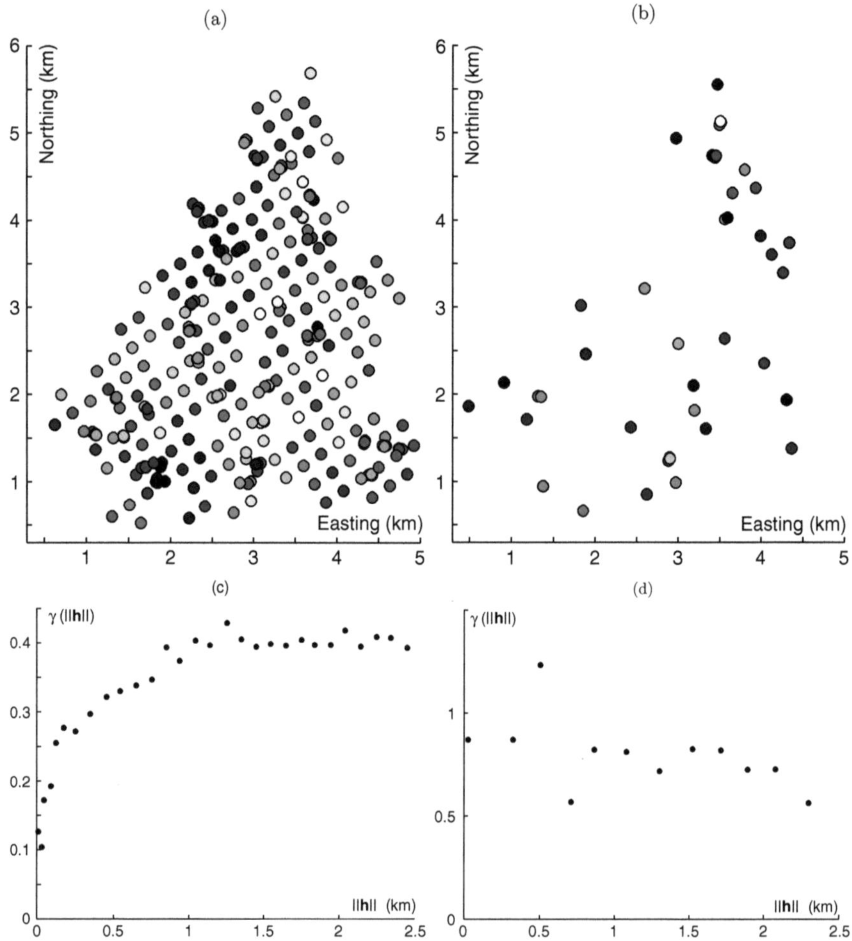

Fig. 5 Locations of the 317 Cadmium values that are below 1 (a) and the 42 values that are above 1 (b), with color proportional to soil concentrations (color scales are different), with corresponding estimated semi-variograms (c and d)

Table 1 Cross-validation results for kriging using raw data (KR) and with marginal transform (KT). First column gives overall Mean Square Errors (MSE), second and third ones give MSE for data below or above threshold set to 1

	MSE		
Method	overall	below 1	above 1
KR	0.746	0.384	3.47
KT	0.723	0.356	3.50

poor predictions are obtained for high values, as expected. This also cast serious doubts about how reliable could be predictions for high values, which are the most important ones to be identified from the environmental viewpoint.

Conclusions and Perspectives

It has been shown how the maxent principle may be useful for estimating marginal pdf's as well as for estimating multivariate pdf's that additionally account for a covariance function. Though we restricted ourselves here to the case of raw moments, the same approach may be used for other moments (e.g., logarithmic moments $E[(\ln Z)^j]$ yielding a lognormal marginal pdf's when $p = 2$) and can be extended for truncated distributions as well, thus offering considerable flexibility. In a marginal transform context, it is already a valuable method; rather than relying on the experimental cdf as advocated in the normal score transform (NST) method, the maxent cdf $F(z)$ can be used instead, thus offering the guarantee of a continuous and differentiable $f(z_0|\mathbf{z}_i)$ at the end (which is not possible using NST of course, as the experimental cdf is badly non differentiable). As multivariate pdf's can be estimated by extending this principle, it is also a valuable alternative for conducting non linear spatial prediction, as it offers the guarantee that it will select a pdf that honors $f(z)$ and $C(\mathbf{h})$ and provides at the same time a safeguard against possible blunders as it is the most entropic one (so containing the lesser amount of extra information that were not explicitly specified by $f(z)$ and $C(\mathbf{h})$).

Though the maxent principle for multivariate pdf can be considered as a sound prediction method in itself, there are some drawbacks linked to computational burden issues, due to the high number of parameters to be estimated. As currently implemented, it is thus restricted to limited neighborhood size, so that additional work may be needed for using it in an fully operational context. However, it is still a promising alternative. Being the most generic solution, it can also be used as an external reference for comparing methods, as it has been illustrated here.

References

Atteia O, Dubois J.-P, Webster R (1994) Geostatistical analysis of soil contamination in the swiss jura. Environ. Pollut 86:315–327

Jaynes E.T. (1982) On the rationale of maximum-entropy methods. Proc IEEE 70:939–952

Jaynes E.T. (2003) Probability theory: the logic of science. Cambridge University Press, New York

Papoulis A (1991) probability, Random Variables and Stochastic Processes. 3rd ed, Mc Graw-Hill, New York

Silverman B.W (1983) Density Estimation for Statistics and Data Analysis. Chapman & Hall

Wu X, (2003) Calculation of maximum entropy densities with application to income distribution. Environ. Pollut 115:347–354

Wu X, Stengos T (2005) Partially adaptive estimation via the maximum entropy densities. Working paper 2005. Department of Economics, University of Guelph, Guelph, Ontario, Canada

Zellner A, Highfiel R.A. (1988) Calculation of maximum entropy distribution and approximation of marginal posterior distributions. J Econometrics 37:195–209

Data Fusion in a Spatial Multivariate Framework: Trading off Hypotheses Against Information

D. Fasbender and P. Bogaert

Abstract Due to the exponential growth in the amount and diversity of data that one may expect to provide greater modeling and predictions opportunities, there is a real need for methods that aim at reconciling them inside a flexible and sound theoretical framework. In a geostatistical prediction context, beside more or less straighforward variations around univariate kriging (e.g. kriging with external drift), the most classical method (i.e. cokriging) is based on a multivariate random field approach of the problem, at the price of strong modeling hypotheses. However, there are expected practical situations where these hypotheses may be hard to fulfill or do not make sense from a modeling viewpoint.

This paper proposes an alternative way of using secondary information for spatial prediction. Based on a data fusion perspective, a general theoretical procedure is proposed. Simple cokriging and Bayesian data fusion are compared both from theoretical and practical viewpoints. Theoretical differences are first emphasized based on the corresponding modeling hypotheses. A case study based on synthetic data subsequently allows to compare both methods from a practical viewpoint. It is shown that, in spite of some simplifying hypotheses required by data fusion, the method is offering quite comparable performances in situations where simple cokriging is expected to be the best possible predictor. Moreover, it offers a much greater flexibility and opens new avenues for incorporating a wide panel of very different and possibly numerous secondary information that, by nature, would not easily fit into a multivariate random field framework, as required by cokriging.

1 Introduction

As the classical (co)kriging predictor relies on the knowledge of first and second-order moments for a given set of random fields (RF's), we will assume here that first and second-order stationarities can be assumed for all variables, and that the

D. Fasbender
Department of Environmental Sciences and Land Use Planning,
Université catholique de Louvain, Belgium
e-mail: fasbender@enge.ucl.ac.be

A. Soares et al. (eds.), *geoENV VI – Geostatistics for Environmental Applications*, 457–466

corresponding functions (i.e. the mean functions and the (cross-)covariance functions) can reasonably be inferred from the data. Without loss of generality and for the sake of simplicity, we will restrict here the discussion to variables with known mean functions so that simple cokriging can be used, though the methodology includes of course the case of non stationary mean and intrinsic stationarity as well (see e.g. Christakos, 1992).

In a first section, a short recall of simple cokriging (SCoK) formulation is presented. The hypotheses involved in this model are emphasized and modeling issues are pointed out with respect to both theoretical and practical viewpoints. Subsequently, the Bayesian Data Fusion (BDF) methodology is presented. For the sake of conciseness and without loss of generality, presentation will be restricted here to the particular case of bivariate Gaussian RFs. Pros and cons of the method are discussed too, and a synthetic case study is presented. The corresponding results indicate that BDF is an interesting alternative in the context of spatial prediction that need to account for additional secondary information which may not fit very well into a multivariate stationary RF framework.

2 Simple Cokriging

In the case of several correlated RFs that are sampled over the same spatial domain, assuming that means are known everywhere, the classical geostatistical method used for conducting multivariate prediction is SCoK (Cressie, 1991). If $\mathcal{Z} = \left\{Z(\mathbf{x}) : \mathbf{x} \in \mathbb{R}^d, Z(\mathbf{x}) \in \mathbb{R}^1\right\}$ and $\mathcal{Y} = \left\{Y(\mathbf{x}) : \mathbf{x} \in \mathbb{R}^d, Y(\mathbf{x}) \in \mathbb{R}^1\right\}$ are two zero-mean second-order stationary RFs and $\mathbf{Z}_\alpha = Z_1, \ldots, Z_n$ and $\mathbf{Y}_\gamma = Y_{n+1}, \ldots, Y_{n+m}$ are corresponding random vectors sampled from them at locations $\mathbf{x}_i$ $(i = 1, \ldots, m+n)$, the SCoK predictor for Z_0 is then

$$Z_0^p = \left(\boldsymbol{\sigma}'_\alpha\ \boldsymbol{\sigma}'_\gamma\right) \begin{pmatrix} \boldsymbol{\Sigma}_{\alpha\alpha} & \boldsymbol{\Sigma}_{\alpha\gamma} \\ \boldsymbol{\Sigma}_{\gamma\alpha} & \boldsymbol{\Sigma}_{\gamma\gamma} \end{pmatrix}^{-1} \begin{pmatrix} \mathbf{z}_\alpha \\ \mathbf{y}_\gamma \end{pmatrix} \tag{1}$$

where the $\boldsymbol{\sigma}$s and $\boldsymbol{\Sigma}$s are obtained from the partitioning of the covariance matrix $\boldsymbol{\Sigma}$ for the whole vector $(Z_0, \mathbf{Z}'_\alpha, \mathbf{Y}'_\gamma)$, with

$$\boldsymbol{\Sigma} = \begin{pmatrix} \sigma_0^2 & \boldsymbol{\sigma}'_\alpha & \boldsymbol{\sigma}'_\gamma \\ \boldsymbol{\sigma}_\alpha & \boldsymbol{\Sigma}_{\alpha\alpha} & \boldsymbol{\Sigma}_{\alpha\gamma} \\ \boldsymbol{\sigma}_\gamma & \boldsymbol{\Sigma}_{\gamma\alpha} & \boldsymbol{\Sigma}_{\gamma\gamma} \end{pmatrix} \tag{2}$$

whereas the corresponding variance of prediction is given by

$$\sigma_0^p = \sigma_0^2 - \left(\boldsymbol{\sigma}'_\alpha\ \boldsymbol{\sigma}'_\gamma\right) \begin{pmatrix} \boldsymbol{\Sigma}_{\alpha\alpha} & \boldsymbol{\Sigma}_{\alpha\gamma} \\ \boldsymbol{\Sigma}_{\gamma\alpha} & \boldsymbol{\Sigma}_{\gamma\gamma} \end{pmatrix}^{-1} \begin{pmatrix} \boldsymbol{\sigma}_\alpha \\ \boldsymbol{\sigma}_\gamma \end{pmatrix} \tag{3}$$

Clearly, SCoK is the best linear predictor, but not necessarily the best predictor. However, assuming multivariate gaussianity for $(Z_0, \mathbf{Z}'_\alpha, \mathbf{Y}'_\gamma)$, the best predictor is linear, and SCoK then corresponds to the result of a linear regression, with

$$Z_0^p = \mathbb{E}\left[Z_0|\mathbf{z}_\alpha, \mathbf{y}_\gamma\right] \quad ; \quad \sigma_0^p = \mathbb{V}ar\left[Z_0|\mathbf{z}_\alpha, \mathbf{y}_\gamma\right] \tag{4}$$

and where $Z_0|\mathbf{z}_\alpha, \mathbf{y}_\gamma \sim N(Z_0^p, \sigma_0^p)$. In other words, SCoK can only be considered as the best predictor when full gaussianity holds. If this is not the case, SCoK is still a valuable predictor, but its prediction variance is not necessarily the smallest possible one and the true conditional probability distribution function (pdf) for $Z_0|\mathbf{z}_\alpha, \mathbf{y}_\gamma$ is not Gaussian in general.

Though SCoK is a straighforward and well-known method that can be easily numerically implemented without difficulties, there are some practical issues that need to be addressed. Clearly, obtaining the covariance matrix $\boldsymbol{\Sigma}$ for arbitrary locations rely on a multivariate modeling of the covariance functions, in order to ensure positive-definiteness of the final results. The most frequently used method is based on the so-called Linear Model of Coreginalization (see e.g. Chilès and Delfiner, 1999 or Goovaerts, 1997), which for a bivariate case is written as

$$\begin{pmatrix} C_{\alpha\alpha}(\mathbf{h}) & C_{\alpha\beta}(\mathbf{h}) \\ C_{\alpha\beta}(\mathbf{h}) & C_{\beta\beta}(\mathbf{h}) \end{pmatrix} = \sum_i \begin{pmatrix} a_{i,\alpha\alpha} & a_{i,\alpha\beta} \\ a_{i,\alpha\beta} & a_{i,\beta\beta} \end{pmatrix} c_i(\mathbf{h}) = \sum_i \mathbf{A}_i c_i(\mathbf{h}) \tag{5}$$

where all matrices $\mathbf{A}_i$ are positive definite and where the same elementary positive definite covariance models $c_i(\mathbf{h})$ must be used for modeling all (cross-)covariance functions. Clearly, this imposes an important (and rarely discussed) conceptual constraint on the method, as any secondary variable that need to be accounted for when predicting Z_0 must fit into the second-order stationary RF paradigm, i.e. the random vector $\mathbf{Y}_\gamma$ must be considered as a sample from a second-order stationary RF that can be characterized by a covariance function $C_{\beta\beta}(\mathbf{x}_i - \mathbf{x}_j)$ that only depends on $\mathbf{x}_i - \mathbf{x}_j$ but neither on $\mathbf{x}_i$ nor on $\mathbf{x}_j$. Unfortunately, it happens frequently that quite useful information does not fit well into this paradigm. As a simple example, in an environmental pollution context, the distance $\mathbf{x} - \mathbf{x}_j$ to a chemical industry located at $\mathbf{x}_j$ is likely to be a quite relevant measure for quantifying atmospheric deposition $Y(\mathbf{x})$, but $Y(\mathbf{x})$ cannot be considered as coming from a RF whose covariance function would be translation invariant, i.e. depending only on $\mathbf{h}$, of course. This in turn may seriously impair the prediction of a related variable of interest $Z(\mathbf{x})$ (e.g. soil pollution) using a cokriging approach, as corresponding covariance function estimates $\widehat{C}_{\beta\beta}(\mathbf{h})$ and $\widehat{C}_{\alpha\beta}(\mathbf{h})$ are meaningless.

3 Bayesian Data Fusion

By the light of the previous example, it appears that relying on a multivariate second-order stationary RF framework is quite arguable in some instances. In order to account for this issue, a new theoretical framework has recently been proposed. Its aim is to alleviate the need of second-order stationarity for secondary variables at the price of mild simplifying hypothesis. Due to space limitations, only main and most important results will be presented here. Additional details can be found and are disscussed at length in Bogaert and Fasbender (2007).

Let us assume that $\mathcal{Z}$ is a zero-mean second-order stationary RF of primary interest, where $\mathbf{Z}$ is random vector sampled from it. Let us define a mapping $g : \mathbb{R} \rightarrow \mathbb{R}$ such that $\mathbf{g}(\mathbf{Z}) = \{g(Z_i)\}$ is a new random vector, along with an arbitrary zero-mean random vector $\mathbf{E}$ of same size and independent from $\mathbf{Z}$. The basic assumption of BDF is to assume that, for any secondary variable, we have $\mathbf{Y} = \mathbf{g}(\mathbf{Z}) + \mathbf{E}$. Clearly, $\mathbf{Z}$ and $\mathbf{Y}$ are collocated random variables but $\mathbf{Y}$ is not longer second-order stationary in general, as its properties also depend on the arbitrary $\mathbf{E}$. According to the independence assumption $\mathbf{E} \perp \mathbf{Z}$, it is also clear that the conditional pdf for $\mathbf{Z}|\mathbf{y}$ is given by

$$f_{\mathbf{z}|\mathbf{y}}(\mathbf{z}|\mathbf{y}) \propto f_{\mathbf{z}} f_{\mathbf{y}|\mathbf{z}}(\mathbf{y}|\mathbf{z}) = f_{\mathbf{z}}(\mathbf{z}) f_{\mathbf{e}}(\mathbf{y} - \mathbf{g}(\mathbf{z})) \tag{6}$$

where $f_{\mathbf{z}}$ is the *a priori* pdf for $\mathbf{Z}$ and $f_{\mathbf{e}}$ is the pdf for $\mathbf{E}$. A more intuitive interpretation of Eq. (6) is to consider that each Y_i can be viewed as an indirect measurements of the true Z_i, as Y_i is a functional $g(Z_i)$ up to an additive error E_i, where it is reasonable in general to assume these errors as independent from the true $\mathbf{Z}$.

Starting from this last very general relation, a straightforward formulation can be proposed in a spatial prediction context. Let us consider that what is sought for is the conditional pdf for Z_0 given a set of observed values $\mathbf{z}_\alpha = z_1, \ldots, z_n$ for the RF of interest $\mathcal{Z}$ along with a set of observed values $\mathbf{y}_\gamma = y_{n+1}, \ldots, y_{n+m}$ for the auxiliary RF $\mathcal{Y}$. Defining additionally the vector $\mathbf{Z}_\beta = Z_{n+1}, \ldots, Z_{n+m}$ of unobserved variables at the same location as $\mathbf{Y}_\gamma$, the conditional pdf for $Z_0|\mathbf{z}_\alpha, \mathbf{y}_\gamma$ is then given by

$$f_{z_0|\mathbf{z}_\alpha,\mathbf{y}_\gamma}(z_0|\mathbf{z}_\alpha, \mathbf{y}_\gamma) \propto \int_{\mathbf{R}^m} f_{z_0,\mathbf{z}_\alpha,\mathbf{z}_\beta}(z_0, \mathbf{z}_\alpha, \mathbf{z}_\beta) f_{\mathbf{e}}(\mathbf{y}_\gamma - \mathbf{g}(\mathbf{z}_\beta)) d\mathbf{z}_\beta \tag{7}$$

This is a very general and nonlinear formula for prediction that requires the knowledge of the joint pdf $f_{z_0,\mathbf{z}_\alpha,\mathbf{z}_\beta}$ as well as the joint pdf of errors $f_{\mathbf{e}}$. A classical approach would be to consider $\mathcal{Z}$ as a Gaussian RF along with a mutual independence hypothesis for the vector $\mathbf{E}$, so that $f_{z_0,\mathbf{z}_\alpha,\mathbf{z}_\beta}$ is multivariate Gaussian and $f_{\mathbf{e}} = \prod_i f_{e_i}$. Using again Baye's theorem, we then have $f_{e_i}(y_i - g(z_i)) \propto f_{z_i|y_i}(z_i|y_i)/f_{z_i}(z_i)$, so that Eq. (7) simplifies to

$$f_{z_0|\mathbf{z}_\alpha,\mathbf{y}_\gamma}(z_0|\mathbf{z}_\alpha, \mathbf{y}_\gamma) \propto \int_{\mathbf{R}^m} f_{z_0,\mathbf{z}_\alpha,\mathbf{z}_\beta}(z_0, \mathbf{z}_\alpha, \mathbf{z}_\beta) \prod_i \frac{f_{z_i|y_i}(z_i|y_i)}{f_{z_i}(z_i)} d\mathbf{z}_\beta \tag{8}$$

As a consequence, Eq. (8) is still a nonlinear expression that requires multivariate integration, but it is now possible to express $f_{z_0,\mathbf{z}_\alpha,\mathbf{z}_\beta}$ from the covariance function $C_{\alpha\alpha}(\mathbf{h})$ as well as to infer the pdf's $f_{z_i|y_i}$ from the data set. A synthetical illustration of this can be found in Bogaert and Fasbender (2007). The real advantage of this formulation is that it is not longer needed to have any stationarity hypothesis about $\mathcal{Y}$ compared to SCoK. Moreover, it can be shown that a similar reasonsing can be used in order to account for multiple secondary information without any difficulties.

4 Comparing Data Fusion and Cokriging

Though BDF and SCoK may appear at the first sight as completely different (and thus difficult to compare) approaches for accounting for secondary information, it can however be shown that close analytical linear relations for expressing the conditional mean and variances can be obtained for BDF if an additional multivariate Gaussian hypothesis is assumed to hold.

It has been reminded that, for jointly Gaussian RF's $\mathcal{Y}$ and $\mathcal{Z}$, the corresponding conditional pdf $f_{z_0|\mathbf{z}_\alpha,\mathbf{y}_\gamma}$ is Gaussian with mean and variance given by Eq. (4), so that SCoK is the best predictor. On the other side, for BDF, one can also remark that assuming that $\mathcal{Y}$ and $\mathcal{Z}$ are jointly zero-mean Gaussian RF's leads to a zero-mean Gaussian vector $(Z_0, \mathbf{Z}_\alpha, \mathbf{Z}_\beta, \mathbf{Y}_\gamma)$ with covariance matrix $\boldsymbol{\Sigma}$ as given by

$$\boldsymbol{\Sigma} = \begin{pmatrix} \sigma_0^2 & \boldsymbol{\sigma}'_\alpha & \boldsymbol{\sigma}'_\beta & \boldsymbol{\sigma}'_\gamma \\ \boldsymbol{\sigma}_\alpha & \boldsymbol{\Sigma}_{\alpha\alpha} & \boldsymbol{\Sigma}_{\alpha\beta} & \boldsymbol{\Sigma}_{\alpha\gamma} \\ \boldsymbol{\sigma}_\beta & \boldsymbol{\Sigma}_{\beta\alpha} & \boldsymbol{\Sigma}_{\beta\beta} & \boldsymbol{\Sigma}_{\beta\gamma} \\ \boldsymbol{\sigma}_\gamma & \boldsymbol{\Sigma}_{\gamma\alpha} & \boldsymbol{\Sigma}_{\gamma\beta} & \boldsymbol{\Sigma}_{\gamma\gamma} \end{pmatrix} \tag{9}$$

Several simplifications will then occur. First, the functional g is now a linear mapping with $\mathbf{g}(\mathbf{z}_\beta) = \frac{\sigma_{yz}}{\sigma_0^2}\mathbf{z}_\beta$, where $\sigma_{yz} = \sigma_{zy} = Cov(Z(\mathbf{x}), Y(\mathbf{x}))$, $\forall \mathbf{x} \in \mathbb{R}^d$. Second, $\mathbf{E}$ is now Gaussian with covariance matrix equal to $\sigma^2_{\gamma|\beta}\mathbf{I}$ where $\sigma^2_{\gamma|\beta}$ is equal to $\sigma_y^2 - \frac{\sigma_{yz}^2}{\sigma_0^2}$ with $\sigma_y^2 = \mathbb{V}ar[Y(\mathbf{x})]$, $\forall \mathbf{x} \in \mathbb{R}^d$. Third, since $(Z_0', \mathbf{Z}_\alpha', \mathbf{Z}_\beta')'$ and $\mathbf{E}$ are Gaussian vectors, the product of pdf's in Eq. (8) is proportional to a Gaussian pdf with mean vector $\mathbf{M}$ and covariance matrix $\mathbf{S}$ given by

$$\begin{cases} \mathbf{S}^{-1} = \begin{pmatrix} \sigma_0^2 & \boldsymbol{\sigma}'_\alpha & \boldsymbol{\sigma}'_\beta \\ \boldsymbol{\sigma}_\alpha & \boldsymbol{\Sigma}_{\alpha\alpha} & \boldsymbol{\Sigma}_{\alpha\beta} \\ \boldsymbol{\sigma}_\beta & \boldsymbol{\Sigma}_{\beta\alpha} & \boldsymbol{\Sigma}_{\beta\beta} \end{pmatrix}^{-1} + \begin{pmatrix} 0 & \mathbf{0} & \mathbf{0} \\ \mathbf{0} & \mathbf{0} & \mathbf{0} \\ \mathbf{0} & \mathbf{0} & \frac{1}{\sigma^2_{\gamma|\beta}}\frac{\sigma_{yz}^2}{\sigma_0^4}\mathbf{I} \end{pmatrix} \\ \mathbf{M} = \frac{1}{\sigma^2_{\gamma|\beta}}\mathbf{S}\begin{pmatrix} 0 \\ \mathbf{0} \\ \frac{\sigma_{yz}}{\sigma_0^2}\mathbf{y}_\gamma \end{pmatrix} \end{cases} \tag{10}$$

Finally, since the integrand of Eq. (8) is proportional to a Gaussian pdf, integrating over $\mathbf{z}_\beta$ and conditioning on $\mathbf{z}_\alpha$ leads to the conclusion that the conditional pdf $Z_0|\mathbf{z}_\alpha, \mathbf{y}_\gamma$ is univariate Gaussian, thus completely characterized by its mean and variance, with

$$\begin{cases} \mathbb{E}[Z_0|\mathbf{z}_\alpha, \mathbf{y}_\gamma] = \frac{1}{\sigma^2_{\gamma|\beta}}\frac{\sigma_{yz}}{\sigma_0^2}\mathbf{s}'_\beta\mathbf{y}_\gamma + \mathbf{s}'_\alpha\mathbf{S}^{-1}_{\alpha\alpha}\left(\mathbf{z}_\alpha - \frac{1}{\sigma^2_{\gamma|\beta}}\frac{\sigma_{yz}}{\sigma_0^2}\mathbf{S}_{\alpha\beta}\mathbf{y}_\gamma\right) \\ \mathbb{V}ar[Z_0|\mathbf{z}_\alpha, \mathbf{y}_\gamma] = s_0 - \mathbf{s}'_\alpha\mathbf{S}^{-1}_{\alpha\alpha}\mathbf{s}_\alpha \end{cases} \tag{11}$$

where

$$\mathbf{S} = \begin{pmatrix} s_0 & \mathbf{s}'_\alpha & \mathbf{s}'_\beta \\ \mathbf{s}_\alpha & \mathbf{S}_{\alpha\alpha} & \mathbf{S}_{\alpha\beta} \\ \mathbf{s}_\beta & \mathbf{S}_{\beta\alpha} & \mathbf{S}_{\beta\beta} \end{pmatrix} \tag{12}$$

The gain of BDF on SCok in terms of inference is non negligible. Indeed, instead of inferring the multivariate covariance model (with LMC namely), we only need to estimate three simple quantities: i) $C_{\alpha\alpha}(\mathbf{h})$ the covariance function of $\mathcal{Z}$, ii) σ_y^2 the variance of $Y(\mathbf{x})$, $\forall \mathbf{x} \in \mathbb{R}^d$ and iii) σ_{yz} the pointwise covariance of $Z(\mathbf{x})$ and $Y(\mathbf{x})$, $\forall \mathbf{x} \in \mathbb{R}^d$.

Comparing Eqs. (1) and (3) with Eq. (11) is not straightforward, but it is worth noting that (i) BDF now provides expressions for the conditional expectation and variance that are linear with respect to the observed values $\mathbf{z}_\alpha$ and $\mathbf{y}_\gamma$ as for SCoK, and (ii) results for BDF and SCoK can be compared too on a common basis, as we fulfill the optimal conditions for using SCoK as the best possible predictor. It is worth remembering however that, in general, BDF and SCoK will obey distinct optimality properties: SCoK is optimal under the hypothesis of jointly Gaussian RF's $\mathcal{Y}$ and $\mathcal{Z}$, whereas BDF only rely on a Gaussian hypothesis for $\mathcal{Z}$ and an independence hypothesis for $\mathbf{E}$, which is a somewhat milder hypothesis than assuming joint Gaussianity for both RF's $\mathcal{Y}$ and $\mathcal{Z}$.

5 A Synthetic Case Study

In order to illustrate the similitudes of the results that are obtained using both approaches in a situation where SCoK is expected to give the best possible results, a synthetic case study is presented. The aim here is to show that, even under optimal conditions for SCoK, using SCoK instead of BDF does not significantly increase the quality of predictions. Stated in other words, the loss of information due to the use of BDF instead of SCoK does not dramatically affect the quality of the predictions.

Let assume a smooth zero-mean unit-variance Gaussian RF $\mathcal{Z}$ for which a realization over a 100×100 regular grid is given in Fig. 1a (covariance function is exponential with sill equal to 1 and range equal to 30). Let assume also a second Gaussian RF $\mathcal{Y}$ (Fig. 1b) defined as a linear combination of the RF $\mathcal{Z}$ and another zero-mean unit-variance Gaussian RF $\mathcal{E}$ with the same covariance function. By taking $\mathcal{Y} = 2\mathcal{Z} + \mathcal{E}$, the resulting RF $\mathcal{Y}$ is thus a zero-mean Gaussian RF with variance equal to 5, and both RF's are jointly Gaussian with pointwise correlation equal to 0.894. Under these conditions, SCoK is thus the best possible predictor. For conducting predictions, two random samples $\mathbf{z}$ and $\mathbf{y}$ are extracted from these simulated grids. In order to be in a situation when SCoK would be interesting compared to simple kriging (i.e. the auxiliary $\mathbf{y}$ conveys valuable extra information compared to what is already known from $\mathbf{z}$), samples size have been chosen equal to 200 and 400 for $\mathbf{z}$ and $\mathbf{y}$, respectively. Predictions of $\mathcal{Z}$ is then conducted at the nodes of the grid using $\mathbf{z}$ and $\mathbf{y}$ as observed values. It is worth noting too that sampling has been conducted so that there are no locations for which $\mathcal{Z}$ and $\mathcal{Y}$ are jointly observed.

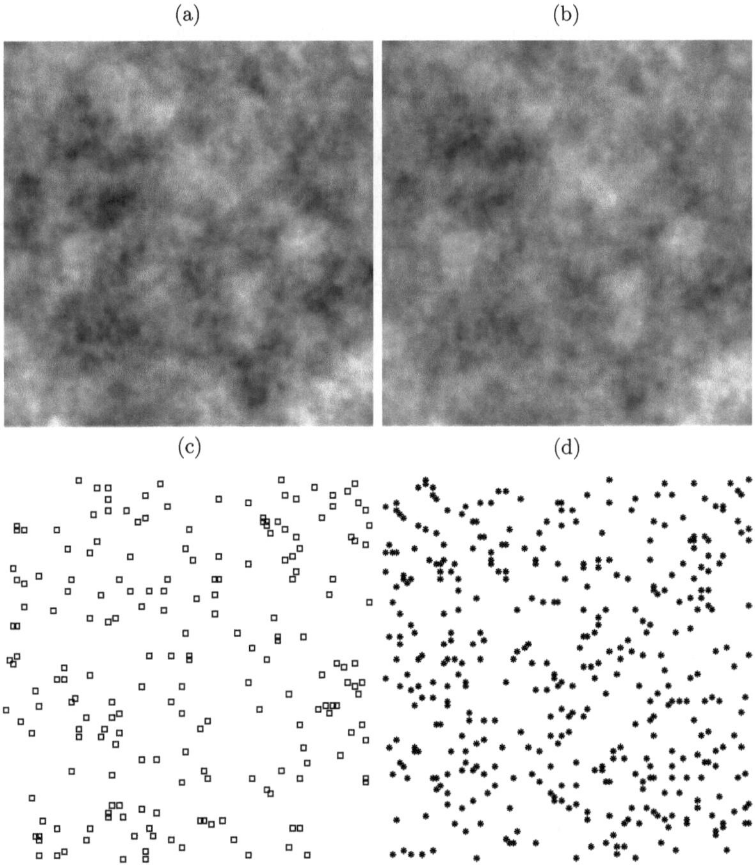

Fig. 1 Simulations over a 100×100 regular grid for RF $\mathcal{Z}$ **(a)** and RF $\mathcal{Y}$ **(b)**, along with the random sample **z** at 200 locations **(c)** and the random sample **y** at 400 locations **(d)**

In order to compare relative differences between BDF and SCoK results compared to simple kriging results, a relative comparison approach has been used here. Clearly, using simple kriging with the n closest observations of **z** (that will be denoted as SK_n) provides a lower bound in terms of prediction quality, as it only makes use of the primary variable. On the other side, using simple kriging with the n closest observations of **z** along with the **z** observations at the n closest locations for the observed **y** (that will be denoted as SK_{2n}) provides an upper bound, as it assumes that the true values for the primary variable are available at locations where only the auxiliary variable is observed. As a consequence, in terms of quality predictions, results for BDF and SCOK will be located somewhere in between these two bounds. This also provides a way to compare the results for BDF and SCoK on a relative scale ranging from SK_n to SK_{2n}.

Figure 2 shows the predictions results obtained using the four previously described methods, namely SCoK (Fig. 2d), BDF (Fig. 2c) and the two extreme simple kriging situations (Figs. 2a and 2b). One can notice that BDF and SCoK provide visually very similar results. This observation is confirmed from the Root Mean Squared Error (RMSE) as computed between simulated and predicted values (see Table 1). As expected, SCoK and BDF gives intermediate results in between the two kriging predictions, with only a slight advantage for SCoK when values are compared on a relative scale.

The above computations can been repeated by keeping everything identical except for the pointwise correlation between the two RF's $\mathcal{Z}$ and $\mathcal{Y}$. It can be seen from Fig. 3 that the relative difference between BDF and SCoK is null for high and

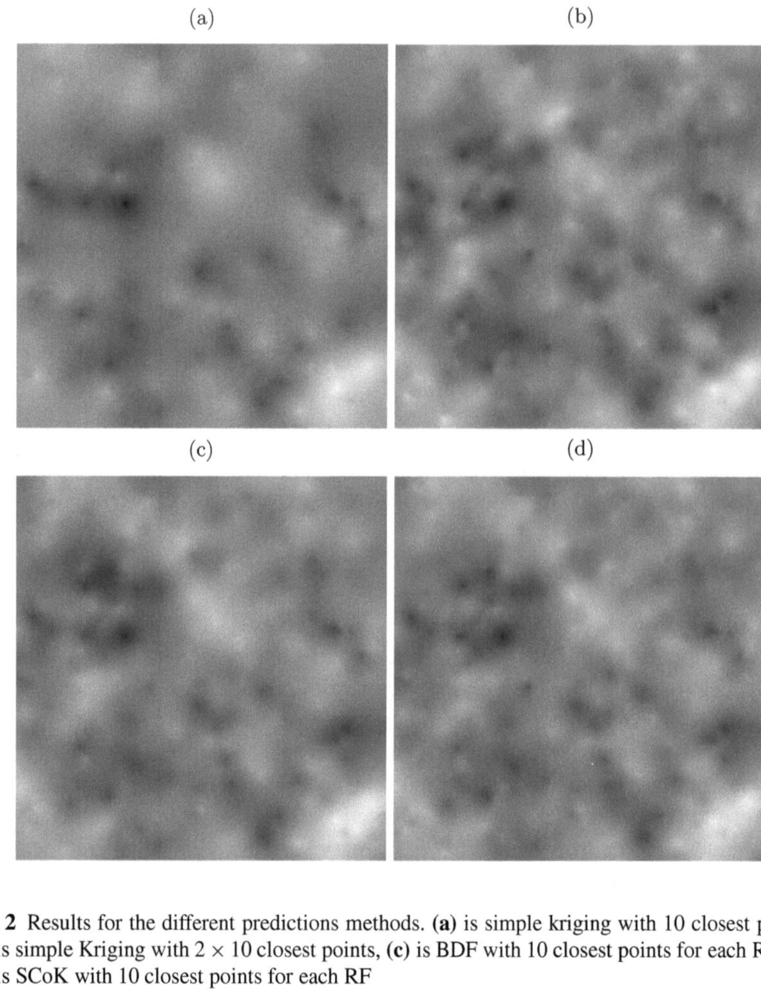

Fig. 2 Results for the different predictions methods. **(a)** is simple kriging with 10 closest points, **(b)** is simple Kriging with 2×10 closest points, **(c)** is BDF with 10 closest points for each RF and **(d)** is SCoK with 10 closest points for each RF

Table 1 Quality assessment for the various prediction methods. RMSEs are computed as the root mean squared differences between simulated and predicted values. Relative RMSEs are RMSEs rescaled between 0 and 1 according to the bounds as provided by SK_n and SK_{2n} (value 1 is thus the best possible result whereas value 0 is the worst possible one)

-	SK_n	SCoK	Bayesian Data Fusion	SK_{2n}
RMSE	0.63	0.53	0.55	0.48
Relative RMSE	0	0.68	0.58	1

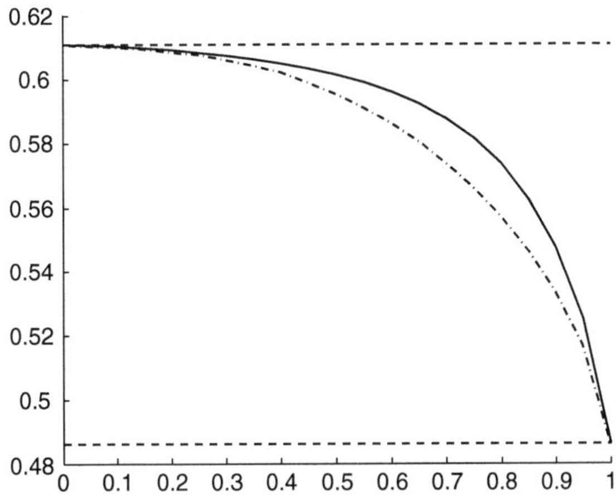

Fig. 3 Evolution of the RMSE with respect to the pointwise correlation between $\mathcal{Z}$ and $\mathcal{Y}$ RF's. Dashed lines correspond to SK_n (higher value) and SK_{2n} (lower value), whereas plain line and dotted line correspond to BDF and SCoK, respectively

low correlations (i.e. when secondary information is useless and when secondary information is equivalent to primary one, respectively), so that biggest differences will be observed for intermediate values.

6 Conclusions

In this paper, a BDF approach has been proposed as an interesting alternative way to account for various secondary information in a spatial prediction context. Indeed, the method does not rely on a classical multivariate second-order stationary RF framework, as it is required for cokriging. Because by nature BDF relies on a different set of assumptions, it is difficult to compare both methods from a general viewpoint in situations where both of them would be relevant. In order to overcome this difficulty, a comparison has been conducted in a situation where SCoK is known to be the best possible predictor, so that the loss of information caused by using BDF instead of SCoK can be assessed on an objective way. Results show that in terms of performances, differences are however quite limited.

Of course, there is no point in using data fusion instead of cokriging when one can reasonably assume that a second-order multivariate RF hypothesis holds, as SCoK is then by definition the right method to be used. However, the application field of BDF is quite more general, as it allows the user to account for secondary information that would not fit into this framework, permitting thus to deal with a much wider panel of situations that could not be meaningfully handled by SCoK. Hence, BDF can be viewed as a robust alternative to cokriging for multivariate prediction where cokriging hypotheses are known to be irrelevant or at least quite arguable.

Finally, it is worth noting that the aim of this paper was not to suggest that sound multivariate spatial prediction methods like cokriging, which have frequently proved to be quite useful, should be discarded or criticized when used under the appropriate hypotheses. It is rather the limited practical pertinency of these hypotheses which is at stake here. It is suggested that in situations where a multivariate modelling does not appear to be conceptually consistent with what is known from data, BDF is then a more reasonable choice by avoiding the need of using a multivariate model (e.g., as the LMC) at the price of mild simplifying hypotheses. Though rather simple in the case of a single auxiliary variable, this modelling problem is expected to become critical in situations where the number and diversity of auxiliary informations sources that need to be accounted for is increasing. This is a typical case where BDF is expected to be a much more flexible way of handling the problem than SCoK.

References

Bogaert P, Fasbender D (2007) Bayesian data fusion in a spatial prediction context: A general formulation. Stochastic Environmental Research and Risk Assessment. Published online

Chilès J.-P, Delfiner P (1999) Geostatistics: modeling spatial uncertainty. John Wiley and Sons, Inc., New York, NY

Christakos G (1992) Random field models in earth sciences. Academic Press, San Diego, CA

Cressie N (1991) Statistics for spatial data. Wiley series in probability and mathematical statistics. John Wiley and Sons, Inc., New York, NY

Goovaerts P (1997) Geostatistics for natural resources evaluation. Oxford University Press, New York, NY

The Challenge of Real-Time Automatic Mapping for Environmental Monitoring Network Management

E. J. Pebesma, G. Dubois and D. Cornford

Abstract The automatic interpolation of environmental monitoring network data such as air quality or radiation levels in real-time setting poses a number of practical and theoretical questions. Among the problems found are (i) dealing and communicating uncertainty of predictions, (ii) automatic (hyper)parameter estimation, (iii) monitoring network heterogeneity, (iv) dealing with outlying extremes, and (v) quality control. In this paper we discuss these issues, in light of the spatial interpolation comparison exercise held in 2004.

1 Introduction

Many environmental variables are monitored in a (semi-)continuous way; examples include air quality and background radiation levels. In order to utilize the network, maps of observed values are usually instantly available to network operators, but maps with interpolated values often need lengthy intervention by (spatial) statisticians before they become available. We believe that spatial interpolation can, and should, be automated to the extent that both in routine and emergency situations interpolated maps can become available in near real-time (i.e., within seconds up to tens of minutes) *without such intervention*. Of course there will always be a role for the spatial statistician in providing in depth analysis of a given data; our focus is on situations where decisions must be made quickly.

In a decision theoretic setting, a map with interpolated (predicted) values, is not sufficient information; knowledge of prediction errors and their probability distributions is necessary for optimal results. We explore some of the issues that the requirement for automatic, probabilistic, real-time, prediction raises.

This paper discusses issues in both algorithm development and their practical implementation in the form of a web service for operational monitoring network management. It will review some of the submissions of the Spatial Interpolation

E. J. Pebesma
Geosciences Faculty, Utrecht University, The Netherlands
e-mail: e.pebesma@geo.uu.nl

A. Soares et al. (eds.), *geoENV VI – Geostatistics for Environmental Applications*, 467–475

Comparison (SIC) 2004 (Dubois and Galmarini, 2005; EUR, 2005). The issues we will address comprise

i. quantifying and communicating exceedance probabilities for given threshold levels, in order to estimate risk of exposure.
ii. the automated estimation of parameters describing the spatial variability in presence of extremes
iii. dealing with heterogeneity of monitoring networks, e.g. across EU member state boundaries
iv. detection of outliers in space and time
v. quality control.

2 Communicating Prediction Error Distributions

Interpolating in two dimensions can be relatively simple. In cases where the variogram is close to an exponential or spherical model, and the nugget variance is small, the inverse distance interpolation algorithm is hard to beat significantly with highly advanced geostatistical models, when the implementation is tuned to have a varying power in the distance weights, or a varying neighbourhood selection. One of the disadvantages of inverse distance methods is that they do not yield interpolation, or prediction errors when no variogram model is assumed. Interpolation errors can be large, and are of importance, if for example someone is faced with the decision whether an area, or how large an area should be evacuated based on the interpolation of measured radiation levels after a radioactive outbreak.

Ideally, an automatic prediction algorithm should provide a user with the full conditional cumulative distribution function (ccdf), which is for a random variable Z at arbitrarily chosen unobserved location s_0 the probability

$$F(Z(s_0), c) = \Pr(Z(s_0) < c \mid z(s_i),\ \ i = 1, ..., n) \tag{1}$$

with $Z(s_i), i \geq 1$ the observed data. Usually s_0 is chosen to be a large number of points (or square blocks) over a regular grid, and $F(Z(s_0), c)$, for a given level c, can be shown as a map. In risk studies, it may be more intuitive to map $1 - F(Z(s_0), c)$, which is the probability of exceeding c, but for the discussion here this is irrelevant. An alternative visualisation is that of the quantile function, obtained by inverting (1), which gives the Z values corresponding to a spatially constant given quantile value $q \in [0, 1]$:

$$F^{-1}(Z(s_0), q) = c \tag{2}$$

such as the median, or the 2.5 and 97.5 percentiles[1].

[1] Although not necessary for the discussion here, we want to note that in a considerable part of the geostatistical literature ccdf's are associated with, or discussed in the context of certain specific

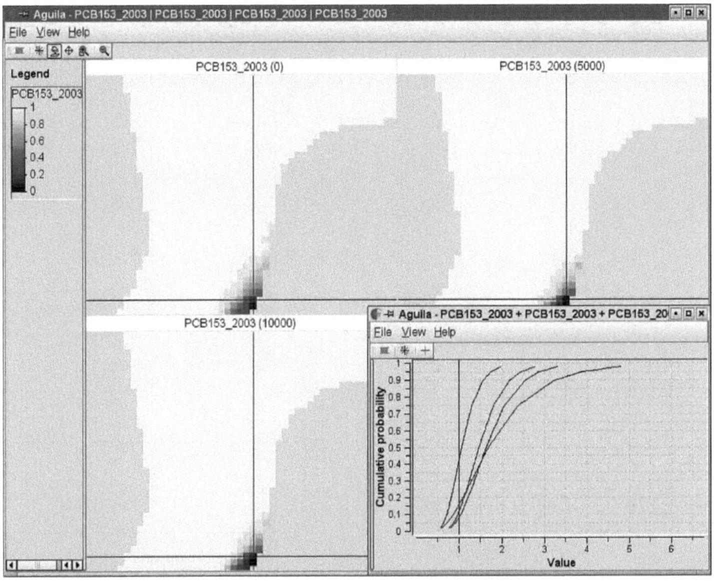

Fig. 1 Screen shot of a tool to visualize the distribution function $F(Z(s), c)$ for PCB-138 concentrations in North Sea floor sediment data, analyzed in Pebesma and Duin (2005). The maps show probabilities for the interpolated values of being below the PCB138 threshold value of 1 ppm (the legend caption misses this point). The scenarios refer to block size: (0) refers to point kriging, (5000) to kriging at 5 km × 5 km block, (10000) to kriging at 10 km × 10 km blocks

For fixed, chosen values of c, the value of $F(Z(s), c)$, or alternatively one minus this value (the probability of exceeding c) may be shown as a static map. Usually but not necessarily, s is a collection of points on a regular grid covering the area studied. Accordingly, for fixed values of q a quantile map for $F^{-1}(Z(s), q)$ can be shown. Choosing these values ahead of time may be guided by regulatory guidelines, e.g. from maximum tolerated or established zero-risk concentrations, but threshold values found there often contain a certain or even considerable amount of arbitrariness in them, and a user may want to change them. An important issue to find out is how much a slight change in c results in a different assessment of the exceedence probability map.

Visual communication of the full functions $F(Z(s), c)$ and $F^{-1}(Z(s), q)$ is area of research. Pebesma et al. (accepted) describe a tool for the dynamic analysis of maps of (1) and (2) under different modelling or interpolation scenarios. In case of the ccdf (Fig. 1) the value of c can be dynamically changed by dragging and dropping the vertical line in the ccdf widget, which is followed by immediate update

forms of kriging, notably indicator kriging and its descendents or generalisations. This is not necessary as ordinary, simple or universal kriging can provide ccdf's whenever a parametric distribution function (e.g. normal, lognormal, normal after Box-Cox transformation) is assumed. Such assumptions may be strong, but so are the assumptions about the identification of tail distributions and their spatial correlation in the indicator and related approaches.

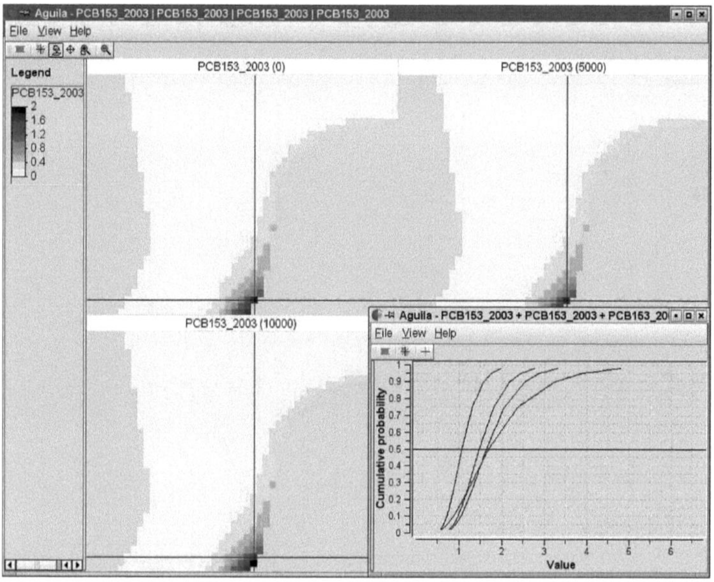

Fig. 2 Screen shot of a tool to visualize the quantile function $F^{-1}(Z(s), q)$ for PCB-138 concentrations in North Sea floor sediment data, analyzed in Pebesma and Duin (2005). The maps show quantiles for the interpolated values for the probability value 0.5 (i.e., the median)

of the corresponding maps of $F(Z(s), c)$. In case of the quantiles plot (Fig. 2), the value of q (horizontal line in the ccdf widget) can be dynamically changed, to be followed by an update of the maps of $F^{-1}(Z(s), q)$.

3 An INTERPOLATE Button, or Web Service?

Ideally, we would like to have a routine (let us say a button in a computer program or web client) which, given a set of measurements, provides near real-time maps of interpolated values, and/or their associated distribution or quantile views. This means that data have to be submitted, interpolated values computed, and ccdf's are returned (Fig. 3).

While the implementation of the interpolation algorithm is clearly very important in determining the accuracy of the predictions, the usefulness of the system also depends on the ease of integration into the overall network management system. An

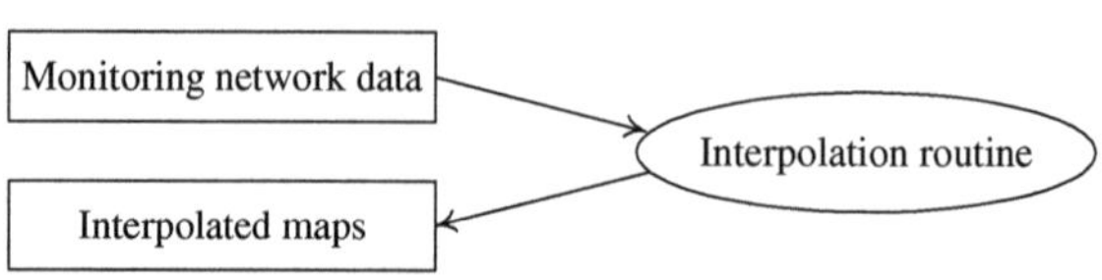

Fig. 3 Idealised data flow for an automatic mapping procedure. Implemented as a web service, the arrows may represent data flow over `http` connections

exciting opportunity is presented by the adoption of a service oriented architecture (often called "web services") with carefully defined interfaces offers an exciting prospect of developing an automatic interpolation service which any user capable of employing web services can utilise. This will necessitate the definition of standards for communicating uncertainty.

4 Monitoring Network Heterogeneity

The idealized situation of Fig. 3 discards much of the information that is usually available for monitoring networks. Besides the measurements themselves, the following, non-exhaustive list may be relevant for the interpolation routine:

- what do the measurements usually look like?
- are these measurements taken from a variable that can only take positive values (e.g. a concentration variable), and does it have an upper boundary?
- are the measurements obtained under identical conditions, or are there differences in measurement device, monitoring network design (e.g. between states or countries), standardization issues, or rules regarding the classification of a monitoring station? (e.g. is an air quality monitoring station classified as *rural* comparable to a likewise classified station in another country?)
- are there variables available to which the monitored variable bears a relationship, that are useful for interpolation? (e.g. ozone may be related to altitude when looking a large region)
- is there any other prior knowledge available that needs to be taken into account in interpolation? (e.g. previous measurements, spatial correlation characteristics, prior beliefs)
- is there historic information that certain measurement stations behave anomalous, more often than others?

A naive interpolation procedure that does not take any of these issues into account may seem fairly easy to implement. When looking at interpolation as a stage in exploratory analysis of monitoring network data, such a naive interpolation procedure should be useful to detect some of the issues mentioned. As a dedicated system for decision support in emergency conditions, the requirements are different. Communicating any (or all) of the information to an interpolation procedure poses another interesting technical problem.

However, even if none of the above information is provided, an automatic interpolation should still be possible. The main issue is then (i) the modelling of the variogram (or covariance function), and (ii) the possible uncertainty about variogram model and/or model parameters. Given measurements and their locations, several issues require careful consideration. If we want to fit models to sample variograms, for example

1. how should we compute the sample variogram (maximum distance, lag interval width, directional or isotropic)?

2. which particular variogram model or group of models do we want to fit?
3. which criteria do we use for the actual fit?
4. which initial values do we provide for the fit, in case it involves non-linear parameters (such as range)?
5. how do we deal with the problem of an ill-fitting model or non-convergence in the fit?

Some of the above questions were discussed, but not typically "solved" by Pebesma (2005). When fitting by ML/REML, questions 2, 4 and 5 are relevant as well. In case of a Bayesian, so-called model-based approach (Diggle et al., 1998), two further questions are

6. which prior distributions should be chosen, automatically, for the variogram fitting procedure, and
7. how do we verify automatically that the Markov chain Monte Carlo algorithm has converged?

In the context of SIC2004, Palaseanu-Lovejoy (2005) has shown that this Bayesian procedure worked when the algorithm was applied to data that matched the prior assumptions, but failed in case of extreme, unexpected outlying data. Clearly further research is required to address these issues in the context of an automatic interpolation method.

5 Outliers in Space and Time

Outliers are of utmost importance, as they either need to be discarded as invalid measurements (monitoring network failure) or indicate extreme conditions, possibly notifying us of an emergency condition. An automatic interpolation routine should never automatically remove outliers in order to remain useful for the second type of situation, but it is useful for network management, to provide a mechanism for deciding which case is true.

Interpolation in the presence of outliers (one or very few highly extreme values) is a major source of trouble for any interpolation procedure (e.g. EUR, 2005). One wonders whether single stationary random fields of whatever kind are useful as models for fields which really include outliers. We might also consider how one field (say, background concentrations) should be distinguished from the second (with outlying measurements) and in addition how the spatial correlation of the outlier field should be characterized on the basis of maybe one or two observations. Cornford (2005) suggested that in case of outliers that arise from real physical processes, we should work to probabilistic models that incorporate the physics of the phenomena modelled, using a data assimilation framework. This is a complex task and it remains to be seen whether one or two outlying observations are sufficient for successful assimilation of the outlying phenomena in absence other information on the magnitude and location of a source, but an integrated space-time analysis does seem indispensable for these cases.

6 Space-time Approaches

One important issue for (near) real-time interpolation is whether past observations should be taken into account for the interpolation based on current observations. If measurements are taken with high frequency, this seems attractive because they may carry additional information when the process is temporally correlated. On the other hand, for certain processes sudden jumps in time (e.g. a radioactive outbreak) may not show up well in interpolated maps if these rely on the regular behaviour that nuclear radiation shows when there is no outbreak. For such emergency cases a space-time model should allow for sudden jumps in time. In any case, when interpolations are needed in near real-time, computation speed is an issue and this may currently be a challenge for space-time approaches, more than for spatial approaches alone.

7 Quality Control and Implementation Issues

As in many other fields, software architectures in Geographic Information Systems are shifting from application oriented to service oriented paradigms. This means that algorithms are not implemented as a button in a stand-alone application, but rather operate as a web service facilitating their use from a client anywhere in the world. We envisage that interpolation is a service that can, and should be served this way. Among the motivations for this are (i) monitoring data are collected in real-time, but are not present in real-time on the client computer, but typically available after a service request, (ii) the network data may not be publicly available, but views on the data or other derivative products may or may not be, or may be available for specific purposes, and (iii) the monitoring data may be served by a varying data base infrastructures and computer architectures.

How can we ensure that the code, or web service, does what it is supposed to do? At the base of software development, one should always build regression tests. Such tests provide input and verified output, such that in an automatic setting the code can be run to verify that it produces output identical to the verified output. As an example, package development in R (R development core team, 2006) stimulates package writers to supply their own tests, which are automatically run when porting packages to a new computing platform, or when R itself is upgraded. Developing regression tests for a wide variety of situations (not only success, but also failure situations) does harden the code, but is no guarantee for quality.

Another aspect is the use of legacy code. Software contains errors, or has undocumented features. Using code leads to errors being found and software that has been maintained for a long time may therefore be expected to contain fewer (unknown) bugs than freshly written code. Use of legacy code may also reflect the environment in which the code is written, e.g. low-level programming languages as C or Fortran, object-oriented languages such as C++ or java, or high-level environments such as R or Matlab. Code written in the latter environments may be easier to verify (by those who can read it), as it is 5–10 times as compact. The underlying numerical algebra is, at least for R, dealt with by legacy linear algebra libraries (lapack/linpack/blas). In addition, every aspect of R is open source, and as such fully verifiable by anyone.

The implementation as a web service facilitates the creation of a web testing client, which can subsequently be used to (automatically) test the performance of *any* other web service that implements the automatic mapping interface. This can give us more confidence in a new implementation since the automated regression tests will be extensive.

8 Lessons Learnt from SIC2004

SIC2004 (Dubois and Galmarini, 2005; EUR, 2005) was a spatial interpolation comparison, especially set up to test automated mapping routines, and to see how they performed in case of unexpected, rare extremes (a simulated local radioactive outbreak). Some lessons learned from this exercise are:

- SIC2004 used overall, average performance criteria. It did not take the probabilistic aspect of prediction (predictive distributions) into account, and did not evaluate as a performance criteria of the area above a certain cut-off value.
- In a comparison of automated mapping routines, one should never reach final conclusions based on comparison experience using a single data set only, and one should use criteria related to the emergency mapping context (Boogaart, 2004).
- All but one of the participants truly applied an automatic interpolation algorithm, meaning that manual intervention took place after discovery of the outliers and before submitting results (Myers, 2004).

Overall, the results of SIC2004 highlighted the need for further research before a truly automatic algorithm can be robustly deployed. The research issues include spatial statistics, algorithmic developments and software implementations, with their practical deployment requiring development of software architecture and standards for interoperability making this a truly inter-disciplinary problem.

9 Discussion

Insurance companies know that knowledge about uncertainties pays off when taking decisions (setting insurance rates). However, they can spread risk because failure happens with some frequency. When taking decisions in environmental emergency conditions (such as treatment or evacuation of populations), the situation is totally different, because taking a wrong decision may worsen (or even cause) a disaster. This does not mean that we do not need the uncertainties, but rather that we (and the decision makers) have to learn how to use probabilistic information optimally.

When there is no direct emergency, the information about the distribution of prediction errors may also be of use for exposure assessment. As an example, it seems that black smoke has health effects for a considerable fraction of the population in parts of Europe and Northern America. Distribution functions obtained from spatial interpolation should be handled with care though; if a spatial interpolation algorithm suggests that in some region the probability of exceeding a critical level is 10%, this does not mean that 10% of the time the level is exceeded, nor that 10% of the population living there is affected. Despite that, interpolation, and analysing error distribution functions may help evaluating monitoring network management (e.g. monitoring network optimization), assess risk of exposure for populations and be instrumental to policy evaluation and development.

Automatic interpolation procedures seem to be far away now, but we expect them to become available, and envisage their use will be adopted by monitoring network management, risk assessments, and policy evaluation instruments.

Acknowledgments This work was funded by the European Commission, under the Sixth Framework Programme, by the Contract N. 033811 with the DG INFSO, action Line IST-2005-2.5.12 ICT for Environmental Risk Management. The views expressed herein are those of the authors and are not necessarily those of the European Commission.

References

Cornford D (2005) Are comparative studies a waste of time? SIC2004 examined. In: Dubois G, Luxembourg (ed) EUR, 2005. Automatic mapping algorithms for routine and emergency monitoring data. Office for official publications of the European Communities. EUR 21595 EN – Scientific and technical research series, ISBN 92-894-9400-X

Diggle PJ, Tawn JA, Moyeed RA (1998) Model-based geostatistics. Appl Stat 47(3): 299–350

Dubois G, Galmarini S (2005) Introduction to the spatial interpolation comparison (SIC) 2004 exercise and presentation of the datasets. Appl GIS 1(2): 09_1–09_11

EUR (2005) Automatic mapping algorithms for routine and emergency monitoring data. In: Dubois G, Luxembourg (ed) Office for official publications of the European communities. EUR 21595 EN – Scientific and technical research series, ISBN 92-894-9400-X

Myers DE (2004) Spatial interpolation comparison exercise 2004: a real problem or an academic exercise. In: Dubois G, Luxembourg (ed) EUR, 2005. Automatic mapping algorithms for routine and emergency monitoring data. Office for official publications of the European communities. EUR 21595 EN – Scientific and technical research series; ISBN 92-894-9400-X

Palaseanu-Lovejoy M (2005) Bayesian automating fitting functions for spatial predictions. Appl GIS 1(2): 14.1–14.14

Pebesma EJ (2005) Mapping radioactivity from monitoring data: automating the classical statistical approach. Appl GIS 1(2): 11.1–11.17

Pebesma EJ, Duin RNM (2005) Spatio-temporal mapping of sea floor sediment pollution in the North Sea. In: Renard PH, Froidevaux R (eds) Proceedings GeoENV 2004–Fifth European conference on geostatistics for environmental applications, Springer

Pebesma EJ, de Jong K, Briggs D (2007) Interactive visualization of uncertain spatial and spatio-temporal data under different scenarios: an air quality example. Int J GIS 21(5): 515–527

R development core team (2006) The R project. `http://www.r-project.org`

Van den Boogaart KG (2005) The comparison of one-click mapping procedures for emergency. In: Dubois G, Luxembourg (ed) EUR, 2005. Automatic mapping algorithms for routine and emergency monitoring data. Office for official publications of the European communities. EUR 21595 EN – Scientific and technical research series, ISBN 92-894-9400-X

Geostatistical Applications of Spartan Spatial Random Fields

S. N. Elogne and D. T. Hristopulos

Abstract Spartan Spatial Random Fields (SSRFs) were recently proposed (Hristopulos 2003) as a new method for modelling spatial dependence. This paper focuses on (i) the inference of Gaussian SSRF model parameters from spatial data using kernel methods and (ii) the identification of geometric anisotropy by means of the covariance tensor identity (CTI) method (Hristopulos 2002). The methods presented are illustrated with the help of synthetic data and a real set of elevation data. Kriging predictions obtained with the Spartan covariance estimator are compared to those obtained with standard estimators. Based on these results, the Spartan estimator provides a useful alterative to parametric covariance estimators, which it may outperform in certain cases.

1 Introduction

Spatial interpolation has many applications in the fields of the earth and environmental sciences. The widely used methodology is based on the kriging algorithm, which employs the spatial continuity structure encoded in the semivariogram or the covariance function. Estimation of the latter from the data is often conducted under the Gaussian and isotropic assumptions. Reliable estimation of the spatial structure is crucial to producing accurate kriging maps.

Here we investigate the problem of estimating the spatial continuity for both isotropic and anisotropic processes, in the framework of Spartan Spatial Random Fields (Hristopulos 2003). The Spartan Spatial Random Fields (SSRFs) aim to provide a versatile, formally and computationally efficient approach for modelling spatial dependence. The SSRFs possess a Gibbs joint probability density function (pdf) that is expressed in terms of physically motivated interactions between the fluctuations, i.e., $f\left[X_\lambda(\mathbf{s})\right] = Z^{-1} \exp\left\{-H\left[X_\lambda(\mathbf{s})\right]\right\}$, where Z is a normalization factor and the *energy functional* H embodies the interactions. The following functional will be used

S. N. Elogne
Department of Mineral Resources Engineering, Technical University of Crete,
Chania 73100, Greece
e-mail: elogne@mred.tuc.gr

A. Soares et al. (eds.), *geoENV VI – Geostatistics for Environmental Applications*, 477–487

$$H = \frac{1}{2\eta_0 \xi^d} \int d\mathbf{s} \left[\left\{ X_\lambda(\mathbf{s}) - m_X(\mathbf{s}) \right\}^2 + \eta_1 \xi^2 \left\{ \nabla X_\lambda(\mathbf{s}) \right\}^2 + \xi^4 \left\{ \nabla^2 X_\lambda(\mathbf{s}) \right\}^2 \right], \tag{1}$$

where $m_X(\mathbf{s}) = E[X_\lambda(\mathrm{s})]$. Equation (1) provides a class of flexible parametric models derived from the same functional. The three terms in the functional can be viewed as physical constraints related to the square of the fluctuations, as well as their gradient and curvature. Such terms can be either physically motivated or simply used as abstract constraints that lead to flexible covariance models.

The *SSRF parameters* include η_0 (the scale parameter), η_1 (the shape parameter), ξ (the characteristic length) and k_c(the frequency cutoff). The isotropic Spartan covariance spectral density is given by the following equation, where $\delta_{\|\mathbf{k}\| \leq k_c} = 1$ for $\| \mathbf{k} \| \leq k_c$ and 0 for $\| \mathbf{k} \| > k_c$:

$$\widetilde{G}_\lambda(\mathbf{k}) = \eta_0 \xi^d \delta_{\|\mathbf{k}\| \leq k_c} \big/ \left(1 + \eta_1 \xi^2 \|\mathbf{k}\|^2 + \xi^4 \|\mathbf{k}\|^4\right). \tag{2}$$

Permissibility conditions follow simply from requiring non-negative values of the spectral density (Hristopulos 2003; Hristopulos & Elogne 2006).

The above covariance spectral density corresponds to differentiable *random fields* (RFs) for finite k_c and non-differentiable ones for infinite k_c (Hristopulos & Elogne 2006). An important issue for practitioners is the ability to differentiate between covariance models (Gorsich & Genton 2000). In the Spartan framework one can distinguish between models based on the values of the SSRF parameters. SSRF model inference is computationally efficient, because it requires the estimation of a small set of parameters (Hristopulos 2003).

This paper is organized as follows: Section 2 focuses on SSRF parameter inference using kernel methods. In Section 3, identification of anisotropic covariance models using the covariance tensor identity method is reviewed, and extensions based on kernel methods are introduced. Section 4 is devoted to the numerical investigations of the proposed methods using simulated data sets. Section 5 presents an application of the methods to real data. Finally, conclusions and some open issues for further research are presented in Section 6.

2 SSRF Parameter Inference

Parameter inference is based on matching sample (experimental) estimates for the variance as well as *generalized gradient and curvature constraints* with respective values of stochastic (model) constraints (Hristopulos 2003; Elogne & Hristopulos, 2006b). The stochastic constraints are as follows:

$$E[S_0] = G_\lambda(0), \quad E[S_1] = 2d\, F_\lambda(a_1)/a_1^2, \tag{3}$$

$$E[S_2] = \left\{ 8d^2 F_\lambda(a_2) - 4d(d-1) F_\lambda(a_2\sqrt{2}) - 2d\, F_\lambda(2a_2) \right\} \Big/ a_2^4, \tag{4}$$

where d is the *spatial dimension*, F_λ the semivariogram, and a_1, a_2 isotropic spatial increments. Using the spectral representation of isotropic covariance models, the stochastic constraints are expressed in terms of one-dimensional integrals that involve the unknown SSRF parameters.

In order to define corresponding sample constraints, a *continuous*, *isotropic* and *compactly supported* kernel$K$ and two bandwidth parameters h_1 and h_2 are introduced. Kernel averages of the field quantities are denoted as follows:

$$\langle X_{i,j}^2 \rangle_h \equiv \sum_{i \neq j} \{X(\mathbf{s}_i) - X(\mathbf{s}_j)\}^2 K((\mathbf{s}_i - \mathbf{s}_j)/h) \Big/ \sum_{i \neq j} K((\mathbf{s}_i - \mathbf{s}_j)/h). \quad (5)$$

If $\overline{\mu_X}$ is the sample mean, the sample constraints are given by

$$\overline{S_0} = \tfrac{1}{n} \sum_i \{X(\mathbf{s}_i) - \overline{\mu_X}\}^2, \quad \overline{S_1} = d \langle X_{i,j}^2 \rangle_{h_1} \Big/ a_1^2, \quad (6)$$

$$\overline{S_2} = \left\{ 4d^2 \mu_1 \langle X_{i,j}^2 \rangle_{h_2} - 2d(d-1)\mu_2 \langle X_{i,j}^2 \rangle_{h_2\sqrt{2}} - d \langle X_{i,j}^2 \rangle_{2h_2} \right\} \Big/ a_2^4. \quad (7)$$

The increments a_1 and a_2 as well as μ_1 and μ_2, are functions of the sampling locations, selected so that the sample constraints are asymptotically unbiased estimators of the stochastic counterparts (Elogne & Hristopulos 2006b).

2.1 Bandwidth Selection

The choice of the bandwidth parameters is a crucial issue. Classical methods are based of minimizing some criterion (e.g., mean square error) which depends on unknown characteristics of the process, such as the true semivariogram model and its second derivative (Garcia-Soidan et al. 2004). In the Spartan framework, the bandwidths are determined from the *consistency principle* $a_p = \left\langle \| \mathrm{s}_i - \mathrm{s}_j \|^{2p} \right\rangle_{h_p}^{1/2p}$, for $p = 1, 2$ (Elogne & Hristopulos 2006b). Under mild regularity conditions on the sampling locations and the kernel, it is proved that $a_p = h_p \left(m_{K,p+1}/m_{K,1} \right)^{1/2p} + o(h_p)$ almost surely for $p = 1, 2$ in d= 2, where $m_{K,j}$ represent moments of the kernel function. Estimates of the increments a_1 and a_2 are derived from the neighbor distance distribution of the sampling network.

2.2 Constraints Fitting

Given the nonlinear dependence of the stochastic constraints, i.e., equations (6) and (7), on the SSRF parameters, the latter need to be determined numerically by solving a system of equations that fit the stochastic constraints to the sample constraints, given by equations (6) and (7). This is accomplished by minimizing a nonlinear *distance functional* Φ, which measures the deviation between the sample and the stochastic constraints. The initial form of the functional given in (Hristopulos 2003) assumed a fixed cutoff frequency. The distance functional has been recently extended (Elogne & Hristopulos 2006b) to allow for direct inference of the cutoff from the data.

The minimization can be implemented using standard optimization algorithms, (e.g., the Nelder-Mead simplex search method). In the cases explored so far the convergence is very fast (for a sample of 100 locations approximately two seconds on a laptop with a Celeron M processor at 1.1GHz and 256Mb RAM, running Matlab under Windows XP; see also Hristopulos 2003). The initial values of the SSRF parameters, except for the shape coefficient, are not a crucial factor in the optimization results. In our experience, all the "solutions" to which the optimization converges lead to similar spatial dependence (covariance function), although sometimes different covariance estimators are obtained, some of which correspond to local minima of Φ (also see Section 4.2).

3 Anisotropy Identification

Spatial data often exhibit continuity properties that depend on the direction in space. Accurate kriging maps require a reliable description of the anisotropic model. Classical approaches for anisotropy estimation are based mostly on empirical methods (Goovaerts 1997). In the Spartan framework, it is possible to formulate the energy functional for general anisotropic dependence. However, this modification increases the number of model parameters: describing *geometric anisotropy* in two dimensions requires an anisotropy ratio ρ and an orientation angle ϕ. If these parameters are known, isotropic coordinate transformations can be applied to obtain a new system, in which the spatial distribution is *statistically isotropic*. The isotropic SSRF model can then be applied in the new system.

Systematic, unsupervised identification of anisotropy parameters, especially if it enables detection of sudden changes in the spatial distribution, is an important aspect of an automated mapping system for environmental monitoring and emergencies warning systems. The *covariance tensor identity* (CTI) approach allows estimating the anisotropic parameters (Hristopulos 2002), and in certain cases it provides explicit solutions for the anisotropic parameters (Hristopulos 2006). For second-order differentiable random fields, if Q_{11}, Q_{22} and Q_{12} denote the elements of the sample gradient tensor $Q_{ij} = \frac{1}{M}\sum_{m=1}^{M} \partial_i X(\mathbf{s}_m)\, \partial_j X(\mathbf{s}_m)$, it follows that

$$Q_{11} = \alpha_{\mathbf{x}} \left\{\rho^2 \left[\sin\phi\right]^2 + \left[\cos\phi\right]^2\right\},\ Q_{22} = \alpha_{\mathbf{x}} \left\{\rho^2 \left[\cos\phi\right]^2 + \left[\sin\phi\right]^2\right\}$$
$$Q_{12} = \alpha_{\mathbf{x}} \left(1-\rho^2\right) \sin\phi \cos\phi, \tag{8}$$

where the coefficient α_{x} is related to the covariance and is independent of ρ and ϕ. Using the *scaled gradient moments* $Z_g = Q_{22}/Q_{11}$ and $Z_f = Q_{12}/Q_{11}$ eliminates $\alpha_{\mathbf{x}}$, and the anisotropic parameter estimates are given by the following equations:

$$\widehat{\phi_{\pm}} = \tan^{-1}\left(\frac{Z_g - 1 \pm \sqrt{\Delta}}{2Z_f}\right) \text{ and } \widehat{\rho_{\pm}} = \left\{\frac{2\left(Z_g - 2Z_f^2 - 1 \pm \sqrt{\Delta}\right)}{\left(Z_g - 1 \pm \sqrt{\Delta}\right)\left(Z_g + 1 \pm \sqrt{\Delta}\right)}\right\}^{1/2}, \tag{9}$$

where $\Delta = (Z_g - 1)^2 + 4Z_f^2$ and the + (−) signs correspond to equivalent solutions with ρ greater (smaller) than unity respectively.

In practice, finite differences are used to estimate the gradient tensor. For an increment b_j in the direction $\overrightarrow{\mathbf{e_j}}$, we define the following quantities in terms of the semivariogram F : $q_{11} = F\left(b_1\overrightarrow{\mathbf{e_1}}\right)$, $q_{22} = F\left(b_2\overrightarrow{\mathbf{e_2}}\right)$ and $q_{12} = F\left(b_1\overrightarrow{\mathbf{e_1}} - b_2\overrightarrow{\mathbf{e_2}}\right)$. We introduce two lag tolerances τ_1 and τ_2, a continuous, and compactly supported kernel K_1 that selects near neighbors in specified directions, and two smoothing parameters h_1 and h_2 (Elogne & Hristopulos 2006a). For conciseness we define $x_{ij} = x_i - x_j$, $y_{ij} = y_i - y_j$. The estimators $\overline{q_{ij}}$ of the q_{ij} are defined as follows:

$$\overline{q_{11}} = \frac{\sum_{i\neq j} K_1\left(x_{ij}/h_1\right)\delta_{|y_{ij}|\leq\tau_1} X_{ij}^2}{\sum_{i\neq j} K_1\left(x_{ij}/h_1\right)\delta_{|y_{ij}|\leq\tau_1}}, \quad \overline{q_{22}} = \frac{\sum_{i\neq j} K_1\left(y_{ij}/h_2\right)\delta_{|x_{ij}|\leq\tau_2} X_{ij}^2}{\sum_{i\neq j} K_1\left(y_{ij}/h_2\right)\delta_{|x_{ij}|\leq\tau_2}},$$
$$\overline{q_{12}} = \frac{\sum_{i\neq j} K_1\left(x_{ij}/h_1\right) K_1\left(y_{ij}/h_2\right)\delta_{x_{ij}\, y_{ij}<0}\, X_{ij}^2}{\sum_{i\neq j} K_1\left(x_{ij}/h_1\right) K_1\left(y_{ij}/h_2\right)\delta_{x_{ij}\, y_{ij}<0}}. \tag{10}$$

The increments b_j are estimated from kernel averages of the distances between sampling points in the respective directions. Bandwidths are linearly related to the increments with coefficients that depend on the kernel moments and follow from asymptotic analysis. The tolerances are taken proportional to the square root of the average area divided by the number of sampling points; the proportionality coefficients are selected to render the tolerance smaller than the bandwidth.

The CTI method allows checking the consistency of the anisotropy estimates by iterative application to the transformed coordinate system. Asymptotic analysis supports the statistical accuracy and reliability of the method.

4 Numerical Investigations

In this section numerical simulations are conducted to evaluate the performance of the methods presented above. The first experiment focuses on the estimation of isotropic spatial dependence from simulated data using an SSRF model. The second experiment concerns the identification of geometric anisotropy from a training data set and cross-validation of the results at a set of prediction points. The triangular kernel is used in all instances of kernel averaging. Unless otherwise specified, the *ordinary kriging* (OK) spatial interpolator is used. The maps are generated on 50×50 square grids.

4.1 First Experiment

One hundred independent samples of size $n = 50$ from a $N(0,1)$ (Gaussian, zero-mean, unit variance) RF are simulated using the Cholesky method on a square of side $L = 2$. The spherical, $\rho_s(||\mathbf{r}||) = \left\{1 - 3||\mathbf{r}||/2b_s + ||\mathbf{r}||^3/2b_s^3\right\} \delta_{||r|| \leq b_s}$, and the exponential, $\rho_e(||\mathbf{r}||) = \exp(-||\mathbf{r}||/b_e)$, covariance models are used with b_s =0.5 and b_e =0.3. For each simulation, the SSRF model parameters are determined as discussed in Section 2. The covariance is calculated by inverting the spectral density (2), which requires an 1-d numerical integral (Hristopulos 2003).

The box plots of the Spartan covariance estimators are displayed in Figure 1 for ten distance lags, uniformly spaced between 0 and three correlation lengths. As shown in the plots, the Spartan estimator captures satisfactorily the spatial dependence of non differentiable processes.

4.2 Second Experiment

One sample of size $n = 400$ from an $N(20,10)$ RF ($m_X = 25$, $\sigma_X = 10$) is generated on a square domain of length $L = 10$. The hole-type covariance $\rho_h(||\mathbf{r}||) = b_h \sin(||\mathbf{r}||/b_h)/||\mathbf{r}||$ (in isotropic coordinates) with $b_h = 1$ and anisotropic parameters $\phi = 20°$, $\rho = 2$ is used. The training set involves $n_{tr} = 100$ randomly selected points, and the prediction set the remaining $n_{pr} = 300$ points.

Figure 2 shows a map derived from all the data using nearest-neighbor interpolation as well as the locations of the training and prediction sets.

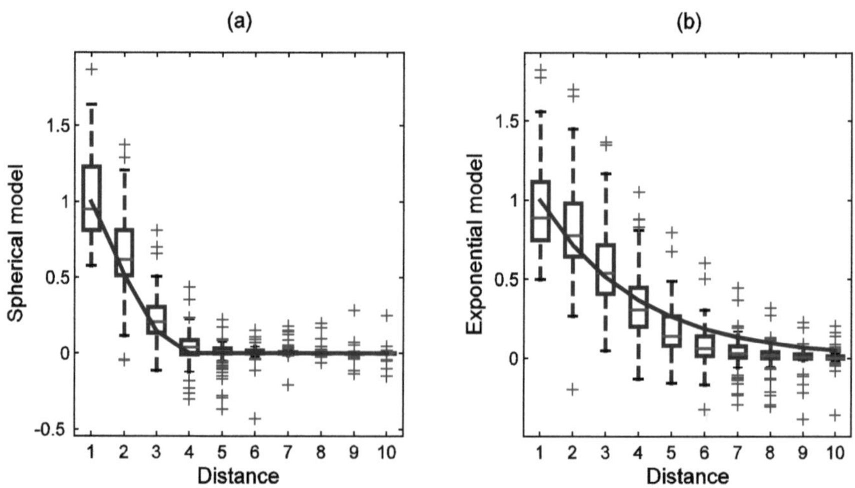

Fig. 1 Box-plots of the Spartan covariance estimator based on 100 independent samples drawn from random fields with spherical (a) and exponential (b) covariance functions, plotted against the theoretical covariances (continuous lines)

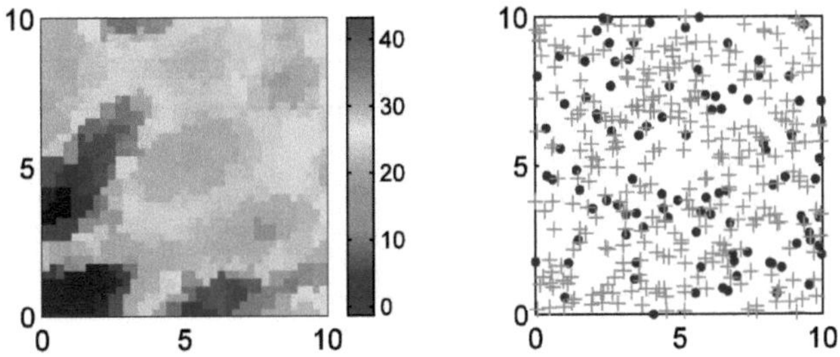

Fig. 2 (Left): Nearest-neighbor interpolation map; (right) Training location set (dots) and prediction location set (crosses)

4.2.1 Model Estimation Under the Isotropic Assumption

The training set data are modeled using the isotropic assumption. The experimental omnidirectional semivariogram is calculated and fitted with three parametric models (hole-type, exponential and spherical). In addition, two Spartan covariance estimators are obtained by constraint fitting. Initial values of η_1 in the range $[-0.5, 1.5]$ lead to the Spartan I model (η_0 =533.59, η_1 =0.56, ξ =0.66, k_c =2.41), while other initial values lead to the Spartan II model (η_0 =533.59, η_1 =1.80, ξ =0.99, k_c =6.51). The latter corresponds to a local minimum of the distance functional, while the Spartan I is the global minimum. Figure 3 displays a plot of the empirical semivariogram, as well as plots of the five estimators. The Spartan I estimator displays an oscillatory dependence, which is also present in the hole-type model used to generate the data. In contrast, the Spartan II model approaches monotonically the sill.

OK predictions are derived at the 300 prediction points and compared with the "actual" values. Table 1 summarizes the performance of the five models, based on the *mean error* (ME), *mean absolute error* (MAE), *root mean square error* (RMSE) and the Pearson *correlation coefficient* (R^2).

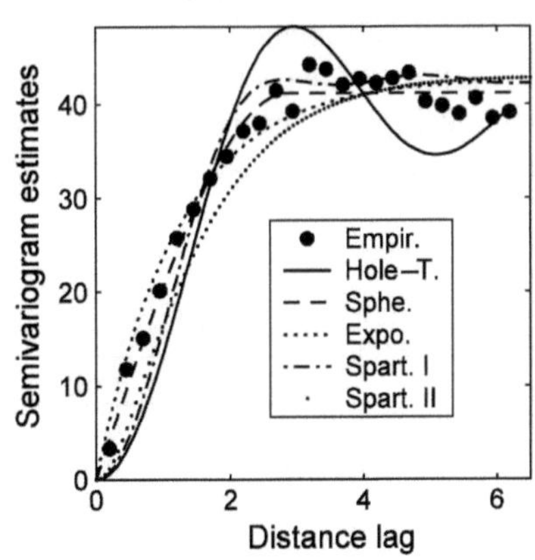

Fig. 3 Plots of the experimental semivariogram and five different isotropic estimators

Table 1 OK cross validation results (isotropic assumption)

	ME	MAE	RMSE	R^2
Spherical	0.98	2.13	3.38	0.90
Exponential	0.99	2.19	3.43	0.90
Hole-Type	1.04	2.99	4.06	0.87
Spartan I	1.00	1.90	3.28	0.91
Spartan II	0.97	2.00	3.31	0.91

All the estimators tested perform similarly with respect to the measures of the Table 1. The hole-type estimator is marginally worse. This may be due partly to the sampling density of the training set not being sufficient to accurately estimate the anti-correlations of the hole-type model. The Spartan I estimator's milder oscillations are in better agreement with the empirical semivariogram than the hole-type estimator. This is probably due to the flexibility provided by the SSRF shape coefficient. The Spartan I estimator has slightly lower MAE and RMSE values than the other estimators, but it gives a marginally higher ME. The R^2values are practically the same for all estimators except for the hole-type.

4.2.2 Model Estimation with Anisotropy Detection

Based on the 100 observations of the training set, the anisotropic parameters obtained by the CTI method are $\widehat{\phi}$=28.75° and $\widehat{\rho}$= 1.71. For comparison, if all 400 locations were used, $\widehat{\phi}$=26.59° and $\widehat{\rho}$=1.76. The increments are equal to $h_1 = 2.20$, $h_2 = 2.08$, and the tolerances are set to $\delta_j = h_j/4$, for $j = 1, 2$. Transformation in the isotropic coordinate system follows (applying CTI in this system yields $\widehat{\phi} = -21.00°$ and $\widehat{\rho}$=1.17).

The spatial dependence is modeled in the isotropic system. The resulting estimators outperform the isotropic counterparts (see Table 2). The hole-type estimator is not shown, because it performs considerably worse than the others. In both the isotropic and anisotropic cases, the Spartan I covariance outperforms the other models, but its advantage is sharpened after the anisotropic correction.

Figure 4 displays the kriging maps obtained with the Spartan I (plot a) and the spherical (plot b) estimators. The spherical model underestimates higher values as evidenced by the ranges of the kriging maps (also compare with Figure 2).

Table 2 OK cross validation results (anisotropic assumption)

	ME	MAE	RMSE	R^2
Spherical	0.85	1.92	2.90	0.94
Exponential	0.87	1.97	2.94	0.94
Spartan I	0.78	1.64	2.72	0.95
Spartan II	0.82	1.78	2.80	0.95

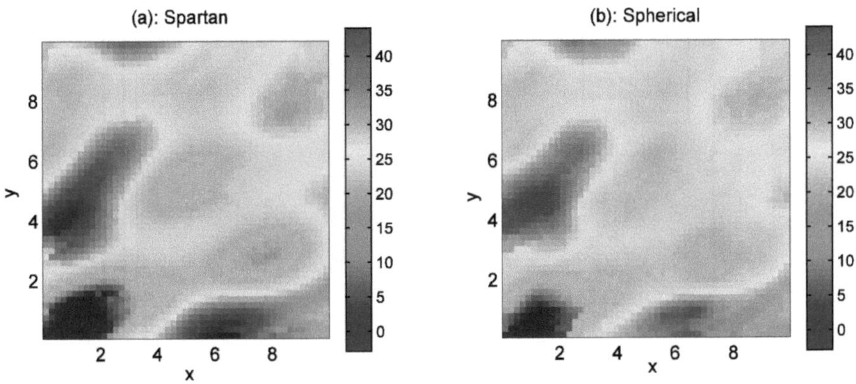

Fig. 4 OK maps obtained with (a) Spartan I and (b) spherical covariance models

5 Application to Elevation Data

We consider a set of elevation data (sample size $n=52$), available online at http://www.maths.lancs.ac.uk/~diggle to illustrate the performance of CTI and the SSRF covariance. CTI application gives $\widehat{\rho}=1.03$ and $\widehat{\phi}=33.11°$ with tolerances taken equal to 25% of the increments. Hence, there is no significant anisotropy.

Next, we perform leave-one-out cross-validation using the Spartan covariance, as well as parametric (spherical, exponential, hole-type, Gaussian and power-law) estimators. The SSRF method yields a single covariance estimator regardless of the initial value of η_1. The best cross-validation results are obtained with the hole-type covariance, but a kriging map based on this model is physically unsatisfactory (it includes negative values). The spherical and the exponential parametric estimators perform also poorly. The Gaussian and SSRF estimators exhibit the best performance as summarized in the Table 3:

The semivariograms (empirical and four estimators) are presented in Fig. 5 (left plot), with the elevation map obtained using the SSRF estimator (right plot). The SSRF estimator misses the clearly non-stationary long-range dependence of the empirical semivariogram, which is either due to mean non-stationarity or long-range fluctuations. Yet, cross validation produces reasonable errors (e.g., mean absolute relative error around 6%). This is due to the greater importance of short-range neighbours in spatial interpolation and the ability of ordinary kriging to capture slowly-changing non-stationarities of the mean.

Table 3 Leave-one-out cross validation OK performance

	ME	MAE	RMSE	MARE	RMSRE	R^2
Gaussian	5.82	53.18	60.85	0.06	0.07	0.39
Spartan	5.17	53.07	60.67	0.06	0.07	0.39

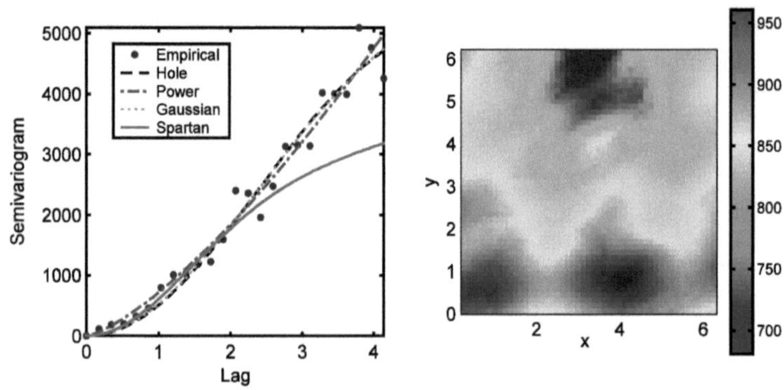

Fig. 5 Elevation semivariograms (left) and OK map based on the SSRF model (right)

6 Conclusions

We present an overview of the SSRF approach, and we investigate its performance by means of simulated and real data. The central idea of the SSRF approach is to use interactions between the field values to model spatial correlations. This viewpoint leads to methodological departures from classical geostatistics, having implications for both *model inference* and *spatial prediction*. Regarding parameter inference, an important advantage of the SSRF approach is the ability to determine the spatial dependence with minimal user involvement.

We also study the application of the CTI method, which is a promising tool for the *automatic detection* of anisotropy, using kernels to estimate the sample averages. The current formulation is based on differentiable covariance models, but extensions to non-differentiable cases are being investigated.

Acknowledgments This research is supported by a Marie Curie Transfer of Knowledge Fellowship (Project SPATSTAT, No. MTKD-CT-2004-014135), and co-funded by the European Social Fund and National Resources – (EPEAEK-II) PYTHAGORAS.

References

Elogne S, Hristopulos D (2006a) Kernel methods for estimating anisotropic parameters by means of the covariance tensor identity. Geophys Res Abstr vol 8, 02170

Elogne S, Hristopulos D (2006b) On the inference of spatial continuity using Spartan random field models, www.arXiv:math.ST/0603430

Garcia-Soidan P, Febrero-Bande M, Gonzalez-Manteiga (2004) Non-parametric kernel estimation of an isotropic semivariogram. J Stat Plan Infer 121:65–92

Goovaerts P (1997) Geostatistics for natural resources evaluation. Oxford, New York

Gorsich D, Genton M (2000) Variogram model selection via non-parametric derivative estimation. Math Geol 27:249–270

Hristopulos D (2002) New anisotropic covariance models and estimation of anisotropic parameters based on the covariance tensor identity. Stoch Env Res Risk A 16:43–62

Hristopulos D (2003) Spartan Gibbs random field models for geostatistical applications. SIAM J Sci Comput 24:2125–2162

Hristopulos D (2006) Identification of spatial anisotropy by means of the covariance tensor identity. In: Dubois (ed) GEUR 21595 EN - Automatic Mapping Algorithms for Routine and Emergency Monitoring Data, Office for Official Publications of the European Communities, Luxembourg, ISBN 92-894-9400-X, pp 103–124

Hristopulos D, Elogne S (2006) Analytic properties and covariance functions of a new class of generalized Gibbs random fields. IEEE T Inform Theory forthcoming (Dec. 2007). Preprint online at: http://arxiv.org/abs/cs.IT/0605073

A New Parallelization Approach for Sequential Simulation

H.S. Vargas, H. Caetano and H. Mata-Lima

Abstract The procedure for spatial sequential simulation – bi-point or multi-point stochastic simulation – of any type of variable starts with the definition of a random path which the simulation should follow in order to generate a structured image of a given attribute. One problem of these algorithms is related to the effort a single processor is required to undertake, especially when applying them to very large grids of nodes.

With the advent of parallel computing and multi-core processors (or multiple execution cores), which are expected to drive a new era of performance and flexibility, providing platforms that can better handle escalating workloads and rapidly evolving usage models, it becomes clear that a scalable parallelization scheme should be developed to allow the usage of such processors to allow for considerable reduction in time spent performing simulations, with clear advantages when used with clusters of multi-processor (or multiple execution core) nodes.

The general idea is to partition the universe in a given number of sections, equal in number to double the number of processors or execution cores, in such a way that the locations to be concurrently simulated are sufficiently apart to be outside search range or multi-point template range. This is only applicable in cases where at least one of the dimensions of the area to be simulated is greater then the chosen range in that direction, which is admitted to be true for cases where parallelization is valuable, particularly for very large fine scale models.

The number of sections, or regions, at which the volume will be segmented is given by an optimization procedure that maximizes the size of each region and minimizes the number of nodes to be sequentially simulated, based on the number of available processors or execution cores.

The results of the proposed parallel simulation method were checked in order to evaluate if they succeeded to reproduce the spatial continuity and spatial patterns of the phenomenon and its distribution function.

H.S. Vargas
CMRP – Centre for Modelling Petroleum Reservoirs, Mining and Georesources Department, Instituto Superior Técnico, Av. Rovisco Pais, 1049-001, Lisboa. Portugal
e-mail: hugo.vargas@total.com

A. Soares et al. (eds.), *geoENV VI – Geostatistics for Environmental Applications*, 489–496

1 Introduction

In recent years, stochastic simulation has increasingly become an indispensable tool for generating equal probability images of a set of random variables with joint probability distribution law, as shown in many references (Goovaerts, 1997, Deutsch and Journel, 1998, Soares, 2006

In earth sciences, the sequential simulation algorithms developed up-to-date, either based on **bi-point** statistics (Sequential Gaussian Simulation – SGS -, Sequential Indicator Simulation – SIS -, Direct Sequential Simulation – DSS -), or **multi-point** statistics (SIMPAT, SNESIM, filterSIM), share a common constraint, imposed by the conditioning data: the simulation of a given node $Z(x_m)$, $1 < m < n$, imposes the knowledge of all the previously simulated values $Z(x_l)$, $l < m$, that became conditioning data and all the experimental data. Therefore, the crucial point of the **sequential simulation** method is knowing the N *ccdf*: $\text{prob}\{Z(x_1) < z|(n)\}$, $\text{prob}\{Z(x_2) < z|(n+1)\}$, $\text{prob}\{Z(x_3) < z|(n+2)\}, \ldots$, and $\text{prob}\{Z(x_N) < z|(n+N-1)\}$. This is usually a problem since as the number of simulated nodes/conditioning data increases, the estimator of the probability of z0 belonging to X becomes unmanageable and inaccurate (Gomez-Hernandez, J., Journel, A., 1993). An approximation to this can be obtained through selecting a limited number of conditioning data $\{n_1\}\subset\{n\}$ with $n_1 \ll n$ so that we have $\text{prob}\{Z(x_0)<z|\,(n_1)\}\approx \text{prob}\{Z(x_0)<z|(n)\}$.

Reduction in time spent performing simulations is a challenge in numerous branches of sciences, particularly in earth sciences in which simulations are related to huge models with millions of cells. In order to mitigate this trouble Dimitrakopoulos, R. (2004) proposed a combination of scales to simulate large grids. But a real parallelization algorithm was yet to be developed.

Reducing simulation effort can be achieved either by sharing global simulation effort or sharing individual simulation effort. In this work we propose a method which allows parallelization of simulation or estimation procedures by sharing the effort of each individual simulation with a great number of processors, exploring the capabilities of newer multi-core processors.

2 Proposed Approach

Using a sequential simulation procedure to simulate a dependent variable in a n-dimensional Cartesian space V with a finite number of nodes N, at location x_1 – $Z(x_1)$ – , the experimental data points and/or previously simulated nodes to be included as simulation constraints are those which fall within variogram range, search range or multi-point template (referred to as range). Any experimental data points or previously simulated nodes outside this range will be ignored. In practice, when working with very large fine-scale grids with fairly small ranges (when compared with image dimensions), if we have spare processing power, it is possible to simultaneously simulate a second dependent variable $Z(x_2)$ (or third, or nth) as long as their ranges do not intersect. There are many ways to ensure this condition; one of them is defining a number of regions of V, according to certain rules, involving the following concepts: Region Definition, Region Association and Node Shuffling (coordinate pair shuffling). We assume that classical ways of defining unique random paths will suffice for our path choosing step of algorithm implementation.

2.1 Region Definition

Considering we have a number of processors p, with shared memory, we are able to simultaneously simulate the same number of dependent variables if we define an optimum number of lists, l, of dependent variable locations (grid nodes) to be simulated for each processor, with each list representing the number of regions in which the initial space will be divided. These lists are ordered sets of nodes of a regular grid to be simulated.

If $V = v_1 \times v_2 \times \ldots \times v_n$, with n being the number of dimensions, we also have a set of ranges R, one for each dimension of space, $R = \{r_1, r_2, \ldots, r_n\}$, taken from the multi-point template or defined by search range.

To verify if we can define regions of space V, the total number of cells to be simulated in one of its dimensions has to allow being divided by its range with a result greater than the number of lists:

$$v_\alpha / r_\alpha \geq l, \qquad \alpha \in \{1, \ldots, \mathrm{n}\}.$$

The number of regions of space is defined by l.

2.2 Region Association

Being that V is composed by an ordered set of contiguous regions $V = \bigcup_{l=1}^{n} w_l$, with l defined above, these have to be associated in proper subsets such as duplets, triplets, ..., p-tuplets, which will then be added to a list (ordered set).

We have to ensure that within any p-tuplet there are only regions which are sufficiently apart to be simultaneously simulated, which, in this context, means being non-contiguous. In this way, a list of p-tuplets, L_p, is defined as an $n(V)/p$ size list of non-repeating proper subsets of p non-contiguous non-repeating regions.

2.3 Node Shuffling

Each region that composes a p-tuplet represents a set of Cartesian coordinates (regular grid nodes) which can be sorted according to any criteria. Each possible sorting represents a path for the simulation of the dependent variable within that list. Consequently, at this stage, there is a list composed of a p-tuplet of lists of nodes that can be simultaneously simulated. This list of lists has to be transformed in a unique shuffled list of p-tuplets of nodes to prevent the complete simulation of any given region before starting the simulation of the following. This can be done by removing a randomly chosen coordinate p-tuplet of coordinates from the list, of initial size a, inserting it into another list and executing this until the second list has reached size a, or, in other words, until the first list is empty.

3 Example

This parallelization approach was tested with a sequential simulation procedure with elevation data from Portugal, on a P4 machine with 2 CPUs. Ninety images of the attribute in a testing image is $121 \times 2440 = 295\,240$ pixels were generated in three groups in the following way: thirty images using two CPUs, thirty images using a single CPU and a path choice algorithm different than the one implemented by the methodology, and thirty simulations using a single CPU and the same path choice algorithm as the two CPU version. The idea is to compare the first two groups of images in order to evaluate if the first group of simulations could be considered as

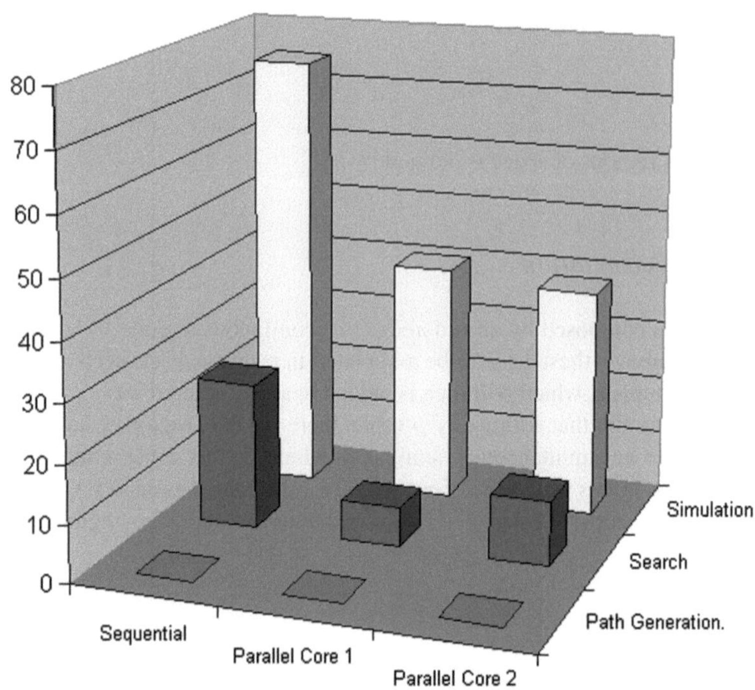

Fig. 1 Graph showing the percentage of time a single-core processor and the dual processor machine used to simulate this example take to execute the simulation

the result of a similar process and if their statistics match. The last group of simulations has the purpose of demonstrating that by using the same well-defined list of simulation nodes with 1 or two CPUs yields the same result. Figure 1 summarizes the results concerning time saving.

The simulation procedure was divided in three steps: Path generation, search for neighboring simulated nodes (or hard data) and simulation of the attribute. Figure 1, on the left, shows the time it takes a single processor to simulate a single image (the sum of path generation, search and simulation). This is our reference. On the right, figure 1 shows the time charge of the each of the two processors relative to the single processor procedure. It is clear that the charge has effectively been divided by the two processors. There is, however, a charge that was not, in this case, divided by the number of processors: the path generation procedure. But it is clear that its effect is negligible in this example.

Figures 2a) and 2b) show a result obtained by using the parallel sequential procedure and the purely sequential procedure, with the same seed value. Figure 1c shows the result of simulating the properties of the attribute using the same path as the one used for the parallel procedure with a pure sequential simulation.

We believe that figure 2a) and 2b) show that the images are the result of a similar process, which figure 2c) reinforces, as does their statistical behavior, shown in fig. 3 and table 1.

In what refers honoring the variograms, figures 4a) through 4h) show the hard data variogram and the variogram of the every type of simulation involved in the testing.

(a) (b) (c)

Fig. 2 a) Simulation result 1 from the group of parallel simulations; b) Result 1 from the group of pure sequential simulations 1; c) Result 1 from the group pure sequential simulations with the same path as the parallel sequential simulations. Elevation values in meters (m)

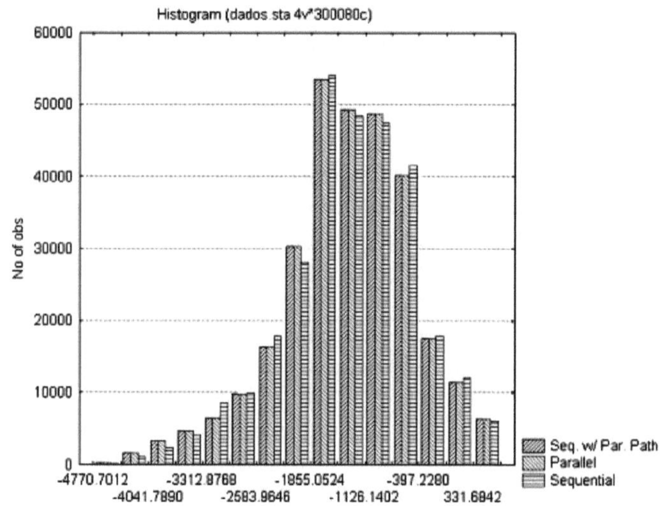

Fig. 3 Global histograms for the images of the attribute shown in Fig. 1

Table 1 Basic statistics of the images shown in Fig. 1

	Seq. w/ Par. Path	Parallel	Sequential
Valid N	300080	300080	300080
Mean	−1360.40	−1360.40	−1353.23
Minimum	−4770.70	−4770.70	−4770.70
Maximum	696.1403	696.1403	696.1403
Variance	785100	785100	770029
Std. Dev	886.059	886.059	877.513

As shown by this figure, the variogram is honored in the parallel procedure, as well as in the classical sequential simulation, as expected.

4 Final Remarks

This preliminary study demonstrates that the methodology is valuable for the purpose of parallelizing the simulation of relatively large images of a given attribute simply by arranging the nodes in such a way that their individual simulations range does not interfere.

The results obtained using this parallelization methodology were tested and verified for consistency in terms of honoring the experimental data points, honoring the experimental histogram and honoring the experimental variogram.

In practice this approach has proven to allow saving time in what concerns, exclusively, the simulation of the dependent variable, reducing time-consumption by a ratio which depends on the number of processors available. The methodology

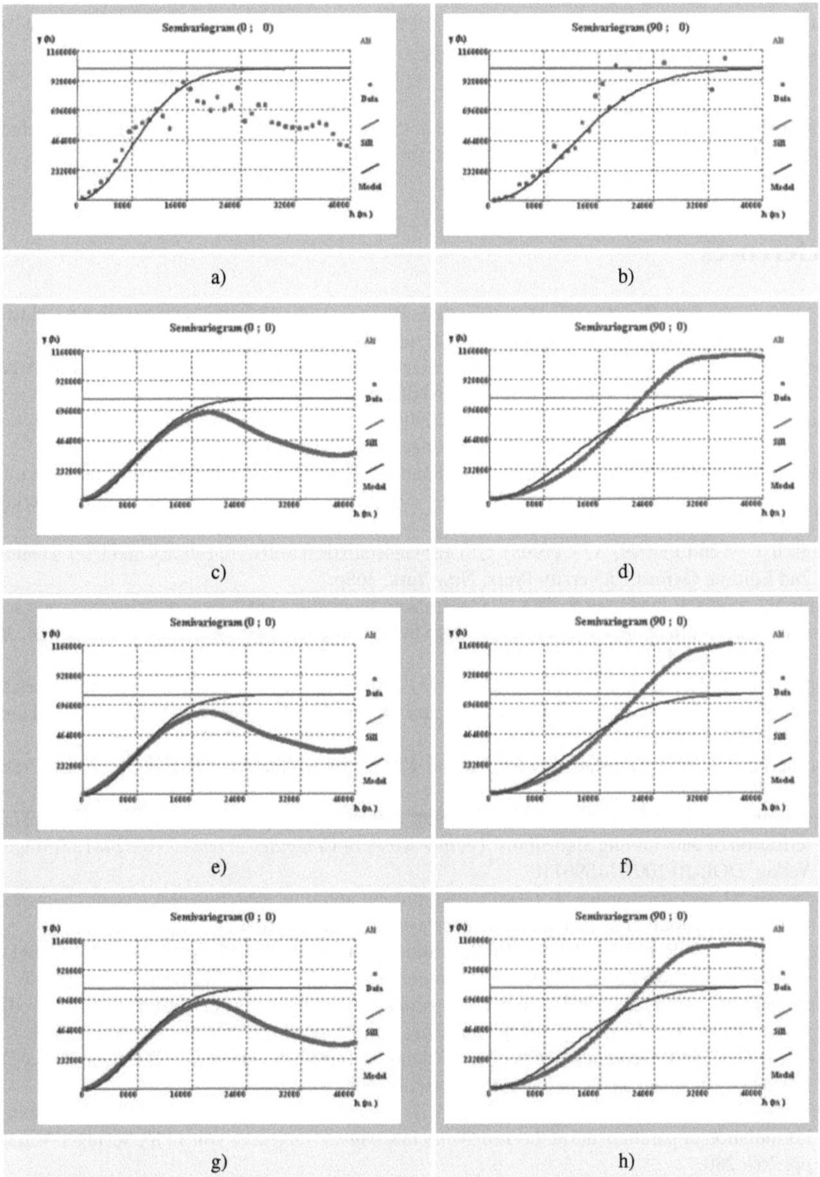

Fig. 4 a) through h) – a) Variogram for direction N-S for the experimental data, b) Variogram for direction W-E for the experimental data, c) Variogram for direction N-S for one of the results of the parallel simulation, d) Variogram for direction W-E for one of the results of the parallel simulation, e) Variogram for direction N-S for one of the results of the sequential simulation, f) Variogram for direction W-E for one of the results of the sequential simulation, g) Variogram for direction N-S for one of the results of the sequential simulation following a simulation path defined by the parallel procedure., h) Variogram for direction W-E for one of the results of the sequential simulation following a simulation path defined by the parallel procedure.

is undergoing complementary studies to be further validated and to verify to what degree, in practice, this can be achieved.

Acknowledgments The authors wish to thank Ana Cristina Marinho da Costa, Miguel Figueiredo Mascarenhas Sousa Filipe, Luis Ponce de Leão for precious discussions.

References

Al-Yamani, A., Sait, S., Youssef, H. and Barada, H. (2002). Parallelizing tabu search on a cluster of heterogeneous workstations. Journal of Heuristics, 8: 277–304.

Benkner, S. and Brandes, F. (2001). High-Level Data Mapping for Clusters of SMPs. *Lecture Notes in Computer Science*, Vol. 2026, Springer-Verlag, p. 1

Chin, W., Khoo, S., Hu, Z., and Takeichi, M. (2000). Deriving parallel codes via invariants. *Lecture Notes in Computer Science*, Vol. 1824, Springer-Verlag, pp. 75–94.

Crauser, A., Mehlhorn, K., Meyer, U. and Sanders, P. (1998). A parallelization of Dijkstra's shortest path algorithm. *Lecture Notes in Computer Science*, Vol. 1450, Springer-Verlag, p. 722.

Deutsch, C.V. and Journel, A.G. (1998). GSLIB: Geostatistical Software Library and User's Guide, 2nd Edition, Oxford University Press, New York, 368p.

Dimitrakopoulos, R. and Luo, X. (2004). Generalized sequential Gaussian simulation on group size ν and screen-effect approximations for large field simulations. Mathematical Geology, Vol. 36, No. 5, pp. 567–591.

Gómez-Hernández, J. and Journel, A.G. (1993) Joint sequential simulation of multi Gaussian fields. In A. Soares, editor, Geostatistics Troia '92, colume 1, pages 85–94. Kluwer Academic Publishers. Dordrecht

Goovaerts, P. (1997). Geostatistics for Natural Resources Evaluation. Oxford University Press, New York. 483p.

Li, X., Jin, R.and Agrawal, G. (2005). Compiler and runtime support for shared memory parallelization of data mining algorithms. *Lecture Notes in Computer Science*, Vol. 2481, Springer-Verlag. DOI: 10.1007/11596110

Nikravesh, M. and Aminzadeh, F. (2001). Past, present and future intelligent reservoir characterization trends. Journal of Petroleum Science and Engineering, 31:67–79

Øye, G. and Hilde Reme, H. (1999). Parallelizations of a compositional simulator with a Galerkin coarse/fine method. *Lecture Notes in Computer Science*, Vol. 1685, Springer-Verlag, p. 586.

Peigin, S. and Epstein, B. (2004). Embedded parallelization approach for optimization in aerodynamic design. *The Journal of Supercomputing*, Vol. 29, pp. 243–263.

Soares, A.O. (2006). Geoestatística para as Ciências da Terra e do Ambiente. 2ª Edição, Colecção Ensino da Ciência e da Tecnologia, IST Press, Lisboa, 206p.

Uchihira, N., Kawata, H. and Tamura, F. (1997). Scenario-based hypersequential programming: Formulation of parallelization. *Lecture Notes in Computer Science*, Vol. 1336, Springer-Verlag, pp. 267–280.

Clustering in Environmental Monitoring Networks: Dimensional Resolutions and Pattern Detection

D. Tuia, C. Kaiser and M. Kanevski

Abstract Monitoring Networks topology and resolutions (spatial and dimensional/ fractal) influence the ability of networks to detect spatial phenomena. In the present paper we consider several fundamental questions related to the clustering of monitoring networks and their ability (1) to detect spatial phenomena and (2) to reproduce spatial patterns using geostatistical simulations. Artificial monitoring networks with known level of clustering characterized by their fractal dimension are sampled on a same reference image with known spatial structure. Subsequently, these networks are used to interpolate using Sequential Gaussian Simulation. Resulting images are compared with several methods. Clustering of networks does not harm global detection of spatial structures (i.e., definition of correct variogram model), but influence heavily the uncertainty related to these maps, especially in tasks of detection of areas-at-risk.

1 Introduction

Design of monitoring networks (MN) is an essential task for correct pattern detection and modelling of environmental phenomena. Non-homogeneous spatial distribution (clustering) of measurement points in space can lead to an over- or underestimation of global parameters such as mean or variance and to nonrepresentative probability distribution functions, which are crucial for conditional stochastic simulations (Deutsch and Journel 1997, Kanevski and Maignan 2004).

Monitoring networks design and optimization have been discussed by several authors (see Christakos 1992, Markus et al. 1999, Caeiro et al. 2003). Traditional spatial design techniques have been recently reviewed in a exhaustive way by De Gruijter et al. (2006).

Most of the studies on MN clustering have been dedicated to the consequences of clustering on distributions (without considering spatial aspect of data) while very

D. Tuia et al.
Institute of Geomatics and Analysis of Risk, University of Lausanne, CH-1015 Switzerland
e-mail: devis.tuia@unil.ch

A. Soares et al. (eds.), *geoENV VI – Geostatistics for Environmental Applications*, 497–509

few of them studied two-point declustering in experimental variogram calculations (Richmond 2002). The present study is a first attempt to characterise the effect of clustering on spatial pattern detection: how clustered networks affect spatial predictions and how the potential losses can be described in terms of spatial patterns. In general, MN can be characterised by spatial and dimensional resolution. Dimensional (fractal) resolution characterises the dimension of the phenomena which can be detected by a particular network: in 2 dimensional space homogeneous networks (no clusters) can detect 2 dimensional phenomena (patterns). Clustered monitoring networks have a dimensional resolution d_f smaller than 2 and are not usually able to detect phenomena having dimension (2-d_f) (Lovejoy et al. 1986). Therefore, a loss of information can occurr, which will cause problems in spatial pattern reconstruction by using interpolations or simulations.

This paper presents synthetic example of simulated spatial patterns sampled with monitoring networks having different level of clustering and different dimensional resolutions.

Section 2 introduces basic notions about dimensional resolution and validity domains that are necessary to characterize the level of clustering of real monitoring networks. Section 3 focuses on the methodology used for the study that is performed in Section 4.

2 Quantitative Description of Network Clustering

There are different measures to quantify MN clustering: topological, statistical and fractal. Each measure characterises different aspects of clustering such as spatial resolution, dimensional resolution or statistical properties of clustering. In general, these measures are connected to each other. In the present study, monitoring networks with dimensional resolutions characterised by fractal dimensions are considered.

2.1 Fractal Dimension of Monitoring Network

Dimensional resolution (ability to detect spatial phenomena) was introduced for characterization of monitoring networks by Lovejoy et al. (1986). By fractals we mean statistically self-similar clustered point objects, whose structure is reproduced throughout the scales and whose dimension is usually not an integer.

In the present paper, fractal dimension d_f is used as a general indicator of clustering, where a dimension lower than 2 can be interpreted as the appearance of clusters at a certain spatial scale. Here, d_f is computed with the box-counting method (Falconer 1990, Peitgen et al. 1992): the area under study is covered by a regular grid of N cells, and the number of cells necessary to cover the whole network, S(L), is computed. Then, the size of boxes L is gradually decreased (accordingly, the number of boxes N is increased). The box-counting operation is repeated *m* times.

For the fractally distributed measurement points, the number of boxes necessary to cover the network points follows a power-law

$$S(L) \quad \sim \quad L^{-d_f} \tag{1}$$

The fractal dimension of the network d_f can be computed as the slope of the regression line after log-transformation of both sides of Eq. (1).

The equations presented above do not take into account real-life situations where different geographical constrains and finite number of measurement points are important. A recent paper by the same authors (Tuia and Kanevski 2006) has shown that a good way to quantify real monitoring networks is to compare them with a reference network generated within the same domains and having predefined fractality.

2.2 Validity Domains and Fractality of Monitoring Networks

Clustering of networks causes incorrect global estimations of mean and variance of the probability distribution function and erroneous spatial predictions over a regular two dimensional space. In geostatistics, a common practice is to interpolate the variable over the whole two-dimensional surface (often a square) and then to clip the results over the area of interest, e.g., with a GIS. These areas of interest, called validity domains (VD), spatially constrain the predictive space. In most cases, fractal dimension of such regions is less than two. In general, VD are related to geographical, political or economical constraints such as political boundaries or topographic barriers. In such cases, even homogeneous monitoring networks have fractal dimension smaller than two. Therefore, in order to quantify clustering of networks within VD, it was proposed to generate reference networks and to compare them with a real measurement network (Kanevski and Maignan 2004, Tuia and Kanevski 2006). Deviations between these networks (real and reference) were used to quantify the degree of clustering. Interpolation techniques have then been applied only on the area of interest, taking into account the irregular shape of the VD.

In order to analyze the effect of clustering on spatial predictions (reconstruction of spatial patterns), a region characterized by heavy geographical constraints has been chosen: the Swiss canton of Graubünden, which is characterized by a validity domain related to its mountainous landscape and to the organization of its inhabited areas into small settlements. The real monitoring network corresponds to indoor radon data measurements network. According to the methodology developed, three MN have been used for the current study (Fig. 1).

A. **Raw network**: a real MN (RMN), related to samples taken during an indoor Radon sampling campaign. The network is composed of $N = 3258$ unique measurements. The RMN is characterized by a high level of clustering corresponding to the fractal dimension $d_f = 1.38$; Two artificial homogeneous monitoring networks (GR network and Pop network) with the same number of sampling points generated within a validity domain of interest:

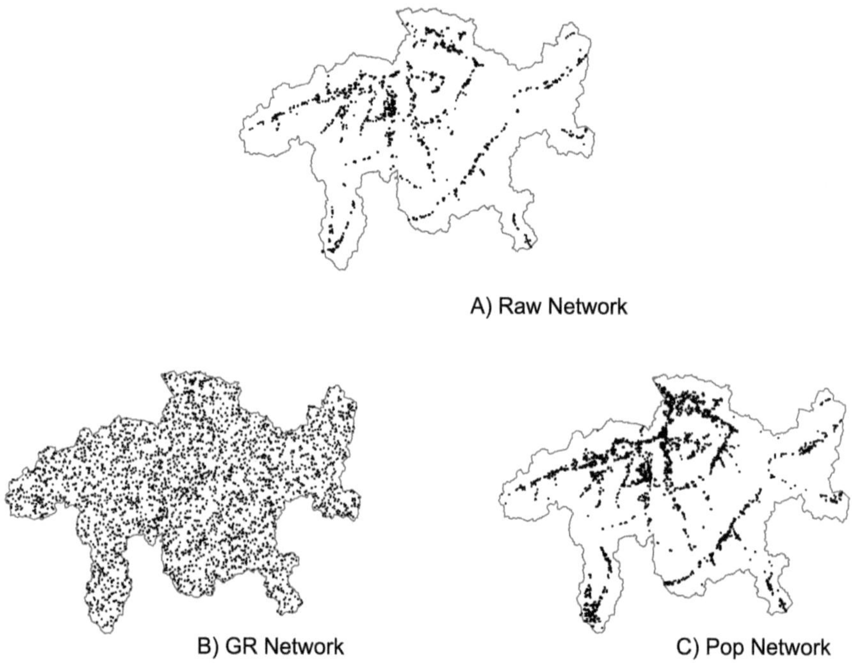

Fig. 1 Monitoring network used for the analysis. (A) RMN; (B) GR network; (C) Pop network

B. **GR network**: 3258 samples homogeneously distributed within the political boundaries of the canton of Graubünden. This VD corresponds to administrative geographical constraints. The dimension d_f of the network is 1.79, where the loss of dimensionality is related to complex boundary effects;
C. **Pop network**: 3258 samples homogeneously distributed within the limits of the populated regions of the canton of Graubünden. This VD corresponds to geographical constraints: the use of populated regions as VD avoids the presence of samples on mountainous regions and provides a distance-related barrier in terms of covariance. The dimension d_f of this network is 1.46.

Thus, the number of observations is constant throughout the networks, and the results should be a function of the network's design, i.e., the level of clustering, as defined by the fractal dimension.

In order to understand and to characterise spatial patterns (and corresponding uncertainties) detected by different networks, a reference model (complete image, CI) was simulated. Then, the CI model was sampled with different MN, which are described above. Finally, conditional simulations were carried out using sampled data and the results were compared with the CI.

3 Simulation of Spatial Patterns

3.1 Sequential Gaussian Simulations of a Reference Pattern

The reference patterns CI have been generated by a nonconditional simulation of a Gaussian random field $Y(u)$ with a given covariance $C_Y(h)$ by using Geostat Office (Kanevski and Maignan 2004):

$$C_Y(h) = \begin{cases} 1 - \frac{3}{2}\frac{|h|}{a} - \frac{1}{2}\frac{|h|^3}{a^3} & h \leq a \\ 1 & h > a \end{cases} \tag{2}$$

Only the results on one reference image called SIM1 and generated according to the isotropic spherical variogram with 20 km correlation range are given (Fig. 2).

Once the reference SIM1 image was generated over the coordinates of Graubünden, it was sampled using three monitoring networks described above. These three artificial "measurement campaigns" were used to reconstruct the original pattern with Sequential Gaussian Simulation algorithm and to make the analysis and comparison between the results. The reconstruction of the patterns has been carried out with complete (including variogram analysis and modelling) conditional SGS based on the three sampling campaigns. The use of conditional SGS gives the possibility not only to compare generated patterns but also to quantify uncertainties and the variability between them.

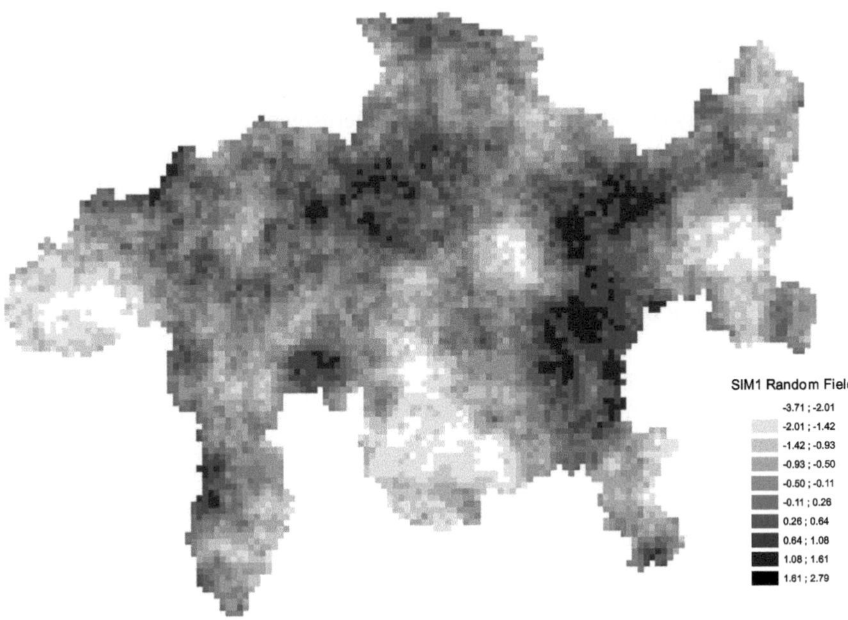

Fig. 2 Artificial phenomenon SIM1 generated by nonconditional simulation over the coordinates of Graubünden

3.2 Tools to Evaluate the Detection of Patterns

Several tools have been applied in order to evaluate the efficiency of pattern reconstruction.

3.2.1 E-type Maps

Conditional SGS provides many realizations of the random field $Z(u)$ governing the phenomenon. The first tool to evaluate the quality of the simulation is the comparison of the maps obtained by computation of the mean value for every simulated node over the M realizations with SIM1. A correct reconstruction of the SIM1 image by the simulations shows the ability of a clustered network to reproduce the underlying spatial structures described, for example, by the γ model.

3.2.2 Probability Isolines

Then, the generation of many realizations of $Z(u)$ allows post-processing analysis, such as the elaboration of maps of probability to exceed a given threshold g:

$$P(X \geq g) = F(g) \tag{3}$$

This procedure allows drawing the isolines corresponding to the same probability to exceed the threshold g. The SIM1 being an artificial reference image, the threshold is defined a priori for the analysis and is not related to a real level of risk.

3.2.3 Spatial Metrics

Finally, the analysis and comparison of the patterns generated by the conditional SGS cannot be made only by simple visual comparison. Several quantitative pattern description metrics coming from landscape ecology (O'Neill et al. 1988, Turner et al. 2001) have been applied on the risk maps discussed above.

3.3 Percentage of Landscape Covered (PLC)

This metric quantifies the percentage of landscape occupied by the patterns. The total area of the pattern is divided by the total area of the landscape (Validity Domain of the political boundaries).

3.4 Land Shape Index (LSI)

The LSI is an indicator of dispersion of a pattern formed by k disconnected patches. It is computed by dividing the total pattern edge length by the edge length of the smallest patch:

$$LSI = \frac{\sum_{i=1}^{k} e_i}{\min e_i} \tag{4}$$

This metric provides a standardized measurement of patches aggregation: the more the LSI increases, the more the patches are disaggregated (McGarigal and Marks 1994).

3.5 Concentration-dependent Fractal Dimension (CDFD, $d_f(Z_{th})$)

CDFD can be estimated with functional box-counting method over simulated nodes exceeding a given threshold Z_{th} for every level of probability tested. The CDFD curve is characterised by the dependence d_{f} (Z_{th}). If the probability level influences the shape of the pattern, then the CDFD curve should decrease with an increase of the level. If the pattern shape is stable, the curve should remain constant or decrease slowly.

4 Discussion

4.1 E-type Maps: Visual Comparison of Results

Fifty stochastic realisations were generated on three sampling networks of the same SIM1 reference phenomenon. In Fig. 3, E-type maps of the realisations are shown.

At a first glance, the networks can reproduce correctly the structure of the phenomenon, i.e., the variogram model. The dependence of the SGS mean results on the clustering of MN is visible by an effect of smoothing of the overall pattern. The GR network gives the best visual result, while the other networks, more clustered, are characterized by smoothed images.

4.2 Probability Isolines: Risk Maps to Draw Pattern Detection

For the SIM1 random field, the value of 1 has been defined as an action threshold for environmental protection (this choice is arbitrary). Then, the SGS simulations allow to draw maps of probability of exceeding that threshold. Figure 4 shows the probability maps related to every set of simulations considered.

The different probability maps studied showed that clustering of the network has a tendency to dilate the regions above a threshold for high uncertainty (i.e., small probabilities). The GR network shows small differences between the regions over the threshold for $P(X \geq 1) = 0.7$ and $P(X \geq 1) = 0.5$, while the Pop, and more clearly the Raw, show increasing differences of patterns between the maps.

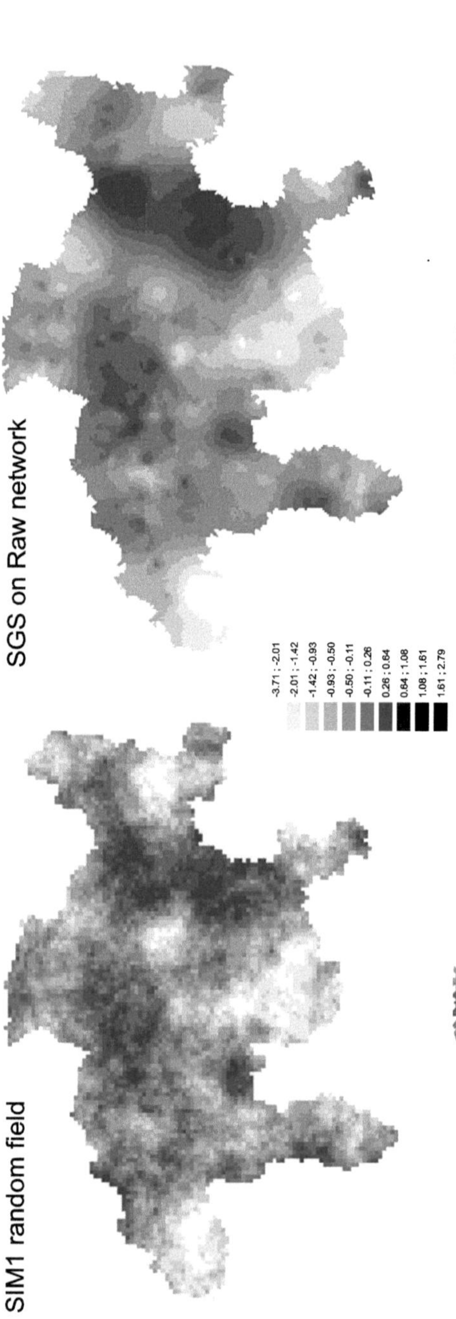

Fig. 3 E-type maps for the simulations obtained on the Graubünden region. Top-right: raw network; bottom-left: GR network; bottom-right: Pop network

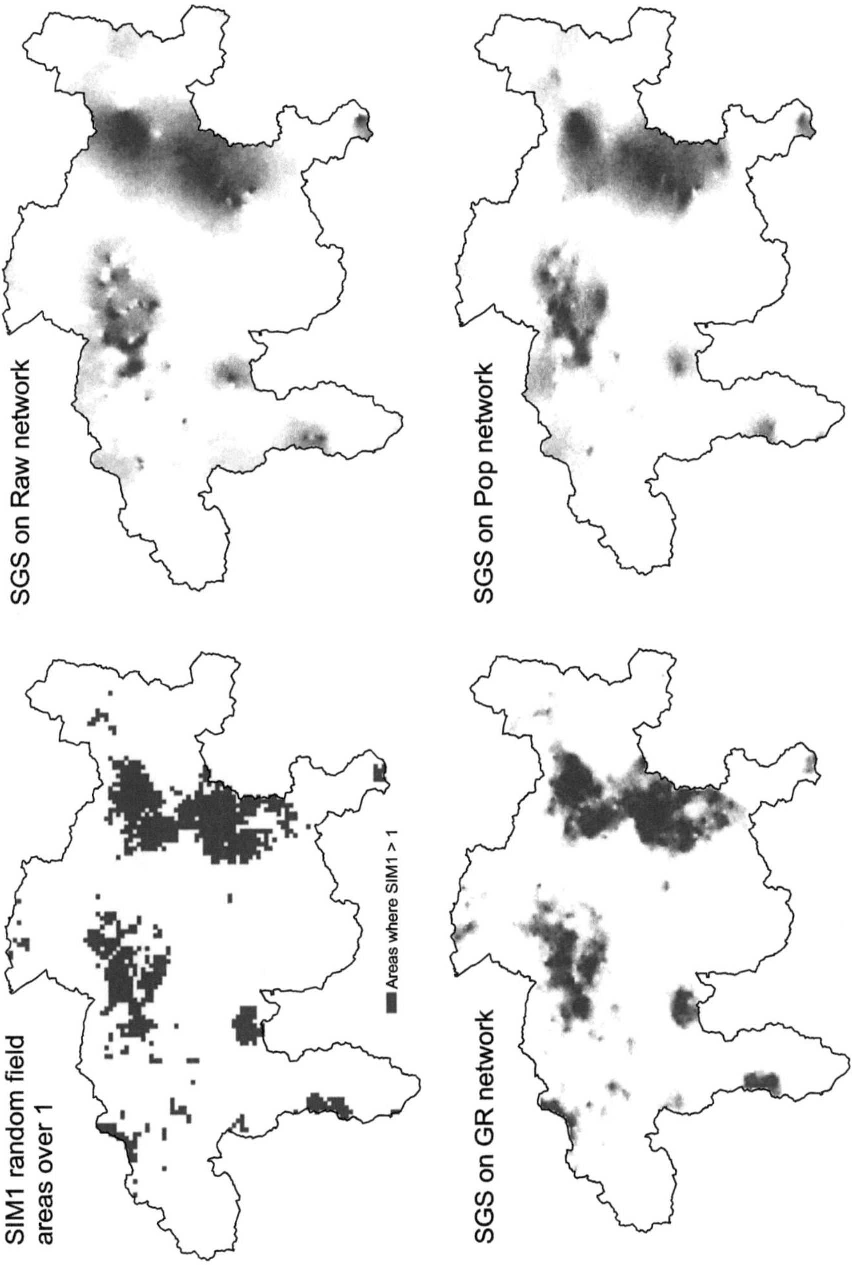

Fig. 4 Probability maps for $P(X \geq 1)$ Bottom-left: GR network; top-right: Raw network; bottom-right: Pop network

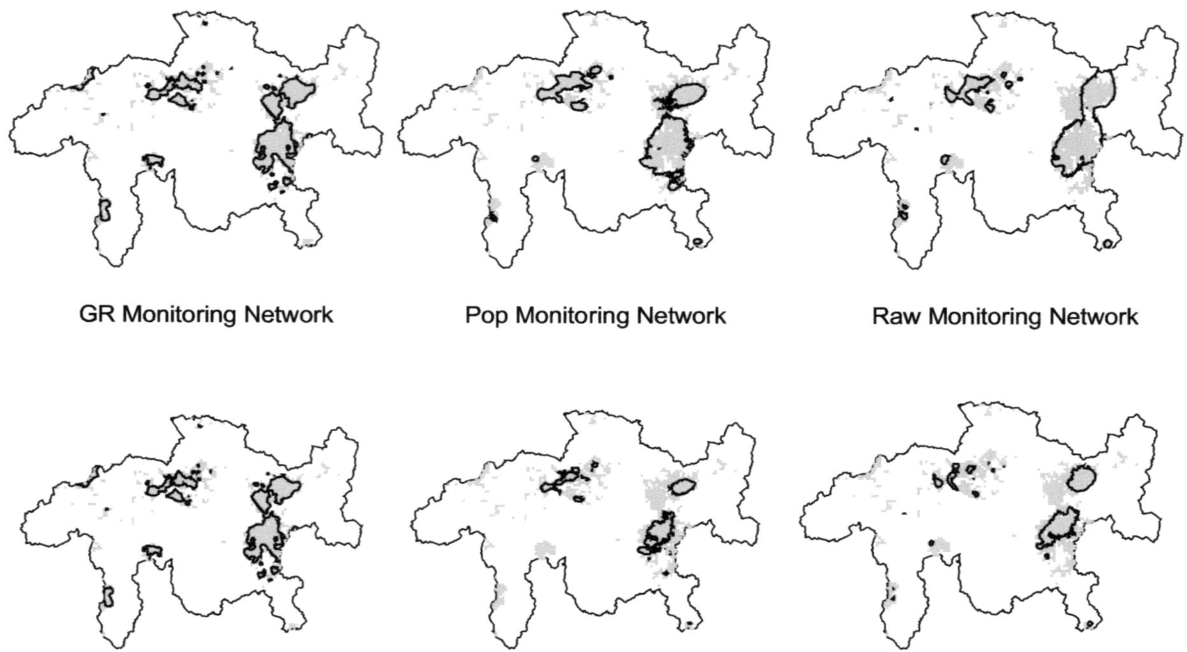

Fig. 5 Probability isolines for $P(X \geq 1) = 0.7$ (top row) and for $P(X \geq 1) = 0.9$ (bottom row). In grey the regions of the random field SIM1 above the threshold. Left: GR network; center: Pop network; right: Raw network

The clustering of networks increases differences between the simulations of $Z(u)$ and destabilizes the shape of the surface at risk for a given probability level.

Comparing only the isolines for probabilities 0.7 and 0.9 (Fig. 5), it can be seen that the clustering of the network leads to a progressive loss of detection of the shape of the area at risk: the GR network allows a correct detection of pattern for small uncertainty levels, while the Pop and Raw networks lose detection on peripheral areas and start having false detections of the phenomenon in areas which are, in reality, below the defined threshold.

4.3 Spatial Metrics

The analysis of the spatial metrics discussed above (Fig. 6) confirmed the observations made following the analysis of probability isolines maps. On one hand, the GR network keeps a higher connectivity level through the probability levels, showed by the slow decrease of PLC (that shows the consistency of small uncertainty levels) and the stability of the CDFD index, which can be explained by the stability of patches keeping their shape and connectivity. On the other hand, clustered networks (Raw and Pop) lead to a faster decrease of PLC, showing a higher uncertainty depending on the probability considered. The CDFD index shows a loss of connectivity of the pattern for small uncertainty (for $F(1) \geq 0.8$) which can even be observed on the map (Fig. 5) by the reduction of the pattern to small patches localized on areas related to high density of samples.

LSI shows the level of aggregation of pattern: the real situation (SIM1) is heavily fragmented, reflecting the complexity of $Z(u)$: the GR network can reproduce partially this disaggregation, which is completely lost with the clustered networks. There are characterized by small values of LSI, i.e., aggregated and smoothed patterns for every probability level.

5 Conclusion

Clustering of monitoring networks has a significant impact on spatial prediction of random fields. Heavily clustered sampling schemes can decrease the quality of definition of areas at risk for environmental and pollution problems. In this study it was shown that even clustered networks can detect correctly the variogram model, but that the realization of the random field cannot provide a correct definition of areas at risk, especially if small uncertainty levels are required.

Patterns created by clustered networks are heavily dependent on the probability level considered. For high levels connectivity is lost, as it is shown by the CDFD analysis. Risk maps can only detect hot spots related to the location of samples.

This study has only used measures of pattern detection based on visual comparison and spatial metrics, which do not analyze patterns in terms of shape or correct reconstruction of the random field. In order to better compare generated patterns,

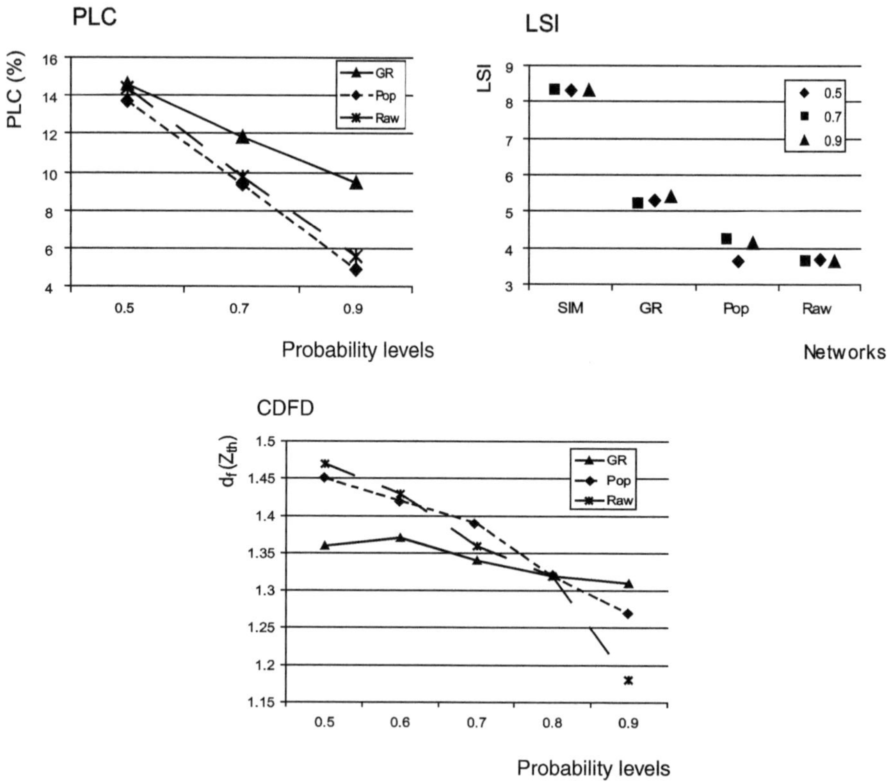

Fig. 6 Spatial metrics. (Top left) Percentage of landscape covered; (Top right) LSI; (bottom) CDFD index

the spatial metrics are being calculated for every independent simulation within our current research. In this way, the point-to-point relationships are preserved and the patterns are coherent for comparison of independent simulations. Approaches based on multiple points statistics and mathematical morphology may be also useful in order to characterize patterns in relation to their shapes and localizations.

Another important issue concerns the fractal dimension used to characterize the clustering of the MN: even if d_f is often used as a global measure of clustering, one must remember that different networks can show the same fractal dimension. The analysis of the robustness of d_f as index of clustering in topologically different situations is central in the process of validation of the index.

Acknowledgments This work has been partially supported by the Swiss National Foundation. Projects "Urbanization Regime and Environmental Impact: Analysis and Modelling of Urban Patterns, Clustering and Metamorphoses" (n.100012–113506) and "GeoKernels": Kernel-Based Methods for Geo- and Environmental Sciences (n.200021–113944).

References

Caeiro, S., Painho, M., Goovaerts, P., Costa, H., Sousa, S. (2003), Spatial sampling design for sediment quality assessment in estuaries. Environmental Modelling & Software, 18:853–859.

Christakos, G. (1992), Random Fields Models in Earth Sciences. San Diego, Academic Press.

De Gruijter J., Brus D., Bierkens M., Knotters M. (2006), Sampling for Natural Resource Monitoring. Berlin Heidelberg, Springer-Verlag.

Deutsch, C., Journel, A. (1997), GSLIB. Geostatistical Software Library and User's Guide. New York, Oxford University Press.

Falconer, K.J. (1990), Fractal Geometry. Mathematical Foundations and Applications. Chichester, John Wiley and Sons.

Kanevski M., Maignan M. (2004), Analysis and Modelling of Spatial Environmental Data. Lausanne, EPFL Press.

Lovejoy S., Schertzer D., Ladoy P. (1986). Fractal characterisation of inhomogeneous geophysical measuring networks. Nature, 319: 43–44.

Markus, A., Welch, W.J., Sacks, J. (1999), Design and analysis for modeling and predicting spatial contamination. *Mathematical Geology*, 31(1):1–22.

McGarigal, L., Marks, B.J. (1994), FRAGSTATS manual: spatial pattern analysis program for quantifying landscape structure. http://www.umass.edu/landeco/research/fragstats/fragstats.html

O'Neill, R.V., Krummel, J.R., Gardner, R.H., Sugihara, G., Jackson, B., DeAngelis, D.L., Milne, B.T., Turner, M.G., Zygmunt, B., Christensen, S.W., Dale, V.H., Graham, R.L. (1988), Indices of landscape pattern. *Landscape Ecology*, 1:153–162.

Peitgen, H.O., Hartmut, J., Saupe, D. (1992), Chaos and Fractals: New Frontiers of Science. New York, Springer-Verlag.

Richmond A. (2002), Two-point declustering for weighting data pairs in experimental variogram calculations. Computers and Geosciences, 28: 231–241.

Tuia, D., Kanevski, M. (2006), Indoor Radon Data Monitoring Networks: Topology, Fractality and Validity Domains, Congress of the International Association of Mathematical Geology (IAMG), Liège, Belgium.

Turner, M.G., Gardner, R.H., O'Neill, R.V. (Eds) (2001), *Landscape Ecology in Theory and Practice: Pattern and Process*. Springer-Verlag, New York.

Quantitative Geology and Geostatistics

1. F.M. Gradstein, F.P. Agterberg, J.C. Brower and W.S. Schwarzacher: *Quantitative Stratigraphy*. 1985 ISBN 90-277-2116-5
2. G. Matheron and M. Armstrong (eds.): *Geostatistical Case Studies*. 1987 ISBN 1-55608-019-0
3. *Cancelled*
4. M. Armstrong (ed.): *Geostatistics*. Proceedings of the 3rd International Geostatistics Congress, held in Avignon, France (1988), 2 volumes. 1989 Set ISBN 0-7923-0204-4
5. A. Soares (ed.): *Geostatistics Traóia* '92, 2 volumes. 1993 Set ISBN 0-7923-2157-X
6. R. Dimitrakopoulos (ed.): *Geostatistics for the Next Century*. 1994 ISBN 0-7923-2650-4
7. M. Armstrong and P.A. Dowd (eds.): *Geostatistical Simulations*. 1994 ISBN 0-7923-2732-2
8. E.Y. Baafi and N.A. Schofield (eds.): *Geostatistics Wollongong* '96, 2 volumes. 1997 Set ISBN 0-7923-4496-0
9. A. Soares, J. Gómez-Hernandez and R. Froidevaux (eds.): *geoENV I - Geostatistics for Environmental Applications*. 1997 ISBN 0-7923-4590-8
10. J. Gómez-Hernandez, A. Soares and R. Froidevaux (eds.): *geoENV II - Geostatistics for Environmental Applications*. 1999 ISBN 0-7923-5783-3
11. P. Monestiez, D. Allard and R. Froidevaux (eds.): *geoENV III - Geostatistics for Environmental Applications*. 2001 ISBN 0-7923-7106-2; Pb 0-7923-7107-0
12. M. Armstrong, C. Bettini, N. Champigny, A. Galli and A. Remacre (eds.): *Geostatistics Rio 2000*. 2002 ISBN 1-4020-0470-2
13. X. Sanchez-Vila, J. Carrera and J.J. Gómez-Hernández (eds.): *geoENV IV - Geostatistics for Environmental Applications*. 2004 ISBN 1-4020-2007-4; Pb 1-4020-2114-3
14. O. Leuangthong and C.V. Deutsch (eds.): *Geostatistics Banff 2004*, 2 volumes. 2005 Set ISBN 1-4020-3515-2
15. A. Soares, M.J. Pereira and R. Dimitrakopoulos (eds.): *geoENV VI - Geostatistics for Environmental Applications*. 2008 ISBN 978-1-4020-6447-0

MIX
Papier aus verantwortungsvollen Quellen
Paper from responsible sources
FSC® C105338
FSC
www.fsc.org

If you have any concerns about our products,
you can contact us on
ProductSafety@springernature.com

In case Publisher is established outside the EU,
the EU authorized representative is:
Springer Nature Customer Service Center GmbH
Europaplatz 3, 69115 Heidelberg, Germany

Printed by Libri Plureos GmbH
in Hamburg, Germany